Subcellular Proteomics

Subcellular Biochemistry
Volume 43

SUBCELLULAR BIOCHEMISTRY

Recent Volumes in this Series

Volume 31 **Intermediate Filaments**
Edited by Harald Herrmann and J. Robin Harris

Volume 32 **alpha-Gal and Anti-Gal: alpha-1,3-Galactosyltransferase, alpha-Gal Epitopes and the Natural Anti-Gal Antibody**
Edited by Uri Galili and Jos-Luis Avila

Volume 33 **Bacterial Invasion into Eukaryotic Cells**
Tobias A. Oelschlaeger and Jorg Hacker

Volume 34 **Fusion of Biological Membranes and Related Problems**
Edited by Herwig Hilderson and Stefan Fuller

Volume 35 **Enzyme-Catalyzed Electron and Radical Transfer**
Andreas Holzenburg and Nigel S. Scrutton

Volume 36 **Phospholipid Metabolism in Apoptosis**
Edited by Peter J. Quinn and Valerian E. Kagan

Volume 37 **Membrane Dynamics and Domains**
Edited by P.J. Quinn

Volume 38 **Alzheimer's Disease: Cellular and Molecular Aspects of Amyloid beta**
Edited by R. Harris and F. Fahrenholz

Volume 39 **Biology of Inositols and Phosphoinositides**
Edited by Lahiri Majumder and B.B. Biswas

Volume 40 **Reviews and Protocols in DT40 Research**
Edited by Jean-Marie Buerstedde and Shunichi Takeda

Volume 41 **Chromatin and Disease**
Edited by Tapas K. Kundu and Dipak Dasgupta

Volume 42 **Inflammation in the Pathogenesis of Chronic Diseases**
Edited by Randall E. Harris

Subcellular Proteomics

From Cell Deconstruction to System Reconstruction

Subcellular Biochemistry Volume 43

Edited by

Eric Bertrand

and

Michel Faupel

Novartis Institutes of Biomedical Research,
Basel,
Switzerland

Springer

A C.I.P. Catalogue record for this book is available from the Library of Congress.

ISBN 978-1-4020-5942-1 (HB)
ISBN 978-1-4020-5943-8 (e-book)

Published by Springer,
P.O. Box 17, 3300 AA Dordrecht, The Netherlands.

www.springer.com

Printed on acid-free paper

To Michel and his speedy recovery. His experience and enthusiasm have been sorely missed in the final stages of preparing this volume.

To my son, Alexandre, far from me now, but still close to my heart.

To my mother, who has been a source of inspiration in many more ways than she could have imagined.

Eric

TABLE OF CONTENTS

ACKNOWLEDGEMENTS

In light of the somewhat eventful circumstances in which the current volume came into being, we are sincerely grateful to the people who contributed to its preparation. Their hard work as well as their patience and kind understanding toward the editors were greatly appreciated. We are most grateful to Mary Johnson, who acted as the publishing editor for this volume. We wish to thank all of the contributors and their staff for submitting their respective chapters as needed, despite busy teaching and research schedules. We would also like to acknowledge several very good friends for their voluntary editorial assistance and even more importantly their unfailing moral support: Eileen Rojo, Hugues Ryckelynck and Claire Mc Donack. Last, but by no mean least, we are particularly grateful to Dr. Robin Harris, who gave us the opportunity to edit this volume for the "Subcellular Biochemistry" book series.

LIST OF CONTRIBUTORS

Bernadette Allinquant
INSERM U573, Centre Paul Broca, Paris, France.

Dario Anselmetti
Experimental Biophysics and Applied Nanoscience, Physics Faculty, Bielefeld University and Bielefeld Institute for Biophysics and Nanoscience (BINAS), Center for Biotechnology (CeBiTec), Bielefeld, Germany.

Fengju Bai
Charles River Laboratories Preclinical Services, Safety Sciences, Worcester, Massachusetts, United States.

Barbara Bakker
BioCentre Amsterdam, Free University Amsterdam, Faculty of Earth and Life Sciences, Department of Molecular Cell Physiology, Amsterdam, The Netherlands.

Anke Becker
Institute for Genome Research, Center for Biotechnology (CeBiTec) Bielefeld University, Bielefeld, Germany.

Eric Bertrand
Novartis Institutes for BioMedical Research, Genome and Proteome Sciences/ Systems Biology, Basel, Switzerland.

Guillaume Boissy
Hybrigenics S.A., Paris, France.

Débora Bonenfant
Novartis Institutes for BioMedical Research, Genome and Proteome Sciences/ Systems Biology, Basel, Switzerland.

Jildau Bouwman
BioCentre Amsterdam, Free University Amsterdam, Faculty of Earth and Life Sciences, Department of Molecular Cell Physiology, Amsterdam, The Netherlands.

Frank J. Bruggeman
BioCentre Amsterdam, Free University Amsterdam, Faculty of Earth and Life Sciences, Department of Molecular Cell Physiology, Amsterdam, The Netherlands and also Manchester Centre for Integrative Systems Biology, Manchester Interdisciplinary BioCentre, University of Manchester, United Kingdom.

Marie-Elaine Caruso
Organelle signaling laboratory, Department of Surgery, McGill University, Montreal, Quebec, Canada.

Queenie W. T. Chan
UBC Centre for Proteomics, Department of Biochemistry and Molecular Biology, University of British Columbia, Vancouver, Canada.

Eric Chevet
Organelle signaling laboratory, Department of Surgery, McGill University, Montreal, Quebec, Canada.

Mark O. Collins
Proteomic Mass Spectrometry, The Wellcome Trust Sanger Institute, Hinxton, Cambridgeshire, United Kingdom.

Vincent Collura
Hybrigenics S.A., Paris, France.

Michel Faupel
Novartis Institutes for BioMedical Research, Genome and Proteome Sciences/ Systems Biology, Basel, Switzerland.

Leonard J. Foster
UBC Centre for Proteomics, Department of Biochemistry and Molecular Biology, University of British Columbia, Vancouver, Canada.

Elvira García Osuna
Center for Bioimage Informatics and Department of Biomedical Engineering, Carnegie Mellon University, Pittsburgh, Pennsylvania, United States.

Seth G. N. Grant
Genes to Cognition, The Wellcome Trust Sanger Institute, Hinxton, Cambridgeshire, United Kingdom.

Dominik Greif
Experimental Biophysics and Applied Nanoscience, Physics Faculty, Bielefeld University and Bielefeld Institute for Biophysics and Nanoscience (BINAS), Center for Biotechnology (CeBiTec), Bielefeld, Germany.

Wibke Hellmich
Experimental Biophysics and Applied Nanoscience, Physics Faculty, Bielefeld University and Bielefeld Institute for Biophysics and Nanoscience (BINAS), Center for Biotechnology (CeBiTec), Bielefeld, Germany.

Deirdre M. Kavanagh
Wellcome Trust Centre for Cell Biology, University of Edinburgh, Edinburgh, United Kingdom.

Markus Kohlmann
Sanofi-Aventis Pharma, Structural & Physical Chemistry, Frankfurt am Main, Germany.

Vassiliki Lazou
Wellcome Trust Centre for Cell Biology, University of Edinburgh, Edinburgh, United Kingdom.

Madalina Maftei
Department of Chemistry, Laboratory of Analytical Chemistry and Biopolymer Structure Analysis, University of Konstanz, Konstanz, Germany.

Poonam Malik
Wellcome Trust Centre for Cell Biology, University of Edinburgh, Edinburgh, United Kingdom.

Joerg Martini
Experimental Biophysics and Applied Nanoscience, Physics Faculty, Bielefeld University and Bielefeld Institute for Biophysics and Nanoscience (BINAS), Center for Biotechnology (CeBiTec), Bielefeld, Germany.

Laurent Mauvieux
Institut d'Hématologie et d'Immunologie, Faculté de Médecine, Université Louis Pasteur and Laboratoire d'Hématologie, Hôpital Hautepierre, Strasbourg, France.

Thomas Merkle
Institute for Genome Research, Center for Biotechnology (CeBiTec) Bielefeld University, Bielefeld, Germany.

Laurent Miguet
Laboratoire de Spectrométrie de Masse Bio-Organique, ECPM, UMR/CNRS 7178, Institut Pluridisciplinaire Hubert CURIEN, Université Louis Pasteur, Strasbourg, France.

Jan Moebius
Protein Mass Spectrometry and Functional Proteomics Group, Rudolf-Virchow-Center for Experimental Biomedicine, University of Wuerzburg, Wuerzburg, Germany.

Dieter R. Mueller
Novartis Institutes for BioMedical Research, Genome and Proteome Sciences /Systems Biology, Basel, Switzerland

Robert F. Murphy
Center for Bioimage Informatics and Departments of Biological Sciences, Biomedical Engineering and Machine Learning, Carnegie Mellon University, Pittsburgh, Pennsylvania, United States.

Christine Olver
Clinical Pathology Section, Colorado State University, Ft Collins, United States.

Noelle Potier
Laboratoire de Spectrométrie de Masse Bio-Organique, ECPM, UMR/CNRS 7178, Institut Pluridisciplinaire Hubert CURIEN, Université Louis Pasteur, Strasbourg, France.

William E. Powell
Wellcome Trust Centre for Cell Biology, University of Edinburgh, Edinburgh, United Kingdom.

Michael Przybylski
Department of Chemistry, Laboratory of Analytical Chemistry and Biopolymer Structure Analysis, University of Konstanz, Konstanz, Germany.

Thierry Rabilloud
CEA-DSV-iRTSV-LBBSI and UMR CNRS 5092, CEA-Grenoble, 17 rue des martyrs, 38054, Grenoble cedex 9.

Ronald Roepman
Department of Human Genetics, Radboud University Nijmegen Medical Centre and Nijmegen Centre for Molecular Life Sciences, Nijmegen, The Netherlands.

Tatiana Rohner
Novartis Institutes for BioMedical Research, Discovery Technologies, Basel, Switzerland.

Robert Ros
Experimental Biophysics and Applied Nanoscience, Physics Faculty, Bielefeld University and Bielefeld Institute for Biophysics and Nanoscience (BINAS), Center for Biotechnology (CeBiTec), Bielefeld, Germany.

Alexandra Ros
Experimental Biophysics and Applied Nanoscience, Physics Faculty, Bielefeld University and Bielefeld Institute for Biophysics and Nanoscience (BINAS), Center for Biotechnology (CeBiTec), Bielefeld, Germany.

Sergio Rossell
BioCentre Amsterdam, Free University Amsterdam, Faculty of Earth and Life Sciences, Department of Molecular Cell Physiology, Amsterdam, The Netherlands.

Sarah Sanglier
Laboratoire de Spectrométrie de Masse Bio-Organique, ECPM, UMR/CNRS 7178, Institut Pluridisciplinaire Hubert CURIEN, Université Louis Pasteur, Strasbourg, France.

Christine Schaeffer
Laboratoire de Spectrométrie de Masse Bio-Organique, ECPM, UMR/CNRS 7178, Institut Pluridisciplinaire Hubert CURIEN, Université Louis Pasteur, Strasbourg, France.

Patrick Schindler
Novartis Institutes for BioMedical Research, Genome and Proteome Sciences/Systems Biology, Basel, Switzerland.

Eric C. Schirmer
Wellcome Trust Centre for Cell Biology, University of Edinburgh, Edinburgh, United Kingdom.

Albert Sickmann
Protein Mass Spectrometry and Functional Proteomics Group, Rudolf-Virchow-Center for Experimental Biomedicine, University of Wuerzburg, Wuerzburg, Germany.

Dieter Staab
Novartis Institutes for BioMedical Research, Discovery Technologies, Basel, Switzerland.

Markus Stoeckli
Novartis Institutes for BioMedical Research, Discovery Technologies, Basel, Switzerland.

Xiaodan Tian
Department of Chemistry, Laboratory of Analytical Chemistry and Biopolymer Structure Analysis, University of Konstanz, Konstanz, Germany.

Katja Toensing
Experimental Biophysics and Applied Nanoscience, Physics Faculty, Bielefeld University and Bielefeld Institute for Biophysics and Nanoscience (BINAS), Center for Biotechnology (CeBiTec), Bielefeld, Germany.

Alain Van Dorsselaer
Laboratoire de Spectrométrie de Masse Bio-Organique, ECPM, UMR/CNRS 7178, Institut Pluridisciplinaire Hubert CURIEN, Université Louis Pasteur, Strasbourg, France.

Karen van Eunen
BioCentre Amsterdam, Free University Amsterdam, Faculty of Earth and Life Sciences, Department of Molecular Cell Physiology, Amsterdam, The Netherlands.

Jan van Oostrum
Novartis Institutes for BioMedical Research, Genome and Proteome Sciences/Systems Biology, Basel, Switzerland.

Michel Vidal
UMR/CNRS 5235, University Montpellier II, Montpellier, France.

Hans Voshol
Novartis Institutes for BioMedical Research, Genome and Proteome Sciences/Systems Biology, Basel, Switzerland.

Annick Waldt
Novartis Institutes for BioMedical Research, Genome and Proteome Sciences/ Systems Biology, Basel, Switzerland.

Hans V. Westerhoff
BioCentre Amsterdam, Free University Amsterdam, Faculty of Earth and Life Sciences, Department of Molecular Cell Physiology, Amsterdam, The Netherlands and also Manchester Centre for Integrative Systems Biology, Manchester Interdisciplinary BioCentre, University of Manchester, United Kingdom.

Brigitte Wiedmann
Novartis Institutes for BioMedical Research, Infectious Disease Area, Cambridge, Massachusetts, United States.

Frank A. Witzmann
Department of Cellular & Integrative Physiology, Indiana University School of Medicine Biotechnology Research & Training Center, Indianapolis, United States.

Uwe Wolfrum
Department of Cell and Matrix Biology, Institute of Zoology, Johannes Gutenberg University of Mainz, Germany.

René P. Zahedi
Protein Mass Spectrometry and Functional Proteomics Group, Rudolf-Virchow-Center for Experimental Biomedicine, University of Wuerzburg, Wuerzburg, Germany.

INTRODUCTION

As proteomics technologies are reaching a plateau in the number of proteins that can be resolved and detected, pre-fractionation steps have become essential to increase the depth of proteomic analysis. So far, many pre-fractionation steps have been based on chromatography methods where the proteins are separated according to their individual physicochemical properties. Subcellular fractionation methods proved to be very potent protein pre-fractionation steps: they allow the representation of low abundance proteins, and they can be combined with chromatography steps. Moreover, as the isolated subcellular components also represent functional units, subcellular fractionation allows the proteomic analysis of protein subsets that are functionally related in a biologically relevant manner. The first three sections of this volume deal with different levels of subcellular organization that also correspond to specific methodological approaches.

In his keynote chapter, Thierry Rabilloud superbly introduces the first section with a thorough definition of membrane proteomics where he pinpoints key theoretical and practical issues of this field, thereby setting the stage for the next contributions. Miguet et al. address the first key issue: the quality of the membrane preparation; they introduce and validate a microparticle strategy for plasma membrane purification. Zahedi et al. deal with the second major issue, the resolving of the hydrophobic proteins found in biological membrane samples, which they solve through two-dimensional BAC/SDS-PAGE gel electrophoresis. To close the first section, Foster and Chan review the proteomics of lipid rafts, membrane structures that are involved in intracellular trafficking and signal transduction. They describe a clever validation scheme based on the sensitivity of lipid rafts to cholesterol disruption.

A central theme in the second section on organelle subproteomes is the variability of their composition and how it can be exploited and interpreted. Kavanagh et al. describe a state-of-the-art substractive proteomics scheme that relies on an in silico purification step based on the comparison of organelle subproteomes. With this approach, they could demonstrate variations in subproteome content across tissues. In the next chapter on synatosome proteomics, Bai and Witzmann review the current efforts to correlate synaptic plasticity and variations in synaptic subproteome content with a special emphasis on post-translational modifications. Finally, Olver and Vidal discuss how the proteomic analysis of exosomes would give clues to the molecular basis of their biogenesis and contribute to a better understanding of their function. Moreover, they propose that the variations observed in exosome protein content are useful for biomarker discovery.

The third section deals with protein complexes, which are considered as the molecular machinery that performs most cell functions. This area is certainly not a trivial one: there are several types of protein complexes and protein-protein interactions and it is not always clear which methodology is most suitable to use in either context. In their chapter, Boissy and Collura sort out for us the concepts and methods encountered in interactomics, guide us through data interpretation issues and share with us their insight on the very nature of interactions. They plead for a systematic integration of interaction maps with functional genomics and molecular genetics data: the potential of such an approach is strikingly demonstrated in the next two chapters. Collins and Grant examine the molecular architecture of membrane associated signalling complexes in the nervous system, they highlight the role scaffolding proteins within these complexes and point out to a few much needed construction rules. Aligning interaction and functional genomics data, they build a case for a modular organisation of large complexes into functional sub-networks. To complete the picture, based on a thorough review on the complexes of the photoreceptor cilia, Roepman and Wolfrum sketch out an approach to organize complexes in functional modules and investigate their interactions.

Assuming that the protein content of an organelle has been inventoried and its protein complexes characterized, the next step is to translate this knowledge into functionally relevant interpretations. This is the purpose of systems biology: if we consider organelles as systems that function and communicate with each other through their protein machinery, it makes sense to apply such an approach at the subcellular level. In their chapter, Caruso and Chevet prove that this concept can actually be applied to reconstruct the stress signalling network of the endoplasmic reticulum. They build an integrative signalling map that qualitatively accounts for the interactions of the stress network with other endoplasmic reticulum machineries and also with other organelles. On the quantitative side, Bruggeman et al. introduce a theoretical framework for subcellular systems biology and thoroughly review the relevant mathematical approaches. Drawing on their extensive experience of metabolic networks, they argue that a direct translation of subcellular units as modules within a mathematical model of the cell can be advantageous both for solving the problem and interpreting the results. Garcia Osuna and Murphy survey the current automated methods for high-throughput determination of protein subcellular location that are used to reconstruct subcellular anatomy at high resolution. These methods provide essential information on the dynamic aspect of subcellular events in individual cells: it would also be extremely interesting to combine them with the molecular switches described by Martini et al. in the next section.

This brings us to the fifth and last section of this volume, where the most recent technological developments in proteomics are reviewed. Martini et al. introduce the emerging field of systems nanobiology that relies on ultrasensitive methods and instruments to investigate cellular processes at the single molecule level. An application, among others, is to monitor the dynamics of protein translocation from one subcellular compartment to another using two-photon laser scanning microscopy and a photoactivable GFP as a molecular switch. Faupel et al. review the current applications of biophotonic technologies to proteomics with a focus on mass spectrometry based molecular imaging. Tian et al. describe a fast-track approach for the characterization of antibody epitopes using Fourier transform ion cyclotron resonance mass spectrometry (FTICR-MS). The application of this method to the Amyloid Precursor Protein has important consequences for the study of intracellular processing pathways relevant to Alzheimer's disease. Finally, Mueller et al. describe the interfacing of LC and MALDI-MS and – MS/MS, discuss its performance, and present selected applications in the proteomics field including the analyses of membrane proteins and protein interactions.

SECTION 1

MEMBRANE PROTEOMICS

CHAPTER 1

KEYNOTES ON MEMBRANE PROTEOMICS

THIERRY RABILLOUD
INSERM U548, France

Table of Contents

1. INTRODUCTION

Before addressing the key issues pertaining to proteomics of membrane proteins, which is an important subject in the wider topic of subcellular proteomics, the first question that must be asked is the definition of a membrane protein. While this might seem trivial at the first glance, the answer is far from being obvious. A commonly accepted definition for a membrane protein is a protein associated with a membrane, that is, a lipid bilayer. However, the meaning of the word "associated" in this case is the subject of almost infinite debate. In fact, two different situations arise. In the first, simple situation, the polypeptide chain spans the lipid bilayer a certain number of times. These proteins are defined as integral or intrinsic membrane proteins, as their association with the lipid bilayer is doubtless. The second, much more complicated case, is when the association with the lipid bilayer is not achieved by transmembrane segments or by transmembrane barrels (e.g. in porins), because many other types of association with membranes are encountered. In one particular type, the association is mediated by a post translational modification of the polypeptide, namely the grafting of an fatty acid or polyisoprenyl chain. Polyisoprenylation has been demonstrated to result in physical association with the membrane, as shown for small G proteins

E. Bertrand and M. Faupel (eds.), Subcellular Proteomics, 3–11.

(Verma et al. 1994). Strong association of proteins with lipid membranes is also achieved by glycolipid anchors such as the glycosylphosphatidyl inositol common in eukaryotic species, or, as in bacterial lipoproteins, by an *N*-acyl diglyceride linkage to an *N*-terminal cysteine residue that becomes available after cleavage of a signal sequence (Cross 1991). Last but not least, the protein can also be acylated by a simple fatty acid chain. In this case, the reality of the anchoring to the lipid bilayer is much less clear.

Protein-membrane association via a post-translational modification introduces the notion of dynamic association and partitioning of proteins between the membrane phase of the cells and the aqueous phase (cytosol or inner phase of organelles). Consequently, such proteins can be found both as membrane-associated and membrane-free, which is not the case with intrinsic membrane proteins which are strictly membrane embedded. Another type of association to membrane is mediated by protein–protein interactions with other membrane proteins. A typical example of this situation is provided by the respiratory complexes. In the case of ubiquinol-cytochrome c oxidoreductase, core proteins 1 and 2 does not show any interaction with the lipid membrane, but only with the protein subunits spanning the membrane (e.g. cytochrome b) (Iwata et al. 1998).

However, the situation is often much less clear than this one, and drives quite fast into the infinite debate of what is a membrane protein apart from the integral ones. In addition to the dynamic view of association to membranes, this complicated situation arises mainly from the operational, biochemical, definition of membranes. When a cell is lysed in an aqueous, detergent-free, medium, the cell-limiting membrane and the network of inner membranes (when applicable) will fragment into vesicles which can be separated from the bulk of cytosol by sedimentation or partition techniques. These techniques are also able to separate membrane-bound organelles (e.g. mitochondria and plasts) from other cellular components, providing good purity. However, not every protein present in such preparations can be considered as a membrane protein. For example, when vesicles will form from larger structures upon lysis, simple physical entrapment will bring soluble proteins in the lumen of the vesicles. Furthermore, it appears more and more clearly that some classes of proteins (e.g. cytoskeletal proteins or ribosomal proteins) are directly or indirectly associated to bona fide transmembrane proteins or to the lipid bilayer. Should these proteins be classified as membrane proteins? Should the non-membrane spanning subunits of respiratory complexes be considered as membrane proteins? This gives an example of the debate that can take place on the notion of membrane proteins.

This could be seen as a very theoretical debate, but this has indeed very practical implications. Let us take the example of the analysis of a reticulum preparation. Endoplasmic reticulum is the place of synthesis of most secreted proteins and of most transmembrane proteins. As such, it contains many ribosomes (the Rough ER) and the reticulum vesicles are known to be associated with cytoskeletal filaments. The problem arises from the fact that the protein content of these cytoskeletal and ribosomal "contaminants" is concentrated in a few proteins. This means in turn that even if these "contaminants" represent a few percent of the total protein mass of the preparation, the few proteins represented will be easily detected by any proteomics analysis method.

Conversely, the "real" reticulum proteins are scattered among almost all the secreted proteins and transmembrane proteins of the cells, plus the resident proteins of the ER. This means in turn that these proteins will appear as less prominent upon proteomics analysis, just because each protein species is more diluted than the few "contaminating" ones.

2. SAMPLE PREPARATION ISSUES FOR MEMBRANE PROTEOMICS

A consequence of the above considerations is that sample preparation is very critical in membrane proteomics. As the volume occupied by the lipid bilayer in a membrane sample is very small in comparison to the volume of the aqueous phase, the transmembrane proteins must be considered as rare components of the sample, although they often drive most of the interest of the researchers. In addition to that, these transmembrane proteins often pose a very difficult solubilization problem. Transmembrane proteins usually contain domains protruding in the extracellular environment or in the cytosol. These domains behave as classical protein domains and are therefore stable in a water-based solvent. However, membrane proteins also contain domains that are embedded more or less deep in the lipid bilayer. Consequently, these domains are stable in a hydrocarbon-like environment. If a membrane protein is to be solubilized prior to its purification, this means that the solvent which is used to this purpose must show at the same time water-like and hydrocarbon-like properties. A water-based solvent will induce aggregation of the membrane domains via hydrophobic interactions, and thus protein precipitation. However, an organic solvent will induce in most cases precipitation of the membrane proteins via their water-soluble domains, which are denatured and coalesce in such a solvent. Some very hydrophobic proteins, however, are soluble in organic solvents (e.g. Molloy et al. 1999; Blonder et al. 2003). This is due to the reduced size of their non-membrane domains, and to the fact that all water-soluble protein domains contain a hydrophobic core, which can be soluble in organic solvents. In some cases, these positive solubilization forces can overcome the precipitation-deriving forces and make the protein soluble in organic solvents.

This is however not a general case, and the general rule for membrane proteins is that they require a solvent that has as the same time strong water-like and strong hydrocarbon-like properties. This is not the situation of mixed solvents, that is mixtures of water and water-miscible solvents, which only offer average properties and not a combination of both properties. Such a combination is only offered by a stable dispersion of hydrocarbon chains in a water-based solvent. This is the definition of lipid membranes, but such assemblies are very difficult to handle along a purification process. Hopefully, this definition is also the one of detergent micelles, which are much easier to handle along a purification process. These constraints explain why detergents are some kind of a universal tool for the disruption of cell membranes and for the solublization of membrane proteins.

The micelle-forming ability of detergents is driven by their chemical structure, which combine a hydrophobic part (the tail), promoting aggregation of the molecules into the micelles, which is linked to a hydrophilic one (the head) promoting water solubility of the individual molecules and of the micelles. By using different structures or heads and tails and by combining various tails with various heads, an almost infinite range of detergents can be generated. Their protein and lipid solubilization properties are of course strongly dependent on the chemical properties of the heads and tails. As a rule of thumb, rigid tails (e.g. steroid-based) and weakly polar heads (e.g. glycosides, oligoethylene glycol) generally lead to "mild" detergents, that is chemicals that have good lipid solubilization properties, but that have weak protein dissociation properties. While these detergents do not denature proteins, they can prove unable to prevent protein–protein interactions promoting precipitation. Consequently, their protein solubilization properties are highly variable, and a lengthy optimization process in the detergent choice is usually needed for various membrane proteins.

Conversely, detergent having a flexible tail and a strongly polar (i.e. ionic) head are viewed as "strong" detergents. In addition of their lipid solubilizing properties, these detergents are able to disrupt the hydrophobic interactions maintaining the structure of the proteins. They are therefore denaturing. However, because of their ionic nature, the bound detergent imparts a net electrical charge to the denatured proteins and induces a strong electrostatic repulsion between protein molecules. Thus, even denatured proteins can no longer aggregate, and these ionic detergents are very powerful protein solubilizing agents.

These protein solubilization conditions have a key impact on the protein fractionation that can be carried out afterwards, in the sense that they will restrict the choice to techniques with which they are compatible. As an example, ionic detergents are not compatible with any technique using protein charge or pI as the fractionation parameter.

All of the above is mainly true when some fractionation is to be carried out at the protein level. In some approaches, the fractionation is carried out only on peptides arising from the digestion of the membrane preparation. In this case too, the choice of the solubilization media will be dictated by the constraints imposed by the subsequent peptide fractionation process. However, the use of chromatographic peptide fractionation tools, and especially those based on reverse-phase chromatography, will make the use of detergents more problematic, leading to specially designed, detergent-free protocols (Wu et al. 2003; Fischer et al. 2006).

3. ISSUES LINKED TO THE SEPARATION PROCESS

Once the membrane proteins have been solubilized, usually in a water-detergent medium, they must be fractionated. As each fractionation method brings its own constraints, this will further restrict the scope of possibilities that can be used for sample solubilization. The following sections will exemplify some constraints linked to the most commonly used fractionation processes in proteomics.

3.1. Constraints in Proteomics Based on Zone Electrophoresis

In these proteomics methods, the separation process is split in two phases (e.g. in Bell et al. 2001). The first phase is a protein separation by denaturing zone electrophoresis, that is in the presence of denaturing detergents, most often sodium dodecyl sulfate (SDS). The second phase is carried out by chromatography on the peptides produced by digestion of the separated proteins. This has no impact on the sample preparation itself, which just needs to be compatible with the initial zone electrophoresis.

This is by far the simplest case. Sample preparation is achieved by mixing the initial sample with a buffered, concentrated solution of an ionic detergent, usually containing a reducer to break disulfide bridges and sometimes an additional nonionic chaotrope such as urea. Ionic detergents are among the most powerful protein denaturing solubilizing agents. Their strong binding to proteins make all proteins to bear an electric charge of the same type, whatever their initial charge may be. This induces in turn a strong electrostatic repulsion between protein molecules, and thus maximal solubility. The system of choice is based on SDS, as this detergent binds rather uniformly to proteins. However, SDS alone at room temperature, even at high concentrations, may not be powerful enough to denature all proteins. This is why heating of the sample in the presence of SDS is usually recommended. The additional denaturation brought by heat synergizes with SDS to produce maximal solubilization and denaturation. However, some hydrophobic transmembrane proteins do not withstand this heating step. In this case, their hydrophobic parts coagulate through the effect of heat much faster than binding of SDS can solubilize them. This leads to precipitation of these proteins.

The use of SDS is not always without drawbacks. One of the most important is encountered when the sample is rich in DNA. A terrible viscosity results, which can hamper the electrophoresis process. Moreover, some protein classes (e.g. glycoproteins) bind SDS poorly and are thus poorly separated in the subsequent electrophoresis. In such cases, it is advisable to use cationic detergents. They are usually less potent than SDS, so that a urea-detergent mixture must be used for optimal solubilization (MacFarlane 1989). Moreover, electrophoresis in the presence of cationic detergents must be carried out at a very acidic pH, which is not technically simple but still feasible (MacFarlane 1989). This technique has however gained recent popularity as a double zone electrophoresis method able to separate even membrane proteins (Hartinger et al. 1996), and showing more separation power than SDS electrophoresis alone.

An important and often overlooked variegation of zone electrophoresis-based separations is the mixed native-denaturing two-dimensional (2D) electrophoresis developed mainly by Schägger and coworkers (Schägger and Von Jagow 1991). Because the first dimension is a native electrophoresis, this approach provides invaluable information upon the assembly of membrane complexes. However, it poses in turn tricky solubilization problems, as the powerful ionic detergents cannot be used because of their denaturing power. The answer to this difficult problem is the combined used of a salt (or a dielectric compound such as aminocaproic acid), a neutral, non-denaturing detergent and a reagent bringing additional electrical charges to the protein complexes to increase their solubility and prevent their aggregation during electrophoresis. This charge-transfer reagent is usually a protein dye such as Coomassie Blue.

Because of the complexity of this solubilization problem, the optimization of the system is usually difficult, and this system has been most often used for mitochondrial and chloroplastic membrane complexes (reviewed in Nijtmans et al. 2002), where it has shown the ability to analyze even very hydrophobic membrane proteins (Devreese et al. 2002). However, this separation approach has been recently applied to whole cell extracts (Camacho-Carjaval et al. 2004) and also to nuclear proteins (Novakova et al. 2006).

3.2. Constraints Proteomics Based on Two-dimensional Gel Electrophoresis

In this scheme, the proteins are first separated by isoelectric focusing (IEF) followed by SDS electrophoresis. The constraints made on sample preparation are thus those induced by the IEF step. One of these constraints is the impossibility to use ionic detergents at high concentrations, as they would mask the protein charge and thus dramatically alter its isoelectric point (pI). Ionic detergents can however be used at low doses to enhance initial solubilization (Wilson et al. 1977), but their amount is limited by the capacity of the IEF system (in terms of ions tolerated) and by the efficiency of the detergent exchange process with takes place during the IEF step. Another major constraint induced by IEF is the requirement for low ionic strength, induced by the high electric fields required for pushing the proteins to their isoelectric points. This means in turn that only uncharged compounds can be used to solubilize proteins, that is neutral chaotropes and detergents. The basic solubilization solutions for IEF thus contain high concentrations of a non-ionic chaotrope, historically urea but now more and more a mixture of urea and thiourea (Rabilloud et al. 1997), together with a reducer and a nonionic detergent. While CHAPS and Triton X-100 are the most popular detergents, it has been recently shown that other detergents can enhance the solubility of proteins and give better performances (Chevallet et al. 1998; Luche et al. 2003). In this solubilization process, detergents play a multiple role. They bind to proteins and help to keep them in solution, but they also break protein-lipids interactions and promote lipid solubilization.

Last but certainly not least, it must be recalled that proteins are at their pI at the end of the IEF process, and the pI is the solubility minimum. This means that solubility problems are very important in IEF. This is especially true with native proteins, and IEF is thus best performed with denatured proteins. However, even in this case, many proteins such as membrane proteins are poorly soluble under IEF conditions. Compared with detergent-based electrophoresis, this problem is further enhanced by the fact that only electrically neutral detergents can be used. Ionic detergents would modify the pI of the proteins and thus prevent any correct separation through this parameter. This however means in turn that the beneficial electrostatic repulsion effect cannot be used in isoelectric focusing, thereby leaving proteins under conditions in which their solubility is minimal. While this is not so much a problem for proteins that are normally water-soluble, this turns to be a major problem for proteins that are poorly water-soluble, and especially transmembrane proteins. This explains why this class of proteins is strongly under-represented in this type of separation (for review see Santoni et al. 2000).

3.3. Constraints in Peptide Separation-Based Approaches

In these methods, the sample preparation process is designed to offer optimal digestion of the proteins into peptides. This means that proteins must be extracted from the sample and denatured to maximize exposure of the protease cleavage sites. This also means that the protease used for peptide production must be active in the separation method. In this case, the robustness of many proteases is a clear advantage. Classical extraction media usually contain either multimolar concentrations of chaotropes or detergents. In the latter case, the sample is usually solubilized and denatured in high concentrations of ionic detergents, and simple dilution is used to bring the detergent concentration down to a point compatible with other steps such as chemical labeling or proteolysis.

The choice between chaotropes and detergent is driven mainly by the constraints imposed by the peptide separation method. In the wide-scope approach based on online two-dimensional chromatography of complex peptide mixtures (Washburn et al. 2001), both the ion exchange and reverse phase steps are very sensitive to detergents. It must be mentioned that these online two-dimensional chromatographic methods are one of the rare cases where the interface between the two separation methods does not bring extra robustness, so that the sample preparation must be compatible with both chromatographic methods. The modification approach (Gevaert et al. 2003), which uses intensively reverse phase chromatography, is also very sensitive to detergent interference. This rules out the use of detergent and favors the use of chatropes, generally nonionic ones because of the ion exchange step. Urea is used for these methods, with possible artefacts induced by urea driven carbamylation of the sample during the lengthy digestion process. Inclusion of thiourea and lowering of the urea concentration could decrease the incidence of carbamylation in these methods. In methods where a detergent-resistant method is used, for example avidin selection of biotinylated peptides (Gygi et al. 1998), extraction by SDS is clearly the method of choice, as the above-mentioned drawbacks are absent.

3.4. Constraints in Protein Chromatography-Based Approaches

In these methods, the separation is carried out by a chromatographic setup. This setup may use various chromatographic principles for protein separation, and this choice will of course alter the possibilities for protein solubilization. For example, the use of an ion-exchange step (Szponarski et al. 2004) will preclude the use of ionic detergents and thus produce a solubilization situation close to the one observed in 2D gels. However, an important advantage of ion exchange over chromatofocusing or IEF is that the proteins do not need to reach their pI in this separation scheme. This allows in turn to increase protein solubility, which is a major concern in transmembrane protein analysis.

Here again, not all chromatographic setups are usable for any proteomics question. The use of protein reverse phase chromatography, which has been advocated for plasma proteomics (Moritz et al. 2005), precludes in turn the use of any detergent of any type. This prevents the use of this chromatographic setup in most subcellular proteomics experiments, where detergents must be used to solubilize the membrane limiting the subcellular compartments.

4. CONCLUDING REMARKS

It should be obvious from the above that membrane proteomics strictly follows Murphy's law. This is due to the fact that there is a mutual exclusion, for physico-chemical reasons, between on the one hand the conditions that must be used to solubilize in water all membrane proteins, including the most hydrophobic ones, and thus give a fair representation of the protein population in the sample, and on the other hand the conditions prevailing in the high resolution peptide separation methods. On top of this problem, there is a second problem linked to the membrane versus aqueous phase volume in many subcellular preparations, which makes transmembrane proteins rare compared to water-soluble proteins.

Thus, everything concurs in making membrane proteomics probably one of the most difficult sub-field in proteomics. Up to now, the proteomics toolbox that we can use has proven quite imperfect to provide a thorough, quantitative and precise (i.e. including post-translational modifications analysis) analysis of membrane proteins.

REFERENCES

Bell, A.W., Ward, M.A., Blackstock, W.P., Freeman, H.N., Choudhary, J.S., Lewis, A.P., Chotai, D., Fazel, A., Gushue, J.N., Paiement, J., Palcy, S., Chevet, E., Lafreniere-Roula, M., Solari, R., Thomas, D.Y., Rowley, A. and Bergeron, J.J. (2001) Proteomics characterization of abundant Golgi membrane proteins. J. Biol. Chem. *276*, 5152–5165.

Blonder, J., Conrads, T.P., Yu, L.R., Terumuma, A., Janini, G.M., Issaq, H.J., Vogel, J. and Veenstra, T.D. (2004) A detergent- and cyanogen bromide-free method for integral membrane proteomics: application to Halobacterium purple membranes and human epidermis. Proteomics *4*, 31–45.

Camacho-Carvajal, M.M., Wollscheid, B., Aebersold, R., Steimle, V. and Schamel, W.W. (2004) Two-dimensional Blue native/SDS gel electrophoresis of multi-protein complexes from whole cellular lysates: a proteomics approach. Mol. Cell. Proteomics *3*, 176–182.

Chevallet, M., Santoni, V., Poinas, A., Rouquié, D., Fuchs, A., Kieffer, S., Rossignol, M., Lunardi, J., Garin, J. and Rabilloud, T. (1998) New zwitterionic detergents improve the analysis of membrane proteins by two-dimensional electrophoresis. Electrophoresis *19*, 1901–1909.

Cross, G.A. (1990) Glycolipid anchoring of plasma membrane proteins. Ann. Rev. Cell. Biol. *6*, 1–39.

Devreese, B., Vanrobaeys, F., Smet, J., Van Beeumen, J. and Van Coster, R. (2002) Mass spectrometric identification of mitochondrial oxidative phosphorylation subunits separated by two-dimensional blue-native polyacrylamide gel electrophoresis. Electrophoresis *23*, 2525–2533.

Fischer, F., Wolters, D., Rogner, M. and Poetsch, A. (2006) Toward the complete membrane proteome: high coverage of integral membrane proteins through transmembrane peptide detection. Mol. Cell. Proteomics *5*, 444–453.

Gevaert, K., Goethals, M., Martens, L., Van Damme, J., Staes, A., Thomas, G.R. and Vandekerckhove, J. (2003) Exploring proteomes and analyzing protein processing by mass spectrometric identification of sorted *N*-terminal peptides. Nat. Biotechnol. *21*, 566–569.

Gygi, S.P., Rist, B., Gerber, S.A., Turecek, F., Gelb, M.H. and Aebersold, R. (1999) Quantitative analysis of complex protein mixtures using isotope-coded affinity tags. Nat. Biotechnol. *17*, 994–999.

Hartinger, J., Stenius, K., Hogemann, D. and Jahn, R. (1996) 16-BAC/SDS-PAGE: a two-dimensional gel electrophoresis system suitable for the separation of integral membrane proteins. Anal. Biochem. *124*, 126–133.

Iwata, S., Lee, J.W., Okada, K., Lee, J.K., Iwata, M., Rasmussen, B., Link, T.A., Ramaswamy, S. and Jap, B.K. (1998) Complete structure of the 11-subunit bovine mitochondrial cytochrome bc1 complex. Science *281*, 64–71.

Luche, S., Santoni, V. and Rabilloud, T. (2003) Evaluation of nonionic and zwitterionic detergents as membrane protein solubilizers in two-dimensional electrophoresis. Proteomics *3*, 249–253.

Macfarlane, D.E. (1989) Two dimensional benzyldimethyl-n-hexadecylammonium chloride-sodium dodecyl sulfate preparative polyacrylamide gel electrophoresis: a high capacity high resolution technique for the purification of proteins from complex mixtures. Anal. Biochem. *176*, 457–463.

Molloy, M.P., Herbert, B.R., Williams, K.L. and Gooley, A.A. (1999) Extraction of *Escherichia coli* proteins with organic solvents prior to two-dimensional electrophoresis. Electrophoresis *20*, 701–704.

Moritz, R.L., Clippingdale, A.B., Kapp, E.A., Eddes, J.S., Ji, H., Gilbert, S., Connolly, L.M. and Simpson, R.J. (2005) Application of 2-D free-flow electrophoresis/RP-HPLC for proteomic analysis of human plasma depleted of multi high-abundance proteins. Proteomics *5*, 3402–3413.

Nijtmans, L.G., Henderson, N.S. and Holt, I.J. (2002) Blue Native electrophoresis to study mitochondrial and other protein complexes. Methods *26*, 327–334.

Novakova, Z., Man, P., Novak, P., Hozak, P. and Hodny, Z. (2006) Separation of nuclear protein complexes by blue native polyacrylamide gel electrophoresis. Electrophoresis *27*, 1277–1287.

Rabilloud, T., Adessi, C., Giraudel, A. and Lunardi, J. (1997) Improvement of the solubilization of proteins in two-dimensional electrophoresis with immobilized pH gradients. Electrophoresis *18*, 307–316.

Santoni, V., Molloy, M.P. and Rabilloud, T. (2000) Membrane proteins and proteomics: un amour imposssible? Electrophoresis *21*, 1054–1070.

Schagger, H. and von Jagow, G. (1991) Blue native electrophoresis for isolation of membrane protein complexes in enzymatically active form. Anal. Biochem. *199*, 223–231.

Szponarski, W., Sommerer, N., Boyer, J.C., Rossignol, M. and Gibrat, R. (2004) Large-scale characterization of integral proteins from Arabidopsis vacuolar membrane by two-dimensional liquid chromatography. Proteomics *4*, 397–406.

Verma, D.P.S., Cheon, C.I. and Hong, Z. (1994) Small GTP-binding proteins and membrane biogenesis in plants. Plant Physiol. *106*, 1–6.

Washburn, M.P., Wolters, D. and Yates, J.R. 3rd. (2001) Large-scale analysis of the yeast proteome by multidimensional protein identification technology. Nat. Biotechnol. *19*, 242–247.

Wilson, D., Hall, M.E., Stone, G.C. and Rubin, R.W. (1977) Some improvements in two-dimensional gel electrophoresis of proteins. Protein mapping of eukaryotic tissue extracts. Anal. Biochem. *83*, 33–44.

Wu, C.C., MacCoss, M.J., Howell, K.E. and Yates, J.R. 3rd. (2003) A method for the comprehensive proteomic analysis of membrane proteins. Nat. Biotechnol. *21*, 532–538.

Corresponding author: Dr. Thierry Rabilloud (thierry.rabilloud@cea.fr)

CHAPTER 2

TWO-DIMENSIONAL BAC/SDS-PAGE FOR MEMBRANE PROTEOMICS

RENÉ P. ZAHEDI, JAN MOEBIUS, and ALBERT SICKMANN
University of Wuerzburg, Germany

Table of Contents

Abstract: Although often used in membrane proteome studies, conventional two-dimensional gel electrophoresis (2-DE) is not well suited for resolving hydrophobic proteins. Nevertheless, an alternative technique, two-dimensional BAC/SDS-PAGE (2-DB) using the cationic detergent benzyldimethyl-*n*-hexadecylammonium chloride (BAC) in the first and the anionic detergent SDS in the second dimension can be utilized as a powerful tool for the separation and analysis of membrane proteins. Systematic studies demonstrated the advantage of 2-DB over one-dimensional SDS-PAGE and 2-DE with regard to membrane proteomics. While in 2-DE gels, in particular proteins with more than one transmembrane domain (TMD) are underrepresented, one-dimensional SDS-PAGE lacks sufficient resolution for large scale analyses. In contrast, 2-DB enabled the identification of extremely hydrophobic proteins like cytochrome-c oxidase subunit I from *S. cerevisiae* with a total of 12 known TMD. Especially the application of tube gels in the first dimension as well as the recent introduction of improved buffer systems hold a great potential for future 2-DB-based membrane studies.

E. Bertrand and M. Faupel (eds.), Subcellular Proteomics, 13–20.

1. INTRODUCTION

The separation of membrane proteins via conventional two-dimensional gel electrophoresis (2-DE) consisting of a first dimension isoelectric focussing step and a subsequent second dimension SDS-PAGE is, although still widely used, strongly biased against hydrophobic proteins. One the one hand, IEF is limited to the usage of zwitterionic or non-ionic detergents, providing weak solubilizing capabilities with regard to hydrophobic proteins in contrast to strong ionic detergents, for example SDS. On the other hand, proteins tend to precipitate upon reaching their isoelectric points (pI) or during the transfer to the second dimension – this is particularly true for membrane proteins. Furthermore, membrane proteins often have pIs in the alkaline region, which in 2-DE is generally characterized by inferior resolution.

Although protocols have been constantly improved in recent years (Molloy 2000; Olsson et al. 2002; Luche et al. 2003), hydrophobic membrane proteins, especially those with multiple transmembrane domains (TMD) are generally underrepresented in 2-DE based proteome studies (Santoni et al. 2000). To account for these inherent limitations alternative electrophoretic techniques have to be applied.

While common one-dimensional SDS-PAGE is a powerful tool for separating hydrophobic membrane proteins (Reinders et al. 2006a), it only provides minor resolution of complex protein samples, and consequently is not suited for differential analyses.

However, alternative 2-DE methodologies, for instance combining SDS-PAGE with PAGE systems based on the usage of cationic detergents like benzyldimethyl-*n*-hexadecylammonium chloride (BAC) (Macfarlane 1983, 1989; Hartinger et al. 1996) or cetyltrimethylammonium bromide (CTAB) (Buxbaum 2003), enabling two-dimensional separation of hydrophobic membrane proteins, have gained more attention in recent years. Thereby, extremely hydrophobic membrane proteins with multiple TMD can be separated with relatively high resolution for subsequent identification by mass spectrometry.

2. TWO-DIMENSIONAL BAC/SDS POLYACRYLAMIDE GEL ELECTROPHORESIS

First introduced by Macfarlane et al. for the separation of base labile protein methylation (Macfarlane, 1983), BAC-PAGE was then improved towards a two-dimensional technique by combination with a subsequent second dimension SDS-PAGE (Macfarlane, 1989). Although, both dimensions comprise a separation according to the molecular weight, slight differences in protein migration properties within the two systems lead to an enhanced resolution, resulting in an elliptical separation area. Already in 1996, Hartinger et al. demonstrated the potential of combined two-dimensional BAC/SDS-PAGE (2-DB) for the separation of membrane proteins (Hartinger et al. 1996). The second dimension SDS-PAGE provides full compatibility with downstream methods like Western Blotting, a broad variety of staining

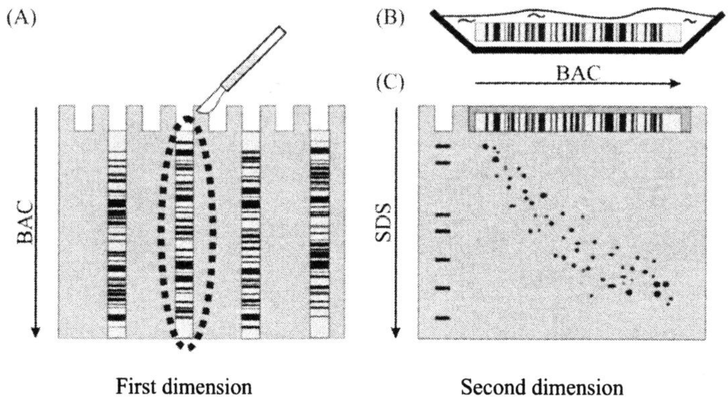

Figure 1. Scheme of a two-dimensional BAC/SDS PAGE using slab gels. (A) Protein samples are separated on the first dimension BAC-PAGE. After visualization of proteins, an entire gel lane is excised. (B) The excised lane is re-buffered in 100 mM Tris, pH 6.8 for 30 min and afterwards incubated in 3× SDS sample buffer for another 5-10 min in order to exchange BAC for SDS. (C) The gel lane is transferred onto the second dimension SDS gel and fixed with a hot agarose solution. After separation staining reveals a characteristic spot pattern within an elliptical area.

techniques and mass spectrometric analysis. Therefore, in recent years several studies have revealed the great potential of this alternative technique for membrane proteomics (Otto et al. 2001; Daub et al. 2002; Diao et al. 2003; Godl et al. 2003; Coughenour et al. 2004; Zahedi et al. 2006).

3. GENERAL WORKFLOW

The general scheme of 2-DB is depicted in Figure 1: Samples are first separated towards the cathode in an acidic PAGE system based on the cationic detergent BAC. Afterwards, protein lanes can be visualized by direct immersion into a colloidal Coomassie staining solution. However, since Coomassie tends to precipitate in presence of cationic detergents which in turn leads to enhanced background staining, an intermediate washing step of at least 2 h is recommended in order to remove BAC from the gel surface.

After visualization whole protein lanes are excised and prepared for the second dimension. Since SDS-PAGE utilizes a more alkaline buffer system than BAC-PAGE, first of all gel lanes have to be re-buffered. Afterwards, BAC is exchanged for SDS by incubation in 3× SDS sample buffer. Finally, protein lanes are transferred onto a second dimension gel for separation and fixed by a hot agarose solution.

4. IMPROVEMENTS

While less complex samples can generally be separated using small gel sizes (approximately 7 × 7 cm) in both dimensions, for samples of higher complexity switching

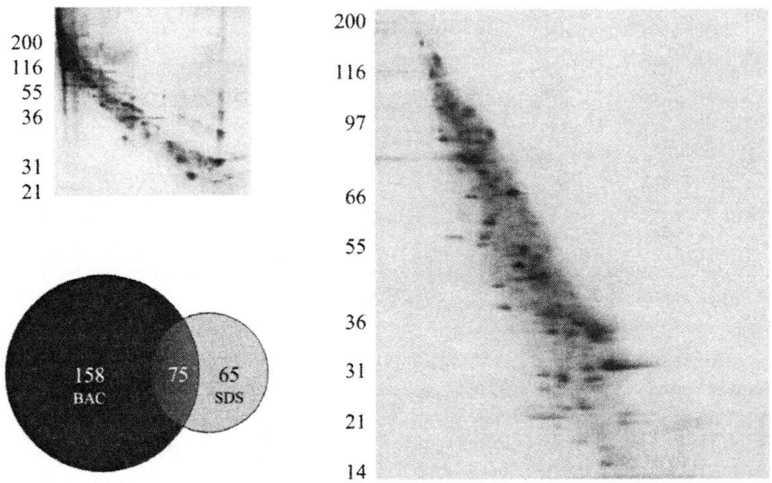

Figure 2. Two-dimensional BAC/SDS-PAGE of platelet membrane proteins. While not only large gels are recommended for complex samples, utilizing tube gels for the first dimension furthermore provides better resolution and more efficient transfer of proteins to the second dimension (less vertical smearing). (A) Small slab gel in the first dimension (7 × 7 cm). (B) Large tube gel in the first dimension (1 mm × 15 cm). (C) Summary of the human platelet membrane proteome study (Moebius et al. 2005). 158 proteins were exclusively identified from 2-DB gels, 65 from 1D-PAGE. An overlap of 75 proteins was identified from both types of gels.

to larger high resolution gels is recommended (Figure 2). Here, slab gels as well as tube gels can be utilized. However, the usage of slab gels is much more complicated in terms of handling (Figure 3).

In general, first dimension BAC-PAGE requires higher voltages and prolonged running times than SDS-PAGE, resulting in an increased heat development – especially in case of large dimension slab gels. Therefore, cooling of running buffers during separation is mandatory.

Besides, an incomplete transfer of proteins from the first to the second dimension can be noticed for slab gels. The usage of tube gels with inner diameters of 1 mm and less totally abolishes this limitation and furthermore provides an improved resolution leading to a higher number of identifications in proteome studies. For this reason, the usage of tube gels is essential for differential studies. Moreover, after first dimension separation, time-consuming staining procedures which might be accompanied by loss of material can be omitted, as the entire gel can be transferred to the second dimension without the need for prior excision.

Resulting from our experience in the separation of a broad variety of samples with 2-DB, resolution strongly depends on the nature of the separated sample. Furthermore, upon long separation times, band broadening during first dimension separation can be observed.

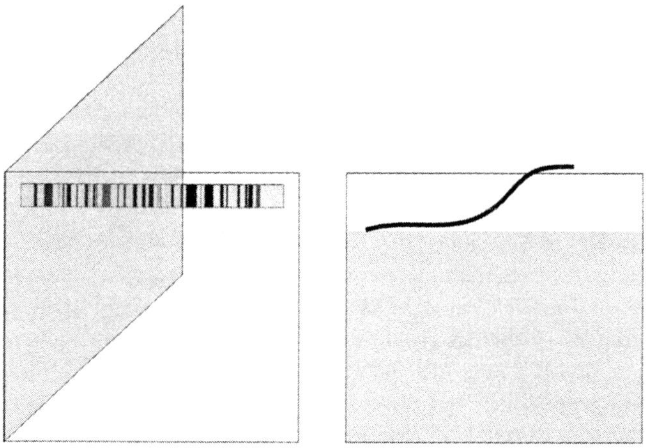

Figure 3. Processing of large slab (left) and tube (right) gels. Since transferring whole gel lanes of approximately 15–20 cm length onto polymerized second dimension gels mostly results in extensive damage of the gel, a more complicated but also safer strategy is recommended here. First the excised gel lane is placed and fixed between two glass plates. Then the separation gel is poured beneath the lane. Finally, a stacking gel is poured on top, enclosing the first dimension gel lane. In case of tube gels, the gel is transferred onto the already polymerized second dimension separation gel and fixed by hot agarose solution.

In general, after membrane purification, it is highly recommended to introduce additional steps like carbonate- (Fujiki et al. 1982) and/or Triton X114-extraction prior to electrophoresis in order to reduce the amount of contaminating soluble proteins. Especially when separating plasma membrane enriched samples, resolution may be impaired in both dimensions. In that case protein precipitation prior to 2-DB may result in an improved separation, however it has to be kept in mind that precipitation procedures are generally not quantitative and might lead to unspecific loss of material.

Recently, an improved BAC-PAGE protocol was introduced, particularly compensating for the inferior efficiency of the stacking gel when compared to common SDS-PAGE systems (Kramer 2006). By systematic studies the composition of gel buffers, running buffers as well as the sample buffer were optimized, resulting in a higher resolution and shorter separation time, comparable to SDS-PAGE (Kramer 2006). Although the impact of these improvements was only investigated for one-dimensional BAC-PAGE, they nevertheless hold a great potential for 2-DB as well. However, this remains to be demonstrated in future studies.

5. POTENTIAL FOR MEMBRANE PROTEOME STUDIES

In systematic studies we demonstrated the potential of 2-DB compared to 2-DE (Zahedi et al. 2005) and SDS-PAGE (Moebius et al. 2005) regarding the separation of membrane proteins.

While 2-DE separation of purified endoplasmatic reticulum membranes from *Canis familiaris* yielded only a few spots after visualization by silver staining, an unequal higher amount of protein spots could be visualized after 2-DB. Among them, Sec61α with a total of ten known TMD and a grand average hydrophobicity (GRAVY) index of 0.558 could be identified by Western blotting as well as mass spectrometry. Furthermore, 54 distinct ribosomal proteins were identified, which cannot be sufficiently resolved by 2-DE since their pIs range from 9 to 12. Furthermore, the separation of mitochondrial membranes from *Saccharomyces cerevisiae* yielded the subsequent identification of the extremely hydrophobic cytochrome-c oxidase subunit I with a total of 12 known TMD and a GRAVY index of 0.74 by mass spectrometry (MS).

In a comprehensive study of the human platelet membrane proteome we demonstrated the need for a combined application of 2-DB and one-dimensional SDS-PAGE for maximizing the amount of identified proteins. Since both techniques address different subproteomes (Figure 2C) they may be utilized in a complementary way. In 2-DB, a higher resolution increases the local protein concentration and facilitates identification. However, in SDS-PAGE, whole lanes can be cut into equidistant slices, eliminating the need for protein visualization prior to excision. Thereby, even proteins, which cannot be visualized by staining procedures and therefore will escape detection after 2-DB, can be identified by subsequent MS.

6. COMPARISON TO OTHER TECHNIQUES

Another two-dimensional PAGE technique, doubled SDS (dSDS), introduced by Rais et al. (2004), is capable of resolving hydrophobic proteins with GRAVY indices of up to 0.86 – however it is has a lower resolution compared to 2-DB due to a smaller accessible separation area.

Since due to the lower resolution in 2-DB gels mostly several proteins co-localize within a single spot, in contrast to conventional 2-DE the usage of LC-MS/MS for protein identification is recommended instead of MALDI-MS.

Table 1 summarizes advantages and properties of the presented electrophoretic methods regarding (membrane) proteome studies.

7. OUTLOOK

Two-dimensional BAC/SDS polyacrylamide gel electrophoresis has been established as further tool in the field of proteome research, especially regarding the separation and analysis of membrane proteins. It is by far more efficient in resolving membrane proteins than common 2-DE and furthermore can be utilized in a complementary way to one-dimensional SDS-PAGE. Therefore, among other techniques, future proteome studies focussing on membrane proteins should include 2-DB as well.

Despite its lower resolution compared to 2-DE, 2-DB can also be applied in combination with the difference gel electrophoresis technique (DIGE) (Unlu et al. 1997; Reinders et al. 2006b), enabling the highly reproducible differential analysis of biological membrane samples.

However, for complex sample mixtures DIGE requires a very high resolution and distinct protein spots. Although the afore mentioned improvements by Kramer further

Table 1. Comparison of different PAGE methods

	1D-PAGE	2-DE	2-DB	dSDS
Resolution	–	++	o	O
Differential studies	–	++	+	O
Proteins are focussed to	Bands	Spots	Spots	Spots
Local concentration	Low	High	High	High
Separation area	Small lane	Entire gel	Elliptical area	Elliptical area
Isoforms	–	+	–	–
Membrane proteins	+	–	+	+
Alkaline proteins	+	–	+	+
Visualization before MS	Not necessary	Necessary	Necessary	Necessary
Amenability to analysis	Entire lane	Only visualized spots	Only visualized spots	Only visualized spots
Recommended MS strategy	LC-MS/MS	MALDI-MS	LC-MS/MS	LC-MS/MS

enhance resolution during first dimension BAC-PAGE, additional prefractionation of complex mixtures might be necessary in order to account for the high demands of differential gel electrophoresis.

ACKNOWLEDGEMENTS

The authors thank the Deutsche Forschungsgemeinschaft (DFG) for continuous support within the SFB 688 and the FZT 82.

REFERENCES

Buxbaum, E. (2003) Cationic electrophoresis and electrotransfer of membrane glycoproteins. Anal. Biochem. *314*, 70–76.

Coughenour, H.D., Spaulding, R.S. and Thompson, C.M. (2004) The synaptic vesicle proteome: a comparative study in membrane protein identification. Proteomics *4*, 3141–3155.

Daub, H., Blencke, S., Habenberger, P., Kurtenbach, A., Dennenmoser, J., Wissing, J., Ullrich, A. and Cotten, M. (2002) Identification of SRPK1 and SRPK2 as the major cellular protein kinases phosphorylating hepatitis B virus core protein. J. Virol. *76*, 8124–8137.

Diao, A., Rahman, D., Pappin, D.J., Lucocq, J. and Lowe, M. (2003) The coiled-coil membrane protein golgin 84 is a novel rab effector required for Golgi ribbon formation. J. Cell Biol. *160*, 201–212.

Fujiki, Y., Hubbard, A.L., Fowler, S. and Lazarow, P.B. (1982) Isolation of intracellular membranes by means of sodium carbonate treatment: application to endoplasmic reticulum. J. Cell Biol. *93*, 97–102.

Godl, K., Wissing, J., Kurtenbach, A., Habenberger, P., Blencke, S., Gutbrod, H., Salassidis, K., Stein-Gerlach, M., Missio, A., Cotten, M. and Daub, H. (2003) An efficient proteomics method to identify the cellular targets of protein kinase inhibitors. Proc. Natl. Acad. Sci. U.S.A. *100*, 15434–15439.

Hartinger, J., Stenius, K., Hogemann, D. and Jahn, R. (1996) 16-BAC/SDS-PAGE: a two-dimensional gel electrophoresis system suitable for the separation of integral membrane proteins. Anal. Biochem. *240*, 126–133.

Kramer, M.L. (2006) A new multiphasic buffer system for benzyldimethyl-*n*-hexadecylammonium chloride polyacrylamide gel electrophoresis of proteins providing efficient stacking. Electrophoresis *27*, 347–356.

Luche, S., Santoni, V. and Rabilloud, T. (2003) Evaluation of nonionic and zwitterionic detergents as membrane protein solubilizers in two-dimensional electrophoresis. Proteomics *3*, 249–253.

Macfarlane, D.E. (1983) Use of benzyldimethyl-*n*-hexadecylammonium chloride ("16-BAC"), a cationic detergent, in an acidic polyacrylamide gel electrophoresis system to detect base labile protein methylation in intact cells. Anal. Biochem. *132*, 231–235.

Macfarlane, D.E. (1989) Two dimensional benzyldimethyl-*n*-hexadecylammonium chloride – sodium dodecyl sulfate preparative polyacrylamide gel electrophoresis: a high capacity high resolution technique for the purification of proteins from complex mixtures. Anal. Biochem. *176*, 457–463.

Moebius, J., Zahedi, R.P., Lewandrowski, U., Berger, C., Walter, U. and Sickmann, A. (2005) The human platelet membrane proteome reveals several new potential membrane proteins. Mol. Cell Proteomics *4*, 1754–1761.

Molloy, M.P. (2000) Two-dimensional electrophoresis of membrane proteins using immobilized pH gradients. Anal. Biochem. *280*, 1–10.

Olsson, I., Larsson, K., Palmgren, R. and Bjellqvist, B. (2002) Organic disulfides as a means to generate streak-free two-dimensional maps with narrow range basic immobilized pH gradient strips as first dimension. Proteomics *2*, 1630–1632.

Otto, H., Dreger, M., Bengtsson, L. and Hucho, F. (2001) Identification of tyrosine-phosphorylated proteins associated with the nuclear envelope. Eur. J. Biochem. *268*, 420–428.

Rais, I., Karas, M. and Schagger, H. (2004) Two-dimensional electrophoresis for the isolation of integral membrane proteins and mass spectrometric identification. Proteomics *4*, 2567–2571.

Reinders, J., Zahedi, R.P., Pfanner, N., Meisinger, C. and Sickmann, A. (2006a) Toward the Complete Yeast Mitochondrial Proteome: Multidimensional Separation Techniques for Mitochondrial Proteomics. J. Proteome Res. *5*, 1543–1554.

Reinders, Y., Schulz, I., Graf, R. and Sickmann, A. (2006b) Identification of novel centrosomal proteins in Dictyostelium discoideum by comparative proteomic approaches. J. Proteome Res. *5*, 589–598.

Santoni, V., Molloy, M. and Rabilloud, T. (2000) Membrane proteins and proteomics: un amour impossible? Electrophoresis *21*, 1054–1070.

Unlu, M., Morgan, M.E. and Minden, J.S. (1997) Difference gel electrophoresis: a single gel method for detecting changes in protein extracts. Electrophoresis *18*, 2071–2077.

Zahedi, R.P., Meisinger, C. and Sickmann, A. (2005) Two-dimensional benzyldimethyl-*n*-hexadecylammonium chloride/SDS-PAGE for membrane proteomics. Proteomics *5*, 3581–3588.

Zahedi, R.P., Sickmann, A., Boehm, A.M., Winkler, C., Zufall, N., Schonfisch, B., Guiard, B., Pfanner, N. and Meisinger, C. (2006) Proteomic analysis of the yeast mitochondrial outer membrane reveals accumulation of a subclass of preproteins. Mol. Biol. Cell *17*, 1436–1450.

Corresponding author: Dr. Albert Sickmann

CHAPTER 3

MICROPARTICLES: A NEW TOOL FOR PLASMA MEMBRANE SUB-CELLULAR PROTEOMIC

LAURENT MIGUET[1], SARAH SANGLIER[1], CHRISTINE SCHAEFFER[1], NOELLE POTIER[1], LAURENT MAUVIEUX[2], and ALAIN VAN DORSSELAER[1]

[1] *Laboratoire de Spectrométrie de Masse Bio-Organique, ECPM, UMR/CNRS 7178, Institut Pluridisciplinaire Hubert CURIEN, Université Louis Pasteur, Strasbourg, France*

[2] *Institut d'Hématologie et d'Immunologie, Faculté de Médecine, Université Louis Pasteur and Laboratoire d'Hématologie, Hôpital Hautepierre, Strasbourg, France*

Table of Contents

1. INTRODUCTION

Membranes are critical components of cellular structure. It has been reported that plasma membrane proteins represent about 30% of all cellular proteins (Wallin and von Heijne 1998). Even if plasma membrane was considered for a long time as a simple biological barrier between the cytosol of the cell and the extra-cellular environment, these membrane proteins have been demonstrated to play a crucial role in the different fundamental biological processes as exchange of component or signal transduction. Also, more than half of all anticipated pharmacological drug targets are predicted to be localized to the plasma membrane (Jang and Hanash 2003). Plasma membrane can then clearly be considered as a sub-cellular compartment of first interest in regard to different diagnosis and/or therapeutic target proteins. Indeed, for each membrane protein there is potentially a specific antibody which can be used

E. Bertrand and M. Faupel (eds.), Subcellular Proteomics, 21–34.

for the diagnosis of several pathologies and also for treatment using armed antibodies (Harris 2004). Therefore, proteomic analysis of plasma membrane proteins is of first importance. Despite the importance of plasma membrane proteins, there is less understanding in this class of proteins due to the difficulty to obtain enriched plasma membrane proteins preparation from eukaryotic cells. Until now many different strategies have been applied but are still laborious and imperfect. In example of these different approaches, biotinylation (Peirce et al. 2004; Zhao et al. 2004) silica coated (Rahbar and Fenselau 2004), partition phase repartition (Qoronfleh et al. 2003) or partial tryptic surface digestion have been tested, but they still remain unsatisfactory. Although some studies of plasma membrane proteins using 2-DE have been reported (Galeva and Altermann 2002; Luche et al. 2003), but the separation of such hydrophobic proteins have been often poor (Santoni et al. 2000). We describe here a new strategy in order to increase the proportions of plasma membrane proteins identified with the highest possible coverage percentage in the different membrane proteins preparations analyzed by mass spectrometry. We will focus on microparticles as a source of plasma membrane proteins.

2. MICROPARTICLES

Each of the two leaflets of the plasma membrane bilayer has a specific lipid composition. Aminophospholipids (phosphatidylserine and phosphatidylethanolamine) are specifically segregated in the inner layer of the membrane, whereas the phosphatidylcholine and the sphingomyelin are enriched in the external leaflet (Bevers et al. 1998). When cells are submitted to various stress conditions as mitogenic activations or apoptosis, the constitutive asymmetry between the inner and the outer leaflet of the plasma membrane is disrupted. The major changes in the plasma membrane constitution will be the delocalization of the phosphatidylserine to the outer leaflet and an augmentation of the Ca^{2+} ion concentration in the cytoplasm. Such changes are going to disrupt the organisation of the cytoskeleton and drive to a blebbing of the plasma membrane and release microvesicles which are named microparticles (MPs) (Figure 1) (Hugel et al. 2005; Miguet et al. 2005).

Such microvesicles have size variable between 50 nm to 1 μm and differ from other vesicles (like exosomes (30–100 nm)). In general, microparticles are phospholipids vesicles derived from eukaryotic cells as a result of different types of stimulation. Microparticles can also be defined as phospholipids microvesicles containing certain membrane proteins originating from the parental cell. Microparticles circulate in the blood and contribute to numerous physiological processes. MPs have been described in various haematopoietic cells as platelets (Heijnen et al. 1999), T-cells (Blanchard et al. 2002), polynuclear neutrophils (Mesri and Altieri 1999) or dendritic cells. After have been considered as cell dust, MPs are now considered to reflect cell activation. Platelet derived microparticles have been the most extensively studied until now. They are now accepted to play an important role in the procoagulant

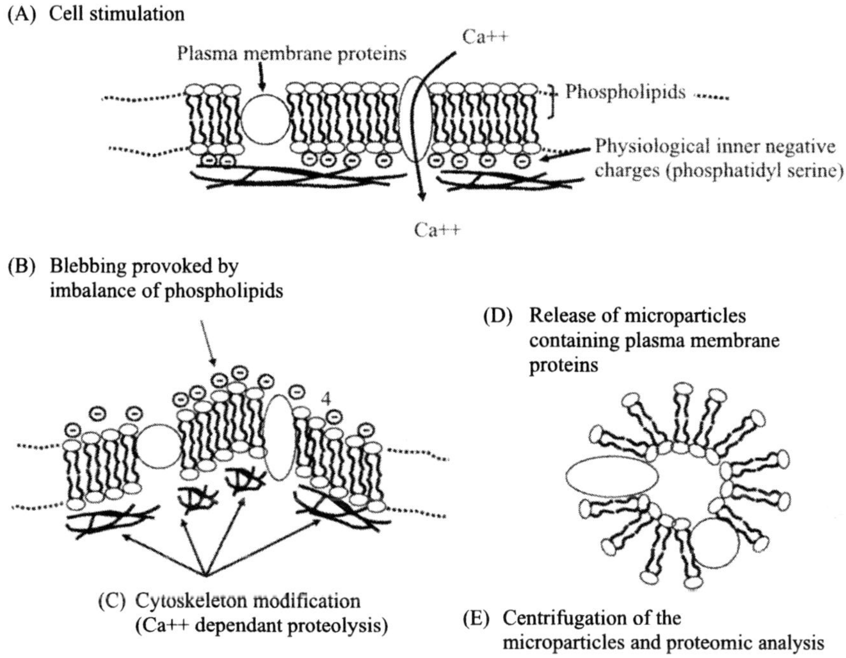

Figure 1. Scheme of microparticles formation. Cell stimulation (A) induces a remodelling of the phosphatidyl serine repartition (B) and an increase of cytosolic calcium concentration (C). Such remodelling induces the MPs formation (D) which are easily separated by centrifugation (E).

process due to the presentation of phospholipids as phosphatidylserine to other platelet which activate them for example (Hugel et al. 2005; Pilzer et al. 2005; Simak and Gelderman 2006).

As microparticles are microvesicles formed directly from plasma membrane, exploration of their protein composition in different pathologies may provide valuable plasma membrane markers specific of the cells they originate from (Pilzer et al. 2005). We then proposed to take the advantage of the physiological production of MPs in order to increase the proportion of plasma membrane proteins and lower the number of irrelevant proteins of the sample.

3. STRATEGY USED FOR MICROPARTICLES PROTEOMIC ANALYSIS

3.1. Presentation of the Strategy

Microparticles are naturally enriched in plasma membrane proteins. They correspond themselves to a sub-proteome, and it is tempting to use them as a substitute to perform plasma membrane proteomics. The methodology described bellow has been developed to perform the proteomic analysis of microparticles. The major problem is linked to the fact that proteins extracted from microparticles are very hydrophobic and therefore cannot be separated on 2-DE (Santoni et al. 2000). Indeed, as a strong ionic detergent must be used for solubilization of the hydrophobic proteins an electro-focalisation step cannot be performed. The protein extract can then only be separated on 1D-SDS-PAGE gel (Galeva and Altermann 2002). The gel lanes obtained are then cut in consecutive pieces of 2 mm. All these pieces (about 100 for a 20 cm gel) are then digested with trypsin. The tryptic peptides are further extracted and analyzed by MALDI-TOF and ESI-nano-LC-MS-MS.

MALDI spectra obtained on these very complex peptide mixtures (several hundred of peptides are expected in a 2 mm slice of 1D gel) only allowed the identification of the few most abundant proteins using MASCOT (http://www.matrixscience.com/), Profound or ProteinProspector (http://prospector.ucsf.edu/).

The peptide mixtures are then analyzed by ESI-nano-LC-MS/MS. The mass data recorded during the different analysis were processed and converted into MassLynx.pkl (Q-TOFII, Waters) or Esquire.mgf (IonTrap HCT+, Bruker) format in order to be submitted to a protein database via the different search engines. This step allowed usually the identification of about 10 to 20 different proteins in each 2 mm gel slices. This strategy could seem to be time-consuming method (a 20 cm gel generates about 100 slices), but this approach allowed the identification of 400–500 proteins for each microparticles preparation. These 400–500 proteins were mainly identified in the 2 mm gel slices regrouped from 20 to 100 kDa (about 50 slices). In each slice an average of ten proteins was identified. This rather low number is probably due to the fact that the protein extracted from each slice is a sub-proteome. The identification of these proteins using nanoLC-MS/MS was often obtained with a low coverage and few peptides, which was probably due to the fact that only major proteins were detected in each 2 mm gel slice. Since sensitivity and chromatographic resolution are probably one of the key elements to improve both the number of proteins identified and the sequence coverage, we have recently used a LC-Chip chromatographic system (Agilent) coupled to an IonTrap (Bruker) on about 10 slices. In these experiments, the number of protein identified was indeed increased (a factor of 30%). Since MALDI-MS analysis performed on the raw peptide extract was very disappointing from the point of view of protein identification, mainly because of the signal suppression effect, we have developed a new MALDI approach. The suppression effect can only be overcome if the number of peptides present in each fraction is reduced. In order to reduce the complexity of the peptide mixture, a chromatographic fractionation step was performed on the tryptic peptides extract. Figure 2 summarizes the strategy and the different analysis performed on each 2 mm 1D gel tryptic peptide extract.

The column used for this separation was adapted in size to the small quantities of peptides extracted from each slice. It was found that micro-chromatography with 0.3 mm diameter columns yield the best compromise for chromatographic resolution

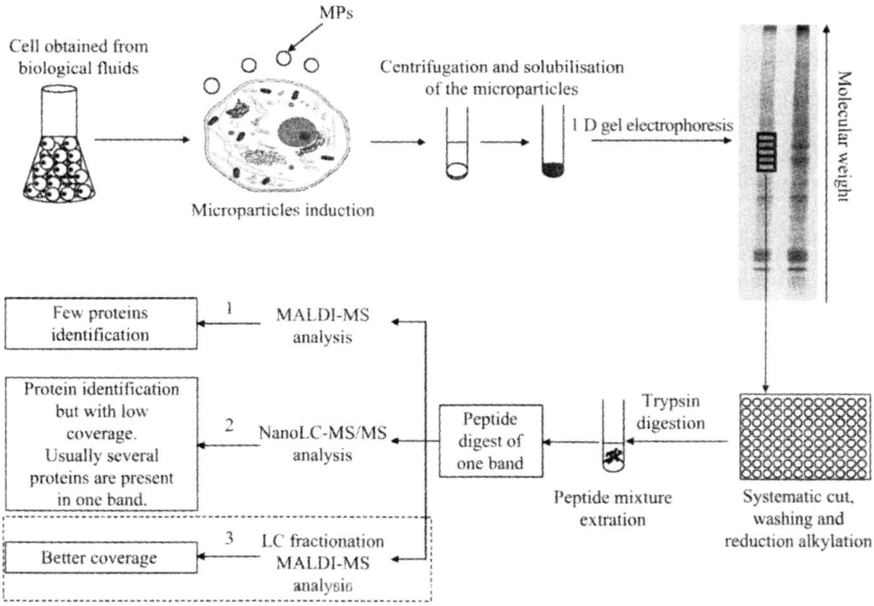

Figure 2. Strategy developed for the microparticles preparation and analysis by mass spectrometry. The step 3 is described in more detail on Figure 3.

and fractionation feasibility. This approach is therefore based on the possibility to collect efficiently fractions of a few μL from a flow rate of about 4 μL/min. In order to speed up the process of data acquisition, a post-column split was used in some cases, allowing to have both ESI-microLC-MS-MS data and fraction collection. Each collected micro-fraction was then submitted to MALDI-MS analysis (0.5 μL used). All MALDI-MS data were than merged to obtain a single peak list showing masses of peptides present in the original extract of one single slice. This single peak list, corresponding to a fictitious MALDI spectrum obtained on the total peptides extract where suppression effect would be absent, allowed to improve very significantly the coverage percentage on the proteins identified by MS-MS data (see bellow).

The "online" or "offline" LC-ESI analysis also generated a list of peptide masses which were usually complementary to the MALDI peaklist. The few microliters left from the collected fractions after MALDI-MS analysis can be used, when necessary, for complementary analysis (MALDI-MS-MS or nano-ESI-MSn).

The complementarity of all the data obtained in this strategy (LC-MALDI-MS and nanoLC-MS/MS) allowed to increase significantly the coverage percentage for all the gel slices. Figure 3 summarizes this strategy and technical details are presented below.

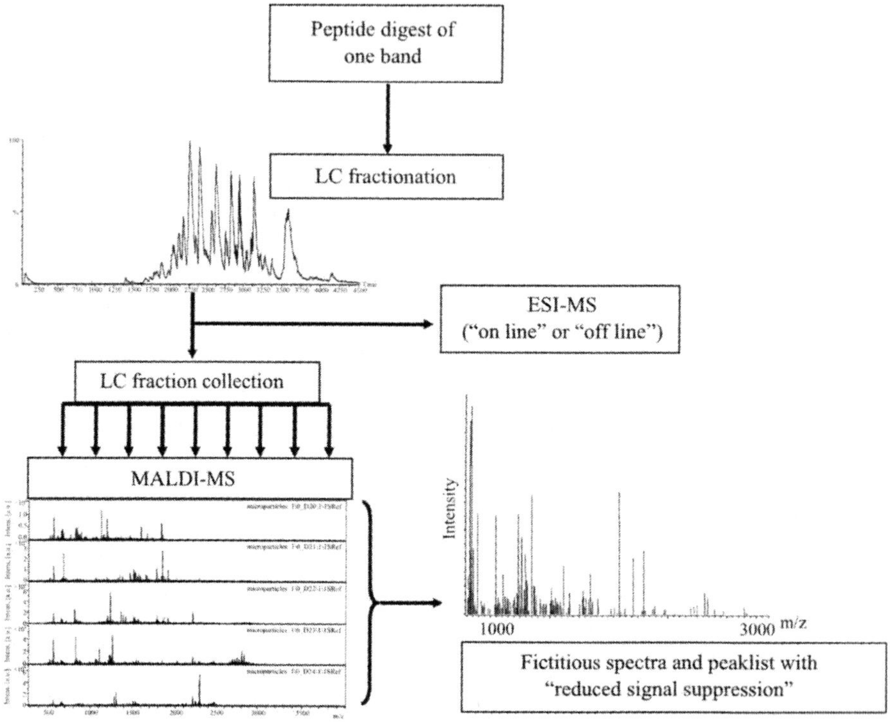

Figure 3. LC-MALDI strategy. Fractionation of the peptide mixture and MALDI-MS analysis.

3.2. Technical Details

The micro-fractionation was realized using a capillary LC system (Agilent 1100 series, Wilmington, DE, USA) equipped with a micro-fraction collector (Agilent 1100 series micro-fraction collector, Wilmington, DE, USA), and all the capillary are optimized for very small column. A total of 15 μl of the samples were injected onto the LC column (Zorbax 300SB-C18, 15 cm × 300 μm, 3.5 μm). Mobile phase solvents were water (A) and acetonitrile (B), containing 0.05 and 0.045% TFA, respectively. Separation was achieved with the following gradient: 5 min at 10% B; from 10 to 80% B in 60 min. The column was maintained at 30°C and the flow rate was set at 4 μL min. Between 3 minutes and 60 minutes, fractions were collected each 2 min in a 96 microwell plate using the automatized micro-fraction collector. Such chromatographic steps generate about 15 fractions containing peptides. All the collected fractions are analyzed by MALDI-MS in order to obtain 20 MALDI-MS spectra. This acquisition is realized on ULTRAFLEX (Bruker) using WARP-LC software (Bruker) for acquisition and treatment of the data (Figure 4).

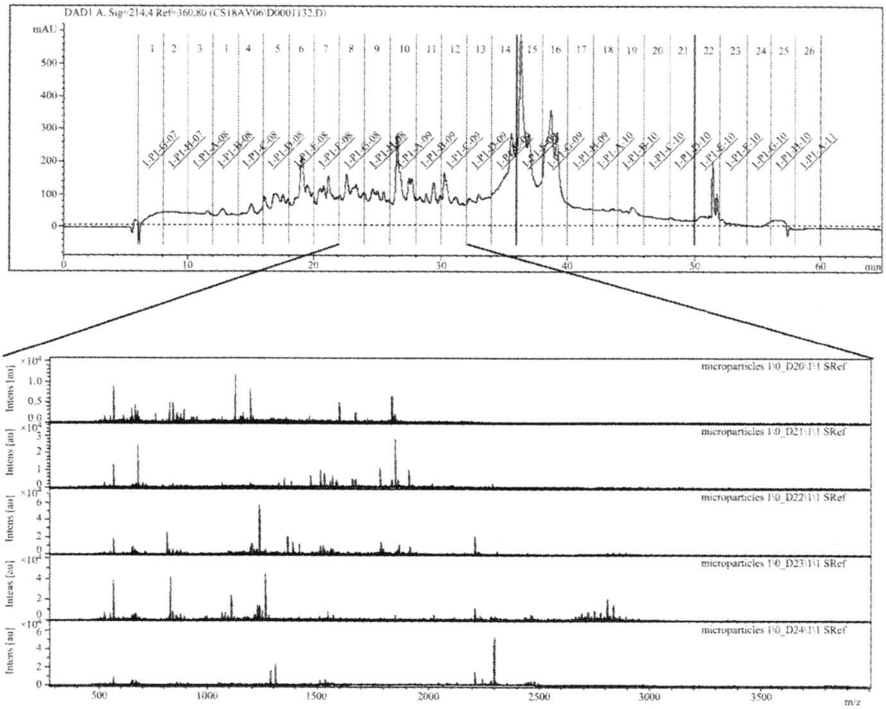

Figure 4. Example of peptide fractionation performed on a 1D gel slice. 26 fractions were collected and analyzed by MALDI-MS. The figure focuses on five of them.

The different spectra are automatically treated in order to generate a single non redundant peaklist. Even if this reconstituted peaklist does not allow new protein identification using peptide mass fingerprint, it combines hundreds of different peptide masses. These masses can be used to obtain a better coverage percentage of the different proteins identified in the different spot. Indeed, in the 1DE LC-MS/MS approach, the identification of the different proteins is performed with few peptides generating low sequence coverage. The different proteins identified by nanoLC-MS/MS are digested *in-silico* in order to generate the theoretical peaklilst. The theoretical and the experimental peaklists are compared together in order to identify the masses corresponding to the tryptic peptides of the identified proteins. The peaks accepted for this compilation must have a signal/noise greater than 5 and a mass error lower than 20 ppm. This strategy allows than a better coverage of the different proteins identified by nanoLC-MS/MS.

Figure 5 summarizes the all strategy, and an example of this approach obtained from the analysis of a single 1D gel slice is reported in Table 1.

Table 1 reports the different proteins identified during the nanoLC-MS/MS analysis of a single 1D gel slice. Peptides obtained from the nanoLC-MS/MS and

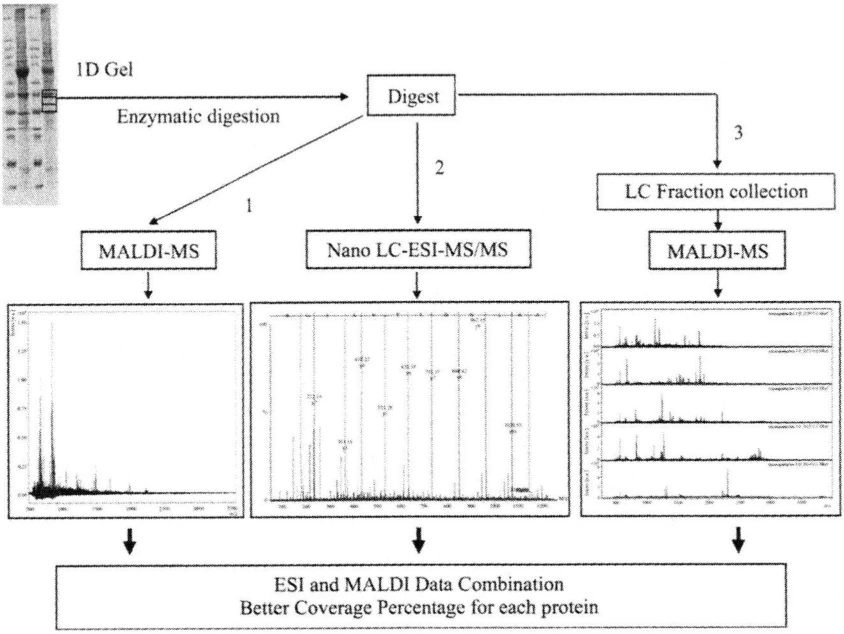

Figure 5. Summary of the strategy applied for a better sequence coverage of the identified proteins in 1D gel. The first step is the analysis of the raw peptide mixture by MALDI. The step 2 corresponds to the nanoLC-MS/MS analysis of the peptide digests, and the step 3 corresponds to the "off-line" fractionation of the peptide digest and MALDI-MS analysis. The combination of these approaches allow a better coverage of the different proteins identified.

LC-MALDI-MS experiments are also reported. The combination of both ESI and MALDI experiments shows well the advantage in term of coverage of this double ionization process. Indeed, some peptides are identified specifically from the ESI ionization process and other from the MALDI process, increasing the whole number of identified peptides.

The improvement of sequence coverage of the proteins is very important first to enhance the confidence of the different identifications and second to have a better characterization of the proteins, particularly in term of post-translational modifications and/or mutations.

Moreover, we have determined the false positive rate for this approach. Many tryptic peptides originated from different proteins can be attributed to a single mass (e.g.: HQHPLQCVMEK 1364.63 Da and EADFINCVIWR 1364.65 Da; $\Delta M < 20$ ppm). Thereby false positive identification may occur. To evaluate the false positive rate, we have selected three common proteins which were not identified during the nanoLC-MS/MS analysis. These proteins (tubulin, actin and myosin) were digested *in-silico*, and the generated mass lists were compared to the LC-MALDI-MS peaklist. A total of only five masses were attributed to the three

Table 1. Summary of the proteins identified by nanoLC-MS/MS from a single 1D gel slice, and the increase of the coverage using LC-MALDI-MS strategy

Accession number	Protein	Peptides identified by nanoLC-MS/MS	Coverage (%)	Peptides identified by LC-MALDI	Combination MALDI/ESI	
					Number of peptides	Final coverage (%)
P08238	Heat shock protein HSP 90-beta	25	31	20	30	47
P07900	Heat shock protein HSP 90-alpha	16	21	27	33	55
P46940	Ras GTPase-activating-like protein	7	5	22	27	22
Q01813	6-phosphofructokinase type C	4	5	7	11	19
P08195	4F2 cell-surface antigen (CD98)	4	9	10	11	30
P23921	Ribonucleoside-diphosphate reductase	3	4	13	14	20
P26639	Threonyl-tRNA synthetase	3	4	13	15	21
P02786	Transferrin receptor (CD71 antigen)	2	2	14	16	29
P16070	CD44 antigen precursor	1	1	3	3	4
Q15758	Neutral amino acid transporter B(0)	1	2	3	4	9

proteins, which is very few compared to the masses attributed to the different proteins in Table 1. This low false positive rate is due to the low tolerance accepted in our strategy (<20 ppm).

To conclude, we have developed a strategy which combines LC-MALDI-MS and nanoLC-MS/MS in order to identify proteins originating from 1D gel with a high coverage compatible with plasma membrane study (CD98, CD71, CD44).

4. VALIDATION OF MICROPARTICLES AS A NEW TOOL FOR PLASMA MEMBRANE PREPARATION

In order to validate the hypothesis that microparticles are structures enriched in plasma membrane proteins, we have analysed the proteome composition of different microparticle preparations obtained from a T-Lymphocytic cell line. Such microparticles can be produced and enriched *in vitro* by mitogenic activation (PHA) or apoptosis induction (Act D1, TNFα) for instance. Microparticles protein mixtures obtained from the two stimulations were separated independently on a 1D gel. The different gels were cut each 2 mm and the different slides were digested and analyzed by nanoLC-MS/MS.

This approach allowed the identification of 390 proteins among which 34% were localized to the plasma membrane. The microparticles obtained from the two different ways of stimulation did not show significant variation in their proteome (Miguet et al. 2005).

These results have been compared to a systematic analysis of a classical plasma membrane protein preparation, generated by ultracentrifugation (100,000 g on sucrose). This plasma membrane isolation has allowed the identification of 292 proteins, and only 69 proteins were located to the plasma membrane (24 vs. 34% for the microparticles preparation). Moreover, 75 plasma membrane proteins were identified specifically in the microparticles preparation, and only 9 were identified only in the plasma membrane preparation. These results are summarized in Figure 6.

On the other hand different groups have started working on the plasma membrane microparticles proteome (Banfi et al. 2005; Garcia et al. 2005) The results obtained by these groups are very different from the results coming from our lab. This might be due to the difference in the microparticles centrifugation step used for MPs isolation. Indeed these two groups used intense centrifugation (between 100,000 and 250,000 g) which is very high in comparison with the theorical microparticles size. This step pellet indeed the microparticles, but also the other vesicles contained in the solution as exosomes which are much smaller. For example, the classical way to prepare exosomes is to centrifuge for half an hour between 100,000 and 200,000 g (Thery et al. 2001; Wubbolts et al. 2003). We believe that this centrifugation step is crucial in order to generate a pure preparation of microparticles. This point was already underlined, even if not discussed, by Jin et al. (2005). In this aim a centrifugation of 20,000 g during 45 min, that we used, is appropriate to generate a homogeneous microparticles preparation (Hugel et al. 2005).

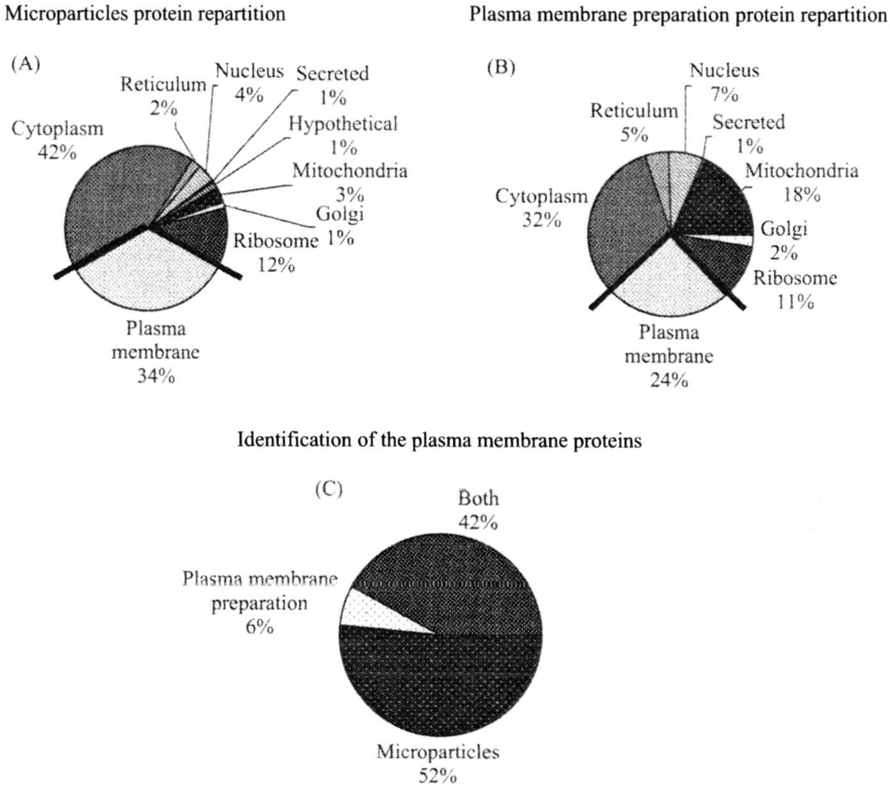

Figure 6. Representation of the protein repartition in function of the preparation. (A) Repartition of the proteins identified in the microparticles. (B) Repartition of the proteins identified in the classical plasma membrane preparation. (C) Comparison of the plasma membrane identified in the microparticles and in the plasma membrane preparation.

5. PERSPECTIVES

As it is potentially possible to generate microparticles from almost all cell types, it could be a new tool for sub-cellular plasma membrane study and biomarker discovery. Even if quantitation is not assessed, the comparison of the microparticles sub-proteome, originating from different pathologies, may enlighten differences in protein expression. As our group is interested in the study of the different B-cell chronic lymphopathies, we have decided to take advantage of the microparticles preparations obtained from various pathologies in order to realize their proteome. In this aim, fresh cells were obtained following informed consent, and subjected to actinomycinD stimulation. The obtained microparticles were first separated on 1D gel before to be systematically analyzed by nanoLC-MS/MS. We analyzed a lymphoma

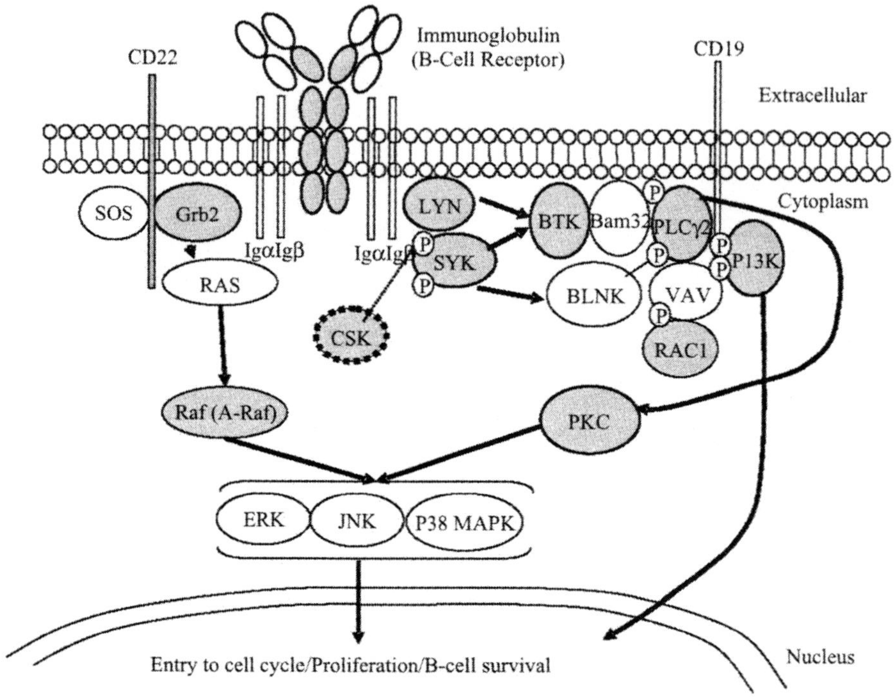

Figure 7. Representation of the B-cell receptor, and the different kinase implied in the signal transduction for the cell proliferation. The proteins in gray were identified in the microparticles obtained from the patient. Most of the proteins implied in the signal transduction were identified using this strategy.

(small cell lymphoma). This analysis has allowed the identification of more than 400 proteins among which 30% were localized to the plasma membrane or associated to the plasma membrane.

Moreover, in this study, a large number of proteins involved in the signal transduction were identified. Indeed B-cell receptor signaling proteins (numerous membrane receptors, kinase proteins) implied in the signal transduction was detected. This complexome identified in the microparticles is reported on Figure 7. The proteins in grey were identified in our study.

In conclusion, plasma membrane microparticles represent a sub-proteome reflecting the plasma membrane composition. These structures have the advantage of not being contaminated by membrane proteins originating from other cell components (golgi, nucleus, mitochondria). As microparticles can be produced by almost all cell lines, they can be useful to discover new potential biomarkers. In this goal we have undertaken the comparison of microparticles originating from different kind of lymphoma. The analytical strategy developed for the characterization of the plasma membrane sub-proteome, based on micro-fractionation, allowed the identification of hundreds proteins with a better coverage percentage than the classical nanoLC-MS/MS approach. The combinations of nanoLC-MS/MS, micro-fractionation and "off-line" MALDI-MS data is now used routinely in general study for the discovery of new specific biomarkers.

REFERENCES

Banfi, C., Brioschi, M., Wait, R., Begum, S., Gianazza, E., Pirillo, A., Mussoni, L. and Tremoli, E. (2005) Proteome of endothelial cell-derived procoagulant microparticles. Proteomics *5*, 4443–4455.

Bevers, E.M., Comfurius, P., Dekkers, D.W., Harmsma, M. and Zwaal, R.F. (1998) Regulatory mechanisms of transmembrane phospholipid distributions and pathophysiological implications of transbilayer lipid scrambling. Lupus *7* (Suppl), S126–S131.

Blanchard, N., Lankar, D., Faure, F., Regnault, A., Dumont, C., Raposo, G. and Hivroz, C. (2002) TCR activation of human T cells induces the production of exosomes bearing the TCR/CD3/zeta complex. J. Immunol. *168*, 3235–3241.

Galeva, N. and Altermann, M. (2002) Comparison of one-dimensional and two-dimensional gel electrophoresis as a separation tool for proteomic analysis of rat liver microsomes: cytochromes P450 and other membrane proteins. Proteomics *2*, 713–722.

Garcia, B.A., Smalley, D.M., Cho, H., Shabanowitz, J., Ley, K. and Hunt, D.F. (2005) The platelet microparticle proteome. J. Proteome Res. *4*, 1516–1521.

Harris, M. (2004) Monoclonal antibodies as therapeutic agents for cancer. Lancet Oncol. *5*, 292–302.

Heijnen, H.F., Schiel, A.E., Fijnheer, R., Geuze, H.J. and Sixma, J.J. (1999) Activated platelets release two types of membrane vesicles: microvesicles by surface shedding and exosomes derived from exocytosis of multivesicular bodies and alpha-granules. Blood *94*, 3791–3799.

Hugel, B., Martinez, M.C., Kunzelmann, C. and Freyssinet, J.M. (2005) Membrane microparticles: two sides of the coin. Physiology (Bethesda) *20*, 22–27.

Jang, J.H. and Hanash, S. (2003) Profiling of the cell surface proteome. Proteomics *3*, 1947–1954.

Jin, M., Drwal, G., Bourgeois, T., Saltz, J. and Wu, H.M. (2005) Distinct proteome features of plasma microparticles. Proteomics *5*, 1940–1952.

Luche, S., Santoni, V. and Rabilloud, T. (2003) Evaluation of nonionic and zwitterionic detergents as membrane protein solubilizers in two-dimensional electrophoresis. Proteomics *3*, 249–253.

Mesri, M. and Altieri, D.C. (1999) Leukocyte microparticles stimulate endothelial cell cytokine release and tissue factor induction in a JNK1 signaling pathway. J. Biol. Chem. *274*, 23111–23118.

Miguet, L., Pacaud, K., Felden, C., Hugel, B., Martinez, M.C., Freyssinet, J.M., Herbrecht, R., Potier, N., van Dorsselaer, A. and Mauvieux, L. (2005) Proteomic analysis of malignant lymphocyte membrane microparticles using double ionization coverage optimization. Proteomics *6*, 153–171.

Peirce, M.J., Wait, R., Begum, S., Saklatvala, J. and Cope, A.P. (2004) Expression profiling of lymphocyte plasma membrane proteins. Mol. Cell Proteomics *3*, 56–65.

Pilzer, D., Gasser, O., Moskovich, O., Schifferli, J.A. and Fishelson, Z. (2005) Emission of membrane vesicles: roles in complement resistance, immunity and cancer. Springer Semin. Immunopathol. *27*, 375–387.

Qoronfleh, M.W., Benton, B., Ignacio, R. and Kaboord, B. (2003) Selective enrichment of membrane proteins by partition phase separation for proteomic studies. J. Biomed. Biotechnol. *2003*, 249–255.

Rahbar, A.M. and Fenselau, C. (2004) Integration of Jacobson's pellicle method into proteomic strategies for plasma membrane proteins. J. Proteome Res. *3*, 1267–1277.

Santoni, V., Molloy, M. and Rabilloud, T. (2000) Membrane proteins and proteomics: un amour impossible? Electrophoresis *21*, 1054–1070.

Simak, J. and Gelderman, M.P. (2006) Cell membrane microparticles in blood and blood products: potentially pathogenic agents and diagnostic markers. Transfus. Med. Rev. *20*, 1–26.

Thery, C., Boussac, M., Veron, P., Ricciardi-Castagnoli, P., Raposo, G., Garin, J. and Amigorena, S. (2001) Proteomic analysis of dendritic cell-derived exosomes: a secreted subcellular compartment distinct from apoptotic vesicles. J. Immunol. *166*, 7309–7318.

Wallin, E. and von Heijne, G. (1998) Genome-wide analysis of integral membrane proteins from eubacterial, archaean, and eukaryotic organisms. Protein Sci. *7*, 1029–1038.

Wubbolts, R., Leckie, R.S., Veenhuizen, P.T., Schwarzmann, G., Mobius, W., Hoernschemeyer, J., Slot, J.W., Geuze, H.J. and Stoorvogel, W. (2003) Proteomic and biochemical analyses of human B cell-derived exosomes. Potential implications for their function and multivesicular body formation. J. Biol. Chem. *278*, 10963–10972.

Zhao, Y., Zhang, W. and Kho, Y. (2004) Proteomic analysis of integral plasma membrane proteins. Anal. Chem. *76*, 1817–1823.

Corresponding author: Pr. Alain Van Dorsselaer (vandors@chimie.u-strasbg.fr)

CHAPTER 4

LIPID RAFT PROTEOMICS: MORE THAN JUST DETERGENT-RESISTANT MEMBRANES

LEONARD J. FOSTER and QUEENIE W. T. CHAN
University of British Columbia, Canada

Table of Contents

Abstract: The fluid mosaic model of membrane bilayers implies that proteins and lipids are homogenously distributed in the 2D surface of a membrane. Numerous lines of biochemical, biophysical and optical evidence now suggest that organized sub-domains of membranes exist, a subset of which are known as lipid rafts. Rafts are enriched in cholesterol, saturated phospholipids, sphingolipids and what is thought to be a specific subset of proteins. Biologically rafts have been implicated in several fundamental processes, including signal transduction, bacterial invasion, apical/basolateral sorting in polarized cells and viral budding; therefore, defining the raft proteome is an attractive goal. Rafts can be enriched biochemically by taking advantage of their buoyant density and resistance to non-ionic detergents so numerous studies have used a fraction so enriched as a starting point for characterizing the proteome of lipid rafts. This review will focus on approaches to lipid raft proteomics with a specific emphasis on the use of quantitative methods to ensure the specificity and/or functionality of raft proteins.

E. Bertrand and M. Faupel (eds.), Subcellular Proteomics, 35–47.

1. INTRODUCTION

1.1. Lipid Raft Biology

The lipid raft theory (Simons and van Meer 1988; Simons and Ikonen 1997) was first proposed as a model for explaining the clustering behavior of cholesterol and sphingolipids in membranes and their role in the polarization of epithelial cells. In short, glycosphingolipids cluster laterally in membranes, likely via weak hydrogen bonding between the carbohydrate head groups, and any voids between the saturated hydrocarbon chains are filled by cholesterol. These assemblies of lipids would exclude other membrane lipids, such as unsaturated phosphatidic acid-based components. Specific classes of proteins were also observed to target to lipid rafts, including glycosylphosphatidylinositol (GPI)-anchored proteins, doubly acylated tyrosine kinases and certain transmembrane domain proteins (Skibbens et al. 1989; Sargiacomo et al. 1993; Casey 1995; Danielsen and van Deurs 1995) and these observations led to the speculation that rafts play a role in several processes that were known to depend on these selected proteins. Several fundamental processes have been implicated, including endocytosis and bacterial invasion, but the two phenomenon with which rafts are most often associated are polarization of epithelial cells and T-cell receptor (TCR) signalling. After development in the Golgi stacks, rafts are targeted specifically to apical membranes of polarized cells and the apical targeting of several other proteins seems to depend on this. Lipid-anchored proteins probably partition into rafts themselves while other proteins may bind raft-specific lectins and come along for the ride (Simons and Ikonen 1997). In TCR signalling, active, phosphorylated CD3ζ is recruited to lipid rafts together with other kinases such as ZAP70 (Qian and Weiss 1997). The localization to rafts not only brings several interacting factors together, it also seems to protect the phosphorylated CD3ζ from deactivating phosphatases (Zeyda and Stulnig 2006).

1.2. Lipid Rafts Versus Caveolae

Caveolae were described in 1953 as flask-shaped invaginations on the inner surface of the plasma membrane (Palade 1953) and they require caveolin-1 protein to maintain their structure. Caveolae have been implicated in numerous processes as well, including endocytosis, but recent work has shown unequivocally that caveolae are rigid structures and do not participate in internalization (Thomsen et al. 2002). The relationship between caveolae and rafts is not completely agreed upon, with the major players on one side believing that caveolae are simply a subset of rafts (Rajendran and Simons 2005) and on the other side stating that caveolae and rafts, while related, are distinct entities (McMahon et al. 2006). Most of proteomic studies that will be discussed here have utilized biochemical purifications that will enrich for both caveolae and rafts with only one specifically targeting caveolae (McMahon et al. 2006) so for simplicity we will consider caveolae to be a subset of rafts.

1.3. Lipid Raft Proteomics to Date

Lipid rafts have been incredibly popular subjects for proteomic investigations, with more published studies that have tried to ascertain the composition of these intriguing domains than any other subcellular fraction of the cell. To date at least 23 papers on lipid raft proteomics have been published (von Haller et al. 2001; Nebl et al. 2002; Bini et al. 2003; Foster et al. 2003; Ledesma et al. 2003; Li et al. 2003; von Haller et al. 2003a, b; Bae et al. 2004; Blonder et al. 2004; Li et al. 2004; Sprenger et al. 2004; Tu et al. 2004; Blonder et al. 2005; Borner et al. 2005; Karsan et al. 2005; Maclellan et al. 2005; Man et al. 2005; McMahon et al. 2006; Paradela et al. 2005; Sanders et al. 2005; Gupta et al. 2006; Nguyen et al. 2006), reflecting the central role that they apparently play in several basic processes. The majority of these studies have taken the traditional proteomics approach: identify the proteins in a sample for the sake of knowing what they are. Some have addressed specific biological questions (e.g. what proteins are recruited to rafts upon activation) (Bini et al. 2003; von Haller et al. 2003a, b; Tu et al. 2004; Maclellan et al. 2005; Gupta et al. 2006) and some have applied quantitative proteomics approaches to overcome some of the limitations of biochemistry in purifying compartments to homogeneity (Foster et al. 2003; Borner et al. 2005). Unfortunately it is difficult to extract a meaningful message from the sum of this plethora of data since, by and large, each study was done in different biological systems with different mass spectrometric approaches. In this review we attempt to coalesce all these disparate data into an overall picture of lipid raft composition and we discuss specific studies that have gone beyond qualitative proteomics to measure dynamic changes in lipid raft proteins or to assess the specificity of lipid raft components.

2. RAFT BIOCHEMISTRY

2.1. Purification by Non-ionic Detergents

The original and most commonly used method for enriching lipid rafts takes advantage of the resistance of rafts to solubilization in ice cold, non-ionic detergents such as Triton X-100 (Brown and Rose 1992). The majority of cellular membrane lipids and proteins are efficiently solubilized in 1% Triton X-100 so that any intact membranes can be recovered by floatation in a sucrose gradient due to the buoyancy of the lipids. Because of this characteristic the biochemical fraction resulting from this purification is referred to as "detergent-resistant membranes" (DRMs), detergent-insoluble glycolipid-rich domains (DIGs), glycolipid-enriched membranes (GEMs) or low-density Triton-insoluble (LDTI) complexes, among others. These acronyms are often used interchangeably with "lipid rafts" but, regardless of the name given, the biochemical fraction purified in this way is distinct from the lipid raft entity since lipid rafts are but a subset of all the buoyant, detergent-insoluble material in a cell (Foster et al. 2003). In this review we will use the term "lipid raft" only to refer to the concept of an ordered lipid subdomain and not the biochemically enriched fraction. Likewise, "DRM" will be used to refer to the biochemical fraction.

Triton X-100 is by far the most popular choice for raft enrichment but the selection of detergent has a marked effect on the resulting preparation (Chamberlain 2004). In a recent study, Veenstra et al. (Blonder et al. 2006) used chemically and enzymatically introduced stable isotope labels to quantify the differences in the proteomics of Brij-96 (another non-ionic detergent) and Triton X-100-insoluble, low density fractions and the findings were very revealing. Triton X-100 extracts contained considerably more proteins than does Brij-96 and while many of the additional proteins were likely to be true components of rafts, the number of non-raft proteins was also much greater. This work further reinforces the need for rigorous controls for specificity, as will be discussed below.

2.2. Purification by High pH and Sonication

The main alternative to detergent insolubility involves physically shearing the cells and membranes by sonication and then floating the rafts on sucrose or Optiprep density gradients under physiological buffer conditions (Smart et al. 1995) or in 500 mM sodium carbonate, pH 11.5 (Song et al. 1996). These detergent-free methods appear to result in preparations that are very similar to detergent-resistant low-density fractions, both from Western blotting for specific markers and from less biased proteomic analysis. Detergent-free methods are considered by some to enrich caveolae specifically, although this remains controversial.

2.3. Pharmacological Disruption

Cholesterol is a crucial component of lipid rafts as it is thought to intercalate between the hydrophobic tails of other membrane lipids to provide rigidity, and presumably detergent-resistance, to rafts. Simons et al. discovered that lipid raft markers such as influenza virus hemagglutinin no longer trafficked correctly and lost their detergent insolubility when cells were treated with methyl β-cyclodextrin (MβCD) (Keller and Simons 1998) or lovastatin. Lovastatin inhibits cholesterol synthesis by blocking 3-hydroxy, 3-methylglutaryl coenzyme A reductase that, over time, depletes membrane cholesterol. MβCD, on the other hand, acts more acutely by chelating cholesterol and extracting it from the membrane. Several other cholesterol-disrupting agents have emerged but sensitivity to MβCD remains the gold standard for true components of lipid rafts.

3. USE OF QUANTITATIVE PROTEOMICS TO INCREASE SPECIFICITY

Lipid rafts had been studied for almost a decade prior to proteomics coming onto the scene and in all that time no evidence emerged that suggested that the mitochondrial proton pump or any of several other mitochondrial enzymes could be found on lipid rafts. Yet all published lipid raft proteomics studies have found components of the

F_1/F_0 ATPase or other highly abundant mitochondrial proteins, in addition to other unexpected proteins from other subcellular locations. Of course, it is likely that no one actually looked for such proteins in rafts before and with the unbiased nature of proteomics one expects (or even hopes!) to find components that had not been previously described. However, any analysis of a biochemical fraction can only be as good as the initial preparation of that fraction and in previous lipid raft studies the minimum standard one has to meet to claim a protein is in rafts is to demonstrate sensitivity to cholesterol disruption.

We have previously reported a proteomic analysis of lipid rafts based on this minimum standard (Foster et al. 2003). By applying Stable Isotope Labelling by Amino acids in Cell culture (SILAC, (Ong et al. 2002, 2003)) we encoded those proteins remaining in DRMs after cholesterol depletion using MβCD with a different mass from DRM proteins isolated from untreated cells (Figure 1; **see colour insert**). Seven hundred and three proteins were identified in DRMs, of which 392 were tested for their susceptibility to MβCD. By this approach a protein had to be present in the DRM but also had to be depleted from the DRM by MβCD treatment to be considered a raft protein. Of the 392 proteins measured, 241 were susceptible to MβCD and thus considered true raft proteins. As was mentioned, rafts had long been theorized to be signaling platforms and our findings provided more supporting evidence for this model by demonstrating that signaling molecules such as kinases and phosphatases were enriched in lipid rafts. Of equal importance though, was the observation that every known marker of raft proteins identified in the study fell into the cholesterol-dependent category, as did several other families of proteins thought to reside in rafts. However, levels of mitochondrial proteins such as voltage-dependent anion-selective channels (VDACs), cytochrome C oxidase subunits and proton pump subunits were all roughly equal between the control and the MβCD-treated cells. This finding provides concrete evidence that such proteins do not reside in cholesterol-dependent membranes and thus are not components of lipid rafts by that definition. Bini et al. (2003) used MβCD in studying the raft proteome but do not comment on the effects of this treatment and Ledesma et al. (2003) quantified the effects of multiple cholesterol disrupting agents on a handful of DRM proteins.

Borner et al. (2005) took a different approach to address the specific localization of DRM proteins to lipid rafts. In this study the relative enrichment of proteins in the detergent-resistant fraction versus total cell membranes was used as a measure of the specificity of putative lipid raft proteins. In their study Borner et al. used *Arabidopsis thaliana* as the model system so a direct comparison with mammalian DRM reports is difficult and perhaps unfair since it is not always clear what the human homologue of a plant protein is. However, it is interesting to note that Borner et al. also found that the common contaminants in DRMs are certain highly abundant mitochondrial and endoplasmic reticulum proteins.

4. USE OF QUANTITATIVE PROTEOMICS TO ASSESS FUNCTION

Lipid rafts are thought to participate in signal transduction by forcing specific types of proteins that partition into the liquid ordered environment into close proximity with

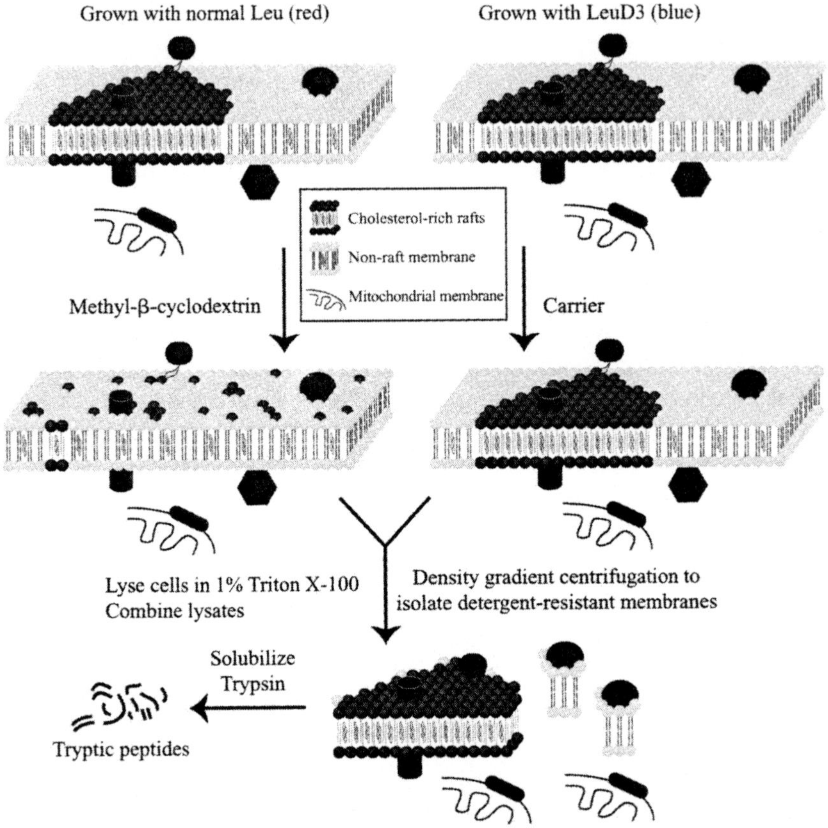

Figure 1. Use of SILAC to distinguish true raft proteins. In order to distinguish true raft proteins from co-purifying contaminants two populations of HeLa cells were grown, one in normal isotopic abundance amino acids (LeuH3, red proteins) and the other with triply deuterated leucine (LeuD3, blue proteins). Cholesterol was depleted from LeuH3 cells by treating with 10 mM methyl β-cyclodextrin (MβCD) while the LeuD3 cells were treated with carrier alone. LeuH3 and LeuD3 cells were lysed and combined prior to isolating detergent resistant membranes by floatation in a discontinuous sucrose gradient. MβCD treatment caused LeuH3 rafts to lose their structural integrity so that they would no longer be detergent-resistant, unlike rafts originating from LeuD3 cells. Conversely, co-purifying non-raft proteins from both cell populations would still enrich equally well from LeuD3 and LeuH3 cells since they would be unaffected by the MβCD treatment. LeuD3/LeuH3 ratios for raft components as measured by the relative ion intensities of tryptic peptides were all greater than 3, while ratios for co-purifying proteins grouped around 1 (reprinted from (Foster et al. 2003)).

one another, leading to the recruitment of other signaling factors to the raft platform. Obviously then, one would expect to find such recruited proteins in DRMs following stimulation. To this end, several recent studies have applied quantitative proteomics to characterize the changes in the DRM proteome following various stimulations. In the theme of this review, these studies, in addition to advancing our knowledge of the signaling pathways, have the added advantage that raft specificity and cholesterol dependence become a moot point since any protein that changes with stimulation is specific by definition.

Activation of the T-cell receptor (TCR) signaling pathway through the clustering of lipid rafts is the prototypic signaling paradigm involving rafts and three studies have used proteomics to demonstrate that TCR activation does indeed lead to massive changes in the DRM proteome. Bini et al. (2003) used two-dimensional gel electrophoresis (2DE) to follow changes in DRMs following antibody stimulation of TCR. While several spots on the 2D gels were observed to change with stimulation, the limitations of MALDI-TOF peptide mass fingerprinting prevented the investigators from identifying more than a handful of these proteins. Of those that were identified by mass spectrometry most were highly abundant mitochondrial proteins but several raft markers that were below the detection limit of MALDI-TOF were subsequently confirmed by Western blotting. von Haller et al. (2003a, b) used isotope-coded affinity tags (ICAT) to quantify the differences in DRM composition following cross-linking/activation of TCR and CD28. Six hundred and eighty five proteins were identified but the authors focus more on the bioinformatic handling of their data and do not draw any biological conclusions from their data. Nonetheless, the technical quality of this study makes the data they have published a tremendous resource for further exploration. More recently, Tu et al. (2004) used classical 2D gel-based quantitation to examine the composition of isolated rafts from unstimulated cells and from TCR/CD28 co-stimulated Jurkat cells. Their analysis revealed that two heat shock proteins, vimentin, calmodulin and a Rho guanine nucleotide dissociation inhibitor were recruited to the rafts after CD3/CD28 stimulation but it was unclear what functional role these proteins might play in CD3/CD28 signalling. However, the authors also show through other means that IκB Kinase complexes with flotillin, a major raft protein, and that this complex is disrupted by MβCD.

The TCR pathway is not the only signaling system thought to involve lipid rafts. Binding of the B-cell receptor (BCR) by cognate antigen also leads to the coalescence of lipid rafts and the initiation of a pathway that leads to activation and proliferation of B cells. Gupta et al. (2006) used ICAT to quantitatively compare the DRM proteome before and after BCR engagement by anti-IgM antibodies. The authors report only four proteins that were observed to change with stimulation: BCR itself and two myosin motor proteins increased with stimulation, while ezrin, a structural protein, decreased with stimulation. The finding that ezrin dissociates from rafts upon stimulation intrigued the authors and they went on to identify a regulated phosphorylation site in the protein and observed that perturbation of ezrin function blocked raft coalescence. This work has led to the reformulation of models for the involvement of the cytoskeleton in antigen presentation and the central role that ezrin plays in this.

Apart from immunological processes, other ligands are also known to act through lipid rafts. Maclellan et al. (2005) have demonstrated quantitative differences in DRMs from primary smooth muscle cells stimulated with platelet-derived growth factor (PDGF) using 2DE and ICAT. Twenty-three DRM proteins were observed to change after 15 min of PDGF stimulation, including several GPI-anchored proteins, cytoskeletal elements and endocytosis machinery. These findings are consistent with the morphological changes PDGF induces in smooth muscle cells and the endocytic down-regulation that is observed for multiple growth factor signals.

The functional studies described here only begin to delve into the role of lipid rafts in signal transduction. Nonetheless, all five studies have clearly advanced our understanding of raft biology beyond what was achievable using more standard cell biology and biochemistry. We expect that many more such studies will be reported, perhaps with more temporal resolution (Blagoev et al. 2004), that will help to clarify the role of these membrane domains.

5. COMPARISON OF RAFT PROTEOMIC STUDIES

As of this writing there were 23 raft proteomic analyses reported in the literature but since virtually all of these were performed in different biological systems a comprehensive picture of rafts is difficult to establish. Indeed, a single unified model of raft protein composition would be meaningless anyway since cell and tissue-type differences in rafts certainly exist. For example, TCR and BCR are expressed only in specific hematopoetic cells and would not be expected in epithelial cells. Nonetheless, the ubiquitousness of rafts suggests that there is likely a common core set of proteins that at least provide structure to the membrane subdomains.

We have carried out a bioinformatic analysis of all the protein lists reported in DRM proteomic studies to gain a better understanding of the consistencies and inconsistencies between studies (Table 1). In order to avoid the minefield of defining orthologous proteins between distantly related species we have focused our attention on primates and rodents, which represent 20 of the 23 studies. We first took the entire DRM protein list reported in each study and attempted to retrieve the amino acid sequences for all accession numbers. Several studies fell out of the analysis at this first stage as they did not report their entire protein list or accession numbers (Ledesma et al. 2003; Bae et al. 2004; Gupta et al. 2006) or they reported untraceable accession numbers from a private database (von Haller et al. 2003a, b). The remaining 15 studies reported widely varying numbers of protein identifications, anywhere from 15 to 738. We first asked what fraction of the proteins reported in each study had been observed to be sensitive to MβCD-mediated cholesterol disruption (Foster et al. 2003). Similar to our original findings, a significant proportion of DRM proteins found by others are likely to by contaminants that co-purify with lipid rafts. These contaminants tended to be highly abundant cytosolic proteins such as chaperones and cytoskeletal elements and mitochondrial proteins such as VDACs and ATP synthase subunits.

The ubiquitousness of lipid rafts suggests, as we have said above, that rafts from different sources should share some similar structural proteins, at the very least. To this end we also asked what fraction of proteins in each DRM analysis were also seen in at least 50% of the other studies (Table 1). We chose to use 50% instead of absolute commonality to all studies since such an analysis would be limited by the study with the fewest number of reported proteins, a bias others have overlooked to claim

Table 1. Summary of raft proteomic studies

Source[a]	Reported IDs[b]	Proteins found[c]	True raft proteins (%)[d]	Common proteins[e]	Reference
RBK-2H3 (rat)	738	738	67	34	(Blonder et al. 2005)
Jurkat T cells (human)	685	0[f]	n/a	n/a	(von Haller et al. 2003a, b)
Vero cells (monkey)	380	375	68	33	(Blonder et al. 2004)
HUVEC (human)	358	354	68	20	(Karsan et al. 2005)
HeLa (human)	241	241[g]	100	20	(Foster et al. 2003)
Liver (mouse)	196	0[h]	n/a	n/a	(Bae et al. 2004)
	105/177[j]	174	60	26	(Man et al. 2005)
	66/122	145	63	25	
Natural killer cells (rat)[i]	48/117	141	68	27	
Skin fibroblasts (human)	98	91	59	12	(McMahon et al. 2006)
Jurkat T cells (human)	72	70	81	11	(von Haller et al. 2001)
THP monocytes (human)	71[k]	87	68	22	(Li et al. 2004)
Smooth muscle (human)	64	63	81	8	(Maclellan et al. 2005)
Placenta (human)	64	63	62	7	(Paradela et al. 2005)
HUVEC (human)	54	48	79	12	(Sprenger et al. 2004)
THP monocytes (human)	52	51	85	14	(Li et al. 2003)
Jurkat T cells (human)	36	32	72	6	(Tu et al. 2004)
Jurkat T cells (human)	23	19	44	8	(Bini et al. 2003)
Intestinal brush border (rat)	15	14	100	2	(Nguyen et al. 2006)
Ramos cells (human)	n/a[l]	n/a	n/a	n/a	(Gupta et al. 2006)
Hippocampal neurons (rat)	n/a[l]	n/a	n/a	n/a	(Ledesma et al. 2003)
Merozoite (Plasmodium falciparum	122	n/a[m]	n/a	n/a	(Sanders et al. 2005)
Callus tissue (Arabidopsis)	63	n/a[m]	n/a	n/a	(Borner et al. 2005)
Neutrophils (cow)	19	n/a[m]	n/a	n/a	(Nebl et al. 2002)

[a] Tissue or cell and species source analyzed.
[b] Number of protein identifications reported by study authors.
[c] Number of protein sequences still retrievable from on-line databases using reported accession numbers.
[d] Fraction of proteins reported in the indicated study that is likely to actually be components of lipid rafts based on comparison to MβCD-sensitive proteins in Foster et al. (2003). For within species comparisons 95% identity was required for two sequences to be considered equivalent and for cross-species comparisons 85% identity was required (Foster et al. 2006).
[e] Number of reported proteins found in at least 50% of all DRM proteomic studies based on BLAST comparison using homology rules described above.
[f] Authors report only protein names and "in-house" accession numbers so data were not comparable to any other study.
[g] 387 proteins were reported in this study but only 241 were claimed to actually reside in rafts.
[h] Protein accession numbers claimed to be in supplementary material but none available.
[i] Results for Brij, carbonate and NP-40 extractions reported separately.
[j] Number of IDs in "2 + 3 + 4" and "5 + 6" pools respectively.
[k] Number reflects collapsed isoforms.
[l] No complete protein list or accession numbers provided.
[m] Bioinformatic analysis performed only on rodent and primate DRM datasets.

that there is no consistency between DRM proteomes (McMahon et al. 2006). The proteins common to at least half of all reports include some (e.g. flotillins 1 and 2, 5'-nucleotidase) but not all (e.g. caveolin) raft markers. There is also a considerable range of common proteins seen in each individual study, which can largely be explained by the depth of coverage in the particular study and the biological system studied. For instance, since there are more reports of DRM analyses in B and T-cells those studies tended to have higher numbers of common proteins (Table 1).

While this correlation of all the published DRM proteomes is very enlightening, perhaps the clearest lesson to be learned from it is that a careful analysis of specificity (e.g. sensitivity to cholesterol disruption) and functionality is required in each biological system before one could attempt to build a global picture of rafts.

6. CONCLUSIONS

Lipid rafts remain a contentious subject, with skeptics citing inconsistencies in parameters such as raft size measured by different methods and discrepancies between model lipid systems and cells as proof that rafts are artifactual. Such criticisms are likely to be addressed with a better understanding of the biophysics of biological membranes but in the meantime cell biologists and proteomicists continue to study them as if they are real. In the heady early days of proteomics there was a big rush to apply the technology to any exciting area of biology, including lipid rafts. As a result, there is a plethora of information available about the protein composition of DRMs. However, with quantitative methodologies maturing (Foster 2006), reducing in price and becoming easier to analyze with open-source tools we believe that the days are past where it is reasonable to simply identify proteins in DRMs and equate them to lipid raft proteins. Given that a third or more of all DRM proteins are likely to be contaminants (Foster et al. 2003) such an approach seems very dangerous! For example, several studies discussed here have concluded that numerous proteins previously thought to be specific to mitochondria also reside in rafts/caveolae. However, the cholesterol-independence of the proteins on which these findings are based indicates that these conclusions are incorrect. We believe that future raft/caveolae proteomics efforts should focus more on specificity and/or functionality. Several reagents that affect cellular cholesterol levels are now available and while none of them are perfect (Pike and Miller 1998), they can be used as a means of cross-validation. On the other hand, functional studies involving the quantitation of DRM proteomics between different cell states bypass the need for direct control for specificity since, by definition, if the level of a protein in DRMs is observed to change after antagonism then it is specific. In the near term we expect that more functional proteomic analysis of DRMs will emerge, following changes induced by the multitude of stimuli that are thought to engage rafts. Further in the future, as high-throughput and more advanced imaging techniques become mainstream, raft proteomics will marry mass spectrometry-based analyses with microscopic analysis of each identified raft protein to arrive at a much more detailed picture of raft dynamics.

ACKNOWLEDGEMENTS

This work is supported in part by grants to LJF from the Natural Sciences and Engineering Research Council of Canada (NSERC) and from the Canadian Institutes of Health Research. QWTC is supported by a NSERC CGS M graduate scholarship and LJF is a Michael Smith Foundation Scholar and the Canada Research Chair in Organelle Proteomics. LJF and QWTC thank Carmen de Hoog for critical reading of the manuscript.

REFERENCES

Bae, T.J., Kim, M.S., Kim, J.W., Kim, B.W., Choo, H.J., Lee, J.W., Kim, K.B., Lee, C.S., Kim, J.H., Chang, S.Y., Kang, C.Y., Lee, S.W. and Ko, Y.G. (2004) Lipid raft proteome reveals ATP synthase complex in the cell surface. Proteomics *4*, 3536–3548.

Bini, L., Pacini, S., Liberatori, S., Valensin, S., Pellegrini, M., Raggiaschi, R., Pallini, V. and Baldari, C.T. (2003) Extensive temporally regulated reorganization of the lipid raft proteome following T-cell antigen receptor triggering. Biochem. J. *369*, 301–309.

Blagoev, B., Ong, S.E., Kratchmarova, I. and Mann, M. (2004) Temporal analysis of phosphotyrosine-dependent signaling networks by quantitative proteomics. Nat. Biotechnol. *22*, 1139–1145

Blonder, J., Hale, M.L., Chan, K.C., Yu, L.R., Lucas, D.A., Conrads, T.P., Zhou, M., Popoff, M.R., Issaq, H.J., Stiles, B.G. and Veenstra, T.D. (2005) Quantitative profiling of the detergent-resistant membrane proteome of iota-b toxin induced vero cells. J. Proteome Res. *4*, 523–531.

Blonder, J., Terunuma, A., Conrads, T.P., Chan, K.C., Yee, C., Lucas, D.A., Schaefer, C.F., Yu, L.R., Issaq, H.J., Veenstra, T.D. and Vogel, J.C. (2004) A proteomic characterization of the plasma membrane of human epidermis by high-throughput mass spectrometry. J. Invest. Dermatol. *123*, 691–699.

Blonder, J., Yu, L.R., Radeva, G., Chan, K.C., Lucas, D.A., Waybright, T.J., Issaq, H.J., Sharom, F.J. and Veenstra, T.D. (2006) Combined chemical and enzymatic stable isotope labeling for quantitative profiling of detergent-insoluble membrane proteins isolated using Triton X-100 and Brij-96. J. Proteome Res. *5*, 349–360.

Borner, G.H., Sherrier, D.J., Weimar, T., Michaelson, L.V., Hawkins, N.D., Macaskill, A., Napier, J.A., Beale, M.H., Lilley, K.S. and Dupree, P. (2005) Analysis of detergent-resistant membranes in Arabidopsis. Evidence for plasma membrane lipid rafts. Plant Physiol. *137*, 104–116.

Brown, D.A. and Rose, J.K. (1992) Sorting of GPI-anchored proteins to glycolipid-enriched membrane subdomains during transport to the apical cell surface. Cell *68*, 533–544.

Casey, P.J. (1995) Protein lipidation in cell signaling. Science *268*, 221–225.

Chamberlain, L.H. (2004) Detergents as tools for the purification and classification of lipid rafts. FEBS Lett. *559*, 1–5.

Danielsen, E.M. and van Deurs, B. (1995) A transferrin-like GPI-linked iron-binding protein in detergent-insoluble noncaveolar microdomains at the apical surface of fetal intestinal epithelial cells. J. Cell Biol. *131*, 939–950.

Foster, L.J., de Hoog, C.L. and Mann, M. (2003) Unbiased quantitative proteomics of lipid rafts reveals high specificity for signaling factors. Proc. Natl Acad. Sci. U.S.A. *100*, 5813–5818.

Foster, L.J., de Hoog, C.L., Zhang, Y., Zhang, Y., Xie, X., Mootha, V.K. and Mann, M. (2006) A mammalian organelle map by protein correlation profiling. Cell *125*, 187–199.

Gupta, N., Wollscheid, B., Watts, J.D., Scheer, B., Aebersold, R. and Defranco, A.L. (2006) Quantitative proteomic analysis of B cell lipid rafts reveals that ezrin regulates antigen receptor mediated lipid raft dynamics. Nat. Immunol. *7*, 625–633.

Karsan, A., Blonder, J., Law, J., Yaquian, E., Lucas, D.A., Conrads, T.P. and Veenstra, T. (2005) Proteomic analysis of lipid microdomains from lipopolysaccharide-activated human endothelial cells. J. Proteome Res. *4*, 349–357.

Keller, P. and Simons, K. (1998) Cholesterol is required for surface transport of influenza virus hemagglutinin. J. Cell Biol. *140*, 1357–1367.

Ledesma, M.D., Da Silva, J.S., Schevchenko, A., Wilm, M. and Dotti, C.G. (2003) Proteomic characterisation of neuronal sphingolipid-cholesterol microdomains: role in plasminogen activation. Brain Res. *987*, 107–116.

Li, N., Mak, A., Richards, D.P., Naber, C., Keller, B.O., Li, L. and Shaw, A.R. (2003) Monocyte lipid rafts contain proteins implicated in vesicular trafficking and phagosome formation. Proteomics *3*, 536–548.

Li, N., Shaw, A.R., Zhang, N., Mak, A. and Li, L. (2004) Lipid raft proteomics: analysis of in-solution digest of sodium dodecyl sulfate-solubilized lipid raft proteins by liquid chromatography-matrix-assisted laser desorption/ionization tandem mass spectrometry. Proteomics *4*, 3156–3166.

Maclellan, D.L., Steen, H., Adam, R.M., Garlick, M., Zurakowski, D., Gygi, S.P., Freeman, M.R. and Solomon, K.R. (2005) A quantitative proteomic analysis of growth factor-induced compositional changes in lipid rafts of human smooth muscle cells. Proteomics *5*, 4733–4742, Published on-line.

Man, P., Novak, P., Cebecauer, M., Horvath, O., Fiserova, A., Havlicek, V. and Bezouska, K. (2005) Mass spectrometric analysis of the glycosphingolipid-enriched microdomains of rat natural killer cells. Proteomics *5*, 113–122.

McMahon, K.A., Zhu, M., Kwon, S.W., Liu, P., Zhao, Y. and Anderson, R.G. (2006) Detergent-free caveolae proteome suggests an interaction with ER and mitochondria. Proteomics *6*, 143–152.

Nebl, T., Pestonjamasp, K.N., Leszyk, J.D., Crowley, J.L., Oh, S.W. and Luna, E.J. (2002) Proteomic analysis of a detergent-resistant membrane skeleton from neutrophil plasma membranes. J. Biol. Chem. *277*, 43399–43409.

Nguyen, H.T., Amine, A.B., Lafitte, D., Waheed, A.A., Nicoletti, C., Villard, C., Letisse, M., Deyris, V., Roziere, M., Tchiakpe, L., Danielle, C.D., Comeau, L. and Hiol, A. (2006) Proteomic characterization of lipid rafts markers from the rat intestinal brush border. Biochem. Biophys. Res. Commun. *342*, 236–244.

Ong, S.E., Blagoev, B., Kratchmarova, I., Kristensen, D.B., Steen, H., Pandey, A., Mann, M. (2002) Stable isotope labeling by amino acids in cell culture, SILAC, as a simple and accurate approach to expression proteomics. Mol. Cell. Proteomics *1*, 376–386.

Ong, S.E., Foster, L.J. and Mann, M. (2003) Mass spectrometric-based approaches in quantitative proteomics. Methods *29*, 124–130.

Palade, C.E. (1953) Fine structure of blood capillaries. J. Applied Phys. *24*, 1424.

Paradela, A., Bravo, S.B., Henriquez, M., Riquelme, G., Gavilanes, F., Gonzalez-Ros, J.M. and Albar, J.P. (2005) Proteomic analysis of apical microvillous membranes of syncytiotrophoblast cells reveals a high degree of similarity with lipid rafts. J. Proteome Res. *4*, 2435–2441.

Pike, L.J. and Miller, J.M. (1998) Cholesterol depletion delocalizes phosphatidylinositol bisphosphate and inhibits hormone-stimulated phosphatidylinositol turnover. J. Biol. Chem. *273*, 22298–22304.

Qian, D. and Weiss, A. (1997) T cell antigen receptor signal transduction. Curr. Opin. Cell Biol. *9*, 205–212.

Rajendran, L. and Simons, K. (2005) Lipid rafts and membrane dynamics. J. Cell Sci. *118*, 1099–1102.

Sanders, P.R., Gilson, P.R., Cantin, G.T., Greenbaum, D.C., Nebl, T., Carucci, D.J., McConville, M.J., Schofield, L., Hodder, A.N., Yates, J.R. 3rd and Crabb, B.S. (2005) Distinct protein classes including novel merozoite surface antigens in raft-like membranes of Plasmodium falciparum. J. Biol. Chem. *280*, 40169–40176.

Sargiacomo, M., Sudol, M., Tang, Z. and Lisanti, M.P. (1993) Signal transducing molecules and glycosylphosphatidylinositol-linked proteins form a caveolin-rich insoluble complex in MDCK cells. J. Cell Biol. *122*, 789–807.

Simons, K. and Ikonen, E. (1997) Functional rafts in cell membranes. Nature *387*, 569–572.

Simons, K. and van Meer, G. (1988) Lipid sorting in epithelial cells. Biochemistry *27*, 6197–6202.

Skibbens, J.E., Roth, M.G. and Matlin, K.S. (1989) Differential extractability of influenza virus hemagglutinin during intracellular transport in polarized epithelial cells and nonpolar fibroblasts. J. Cell Biol. *108*, 821–832.

Smart, E.J., Ying, Y.S., Mineo, C. and Anderson, R.G. (1995) A detergent-free method for purifying caveolae membrane from tissue culture cells. Proc. Natl Acad. Sci. U.S.A. *92*, 10104–10108.

Song, K.S., Li, S., Okamoto, T., Quilliam, L.A., Sargiacomo, M. and Lisanti, M.P. (1996) Co-purification and direct interaction of Ras with caveolin, an integral membrane protein of caveolae microdomains. Detergent-free purification of caveolae microdomains. J. Biol. Chem. *271*, 9690–9697.

Sprenger, R.R., Speijer, D., Back, J.W., De Koster, C.G., Pannekoek, H. and Horrevoets, A.J. (2004) Comparative proteomics of human endothelial cell caveolae and rafts using two-dimensional gel electrophoresis and mass spectrometry. Electrophoresis *25*, 156–172.

Thomsen, P., Roepstorff, K., Stahlhut, M. and van Deurs, B. (2002) Caveolae are highly immobile plasma membrane microdomains, which are not involved in constitutive endocytic trafficking. Mol. Cell Biol. *13*, 238–250.

Tu, X., Huang, A., Bae, D., Slaughter, N., Whitelegge, J., Crother, T., Bickel, P.E. and Nel, A. (2004) Proteome analysis of lipid rafts in Jurkat cells characterizes a raft subset that is involved in NF-kappaB activation. J. Proteome Res. *3*, 445–454.

von Haller, P.D., Donohoe, S., Goodlett, D.R., Aebersold, R. and Watts, J.D. (2001) Mass spectrometric characterization of proteins extracted from Jurkat T cell detergent-resistant membrane domains. Proteomics *1*, 1010–1021.

von Haller, P.D., Yi, E., Donohoe, S., Vaughn, K., Keller, A., Nesvizhskii, A.I., Eng, J., Li, X.J., Goodlett, D.R., Aebersold, R. and Watts, J.D. (2003a) The application of new software tools to quantitative protein profiling via ICAT and tandem mass spectrometry: I. Statistically annotated data sets for peptide sequences and proteins identified via the application of ICAT and tandem mass spectrometry to proteins co-purifying with T cell lipid rafts. Mol. Cell Proteomics *2*, 426–427.

von Haller, P.D., Yi, E., Donohoe, S., Vaughn, K., Keller, A., Nesvizhskii, A.I., Eng, J., Li, X.J., Goodlett, D.R., Aebersold, R. and Watts, J.D. (2003b) The application of new software tools to quantitative protein profiling via ICAT and tandem mass spectrometry: II. Evaluation of tandem mass spectrometry methodologies for large-scale protein analysis and the application of statistical tools for data analysis and interpretation. Mol. Cell Proteomics *2*, 428–442.

Zeyda, M. and Stulnig, T.M. (2006) Lipid Rafts & Co.: an integrated model of membrane organization in T cell activation. Prog. Lipid Res. *45*, 187–202.

Corresponding author: Dr. Leonard J. Foster (ljfoster@interchange.ubc.ca)

SECTION 2

ORGANELLE SUBPROTEOMES

CHAPTER 5

ORGANELLE PROTEOME VARIATION AMONG DIFFERENT CELL TYPES: LESSONS FROM NUCLEAR MEMBRANE PROTEINS

DEIRDRE M. KAVANAGH, WILLIAM E. POWELL, POONAM MALIK, VASSILIKI LAZOU, and ERIC C. SCHIRMER
University of Edinburgh, UK

Table of Contents

Abstract: Most subcellular organelles are expected to be similar among different cell types; however, a recent study suggests a surprising amount of variation in the protein composition at the nuclear envelope. Therefore, to comprehensively identify proteins in subcellular organelles proteomics datasets may need to be generated from multiple cell types. In this chapter we describe a proteomics study that expanded the number of nuclear membrane proteins by 5-fold using a "subtractive" methodology in which a subcellular organelle is partially

E. Bertrand and M. Faupel (eds.), Subcellular Proteomics, 51–76.

purified biochemically and partially purified *in silico*. The biochemical fraction of interest and a separate fraction enriched in proteins known to contaminate it, in this case nuclear envelopes and microsomes respectively, are first isolated and separately analyzed by mass spectrometry. For *in silico* purification, proteins appearing in both fractions are subtracted from the dataset in order to identify proteins that are unique to the organelle being investigated. This approach identified 67 novel putative nuclear envelope transmembrane proteins in rodent liver. Further analysis of their expression levels in other tissues indicates that several are preferentially expressed in liver cell types, which in turn predicts considerable variation in the nuclear envelope proteome among different cell types. Finally, we discuss several issues associated with confirming that these peptide-based identifications represent proteins that truly localize to the nuclear envelope. These studies have complicated rather than simplified our view of the nuclear envelope, but proteomics has set the stage for beginning to understand this highly complex subnuclear organelle.

1. INTRODUCTION

It is clear that each cell type expresses only a fraction of the proteins encoded by the genome, with some unique and some ubiquitous proteins. It is less clear how these unique and ubiquitous proteins are distributed at a subcellular level. Cell-type specific proteins have long been demonstrated in the nucleus where distinctive transcription factors regulate which proteins are made, at the plasma membrane where many signalling cascades are instigated, and in the cytoplasm where many unique metabolic events occur. In contrast, the view is widely held that proteins in most other organelles such as mitochrondria, endoplasmic reticulum (ER), vacuoles, and nucleoli would be largely similar from one cell type to another. For some organelles, recent studies have shown that protein variation is greater than previously thought. For example, cell-type specific differences in the nucleus go well beyond transcription factors with certain histone H1 variants exchanged at specific stages in development and in differentiated cell types, presumably to faciliate unique patterns of gene expression (Brown 2001; Lee et al. 2004). Historically the nuclear subcompartment termed the nuclear envelope (NE) was viewed as having only "structural" roles, serving as a barrier between the nucleus and cytoplasm; however analysis of a recent proteomics dataset suggests that its organization is quite intricate, with an unexpected degree of variation in the NE proteins present in different cell types (Schirmer et al. 2003; Schirmer and Gerace 2005; Schirmer et al. 2005). Moreover, this variation in the NE proteome may provide a resolution to the conundrum of how mutations in a subset of ubiquitous NE proteins can cause tissue-specific diseases (Worman and Courvalin, 2002; Mounkes and Stewart 2004).

The NE is an elaborate structure that can be divided into several distinct subdomains: the nuclear pore complexes (NPCs), the lamin polymer, and a double membrane system consisting of the outer nuclear membrane (ONM), inner nuclear membrane (INM), lumen, and pore membrane (PoM) together with their integral proteins (Figure 1; **see colour insert**). The ONM is not only continuous with the ER, but is also studded with ribosomes indicating that in addition to being the outermost layer of the nucleus it is also a subcompartment of the ER. How much of its complement of integral membrane proteins is unique from more distal ER

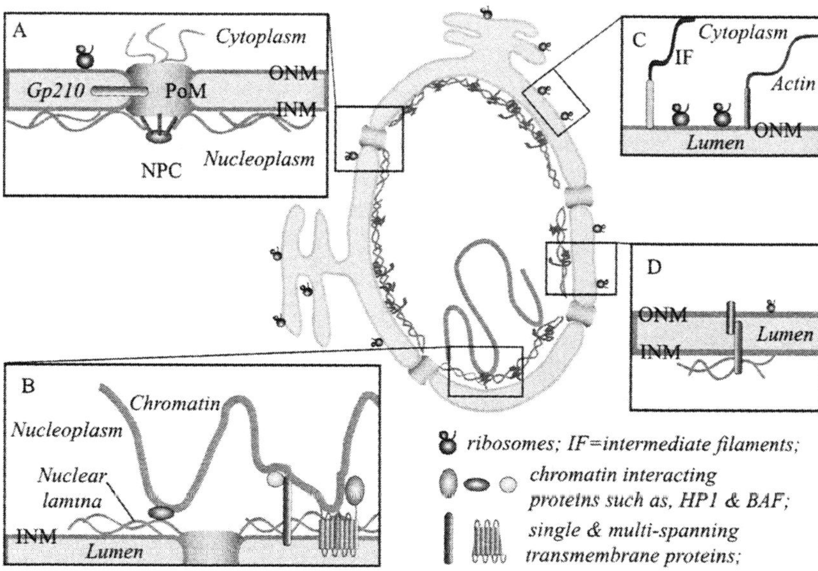

Figure 1. The nuclear envelope and its connections to other cellular compartments. (A) The nuclear envelope is made of outer and inner nuclear membranes (ONM and INM), which are connected at the pore membrane (PoM) that encircles the nuclear pore complexes (NPCs). The NPCs regulate directional trafficking of soluble macromolecules in and out of the nucleus. On the cytoplasmic and nuclear side of the pore respectively are the cytoplasmic filaments and nuclear basket of the NPC. Nup153 of the nuclear basket interacts with lamins. NPCs are tethered to the PoM by two integral proteins, gp210 and POM121. The ONM is continuous with the endoplasmic reticulum (ER) and studded with ribosomes. Thus proximal ER is a major contaminant of NEs. (B) Underneath the INM is the nuclear lamin polymer. Together with associated integral membrane proteins of the INM this polymer is called the nuclear lamina and some unique integral single or multi-spanning transmembrane proteins of the INM interact with chromatin and provide attachment points for the lamina. There are also integral proteins of the INM, PoM, and ONM that are not associated with the lamin polymer. Chromatin and soluble nucleoplasmic proteins may have specific interactions at the NE, but it is hard to distinguish these from contaminants because of the particular "stickiness" of many highly charged DNA and RNA-binding proteins. (C) There are also integral proteins of the ONM that are not associated with the lamin polymer and some ONM-specific proteins bind to cytoplasmic filament systems, providing another major contaminant of NEs. Nesprin proteins appear to be able to link the NE to cytoplasmic intermediate filaments, while SUN proteins link the NE to cytoplasmic actin. (D) Finally, the luminal space between the ONM and INM, a soluble compartment that is also continuous with the ER is largely unexplored territory but some unidentified proteins may make contacts across the lumen in order to maintain its highly regular spacing.

remains unclear, but it has now been clearly demonstrated that proteins involved in tethering the nucleus to the cytoskeleton are distinctive components of the ONM (Starr and Han 2002; Wilhelmsen et al. 2005; Crisp et al. 2006). The lumenal space between the ONM and INM is largely unexplored territory, but recent studies suggest some proteins may make contacts across the lumen in order to maintain its highly regular spacing (Crisp et al. 2006). Such proteins may sterically interfere with lateral diffusion of membrane proteins with large lumenal domains and, indeed, transport from the ER to the INM of transmembrane protein chimeras is inhibited when they carry a large lumenal domain (Ohba et al. 2004). The ONM and INM are joined where NPCs perforate the membrane. The PoM effetively wraps around these approximately 125 MDa (MegaDalton) macromolecular complexes in vertebrates. The INM has a more clearly defined set of integral membrane proteins compared to the ONM because many are physically associated with the nuclear lamin polymer that lies underneath it, and these proteins remain associated with lamins in a biochemical fraction when the lipid of the membranes is extracted with detergent. Lamins are type V intermediate filament proteins that maintain very stable homodimers through the formation of a 52 nm coiled-coil rod by their central alpha-helical domain. Flanking the rod is a short amino-terminal head domain, and a large globular tail that contains an (immunglobulin) Ig fold (Stuurman et al. 1998; Dhe-Paganon et al. 2002; Krimm et al. 2002). While the membrane forms a barrier to diffusion of soluble macromolecules between the nucleus and cytoplasm, the lamins provide mechanical stability to the structure (Liu et al. 2000; Schirmer et al. 2001; Lammerding et al. 2004).

2. THE HISTORY OF NETs

Before the application of proteomics to the NE, proteins of this structure were identified through a wide variety of approaches. Lamins were the first NE proteins characterized (roughly 30 years ago) because their solubility characteristics as intermediate filament proteins and their abundance facilitated easy biochemical enrichment (Aaronson and Blobel 1975; Gerace et al. 1978). It is estimated that there are ~3 million lamin molecules in the average mammalian nucleus (Gerace and Burke 1988).

Individually, the integral membrane proteins are much less abundant; thus, it was over a decade later that the first NE transmembrane proteins (NETs) were discovered. The lamin B receptor (LBR) was identified by its binding to lamin B1 (Worman et al. 1988), while lamina-associated polypeptides 1 and 2 (LAP1 and LAP2) were identified in a screen for monoclonal antibodies using the large number of proteins in a biochemically isolated NE fraction (Foisner and Gerace 1993; Senior and Gerace 1988). Consistent with their identification by this means, the two LAP proteins are likely to be among the most abundant of the NETs. This is further supported by the high relative abundance of LAP peptides recovered by proteomics (Schirmer et al. 2003). Two NETs were identified as part of the NPC, gp210 and POM121 (pore membrane 121) (Greber et al. 1990; Hallberg et al. 1993). Some NETs were identified by microscopy studies: MAN1 was identified fortuitously from an autoimmune serum that produced the nuclear "rim" staining characteristic of lamins (Paulin-Levasseur et al. 1996) while nurim (for nuclear rim) was identified by the screening of a (green fluorescent protein) GFP-cDNA fusion library for proteins that yielded a similar localization (Rolls et al. 1999). Human genetics identified emerin as the gene responsible for Emery-Dreifuss Muscular Dystrophy (Bione et al. 1994) and subsequent work to characterize its protein product determined that it was also a NET (Manilal et al. 1996).

Characterization of a protein identified in a 2-hybrid screen for partners of a kinase of the muscle postsynaptic membrane provided the first evidence of a new family of proteins at the NE (Apel et al. 2000). Originally named Syne for synaptic nuclear envelope (Apel et al. 2000), this family is becoming increasingly referred to as Nesprins for nuclear envelope spectrin repeat because each member contains multiple copies of this motif (Zhang et al. 2001). Two separate genes were originally identified (Apel et al. 2000) that share a common sequence motif termed the KASH domain (for Klarsicht, ANC-1, and Syne-1 homology after the founding members from Drosophila (Mosley-Bishop et al. 1999), *Caenorhabditis elegans* (Starr and Han 2002), and mouse (Apel et al. 2000). A second protein family, denoted SUN (for Sad-1 and UNC domain), was originally identified at the NE in *C.elegans* (Unc84; Malone et al. 1999). Its first mammalian homolog, Unc84A/ SUN1, was discovered in the first proteomics study for NETs (see below; Dreger et al. 2001). SUN2 was identified in an effort to isolate genes with differential expression in heart tissues (Sun et al. 2002).

That UNCL was found to target to the NE was surprising because it was uncovered in a study to identify chaperones that function in assembly of nicotinic acetylcholine receptors (Fitzgerald et al. 2000). Although little note was taken of the fact at the time, its discovery provided the first indication that NET composition differs among different cell types: the study was predicated on the observation that these receptors do not properly assemble in many mammalian cell lines and, indeed, UNCL is absent from most cell types.

2.1. The Genomics Era

This wide range of approaches has been replaced by proteomics in the post-genomics era. Interestingly, genomics itself was inadequate for the identification of NETs. For example, a weak homology LAP2, emerin and MAN1 "LEM" domain was identified when MAN1 was cloned and sequenced (Lin et al. 2000) because this domain was also found to occur in LAP2 and emerin. However, not all proteins containing LEM domains are concentrated at the NE (Dechat et al. 1998; Raju et al. 2003) and the presence of a LEM domain is not always maintained within a protein family, for example, the Drosophila Syne/Nesprin homolog (Klarsicht) contains a LEM domain while the mammalian members of this family do not (Wagner et al. 2004).

There were three initial proteomics studies of NE proteins. Two of these focused on the NPCs and will not be detailed here other than to mention that the approaches were geared towards isolating its structural core (Cronshaw et al. 2002, Rout et al. 2000). The third study focussed on identifying novel INM proteins, using a comparative proteomic approach where different NE subfractions isolated from cultured neuroblastoma cells were compared based on the characteristics of previously known NETs

(Dreger et al. 2001). Integral proteins are enriched after extraction with chaotropes (in this case 4 M urea, 0.1 M Na_2CO_3); however NEs so extracted yield a pellet containing membrane proteins not only of the INM, but also of the ONM and ER. Since known INM proteins remain pelletable after detergent extraction (in this case 1% Triton X-100), they considered proteins of the chaotrope pellet that were also found in the detergent pellet to be good candidates for novel INM proteins. Each subfraction was separated on 2-D gels, the protein spots were excised, and each spot was analyzed by matrix-associated laser desorption ionization (MALDI) mass spectrometry. This analysis identified most expected previously characterized INM proteins in both the chaotrope and detergent fractions; however LAP1 was not identified, nor were the later-identified SUN2 and nesprins. (UNCL was not identified as expected due to its limited tissue expression.) Moreover, LBR and emerin, though identified, did not behave as expected: LBR only appeared in the detergent-resistant fraction and emerin only appeared in the chaotrope-resistant fraction. Nonetheless, the approach was successful in identifying four novel proteins that appeared in both fractions: two additional splice variants of LAP2 (previously predicted from mRNA; Berger et al. 1996), the first mammalian SUN family protein (Unc84A/SUN1), and a completely novel protein with no predicted functions that was named LUMA (Dreger et al. 2001). Several additional putative novel NETs were identified that only occurred in one or the other fraction.

Though both creative and productive, in retrospect there were certain limitations to this approach. The assumption that the membrane extraction characteristics of the proteins would be conserved was in practice not absolute. Moreover, the inherent problems of working with gel-extracted bands were exemplified by the inability to make identifications for roughly 25% of the gel spots (Dreger et al. 2001).

3. SUBTRACTIVE PROTEOMICS

Like the Dreger et al. study, the fourth NE proteomics study took the view that it is impossible to truly purify the NE biochemically because of its many connections to ER, cytoplasmic filament systems, and chromatin. It also assumed that any chemical extraction chosen to improve purity might remove true components as well as contaminants. The design of this fourth approach renders it much better for identification of NE-specific transmembrane proteins than for soluble NE-specific components. The principle is that transmembrane proteins in isolated NEs should derive only from the INM, PoM, ONM, or ER. In contrast, microsomal membranes (MMs) are rich in ER transmembrane proteins, but are free of INM, PoM, or ONM because the large dense nuclei are readily pelletable away from the smaller membrane vesicles obtained by disruption of ER and other membrane systems. Therefore, while isolated nuclei will have contamination from ER, MMs will be free of nuclear contamination: by separately analyzing NEs and MMs and subtracting proteins identified in both fractions, NE-specific transmembrane proteins can be identified (Figure 2). In contrast, soluble contaminants could derive from "sticky" chromatin proteins released during NE purification, collapsed cytoplasmic filaments and associated proteins, or

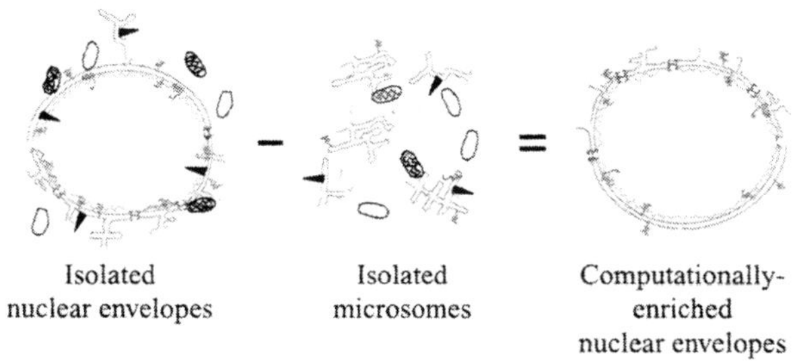

Figure 2. Subtractive proteomics. It is impossible to purify NEs to homogeneity because of the many connections to both the nucleoplasm and the cytoplasm. Thus, biochemically "purified" NEs are expected to be contaminated with chromatin and cytoskeletal proteins and with vesicles from organelles such as mitochodria and ER. In contrast, some of these expected contaminants can be purified free of NE contamination. One such contaminant is ER, which can be isolated as microsomes. Another is mitochondria, which has a well characterized protein complement. Therefore NE and microsomal membrane fractions are separately isolated and analyzed for protein content by MudPIT. All proteins appearing in both fractions are removed from the NE dataset because they could be due to ER vesicles sticking to the isolated nuclear NEs. Similarly, known mitochondrial proteins are removed. Because ER and mitochondria are the only expected membrane contaminants of NEs, all remaining integral membrane proteins in the NE fraction should be NE-specific in theory. After prediction by computer algorithm for membrane-spanning segments, an *in silico* purified NE transmembrane protein list is obtained. A limitation of this approach is that it discounts any proteins that are found both within the ER and the NE membranes (e.g. solid black triangles).

proteins caught in transit through the NPC at the time of purification. Although well-characterized contaminants such as cytoplasmic filaments can easily be identified and discarded, it is difficult to distinguish a separate fraction to subtract in order to identify which other soluble proteins are normally enriched at the NE.

3.1. Biochemical Fractionation

To purify NEs (Blobel and Potter, 1966; Dwyer and Blobel 1976), cells are first lysed by dounce homogenization to release nuclei. Specific protocols for this step vary considerably from one cell type to another based on such criteria as nuclear:cytoplasmic volume ratio, types and abundance of cytoplasmic filament systems, and amount of connective tissue. For example, soft tissues with low nuclear:cytoplasmic ratios such as liver can be directly homogenized in a Potter-Elvehjem homogenizer (Blobel and Potter 1966), while lymphocytes that have much higher nuclear:cytoplasmic ratios must be first hypotonically swollen and then dounce homogenized (Fields et al. 1988). Some protocols call for treatment with filament destabilizing drugs (such as cytochalasin B) because cytoskeletal filaments have been shown to collapse on the nuclear surface when cells are homogenized (Staufenbiel and Deppert 1982), but the tightly banded filaments in muscle are impervious to such treatments. To remove these filaments, the muscle must be first minced or treated with a blade homogenizer (such as a polytron) prior to dounce homogenization. Care must be taken to break the muscle fibers into pieces that are much larger than nuclei because pieces of similar density will be difficult to separate from nuclei in subsequent steps (Held et al. 1977; Kuehl 1977). It is noteworthy that many protocols for isolation of nuclei use detergent in buffers, but this must be avoided as detergents also remove most of the NE.

Because the intact nucleus is larger and denser than any other organelle in most cells, it is readily separated from other organelles by pelleting centrifugation. The speed and duration of this pelleting step should be appropriately modified for different tissues in order to separate the nuclei as much as possible from other cellular structures before subsequent steps (e.g. an initial step at slow speed to pellet large debris without nuclei, followed by a step at intermediate speed to pellet nuclei away from smaller cell debris). Many other membranes and contaminants that co-sediment with the nuclei in this first step are separated by floating on sucrose cushions that the intact nuclei are sufficiently dense to penetrate. Many membranes will float in 1.9 M sucrose; so sucrose is mixed with the homogenate to this density and underlayed with a sucrose cushion of higher density. If too much cell debris is loaded onto the sucrose cushions, it will accumulate at the interface and can prevent penetration by the nuclei. The size, shape, and density of nuclei vary considerably among different cell types and thus the density of sucrose cushions must be accordingly varied for effective recovery. It is also possible to isolate nuclei using other density gradient systems such as percoll (Hahn and Covault 1990).

Isolated nuclei are then treated with DNase and RNase to digest DNA, which is then washed out of the nucleus along with a significant percentage of the nucleoplasmic content using high salt buffers (~half molar). This released chromatin material is then separated from NEs by floating on another sucrose cushion. Through this disruptive process NEs may fracture, but their structure is largely maintained because the membranes and their integral protein complement are connected to the salt-resistant lamin polymer. The resulting nuclear envelopes are thus largely, but not completely, washed clean of both cytoplasmic and nucleoplasmic contaminants. Besides ER membranes (observed as single membrane vesicles by EM in NE preparations; see supplemental material in Schirmer et al. 2003), expected contaminants include mitochondria, cytoplasmic filaments, and some highly charged chromatin proteins that are very "sticky" and resistant to extraction with high salt.

To prepare MMs (Gerace et al. 1982; Scheele 1983), nuclei are first removed from a cell homogenate by low-speed centrifugation. Due to the fact that many mitochondria will saturate the subsequent sucrose cushions, the post-nuclear supernatant is subjected to an intermediate-speed centrifugation step to pellet most mitochondria. The resulting supernatant is mixed with sucrose and layered to float membranes during the next centrifugation step. Different types of membranes will separate into the different density layers of sucrose: the MM fraction is recovered, the sucrose diluted, and the membranes pelleted at high speed.

3.2. MudPIT Proteomics

Multi-dimensional Protein Identification Technology (MudPIT) incorporates several liquid chromatography steps with tandem mass spectrometry to identify proteins in a complex biochemical fraction (Washburn et al. 2001; Wolters et al. 2001). Alternating reverse-phase and cation exchange steps provides separation of the enormous complexity of peptides generated by direct digestion of the fraction without prior separation of individual proteins. Among the advantages that this provides is that proteins which migrate too closely on gels or other isolation platforms can still be identified. Moreover, proteins of low abundance are more likely to be distinguished. These benefits are especially important for membrane proteins, which often have extreme isoelectric points that limit effective resolution on 2-D gels (Santoni et al. 2000). As peptides are not linked to their corresponding protein band when using MudPIT, protein identifications must be made for each individual peptide. To accomplish this, tandem mass spectrometry is utilized in which each analyzed peptide is subsequently fragmented and the masses of the smaller peptides generated are also measured. In many cases this combined footprint can be used to match all the fragments within the full-length peptide with a unique sequence in the genome databases, thus enabling protein identification from a single peptide. Since a significant number of peptides analyzed do not produce identifications, the method is not quantitative alone; however, combining this method with analysis of metabolically labeled fractions can add information on abundance (MacCoss et al. 2005).

3.3. Subtraction

Rodent liver NEs and MMs were extracted with 0.1 M NaOH to enrich for membrane proteins. In the NE fraction 566 proteins were identified, while 652 proteins were identified in the MMs (Schirmer et al. 2003). 41% of proteins in the NE fraction were also observed in the MM fraction; therefore, according to the subtractive methodology, these proteins were disregarded. Because the well-characterized NETs observed in the earlier proteomics study did not all fractionate in extractions as expected (Dreger et al. 2001), a separate NE fraction was extracted with salt and detergent to enrich for proteins associated with the lamin polymer. This identified 1830 proteins, and again any also appearing in the MM fraction were disregarded. Proteins remaining in the NE dataset were further analyzed for probability of transmembrane spanning segments by computer algorithm. Most algorithms used for predicting transmembrane spanning segments measure alpha helices with a hydrophobic character. However, membrane-spanning segments could also be simple hydrophobic stretches, multimers of beta barrels (Tamm et al. 2004), or hydrophobic "domes" (McKinney and Cravatt 2005). Algorithms for such predictions have evolved considerably in the last few years. At the time of the study, the dataset was only interrogated for hydrophobic helices and the available algorithms yielded different results. Because of this, an algorithm was selected that compares proteins to a database of well-characterized membrane-spanning segments (TMPred: www.ch.embnet.org/software/TMPRED_form.html) and parameters were selected that resulted in correct predictions for the previously characterized NETs.

After eliminating previously characterized proteins, 67 putative new NETs were identified between the NaOH extracted and salt/detergent extracted NE fractions.

These included several novel proteins related to previously characterized NETs: the third member of the Syne/ Nesprin family, NET53 (later called Nesprin 3; Wilhelmsen et al. 2005); a homolog of LAP1, NET9 (later called LULL1; Goodchild and Dauer 2005); and two new LEM domain proteins, NET25 (later called Lem2; Brachner et al. 2005; Lee and Wilson 2004) and NET66.

4. CONFIRMATION OF NUCLEAR ENVELOPE RESIDENCE

For simplicity we will refer to these proteins collectively as NETs, although until they are confirmed at the NE they are in fact "putative" NETs. However, determination of targeting to the NE is not always trivial.

4.1. Determination of Subcellular Localization

The simplest method to test for NE residence is to transfect the NET as a fusion with an epitope tag into tissue culture cells and visualize by fluorescence microscopy. Concentration at the nuclear periphery would be considered confirmation of NE targeting. However, overexpression tends to result in accumulation in the ER as well as the NE. This is the case even for well-characterized NETs for which antibodies recognize only the nuclear rim, and is presumably due to saturation of binding sites at the NE (Soullam and Worman 1993; Ellenberg et al. 1997). If a putative NET is normally expressed at a low level and requires a limiting NE binding partner for targeting, then distinguishing the staining at the nuclear periphery against the high background in the ER may be difficult.

Many characterized INM proteins have been shown to interact with lamins (Ye et al. 1998). For the subset of putative NETs that share this characteristic, a more definite result can be achieved. As previously discussed, the lamin polymer is insoluble to extraction with detergent and salt; therefore retention of NETs in cells extracted with detergent (e.g. 0.5% Triton X-100) prior to fixation for microscopy confirms their direct, or indirect, association with the lamin polymer and NE localization. However, loss of the protein to a pre-extraction with detergent would occur for proteins not tethered to the lamin polymer whether they are normally localized to the INM, ONM, or ER. All eight of the putative NETs originally tested targeted to the NE, but only five remained at the NE after the detergent pre-extraction (Schirmer et al. 2003).

Epitope tags produce false-negatives when fusion to the tag blocks access to sequences required for proper targeting. Many NE proteins have large nucleoplasmic regions that bind to both chromatin proteins and lamins and this binding appears to be responsible for their accumulation in the NE. Thus fusion of a large tag in a nucleoplasmic region could block access to these binding partners and cause mistargeting. This appears to be the case for LAP1 and LAP2, which both have large nucleoplasmic domains: chimeric proteins failed to target properly when GFP was fused to the amino-terminus of either protein, whilst GFP fusions to the carboxyl-terminus were correctly localized to the NE.

Although epitope tags may produce artifacts, so may antibodies and several NE proteins have been reported to have significant problems from epitope masking. In a study using three different lamin B1 antibodies on heart and hippocampus sections, one stained just cardiomyoctyes, another stained just hippocampus, and the third stained both (Tunnah et al. 2005). As each antibody recognized a different part of the protein, this argues that different epitopes are masked in different tissues. In the case of emerin, six different monoclonal antibodies that recognize the protein by immunofluorescence in other cell types failed to recognize emerin in spleen, even though it was clearly present by Western blot analysis (Tunnah et al. 2005). Such extreme epitope masking for emerin may be understandable in light of the many partners that have been mapped to bind throughout its length (Bengtsson and Wilson 2004).

4.2. Determination of Membrane Insertion

Soluble binding partners of NETs or lamins could also concentrate at the NE. Therefore full confirmation of these proteins as NETs requires also clear determination of membrane insertion. As mentioned above (Section 3.3), prediction algorithms have improved considerably since this computational analysis was undertaken, yet different algorithms still produce contradictory predictions. Thus, some of the putative NETs may not be integral membrane proteins and all should be directly tested. A simple test is to parallel extraction characteristics with well-defined integral membrane proteins; however this proved problematic in the comparative proteomics study (Section 2.1; Dreger et al. 2001). Moreover, the NETs identified in the subtractive liver study had differing extraction characteristics: of the eight new NETs for which targeting to the NE was initially determined, only one appeared in both NaOH and salt/detergent extracted fractions, and, although the NaOH-extraction is considered better for isolation of transmembrane proteins, all three proteins that appeared in only the salt/detergent extracted fraction targeted to the NE (Schirmer et al. 2003). The best test for membrane insertion is the protection provided by membranes against proteases for example, protein sequences in the lumen are resistant to proteolytic digestion. To test this, however, requires either multiple versions of the protein with different epitope tags or a panel of antibodies to the protein, which may not be readily available for the large set of proteins identified in a proteomics dataset.

Although the predictions generated by computer algorithm must be confirmed by direct testing for individual NETs, in aggregate they can provide important insights into a large dataset. For example, it is noteworthy that eight of the 67 putative NETs were predicted to have signal peptides, yet nuclear retention signals are frequently predicted to be signal peptides (for more information see http://www.cbs.dtu.dk/services/SignalP/). Three of these eight were tested for targeting with a small N-terminal hemagglutinin (HA) epitope tag: all three targeted to the NE, indicating that the predicted signal peptides were not cleaved and are instead NE retention signals or regular transmembrane segments. Moreover, all three resisted a pre-extraction with detergent, indicating that they occur in the INM.

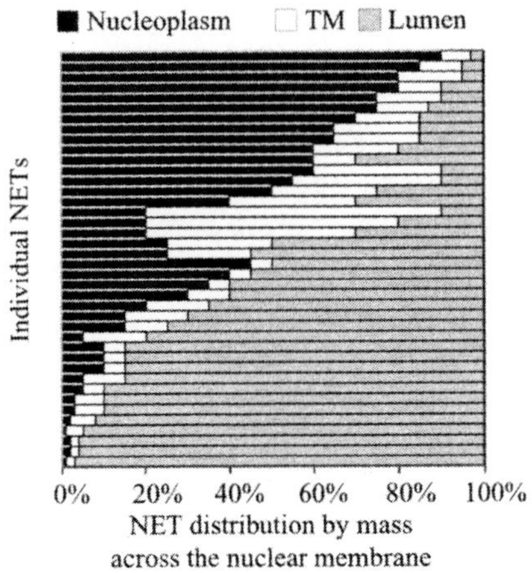

Figure 3. Predicted nuclear membrane topologies of new NETs. The TMHMM algorithm (http://www.cbs.dtu.dk/services/TMHMM-2.0/) was used to predict transmembrane helices for the NETs (plus splice variants for 5 NETs). This algorithm does not predict transmembrane segments for all the NETs that had been predicted by TMPred (Section 3.3); nonetheless, it was chosen for this analysis because it is considered to be more stringent. For those NETs with high probability predictions, the assigned topologies were used to calculate the percentage of the mass of each NET that should occur in the lumen, membrane, or nucleoplasm (nucleoplasm could be cytoplasm depending on if they reside in the INM or ONM). The percentage of the mass of each protein in each of the three domains is depicted graphically. Intriguingly, there were roughly as many proteins with the majority of their mass in the lumen as in the nucleoplasm. Several NETs also had a majority of their mass in the membrane.

This finding is particularly interesting as it suggests most or all NETs are type II transmembrane proteins.

Another interesting aspect of this dataset comes from the topology predictions of the transmembrane algorithms. With the exception of gp210, the principal mass of the originally characterized NETs occurs in the nucleoplasm. However, many among this original set of NETs were identified on the basis of their interactions with the lamina; so a concentration in the nucleoplasm is not surprising. Within the set of novel putative NETs, where identification was based separately on membrane (NaOH-extracted fraction) and on lamina association (salt/ detergent extracted fraction), the topological predictions indicate that there are as many NETs with their principal mass in the lumen as in the nucleoplasm (Figure 3).

4.3. Additional Complications

Several other issues may further complicate determination of NE targeting. One of these is the existence of splice variants for many NETs. For example, the LAP2 gene produces at least seven separate mRNAs (Berger et al. 1996): most have a transmembrane segment, but in what cell types each is produced and whether all target to the NE has not been clarified. The alpha splice variant, which is soluble, concentrates both at the nuclear rim and in specific locations in the nuclear interior (Dechat et al. 1998). The Syne/ Nesprin family also has many splice variants, some of which do not target to the NE (Gough et al. 2003). Evidence of multiple splice isoforms for 18 out of the 67 novel putative NETs is present in the cDNA and EST databases. Some splice variants have different numbers of predicted membrane-spanning segments and large sequence blocks from other exons, which could alter accessibility to or affinity for targeting sites in the NE. Thus, some splice variants could target to other cellular locations.

NE retention in most studied cases has been found to require a lamin-binding site (Soullam and Worman 1993). However, NETs such as emerin have been shown to bind other NETs (Bengtsson and Wilson 2004) and in *C.elegans* the NET Unc83 is dependent upon the NET Unc84 for its targeting to the NE (Malone et al. 1999). Unc84 itself is dependent upon lamin A (Lee et al. 2002). Thus if a particular partner protein is absent from a specific cell type, a NET could target to the NE in one cell type but not another.

5. VARIATION IN THE NUCLEAR ENVELOPE PROTEOME

Nearly all of the NETs identified prior to the subtractive study appeared to be ubiquitously expressed: the only exception was UNCL (Fitzgerald et al. 2000). As no message for UNCL was detected in liver by Northern analysis, it was not surprising that UNCL did not appear in the liver proteomics dataset (Schirmer et al. 2003; Schirmer and Gerace 2005). The surprising number of NETs identified in the subtractive proteomics study raises the possibility that some might be cell-type specific because liver is a composite of several different cell types including hepatocytes, kuppfer cells, lipocytes, and endothelial cells contributed by its extensive vasculature.

5.1. The Transcriptome Database

The relative expression levels of NET mRNAs in different tissues were obtained from the transcriptome database of the Genome Institute of the Novartis Research Foundation. This DNA array dataset was generated using mRNAs isolated from 72 human and 61 mouse tissues and cell types (Su et al. 2002). Experiments were standardized within this large study so that the level of expression in one tissue could be compared to another through the numerical values given for each message in their respective tissues. When the highest numerical value given for a particular NET across the range of tissues was divided by the lowest numerical value given for that NET, a low ratio reflects ubiquitous expression of the NET at similar levels: a high ratio either indicates a wide range of expression levels or its absence in a subset of tissues. Of the 53 putative NETs that were encoded on the Affymetrix chips used, less than one fifth were uniformly expressed with lower than 6-fold variation among different tissues (Figure 4A).

When the highest value was instead divided by the median value, high ratios were also observed (Figure 4B). Here a high ratio is inferred to indicate a significant

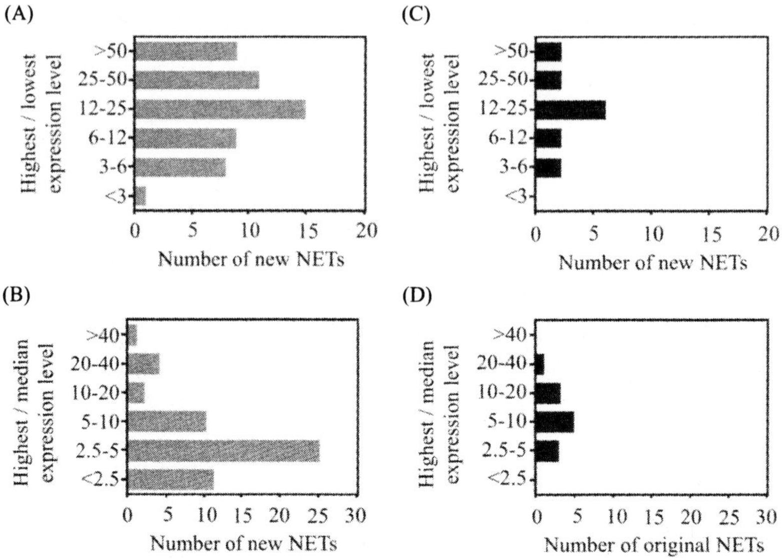

Figure 4. Histogram of NET expression variation. Expression data was generated by the laboratory of John Hogenesch at the Genomics Institute of the Novartis Research Foundation as part of their "transcriptome" database (available online at http://symatlas.gnf.org/SymAtlas/; Su et al. 2002). Maximum, median, and minimum values for NETs among the different tissues were determined and their ratios calculated. (A) The ratio of the highest expression level to the lowest is presented for the new NETs identified in the subtractive study (Schirmer et al. 2003). If a NET is absent from some tissues and thus the lowest value is at the level of background, a high ratio would be produced. Very few NETs had low ratios. (B) The ratio of the highest expression level to the median expression level for the new NETs. In this case a high ratio indicates a preference in expression for a subset of cell types. Though the distribution changes compared to A, there are still several NETs with very high ratios. (C and D) The ratio of the highest expression level to the lowest level, C, and to the median level, D, is presented for the previously characterized NETs. Surprisingly, these also exhibited considerable variation.

preference for expression in a subset of tissues or cell types. One third of the NETs yielded ratios greater than 5; thus, the tissues exhibiting highest expression have levels at least 5-fold higher than the majority of tissues. The cell types where the highest expression levels occurred for NETs from the rodent liver dataset tended to be those present in the starting material: that is liver, adipocytes (similar to lipoctyes), and blood cells (Schirmer et al. 2005; Schirmer and Gerace 2005). These observations support the idea that a subset of NE proteins will be ubiquitous while another subset will be unique to each tissue investigated: for example, if each major cell type in liver contributed five unique NETs, there would be 50 core NETs and 30 tissue-specific NETs among the 80 (67 novel + 13 previously characterized) identified in rodent liver. This may also help to explain why the comparative study that used a single cell line (Dreger et al. 2001) identified a much smaller number of NETs compared to the subtractive liver study (Schirmer et al. 2003).

Surprisingly, the transcriptome data also indicates considerable variation in expression levels among different cell types for several of the previously characterized NETs (Figure 4C and D). Nonetheless these data suggest that the original NETs are more ubiquitous in their expression than the new NETs because their average levels of expression are much higher: using numbers from the relative values given for expression in the transcriptome data the average lowest levels of expression were roughly 2-fold higher in the original NETs compared to the new NETs (161 and 83 respectively), while the average highest levels were more than 4-fold higher (4,797 vs. 1,110). This is also consistent with the idea that their earlier identification was facilitated by their relative abundance. Although apparently ubiquitously expressed, LBR was so highly expressed in hematopoietic cells that its ratio for variation from highest to lowest expressing tissue was in excess of 100.

In addition to this variation between different tissues within an organism, the NE proteome appears to be varied between organisms. Blast searches were performed on all 80 NETs, and those with detectable homologs in several completely sequenced genomes are shown in Figure 5. Although the frequent annotations to the genome databases can cause fluxuations in the results for individual NETs, in aggregate the results are striking. Eight NETs appeared to be universally conserved from mammals to yeast. In contrast, seven other NETs were not conserved even between mouse, rat, and human (though this may reflect mistakes in the databases or non-equivalent prediction of hypothetical orfs). Moreover, there is no particular phylogenetic conservation for different NETs as some had yeast homologs, but neither plant nor fly homologs. Some observations were nonetheless striking and may reflect some phylogenetic grouping. For example, 17 NETs were conserved between mammals and flies, but were not found in worms; yet only 6 NETs conserved between mammals and worms did not appear in flies. NET variation between organisms was first highlighted because two proteins identified in Drosophila, Otefin (Ashery-Padan et al. 1997) and YA (Goldberg et al. 1998), have no mammalian homologs and most of the 13 previously characterized mammalian NETs do not have homologs in *Drosophila* (Gruenbaum et al. 2003).

6. NETS IN MULTIPLE CELLULAR LOCATIONS

The central limitation of the subtractive approach is that it identifies only proteins that are unique to the NE: it will disregard proteins that normally occur in both of the fractions analyzed. One such protein is Sec13, an ER trafficking protein that also has a demonstrated function in NPC mediated transport (Siniossoglou et al. 1996). Sec13 thus appeared in both datasets and was accordingly discarded from the NE-unique *in silico* purified protein set. TorsinA also appeared in both datasets and was discarded, but it has since been shown to shuttle between the INM and the ER (Goodchild and Dauer 2004). As 41% of the 566 proteins identified in the membrane-enriched NE fraction were similarly discarded, there may be many other proteins listed in the pre-subtraction dataset that are truly NE proteins, though they are not unique to the NE.

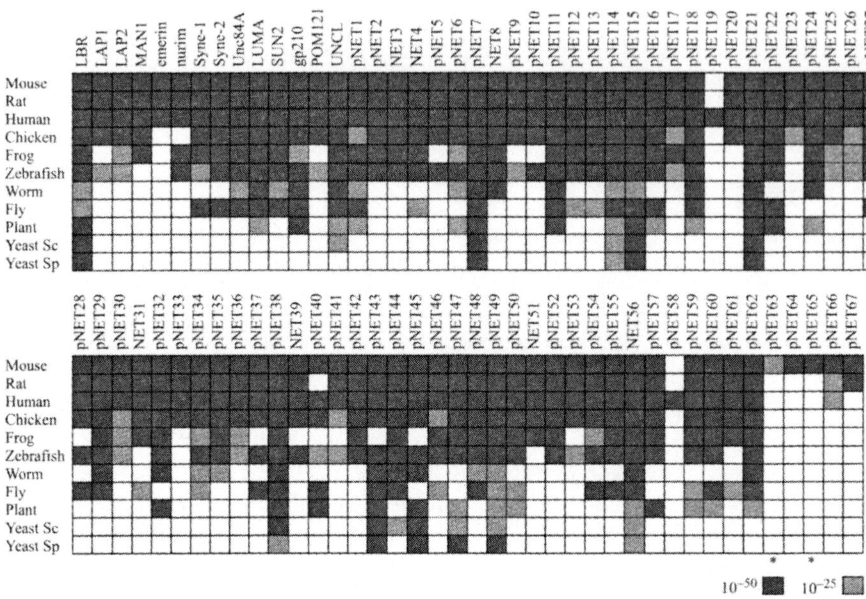

Figure 5. Evolutionary conservation of NETs. Strong homologs do not exist in the Genbank sequence database for many NETs in ten wide-ranging eukaryotic organisms that have completely sequenced genomes: Mouse, Mus musculus; Rat, Rattus norvegicus; Human, *Homo sapiens*; Chicken, *Gallus gallus*; Frog, *Xenopus laevis*; Zebrafish, *Danio rerio*; Worm, *Caenorhabditis elegans*; Fly, *Drosophila melanogaster*; Plant, *Arabidopsis thaliana*; Yeast *Sc*, *Saccharomyces cerevisiae*; Yeast *Sp Schizosaccharomyces pombe*. Dark boxes indicate probability scores lower than 10^{-50} and lighter shaded boxes indicate scores between 10^{-50} and 10^{-25}. White boxes indicate no homologs. The data is based on blast searches using NET protein sequences at different times, most recently updated on 27 April 2006. In cases where database sequences obtained from an earlier blast search (June 2005) had been removed pending sequence corrections, the homologs are maintained in this list. Some NETs that do not appear in all mammalian genomes may be absent because of sequence errors or failure to predict hypothetical orfs in those organisms; however, this seems unlikely for the set of original NETs which have been the subject of extensive searches over the years yet failed to identify homologs in flies (Gruenbaum et al. 2003). Asterisks indicate NETs that are included, yet had low probability scores (below the listed cut-offs) because the annotated sequence had been removed but the exons were still present in the genome sequence.

NETs might also share cellular locations besides ER. The subtractive study focussed on identifying novel NETs, but several transmembrane proteins were identified that had been previously characterized in other cellular compartments (see supplemental material in Schirmer et al. 2003). These range from a plasma membrane receptor (Klein et al. 2004) to an ion transport pump (Arteaga et al. 2004) to mediators of signalling cascades such as the inositol triphosphate receptor (Cruttwell et al. 2005) and a cytosolic phospholipase A2 variant that relocates from the ER to the NE under certain conditions. Several of these have been shown by various means (immuno-EM, immunofluorescence microscopy, Western blotting analysis of subcellular fractions) to target to the NE. Although all of those shown in Table 1 have been tested for

Table 1. Proteins from other cellular compartments also reported at the NE

Protein	Other locations	System tested	Cells/Tissue tested	Detected at NE in subtractive Study	Reference
TorsinA	ER	Exogenous expression of tagged protein	Gli36, BHK cells	Yes	(Schirmer et al. 2003; Goodchild and Dauer 2004)
P-glycoprotein (multi-drug resistance-1 ATP-binding cassette subfamily B)	Plasma membrane, caveolae, and coated vesicles	Immunogold antibody labeling by EM.	Astrocytes	Yes	(Schirmer et al. 2003 Ronaldson et al. 2004)
CPLA2-a (cytosolic phospholipase A_2-a, 14-3-3 zeta)	ER	Determined with polyclonal antibodies	A23187 HeLa and A549 cells,	Yes	(Schirmer et al. 2003; Grewal et al. 2005)
EGFR (epidermal growth factor receptor)	Plasma membrane	At nuclear membrane by biochemical fractionization	*M. Musculus* liver	Yes	(Schirmer et al. 2003; Klein et al. 2004)
InsP 3R (inositol 1, 4, 5-triphosphate receptor type 2)	ER and cell periphery	Exogenous expression of tagged protein	MDCK cells	Yes	(Schirmer et al. 2003; Cruttwell et al. 2005)
Na+/K+ ATPase ß1 subunit	Plasma membrane	Subtype specific antibody labeling	Neurons, satellite cells in *R.norvegicus*	Yes	(Schirmer et al. 2003; Arteaga et al. 2004)

(*Continued*)

Table 1. (Continued)

Protein	Other locations	System tested	Cells/Tissue tested	Detected at NE in subtractive Study	Reference
Ryanodine receptor	Various membranes	Determined by functional effect	NPCs in *X.laevis*	No	(Erickson et al. 2004)
NCX1 (Na+/ Ca++ exchanger membrane-1, solute carrier family 8)	Plasma membrane	Immunoblottting using various antibodies	*H.sapiens* brain and lymphocytes, not at NE in other tissues	No	(Xie et al. 2004)
PV-1 (plasmalemma vesicle protein-1)	ER	Exogenously expressed protein	Bovine aortic endothelial cells	No	(Hnasko and Ben-Jonathan 2005)
NHE-1 (sodium/hydrogen exchanger)	Depends on cell type and species	Fluorescent labeling, found at nuclear membrane	Liver and cardiomyocytes in *R. norvegicus*	No	(Bkaily et al. 2004)
Atypical type IV P-type ATPase	PM, post-mitochondrial supernatant	Immunoelectron microscopy and biochemical fractionization	*O.cuniculus* endometrium	Other family members were identified, but not this variant	(Mansharamani et al. 2001)

localization, typically only one antibody was used so the possibility of cross-reactivity cannot be ruled out. That not all the proteins included in the table were found in the subtractive liver NE study may be explained if they only occur at the NE in certain cell types. This was clearly shown in the case of the NCX1 ion transporter which was detected in brain and lymphocytes, but not in other tissues (Xie et al. 2004). The appearance of these proteins in the NE, though unexpected, makes sense. For example, in the case of ion transporters, the nucleus needs to maintain its internal ion concentration as much as the cytoplasm, and the critical ion concentrations needed for chromatin-modifying enzymes are not the same as those for cytoplasmic enzymes. It is also logical for the NE to have additional mechanisms for signalling between the nucleus and cytoplasm besides the NPC. Thus clarification that other proteins in this proteomics dataset share multiple localizations may add more varied functions to the NE milieu.

7. NUCLEAR ENVELOPE DISEASES AND FUNCTION

Nearly 20 diverse inherited diseases and syndromes are now clearly linked to the NE (Worman and Courvalin 2002; Mounkes and Stewart 2004). These include muscular dystrophies (Bione et al. 1994; Bonne et al. 1999; Muchir et al. 2000; Raffaele Di Barletta et al. 2000; Hanisch et al. 2002), lipodystrophies (Cao and Hegele 2000; Shackleton et al. 2000), neuropathy (De Sandre-Giovannoli et al. 2002), cardiomyopathies (Fatkin et al. 1999), dermopathies (Navarro et al. 2004), bone diseases (Waterham et al. 2003; Hellemans et al. 2004), and the aging disease progeria (De Sandre-Giovannoli et al. 2003; Eriksson et al. 2003). NE proteins mutated in these diseases include the intermediate filament lamin A/C protein and multiple NETs. In addition to diseases caused by proteins unique to the INM, mutations in torsinA which shuttles between the ER and INM cause the disease early-onset torsion dystonia and result in its accumulation in the NE (Goodchild and Dauer 2004).

The combination of these diverse functions and the wide range of diseases associated with the NE suggest that fundamental regulatory mechanisms reside at the NE, the perturbation of which can have profound consequences for the cell. There are several hypotheses explaining how NE proteins could cause these diseases based on known functions for NE proteins. (1) Misregulation of gene expression could result in pathology because both lamins and NETs have been shown to influence transcription (Ellis et al. 1997; Spann et al. 1997; Moir et al. 2000; Nili et al. 2001). (2) Cell cycle misregulation could produce pathology in several ways. Both withdrawal from the cell cycle of a regenerating stem cell population and activation of the cell cycle in a differentiated population could produce pathology over time. Lamin interactions with retinoblastoma (Rb) (Ozaki et al. 1994), DNA replication (Kennedy et al. 2000; Martins et al. 2003), and signal cascades (Steen et al. 2000) as well as NET influences on centriole positioning (Malone et al. 1999) could all affect the cell cycle. (3) Loss of mechanical/ structural stability in the lamina could weaken cells and produce pathology. Lamins and NETs have both been shown to impact on nuclear morphology

(Liu et al. 2000; Schirmer et al. 2001; Hoffmann et al. 2002) and loss of lamin A has been shown to affect the mechanical stability of the lamina *in vivo* (Lammerding et al. 2004), *in vitro* (Schirmer and Gerace 2004), and in disease (Fidzianska et al. 1998). None of these hypotheses can satisfactorily explain the wide range of tissue specificities for NE diseases. For example, while the mechanical stability hypothesis makes sense for muscular dystrophy, it is hard to see how it could apply to diseases affecting neurons or fat cells. These hypotheses also fail to explain how two proximally located mutations in the same protein can cause in one case neuropathy and in another muscular dystrophy.

The data presented above indicating that many NETs are preferentially expressed in certain tissues suggests an alternative hypothesis: the tissue specificity of NE diseases derives from tissue-specific NETs having specifically altered interactions with the ubiquitous NE proteins mutated in disease. This would result in selective dysfunction of multiprotein complexes in those tissues, which could themselves have functions in replication, gene expression, cell cycle, etc. The case of the NETs Unc83 and Unc84 (Section 4.3) provides an example of this. Unc83 is specific to *C.elegans* tissues where nuclear migration occurs, and it must interact with the widely expressed Unc84 to accumulate in the NE (Starr et al. 2001). Mutations in either interacting protein result in the same uncoordinated (*unc*) phenotype. Furthermore, Unc84 requires lamins for its targeting to the NE. This resembles human lamin A and emerin in that emerin requires lamin A to accumulate at the NE and mutations in either protein cause variants of Emery-Dreifuss muscular dystrophy. However both proteins are widely expressed and the relevant tissue-specific partner (such as Unc83) has not yet been clearly identified, though new binding partners of emerin are being rapidly discovered (Section 4.1). The recent explosion in the number of proteins assigned to the NE opens a wide field of possible candidate partners, some of which may themselves cause additional NE diseases.

8. CONCLUSIONS

Many more proteomics datasets of subcellular organelles and compartments generated from different cell types will be required before the full extent of variability for each organelle can be determined. The unexpectedly high degree of variation indicated in the liver NE suggests that other organelles may also exhibit greater variation than previously thought. Currently our laboratory is in the process of analyzing nuclear envelopes from other cell types in an effort to address the former question. The latter question will require analysis of other organelles from multiple cell types.

ACKNOWLEDGEMENTS

We thank Dr. Gavin Wilkie for critical reading of the manuscript. This work was supported by a Dorothy Hodgkin Royal Society Fellowship to P. Malik and a Senior Research Fellowship to E. Schirmer from the Wellcome Trust.

REFERENCES

Aaronson, R.P. and Blobel, G. (1975) Isolation of nuclear pore complexes in association with a lamina. Proc. Natl Acad. Sci. U.S.A. *72*, 1007–1011.

Apel, E.D., Lewis, R.M., Grady, R.M. and Sanes, J.R. (2000) Syne-1, a dystrophin- and Klarsicht-related protein associated with synaptic nuclei at the neuromuscular junction. J. Biol. Chem. *275*, 31986–31995.

Arteaga, M.F., Gutierrez, R., Avila, J., Mobasheri, A., Diaz-Flores, L. and Martin-Vasallo, P. (2004) Regeneration influences expression of the Na+, K+-atpase subunit isoforms in the rat peripheral nervous system. Neuroscience *129*, 691–702.

Ashery-Padan, R., Weiss, A.M., Feinstein, N. and Gruenbaum, Y. (1997) Distinct regions specify the targeting of otefin to the nucleoplasmic side of the nuclear envelope. J. Biol. Chem. *272*, 2493–2499.

Bengtsson, L. and Wilson, K.L. (2004) Multiple and surprising new functions for emerin, a nuclear membrane protein. Curr. Opin. Cell Biol. *16*, 73–79.

Berger, R., Theodor, L., Shoham, J., Gokkel, E., Brok-Simoni, F., Avraham, K.B., Copeland, N.G., Jenkins, N.A., Rechavi, G. and Simon, A.J. (1996) The characterization and localization of the mouse thymopoietin/lamina-associated polypeptide 2 gene and its alternatively spliced products. Genome Res. *6*, 361–370.

Bione, S., Maestrini, E., Rivella, S., Mancini, M., Regis, S., Romeo, G. and Toniolo, D. (1994) Identification of a novel X-linked gene responsible for Emery-Dreifuss muscular dystrophy. Nat. Genet. *8*, 323–327.

Bkaily, G., Nader, M., Avedanian, L., Jacques, D., Perrault, C., Abdel-Samad, D., D'Orleans-Juste, P., Gobeil, F. and Hazzouri, K.M. (2004) Immunofluorescence revealed the presence of NHE-1 in the nuclear membranes of rat cardiomyocytes and isolated nuclei of human, rabbit, and rat aortic and liver tissues. Can. J. Physiol. Pharmacol. *82*, 805–811.

Blobel, G. and Potter, V.R. (1966) Nuclei from rat liver: isolation method that combines purity with high yield. Science *154*, 1662–1665.

Bonne, G., Di Barletta, M.R., Varnous, S., Becane, H.M., Hammouda, E.H., Merlini, L., Muntoni, F., Greenberg, C.R., Gary, F., Urtizberea, J.A., Duboc, D., Fardeau, M., Toniolo, D. and Schwartz, K. (1999) Mutations in the gene encoding lamin A/C cause autosomal dominant Emery-Dreifuss muscular dystrophy. Nat. Genet. *21*, 285–288.

Brachner, A., Reipert, S., Foisner, R. and Gotzmann, J. (2005) LEM2 is a novel MAN1-related inner nuclear membrane protein associated with A-type lamins. J. Cell Sci. *118*, 5797–5810.

Brown, D.T. (2001) Histone variants: are they functionally heterogeneous? Genome Biol. *2*, REVIEWS0006.

Cao, H. and Hegele, R.A. (2000) Nuclear lamin A/C R482Q mutation in Canadian kindreds with Dunnigan-type familial partial lipodystrophy. Hum. Mol. Genet. *9*, 109–112.

Crisp, M., Liu, Q., Roux, K., Rattner, J.B., Shanahan, C., Burke, B., Stahl, P.D. and Hodzic, D. (2006) Coupling of the nucleus and cytoplasm: role of the LINC complex. J. Cell Biol. *172*, 41–53.

Cronshaw, J., Krutchinsky, A., Zhang, W., Chait, B. and Matunis, M. (2002) Proteomic analysis of the mammalian nuclear pore complex. J. Cell Biol. *158*, 915–927.

Cruttwell, C., Bernard, J., Hilly, M., Nicolas, V., Tunwell, R.E. and Mauger, J.P. (2005) Dynamics of the inositol 1,4,5-trisphosphate receptor during polarisation of MDCK cells. Biol. Cell *97*, 699–707.

De Sandre-Giovannoli, A., Bernard, R., Cau, P., Navarro, C., Amiel, J., Boccaccio, I., Lyonnet, S., Stewart, C., Munnich, A., Le Merrer, M. and Levy, N. (2003) Lamin A truncation in Hutchinson-Gilford progeria. Science *300*, 2055.

De Sandre-Giovannoli, A., Chaouch, M., Kozlov, S., Vallat, J., Tazir, M., Kassouri, N., Szepetowski, P., Hammadouche, T., Vandenberghe, A., Stewart, C.L., Grid, D. and Levy, N. (2002) Homozygous defects in LMNA, encoding lamin A/C nuclear-envelope proteins, cause autosomal recessive axonal neuropathy in human (Charcot-Marie-Tooth disorder type 2) and mouse. Am. J. Hum. Genet. *70*, 726–736.

Dechat, T., Gotzmann, J., Stockinger, A., Harris, C.A., Talle, M.A., Siekierka, J.J. and Foisner, R. (1998) Detergent-salt resistance of LAP2alpha in interphase nuclei and phosphorylation-dependent association with chromosomes early in nuclear assembly implies functions in nuclear structure dynamics. EMBO J. *17*, 4887–4902.

Dhe-Paganon, S., Werner, E.D., Chi, Y.-I. and Shoelson, S.E. (2002) Structure of the globular tail of nuclear lamin. J. Biol. Chem. *277*, 17381–17384.

Dreger, M., Bengtsson, L., Schoneberg, T., Otto, H., and Hucho, F. (2001) Nuclear envelope proteomics: novel integral membrane proteins of the inner nuclear membrane. Proc. Natl Acad. Sci. U.S.A. *98*, 11943–11948.

Dwyer, N. and Blobel, G. (1976) A modified procedure for the isolation of a pore complex-lamina fraction from rat liver nuclei. J. Cell. Biol. *70*, 581–591.

Ellenberg, J., Siggia, E.D., Moreira, J.E., Smith, C.L., Presley, J.F., Worman, H.J. and Lippincott-Schwartz, J. (1997) Nuclear membrane dynamics and reassembly in living cells: targeting of an inner nuclear membrane protein in interphase and mitosis. J. Cell. Biol. *138*, 1193–1206.

Ellis, D.J., Jenkins, H., Whitfield, W.G. and Hutchison, C.J. (1997) GST-lamin fusion proteins act as dominant negative mutants in Xenopus egg extract and reveal the function of the lamina in DNA replication. J. Cell Sci. *110*, 2507–2518.

Erickson, E.S., Mooren, O.L., Moore-Nichols, D. and Dunn, R.C. (2004) Activation of ryanodine receptors in the nuclear envelope alters the conformation of the nuclear pore complex. Biophys. Chem. *112*, 1–7.

Eriksson, M., Brown, W.T., Gordon, L.B., Glynn, M.W., Singer, J., Scott, L., Erdos, M.R., Robbins, C.M., Moses, T.Y., Berglund, P., Dutra, A., Pak, E., Durkin, S., Csoka, A.B., Boehnke, M., Glover, T.W. and Collins, F.S. (2003) Recurrent de novo point mutations in lamin A cause Hutchinson-Gilford progeria syndrome. Nature *423*, 293–298.

Fatkin, D., MacRae, C., Sasaki, T., Wolff, M., Porcu, M., Frenneaux, M., Atherton, J., Vidaillet, H.J. Jr., Spudich, S., De Girolami, U., Seidman, J.G., Seidman, C., Muntoni, F., Muehle, G., Johnson, W. and McDonough, B. (1999) Missense mutations in the rod domain of the lamin A/C gene as causes of dilated cardiomyopathy and conduction-system disease. N. Engl. J. Med. *341*, 1715–1724.

Fidzianska, A., Toniolo, D. and Hausmanowa-Petrusewicz, I. (1998) Ultrastructural abnormality of sarcolemmal nuclei in Emery-Dreifuss muscular dystrophy (EDMD). J. Neurol. Sci. *159*, 88–93.

Fields, A.P., Pettit, G.R. and May, W.S. (1988) Phosphorylation of lamin B at the nuclear membrane by activated protein kinase C. J. Biol. Chem. *263*, 8253–8260.

Fitzgerald, J., Kennedy, D., Viseshakul, N., Cohen, B.N., Mattick, J., Bateman, J.F. and Forsayeth, J.R. (2000) UNCL, the mammalian homologue of UNC-50, is an inner nuclear membrane RNA-binding protein. Brain Res. *877*, 110–123.

Foisner, R. and Gerace, L. (1993) Integral membrane proteins of the nuclear envelope interact with lamins and chromosomes, and binding is modulated by mitotic phosphorylation. Cell *73*, 1267–1279.

Gerace, L., Blum, A. and Blobel, G. (1978) Immunocytochemical localization of the major polypeptides of the nuclear pore complex-lamina fraction. Interphase and mitotic distribution. J. Cell. Biol. *79*, 546–566.

Gerace, L. and Burke, B. (1988) Functional organization of the nuclear envelope. Annu. Rev. Cell Biol. *4*, 335–374.

Gerace, L., Ottaviano, Y. and Kondor-Koch, C. (1982) Identification of a major polypeptide of the nuclear pore complex. J. Cell Biol. *95*, 826–837.

Goldberg, M., Lu, H., Stuurman, N., Ashery-Padan, R., Weiss, A.M., Yu, J., Bhattacharyya, D., Fisher, P.A., Gruenbaum, Y. and Wolfner, M.F. (1998) Interactions among Drosophila nuclear envelope proteins lamin, otefin, and YA. Mol. Cell Biol. *18*, 4315–4323.

Goodchild, R.E. and Dauer, W.T. (2004) Mislocalization to the nuclear envelope: an effect of the dystonia-causing torsinA mutation. Proc. Natl Acad. Sci. U.S.A. *101*, 847–852.

Goodchild, R.E. and Dauer, W.T. (2005) The AAA+ protein torsinA interacts with a conserved domain present in LAP1 and a novel ER protein. J. Cell Biol. *168*, 855–862.

Gough, L.L., Fan, J., Chu, S., Winnick, S. and Beck, K.A. (2003) Golgi localization of Syne-1. Mol. Biol. Cell *14*, 2410–2424.

Greber, U.F., Senior, A. and Gerace, L. (1990) A major glycoprotein of the nuclear pore complex is a membrane-spanning polypeptide with a large lumenal domain and a small cytoplasmic tail. EMBO J. *9*, 1495–1502.

Grewal, S., Herbert, S.P., Ponnambalam, S. and Walker, J.H. (2005) Cytosolic phospholipase A2-alpha and cyclooxygenase-2 localize to intracellular membranes of EA.hy.926 endothelial cells that are distinct from the endoplasmic reticulum and the Golgi apparatus. FEBS J. *272*, 1278–1290.

Gruenbaum, Y., Goldman, R.D., Meyuhas, R., Mills, E., Margalit, A., Fridkin, A., Dayani, Y., Prokocimer, M. and Enosh, A. (2003) The nuclear lamina and its functions in the nucleus. Int. Rev. Cytol. *226*, 1–62.

Hahn, C.G. and Covault, J. (1990) Isolation of transcriptionally active nuclei from striated muscle using Percoll density gradients. Anal. Biochem. *190*, 193–197.

Hallberg, E., Wozniak, R.W. and Blobel, G. (1993) An integral membrane protein of the pore membrane domain of the nuclear envelope contains a nucleoporin-like region. J. Cell Biol. *122*, 513–521.

Hanisch, F., Neudecker, S., Wehnert, M. and Zierz, S. (2002) Hauptmann-Thannhauser muscular dystrophy and differential diagnosis of myopathies associated with contractures. Nervenarzt *73*, 1004–1011.

Held, I.R., Rodrigo, R.T., Yeoh, H.C. and Tonaki, H. (1977) Isolation of nuclei from red and white skeletal muscles of the adult rat. Exp. Cell Res. *105*, 191–197.

Hellemans, J., Preobrazhenska, O., Willaert, A., Debeer, P., Verdonk, P.C., Costa, T., Janssens, K., Menten, B., Van Roy, N., Vermeulen, S.J., Savarirayan, R., Van Hul, W., Vanhoenacker, F., Huylebroeck, D., De Paepe, A., Naeyaert, J.M., Vandesompele, J., Speleman, F., Verschueren, K., Coucke, P.J. and Mortier, G.R. (2004) Loss-of-function mutations in LEMD3 result in osteopoikilosis, Buschke-Ollendorff syndrome and melorheostosis. Nat. Genet. *36*, 1213–1218.

Hnasko, R. and Ben-Jonathan, N. (2005) Developmental regulation of PV-1 in rat lung: association with the nuclear envelope and limited colocalization with Cav-1. Am. J. Physiol. Lung Cell Mol. Physiol. *288*, L275–284.

Hoffmann, K., Dreger, C.K., Olins, A.L., Olins, D.E., Shultz, L.D., Lucke, B., Karl, H., Kaps, R., Muller, D., Vaya, A., Aznar, J., Ware, R.E., Sotelo Cruz, N., Lindner, T.H., Herrmann, H., Reis, A. and Sperling, K. (2002) Mutations in the gene encoding the lamin B receptor produce an altered nuclear morphology in granulocytes (Pelger-Huet anomaly. Nat. Genet. *31*, 410–414.

Kennedy, B.K., Barbie, D.A., Classon, M., Dyson, N. and Harlow, E. (2000) Nuclear organization of DNA replication in primary mammalian cells. Genes. Dev. *14*, 2855–2868.

Klein, C., Gensburger, C., Freyermuth, S., Nair, B.C., Labourdette, G. and Malviya, A. N. (2004) A 120 kDa nuclear phospholipase Cgamma1 protein fragment is stimulated *in vivo* by EGF signal phosphorylating nuclear membrane EGFR. Biochemistry *43*, 15873–15883.

Krimm, I., Ostlund, C., Gilquin, B., Couprie, J., Hossenlopp, P., Mornon, J., Bonne, G., Courvalin, J., Worman, H. and Zinn-Justin, S. (2002) The Ig-like structure of the C-terminal domain of lamin A/C, mutated in muscular dystrophies, cardiomyopathy, and partial lipodystrophy. Structure (Camb) *10*, 811–823.

Kuehl, L. (1977) Isolation of skeletal muscle nuclei. Methods Cell Biol. *15*, 79–88.

Lammerding, J., Schulze, P., Takahashi, T., Kozlov, S., Sullivan, T., Kamm, R., Stewart, C. and Lee, R. (2004) Lamin A/C deficiency causes defective nuclear mechanics and mechanotransduction. J. Clin. Invest. *113*, 370–378.

Lee, H., Habas, R. and Abate-Shen, C. (2004) MSX1 cooperates with histone H1b for inhibition of transcription and myogenesis. Science *304*, 1675–1678.

Lee, K., Starr, D., Cohen, M., Liu, J., Han, M., Wilson, K. and Gruenbaum, Y. (2002) Lamin-dependent localization of UNC-84, a protein required for nuclear migration in Caenorhabditis elegans. Mol. Biol. Cell *13*, 892–901.

Lee, K.K. and Wilson, K.L. (2004) All in the family: evidence for four new LEM-domain proteins Lem2 (NET-25), Lem3, Lem4 and Lem5 in the human genome. Symp. Soc. Exp. Biol. *56*, 329–339.

Lin, F., Blake, D.L., Callebaut, I., Skerjanc, I.S., Holmer, L., McBurney, M.W., Paulin-Levasseur, M. and Worman, H.J. (2000) MAN1, an inner nuclear membrane protein that shares the LEM domain with lamina-associated polypeptide 2 and emerin. J. Biol. Chem. *275*, 4840–4847.

Liu, J., Ben-Shahar, T., Riemer, D., Treinin, M., Spann, P., Weber, K., Fire, A. and Gruenbaum, Y. (2000) Essential roles for Caenorhabditis elegans lamin gene in nuclear organization, cell cycle progression, and spatial organization of nuclear pore complexes. Mol. Biol. Cell *11*, 3937–3947.

MacCoss, M.J., Wu, C.C., Matthews, D.E. and Yates, J.R., 3rd (2005) Measurement of the isotope enrichment of stable isotope-labeled proteins using high-resolution mass spectra of peptides. Anal. Chem. *77*, 7646–7653.

Malone, C.J., Fixsen, W.D., Horvitz, H.R. and Han, M. (1999) UNC-84 localizes to the nuclear envelope and is required for nuclear migration and anchoring during C. elegans development. Development *126*, 3171–3181.

Manilal, S., Nguyen, T.M., Sewry, C.A. and Morris, G.E. (1996) The Emery-Dreifuss muscular dystrophy protein, emerin, is a nuclear membrane protein. Hum. Mol. Genet. *5*, 801–808.

Mansharamani, M., Hewetson, A. and Chilton, B.S. (2001) Cloning and characterization of an atypical Type IV P-type ATPase that binds to the RING motif of RUSH transcription factors. J. Biol. Chem. *276*, 3641–3649.

Martins, S., Eikvar, S., Furukawa, K. and Collas, P. (2003) HA95 and LAP2 beta mediate a novel chromatin-nuclear envelope interaction implicated in initiation of DNA replication. J. Cell Biol. *160*, 177–188.

McKinney, M.K. and Cravatt, B.F. (2005) Structure and Function of Fatty Acid Amide Hydrolase. Annu. Rev. Biochem. *74*, 411.

Moir, R., Spann, T., Herrmann, H. and Goldman, R. (2000) Disruption of nuclear lamin organization blocks the elongation phase of DNA replication. J. Cell Biol. *149*, 1179–1192.

Mosley-Bishop, K.L., Li, Q., Patterson, L. and Fischer, J.A. (1999) Molecular analysis of the klarsicht gene and its role in nuclear migration within differentiating cells of the Drosophila eye. Curr. Biol. *9*, 1211–1220.

Mounkes, L. and Stewart, C.L. (2004) Structural organization and functions of the nucleus in development, aging, and disease. Curr. Top. Dev. Biol. *61*, 191–228.

Muchir, A., Bonne, G., van der Kooi, A.J., van Meegen, M., Baas, F., Bolhuis, P.A., de Visser, M. and Schwartz, K. (2000) Identification of mutations in the gene encoding lamins A/C in autosomal dominant limb girdle muscular dystrophy with atrioventricular conduction disturbances (LGMD1B). Hum. Mol. Genet. *9*, 1453–1459.

Navarro, C.L., De Sandre-Giovannoli, A., Bernard, R., Boccaccio, I., Boyer, A., Genevieve, D., Hadj-Rabia, S., Gaudy-Marqueste, C., Smitt, H.S., Vabres, P., Faivre, L., Verloes, A., Van Essen, T., Flori, E., Hennekam, R., Beemer, F.A., Laurent, N., Le Merrer, M., Cau, P. and Levy, N. (2004) Lamin A and ZMPSTE24 (FACE-1) defects cause nuclear disorganization and identify restrictive dermopathy as a lethal neonatal laminopathy. Hum. Mol. Genet. *13*, 2493–2503.

Nili, E., Cojocaru, G.S., Kalma, Y., Ginsberg, D., Copeland, N.G., Gilbert, D.J., Jenkins, N.A., Berger, R., Shaklai, S., Amariglio, N., Brok-Simoni, F., Simon, A.J. and Rechavi, G. (2001) Nuclear membrane protein LAP2beta mediates transcriptional repression alone and together with its binding partner GCL (germ-cell-less). J. Cell Sci. *114*, 3297–3307.

Ohba, T., Schirmer, E.C., Nishimoto, T. and Gerace, L. (2004) Energy- and temperature-dependent transport of integral proteins to the inner nuclear membrane via the nuclear pore. J. Cell Biol. *167*, 1051–1062.

Ozaki, T., Saijo, M., Murakami, K., Enomoto, H., Taya, Y. and Sakiyama, S. (1994) Complex formation between lamin A and the retinoblastoma gene product: identification of the domain on lamin A required for its interaction. Oncogene *9*, 2649–2653.

Paulin-Levasseur, M., Blake, D.L., Julien, M. and Rouleau, L. (1996) The MAN antigens are non-lamin constituents of the nuclear lamina in vertebrate cells. Chromosoma *104*, 367–379.

Raffaele Di Barletta, M., Ricci, E., Galluzzi, G., Tonali, P., Mora, M., Morandi, L., Romorini, A., Voit, T., Orstavik, K.H., Merlini, L., Trevisan, C., Biancalana, V., Housmanowa-Petrusewicz, I., Bione, S., Ricotti, R., Schwartz, K., Bonne, G. and Toniolo, D. (2000) Different Mutations in the LMNA Gene Cause Autosomal Dominant and Autosomal Recessive Emery-Dreifuss Muscular Dystrophy. Am. J. Hum. Genet. *66*, 1407–1412.

Raju, G.P., Dimova, N., Klein, P.S. and Huang, H.C. (2003) SANE, a novel LEM domain protein, regulates bone morphogenetic protein signaling through interaction with Smad1. J. Biol. Chem. *278*, 428–437.

Rolls, M.M., Stein, P.A., Taylor, S.S., Ha, E., McKeon, F. and Rapoport, T.A. (1999) A visual screen of a GFP-fusion library identifies a new type of nuclear envelope membrane protein. J. Cell Biol. *146*, 29–44.

Ronaldson, P.T., Bendayan, M., Gingras, D., Piquette-Miller, M., and Bendayan, R. (2004) Cellular localization and functional expression of P-glycoprotein in rat astrocyte cultures. J. Neurochem. *89*, 788–800.

Rout, M.P., Aitchison, J.D., Suprapto, A., Hjertaas, K., Zhao, Y. and Chait, B.T. (2000) The yeast nuclear pore complex: composition, architecture, and transport mechanism. J. Cell Biol. *148*, 635–651.

Santoni, V., Molloy, M. and Rabilloud, T. (2000) Membrane proteins and proteomics: un amour impossible? Electrophoresis *21*, 1054–1070.

Scheele, G. (1983) Methods for the study of protein translocation across the RER membrane using the reticulocyte lysate translation system and canine pancreatic microsomal membranes. Methods Enzymol. *96*, 94–111.

Schirmer, E.C., Florens, L., Guan, T., Yates, J.R. 3rd and Gerace, L. (2003) Nuclear membrane proteins with potential disease links found by subtractive proteomics. Science *301*, 1380–1382.

Schirmer, E.C., Florens, L., Guan, T., Yates, J.R. 3rd and Gerace, L. (2005) Identification of novel integral membrane proteins of the nuclear envelope with potential disease links using subtractive proteomics, In: Novartis Found Symp No. 264 Nuclear Organization in Development and Disease. D.J. Chadwick and J.A. Goode, eds. (Chichester and New York: John Wiley & Sons Ltd.), pp. 63–76; discussion 76–80, 227–230.

Schirmer, E.C. and Gerace, L. (2004) The stability of the nuclear lamina polymer changes with the composition of lamin subtypes according to their individual binding strengths. J. Biol. Chem. *279*, 42811–42817.

Schirmer, E.C. and Gerace, L. (2005) The nuclear membrane proteome: extending the envelope. Trends Biochem. Sci. *30*, 551–558.

Schirmer, E.C., Guan, T. and Gerace, L. (2001) Involvement of the lamin rod domain in heterotypic lamin interactions important for nuclear organization. J. Cell Biol. *153*, 479–489.

Senior, A. and Gerace, L. (1988) Integral membrane proteins specific to the inner nuclear membrane and associated with the nuclear lamina. J. Cell Biol. *107*, 2029–2036.

Shackleton, S., Lloyd, D.J., Jackson, S.N., Evans, R., Niermeijer, M.F., Singh, B.M., Schmidt, H., Brabant, G., Kumar, S., Durrington, P.N., Gregory, S., O'Rahilly, S. and Trembath, R.C. (2000) LMNA, encoding lamin A/C, is mutated in partial lipodystrophy. Nat. Genet. *24*, 153–156.

Siniossoglou, S., Wimmer, C., Rieger, M., Doye, V., Tekotte, H., Weise, C., Emig, S., Segref, A. and Hurt, E.C. (1996) A novel complex of nucleoporins, which includes Sec13p and a Sec13p homolog, is essential for normal nuclear pores. Cell *84*, 265–275.

Soullam, B. and Worman, H.J. (1993) The amino-terminal domain of the lamin B receptor is a nuclear envelope targeting signal. J. Cell Biol. *120*, 1093–1100.

Spann, T.P., Moir, R.D., Goldman, A.E., Stick, R. and Goldman, R.D. (1997) Disruption of nuclear lamin organization alters the distribution of replication factors and inhibits DNA synthesis. J. Cell Biol. *136*, 1201–1212.

Starr, D.A. and Han, M. (2002) Role of ANC-1 in tethering nuclei to the actin cytoskeleton. Science *298*, 406–409.

Starr, D.A., Hermann, G.J., Malone, C.J., Fixsen, W., Priess, J.R., Horvitz, H.R. and Han, M. (2001) unc-83 encodes a novel component of the nuclear envelope and is essential for proper nuclear migration. Development *128*, 5039–5050.

Staufenbiel, M. and Deppert, W. (1982) Intermediate filament systems are collapsed onto the nuclear surface after isolation of nuclei from tissue culture cells. Exp. Cell Res. *138*, 207–214.

Steen, R.L., Martins, S.B., Tasken, K. and Collas, P. (2000) Recruitment of protein phosphatase 1 to the nuclear envelope by A-kinase anchoring protein AKAP149 is a prerequisite for nuclear lamina assembly. J. Cell Biol. *150*, 1251–1262.

Stuurman, N., Heins, S. and Aebi, U. (1998) Nuclear lamins: their structure, assembly, and interactions. J. Struct. Biol. *122*, 42–66.

Su, A.I., Cooke, M.P., Ching, K.A., Hakak, Y., Walker, J.R., Wiltshire, T., Orth, A.P., Vega, R.G., Sapinoso, L.M., Moqrich, A., Patapoutian, A., Hampton, G.M., Schultz, P.G. and Hogenesch, J.B. (2002) Large-scale analysis of the human and mouse transcriptomes. Proc. Natl Acad. Sci. U.S.A. *99*, 4465–4470.

Sun, G., Yuen Chan, S., Yuan, Y., Wang Chan, K., Qiu, G., Sun, K. and Ping Leung, M. (2002) Isolation of differentially expressed genes in human heart tissues. Biochim. Biophys. Acta. *1588*, 241–246.

Tamm, L.K., Hong, H. and Liang, B. (2004) Folding and assembly of beta-barrel membrane proteins. Biochim. Biophys. Acta. *1666*, 250–263.

Tunnah, D., Sewry, C.A., Vaux, D., Schirmer, E.C. and Morris, G.E. (2005) The apparent absence of lamin B1 and emerin in many tissue nuclei is due to epitope masking. J. Mol. Histol. *36*, 337–344.

Wagner, N., Schmitt, J. and Krohne, G. (2004) Two novel LEM-domain proteins are splice products of the annotated Drosophila melanogaster gene CG9424 (Bocksbeutel). Eur. J. Cell Biol. *82*, 605–616.

Washburn, M.P., Wolters, D. and Yates, J.R. 3rd (2001) Large-scale analysis of the yeast proteome by multidimensional protein identification technology. Nat. Biotechnol. *19*, 242–247.

Waterham, H., Koster, J., Mooyer, P., Noort, G.G., Kelley, R., Wilcox, W., Wanders, R., Hennekam, R. and Oosterwijk, J. (2003) Autosomal Recessive HEM/Greenberg Skeletal Dysplasia Is Caused by 3beta-Hydroxysterol Delta14-Reductase Deficiency Due to Mutations in the Lamin B Receptor Gene. Am. J. Hum. Genet. *72*, 1013–1017.

Wilhelmsen, K., Litjens, S.H., Kuikman, I., Tshimbalanga, N., Janssen, H., van den Bout, I., Raymond, K. and Sonnenberg, A. (2005) Nesprin-3, a novel outer nuclear membrane protein, associates with the cytoskeletal linker protein plectin. J. Cell Biol. *171*, 799–810.

Wolters, D.A., Washburn, M.P. and Yates, J.R. 3rd (2001) An automated multidimensional protein identification technology for shotgun proteomics. Anal. Chem. *73*, 5683–5690.

Worman, H.J. and Courvalin, J.C. (2002) The nuclear lamina and inherited disease. Trends Cell. Biol. *12*, 591–598.

Worman, H.J., Yuan, J., Blobel, G. and Georgatos, S.D. (1988) A lamin B receptor in the nuclear envelope. Proc. Natl Acad. Sci. U.S.A. *85*, 8531–8534.

Xie, X., Wu, G., Lu, Z.H., Rohowsky-Kochan, C. and Ledeen, R.W. (2004) Presence of sodium-calcium exchanger/GM1 complex in the nuclear envelope of non-neural cells: nature of exchanger-GM1 interaction. Neurochem. Res. *29*, 2135–2146.

Ye, Q., Barton, R.M. and Worman, H.J. (1998) Nuclear lamin-binding proteins. Subcell. Biochem. *31*, 587–610.

Zhang, Q., Skepper, J.N., Yang, F., Davies, J.D., Hegyi, L., Roberts, R.G., Weissberg, P.L., Ellis, J.A. and Shanahan, C.M. (2001) Nesprins: a novel family of spectrin-repeat-containing proteins that localize to the nuclear membrane in multiple tissues. J. Cell Sci. *114*, 4485–4498.

Corresponding author: Dr. Eric C. Schirmer (e.schirmer@ed.ac.uk)

CHAPTER 6

SYNAPTOSOME PROTEOMICS

FENGJU BAI[1] and FRANK A. WITZMANN[2]
[1] *Safety Sciences, Charles River Laboratories Preclinical Services, USA*
[2] *Indiana University School of Medicine Biotechnology Research & Training Center, USA*

Table of Contents

Abstract: Our knowledge of the complex synaptic proteome and its relationship to physiological or pathological conditions is rapidly expanding. This has been greatly accelerated by the application of various evolving proteomic techniques, enabling more efficient protein resolution, more accurate protein identification, and more comprehensive characterization of proteins undergoing quantitative and qualitative changes. More recently, the combination of the classical subcellular fractionation techniques for the isolation of synaptosomes from the brain with the various proteomic analyses has facilitated this effort. This has resulted from the enrichment of many low abundant proteins comprising the fundamental structure and molecular machinery of brain neurotransmission and neuroplasticity. The analysis of various subproteomes obtained from the synapse, such as synaptic vesicles, synaptic membranes, presynaptic particles, synaptodendrosomes, and postsynaptic densities (PSD) holds great promise for improving our understanding of the temporal and spatial processes that coordinate synaptic proteins in closely related complexes under both normal and diseased states. This chapter will summarize a selection of recent studies that have drawn upon established and emerging proteomic technologies, along with fractionation techniques that are essential to the isolation and analysis of specific synaptic components, in an effort to understand the complexity and plasticity of the synapse proteome.

E. Bertrand and M. Faupel (eds.), Subcellular Proteomics, 77–98.

1. INTRODUCTION

Neuroscience is a discipline in which proteomics is having a growing impact. Many neuropsychiatric and neurodegenerative diseases, such as Alzheimer's, are thought to involve altered expression of multiple structural and/or metabolic genes and proteins, and therefore are well-suited for proteomic analysis (Kim et al. 2004). The study of other conditions, such as addiction and mood disorders that likely are secondary to altered expression of proteins involved in neurotransmission or neuroplasticity, can also take advantage of the power of global and narrow protein profiling that proteomics offers, for example, to examine the role of synaptic proteins in different disease states.

However, when using proteomics to study central nervous system (CNS) function and pathology, one is faced with a task complicated by diverse regional specialization that is compounded by intricate cellular complexity (neurons, glia and cell projections) and further exacerbated by synaptic heterogeneity and a huge dynamic range of protein expression. The human brain is composed of an estimated $\sim 10^{12}$ heterogeneous neurons that communicate by way of $\sim 10^{15}$ synapses (Pocklington et al. 2006). Whole brain tissues are variably composed of neurons and glial cells, the latter comprising up to 90–95% of the cells (Williams et al. 1988). Because most of the glia are astrocytes (Hansson et al. 2003), protein expression analyzed in a whole brain sample may tell us little about neuronal function, in the classic sense. Numerous efforts have been made to establish reference proteomes for brain tissue from various species by surveying the whole brain or gross brain areas (Edgar et al. 1999; Fountoulakis et al. 1999; Gauss et al. 1999; Langen et al. 1999; Beranova-Giorgianni et al. 2002). Our laboratory demonstrated that the expressed proteome can vary in various brain regions based on genetic selection for alcohol preference, and, within these genetic lines, by functional nuclei (Witzmann et al. 2003). Despite these documented differences in whole brain tissue, it is likely that many of the proteins previously identified by us and by others in whole brain tissue preparations are of glial, not neuronal origin. For proteomics, a meaningful analysis obligates one to by-pass the whole brain, brain region, and even the micropunch (Leng et al. 2004) or laser-capture (Mouledous et al. 2003; Nazarian et al. 2005) sample. Instead, one must opt for the business-end of the CNS, the synapse.

Synapses are electrical or chemical communicative contacts between neurons. Electrical synapses (neuronal gap junctions) function by the propagation of electrical impulses from one cell to another (and vice versa) via direct, physical contact. As a consequence, these synapses are characterized by a relatively simple organization of membrane structure and associated organelles (Zoidl et al. 2002). Electrical synapses are also less mutable, in terms of their function and molecular characteristics, and thus exhibit little of the plasticity that typifies the chemical synapse.

Characteristically, chemical synapses contain a broad range of chemical neurotransmitters and neuropeptides for intercellular communication, in addition to localized translational machinery that is tightly coupled to signaling (Steward et al. 2003). The latter components make these neuronal junctions particularly relevant to proteomic

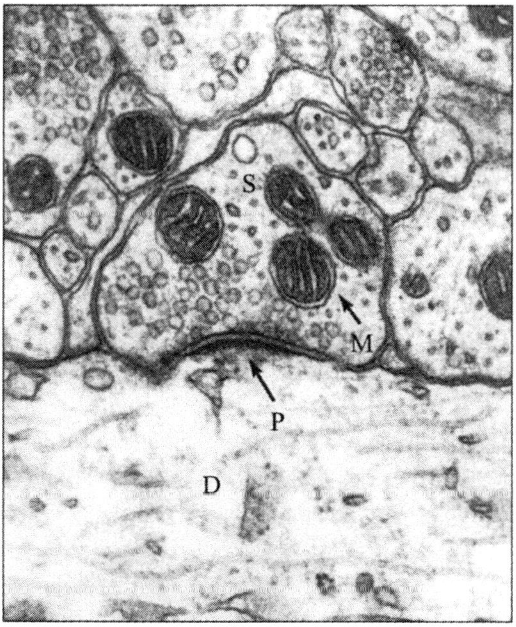

Figure 1. Electron photomicrograph a synapse (56,000X) illustrating the synaptic knob (S) as it ends on the shaft of a dendrite (D) in the central nervous system. P, postsynaptic density; M, mitochondrion. (From: Review of Medical Physiology, 22nd Edition, The McGraw-Hill Co., 2005)

analysis. Cell–cell communication that occurs by chemical transmission is characterized by complex protein-driven molecular mechanisms of synthesis, delivery, storage, docking, fusion, neurotransmitter release, reuptake, etc. (Purves 2004). In general, synapses are composed of three main constituents: a presynaptic component (presynaptic ending, axon terminal), a synaptic cleft, and a postsynaptic component (dendritic spine). The pre- and the postsynaptic membranes are uniquely distinguishable by visible densities along their corresponding plasma membranes. Together with the synaptic cleft, they are collectively referred to as the synapse (see Figure 1).

Typically, the presynaptic ending is further distinguished from the postsynaptic component by the conspicuous presence of neurotransmitter-filled vesicles. In response to presynaptic membrane depolarization, the vesicles exocytose their contents into the cleft through complicated membrane-trafficking events. The presynaptic axon terminal (bouton) of the presynaptic component also contains other organelles such as mitochondria, smooth endoplasmic reticulum, microtubules, and neurofilaments. The presynaptic membrane is variably populated by docking/fusion apparatus, ion channels, and other protein constituents. The 20–30 nM wide synaptic cleft separates the pre- and postsynaptic membranes and generally contains a dense plaque of intercellular material that includes microfilaments.

The postsynaptic membrane, particularly at the dendritic spine, is recognizable by a collection of dense material visible by electron microscopy on its cytoplasmic surface. This so-called post-synaptic density (PSD) (Palay 1958; Gray 1959) is a specialization of the nerve cell's sub-membrane cytoskeleton composed of granular/filamentous material, contains cisternae of smooth endoplasmic reticulum, and its existence seems to be dependent on the presence of the presynaptic ending. A subcellular fraction enriched in structures with PSD-like morphology has been shown to contain signal-transduction molecules thought to regulate receptor localization and function in the CNS (Kennedy 1993).

The morphological characteristics of synapses mentioned above reflect the complex functional dynamics of neurotransmission and its plasticity. Though well-studied for many decades, the synaptic molecular anatomy and clues to its complex function in learning/memory, injury, and disease, are just now being unraveled through the application of emerging protein analytical techniques. Four compelling reasons have been identified that underscore the importance of understanding the protein molecular machinery of the synapse (Pocklington et al. 2006):

1. It is the most important structure for communication between nerve cells.
2. The neurotransmitter receptors and signal transduction machinery within the synapse respond to patterns of electrical activity and instigate biochemical changes in the nerve cell and in so doing modify the brain in response to behavioral experience.
3. Synapse proteins correspond to many human disease genes and drug targets for therapeutics that modulate cognitive illnesses.
4. Synapse proteomic studies have compiled a first draft of the protein composition of the synapse, revealing an unexpectedly high degree of molecular complexity.

The fundamental role of the synapse in neurotransmission and plasticity has fueled increasing proteomic efforts for the identification synaptic protein constituents, leading to the establishment of increasingly comprehensive mapping and profiling of the synaptic proteome. This chapter will summarize a selection of recent studies that have drawn upon established and emerging proteomic technologies, along with fractionation techniques that are essential to the isolation and analysis of specific synaptic components, in an effort to understand the complexity and plasticity of the synapse proteome.

2. SYNAPTOSOME FRACTIONATION

Sample complexity reduction strategies are required to enable neuroscientists to address the issues mentioned above, and facilitate meaningful applications of proteomic analyses. The feasibility of protein enrichment by subcellular fractionation has been demonstrated through the analysis of the rat brain sub-proteomes of cytosolic, mitochondrial and microsomal fractions (Krapfenbauer et al. 2003). Another well-established subcellular fractionation technique, synaptosomal isolation, has recently been applied in various proteomic studies, providing an enrichment of cellular components found at the synapse. By further fractionating the synaptosomes, one can use proteomic approaches to study even more specific synaptic regions.

2.1. Synaptosomes

The term "synaptosome" was first mentioned in a paper published in 1964 by Whittaker's group (Whittaker et al. 1964). At that time, the aim was to explore the synaptosomal localization of several known and putative neurotransmitters and their synthesizing enzymes, and to subfractionate the disrupted synaptosomes to obtain homogeneous fractions of synaptic vesicles and other synaptic components. In this context, rather than organelles, synaptosomes are artificial, membranous sacs that contain synaptic components and are generated by subcellular fractionation of homogenized or ground-up nerve tissue. They are often referred to as "pinched-off nerve endings," because the lipid bilayers naturally reseal together after the axon terminals are torn off by the physical shearing force of homogenization. Synaptosomes contain the complete presynaptic terminal, including mitochondria and synaptic vesicles, along with the postsynaptic membrane and the postsynaptic density (PSD). This typical morphology can be observed clearly via electron microscopy (Figure 2, by permission from (Schrimpf et al. 2005)). Because all the molecular machinery for the release, uptake and storage of neurotransmitters remain intact, synaptosomes are often used to study synaptic transmission.

2.2. Synaptosome Preparation

In a manner similar to the isolation of other subcellular fractions containing relatively pure cellular organelles, differential centrifugation or sucrose (Gray and Whittaker 1962; Whittaker et al. 1964), Ficoll/sucrose (Booth et al. 1978), or Percoll (Nagy et al. 1984) density-gradient centrifugation techniques are used for the preparation of synaptosomes. Though the reagents for buffering and for the formation of the density gradient may differ from lab to lab, depending on personal preferences or specific applications, all follow the steps involved in purifying a crude synaptosomal fraction from their mitochondrial and microsomal contamination. For instance, here we describe briefly a protocol used in our laboratory for the routine preparation of synaptosomes from fresh brain tissue. Whole brain or a grossly dissected brain region is homogenized in buffer (0.32 M sucrose, 20 mM HEPES, pH 7.4, with proper reducing reagent and protease inhibitors) and centrifuged at 1000 × g for 10 min to pellet the membrane fragments and nuclei. The supernatant is then further centrifuged at 17,000 × g for 15 min to obtain the pellet containing synaptosomes contaminated with mitochondria and microsomes. This crude synaptosome fraction is then further purified using a discontinuous sucrose density gradient consisting of a 0.8 M sucrose layer on the top and a 1.2 M sucrose layer on the bottom. The crude

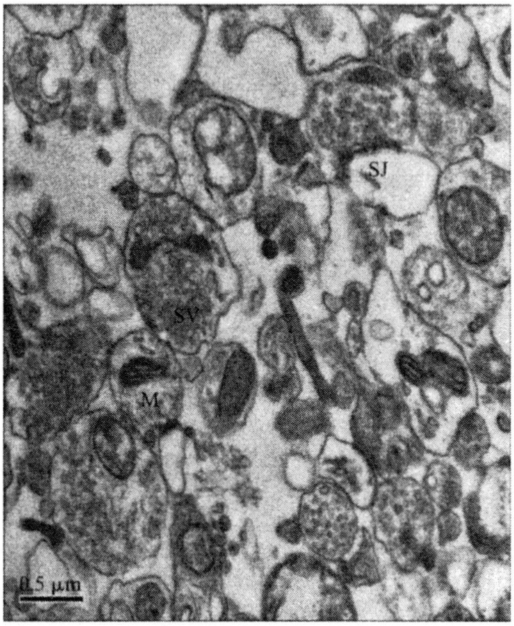

Figure 2. Electron micrograph of a synaptosome fraction isolated from mouse brain by Ficoll density-gradient centrifugation. SV, synaptic vesicles; M, mitochondria; SJ, synaptic junction with attached PSyD. Bar = 0.5 mm. (From: Schrimpf *et al.* 2005)

synaptosome suspension is layered on top of the 0.8 M sucrose layer and centrifuged at 54,000 × g for 90 min. The pure synaptosomal fraction can be obtained from the interface of 0.8 M sucrose and 1.2 M sucrose. The purity of the isolated synaptosomes can be inspected morphologically under an electron microscope (Figure 2) or analyzed biochemically using enzymatic markers.

3. THE SYNAPTOSOMAL PROTEOME

Comprehensive studies on the synaptic proteome have been rare. Not until quite recently has the mass spectrometric technical momentum developed for detecting and documenting a comprehensive and coherent map of synaptic proteins, currently numbering approximately 1000 unique proteins (Grant 2006). This momentum has been driven by an urgent need in the neuroscientific community for molecular markers of neuropsychiatric and neurodegenerative disorders, as well as a desire to understand the molecular mechanisms underlying synaptic neurotransmission and plasticity. Table 1 presents a compact list of recent proteomic efforts where protein fractions were derived from synapses, and the proteomic methods used to study them. These studies are described further in the text that follows.

Table 1. Protein fractions derived from synapses and the proteomic methods used to study them

Synaptic Fraction Source	Proteomic Platform	Proteome Analyzed	Reference
Synaptosomes			
Rat Brain	1D-PAGE, 2-DE; LC-MS/MS; Western blot	O-GlcNAc modified proteins	(Cole et al. 2001)
Human Cortex	IMAC; LC-MS/MS	Phosphoproteome	(DeGiorgis et al. 2005)
Gerbil cortex	2-DE; MALDI-TOF; LC-MS/MS; Western blot	Redox proteome	(Boyd-Kimball et al. 2005)
Mouse Brain	IMAC; LC-MS/MS	Phosphoproteome	(Collins et al. 2005a, b)
Mouse Brain	ICAT; LC-MS/MS	Protein ID	(Schrimpf et al. 2005)
Rat forebrain	Affinity enrichment; 2-DE; MALDI-TOF; LC-MS/MS	Protein ID, shotgun, and PTMs	(Witzmann et al. 2005)
Synaptic Vesicles			
Rat brain	2-DE, 16-BAC/SDS-PAGE; LC-MS/MS	Protein ID	(Coughenour et al. 2004)
Rat brain	16-BAC/SDS-PAGE; MALDI-TOF	Protein ID	(Morciano et al. 2005)
Synaptic membranes			
Rat forebrain	SDS-PAGE, 2-DE, SCX, HPLC; MALDI-TOF, LC-MS/MS	Protein ID	(Stevens et al. 2003)
Rat forebrain	2-DE; LC-MS/MS	Protein ID	(Li et al. 2004)
Rat brain	ICAT; LC-MS/MS	Differential expression	(Prokai et al. 2005)
Rat brain (SV & presyn. mem.)	6-BAC/SDS-PAGE; MALDI-TOF	Protein ID	(Morciano et al. 2005)

(Continued)

Table 1. (Continued)

Synaptic Fraction Source	Proteomic Platform	Proteome Analyzed	Reference
Postsynaptic Density (PSD)			
Rat forebrain	SDS-PAGE; MALDI-TOF	Protein ID	(Walikonis et al. 2000)
Mouse forebrain	2-DE; LC-MS/MS	Protein ID and differential expression	(Satoh et al. 2002)
Rat forebrain	2-DE; MALDI-TOF; ICAT; LC-MS/MS	Protein ID	(Li et al. 2004)
Rat forebrain	MudPIT	Protein ID	(Yoshimura et al. 2004)
Rat & mouse brain	SDS-PAGE; LC-MS/MS; Western blot	Protein ID	(Jordan et al. 2004)
Rat forebrain	SDS-PAGE; LC-MS/MS	Protein ID and semi-quantification	(Peng et al. 2004)
Rat brain	ICAT, SCX; LC-MS/MS; Immunoelectron microscopy	Protein ID and quantification	(Li et al. 2005)
Rat cortex	MudPIT	Protein ID	(Phillips et al. 2005)
Rat forebrain & cerebellum	ICAT; LC-MS/MS	Relative and absolute quantification	(Cheng et al. 2006)
Mouse brain	IP, affinity purification; 2-DE; LC-MS/MS; Western blot	Protein ID	(Collins et al. 2005b)
Rat hippocampus	MUDPIT	Protein ID	(Dosemeci et al. 2006)
Mouse brain	SCX, IMAC; LC-MS/MS; Western blot	Phosphoproteome	(Trinidad et al. 2006)
Mouse brain	lectin weak affinity chromatography; O-GlcNAc LC-MS/MS; MALDI-TOF	modified proteins	(Vosseller et al. 2006)
Presynaptic Particle Fraction (PPF)			
Rat cortex	MudPIT	Protein ID	(Phillips et al. 2005)

3.1. Sample Complexity Reduction and Protein Enrichment

In addition to the direct analysis of the synaptosomal proteome (Boyd-Kimball et al. 2005; Schrimpf et al. 2005; Witzmann et al. 2005), protein subsets have been extracted and examined either as a subcompartment or subfraction of synaptosomes, such as synaptic vesicles (Coughenour et al. 2004; Morciano et al. 2005), or from a particular structural component, such as synaptic membranes (Stevens et al. 2003), presynaptic particles (Phillips et al. 2005), synaptodendrosomes (dendritic spine preparations) (Rao et al. 1993; Leski et al. 1996; Villanueva et al. 2001; Jiang et al. 2002), and PSD (Walikonis et al. 2000; Satoh et al. 2002; Li et al. 2004; Peng et al. 2004; Yoshimura et al. 2004; Phillips et al. 2005; Cheng et al. 2006; Dosemeci et al. 2006; Jordan et al. 2006). The analysis of these subproteomes holds great promise for improving our understanding of the temporal and spatial processes that coordinate synaptic proteins in closely related complexes under both normal physiological or pathological conditions.

Synaptosomal proteins also can be enriched by immunoprecipitation and/or affinity purification. Immunoprecipitation uses highly specific antibodies to remove individual target proteins or groups of related proteins (with similar antigenicity) from a complex mixture. Often, those proteins closely interacting with the target protein(s) are also recovered, making this a commonly used method to study protein-protein interaction (Phizicky and Fields 1995). Affinity purification purifies the protein of interest, such as antibodies, fusion-tagged proteins, and biotinylated proteins by virtue of its specific binding properties to an immobilized ligand. For example, Concanavalin A (Con A) is a tetrameric metalloprotein that binds sugar moieties with C-3, C-4 and C-5 hydroxyl groups. Con A coupled to Sepharose is routinely used for the separation and purification of glycoproteins, specifically for proteomic analysis. The power of sample complexity reduction and protein enrichment has been demonstrated by the analysis of PSD components using the combination of subcellular fractionation, immunoaffinity purification and LC-MS/MS, where more than 400 PSD components (Walikonis et al. 2000; Jordan et al. 2004; Li et al. 2004; Peng et al. 2004; Yoshimura et al. 2004; Jordan et al. 2006) and 186 NMDA receptor-associated proteins (Husi et al. 2000) have been isolated and identified.

Another subsynaptic fraction derived directly from presynaptic specializations, the presynaptic particle fraction (PPF) can be separated from PSD by adjusting the pH of Triton X-100 extraction of isolated trans-synaptic scaffolds. As one might expect, it has been shown that the major proteins of the PPF, clathrin and dynamin, are concentrated in the presynaptic compartment (Phillips et al. 2005). These investigators used multidimensional protein identification technology (MudPIT) to compare the PPF and the PSD fraction. Of 341 proteins identified, 50 localized in the PPF, 231 in the PSD fraction, and 60 were found to be common to both fractions. The PPF was also characterized by a low proportion of actin and actin-associated proteins along with a high proportion of vesicle proteins. The authors concluded that the PPF consists of presynaptic proteins not connected to the actin-based synaptic framework and that clathrin may be an anchorage scaffold for many presynaptic proteins.

3.2. Proteomic Analysis Platforms

The various technical approaches used in proteomics for synaptosomal proteome analysis are addressed throughout this volume and therefore will not be formally described or listed here. Whether the approach is gel-based, mass spec-based, uses stable-isotopic labels, or is label-free; each has its own unique merits and deficiencies, and many of the latter are shared among them all. Consequently, the suitability of a chosen technique can dramatically affect the outcome of a study and experimental design must therefore be done with great care. The combination of specific protein enrichment/complexity reduction strategies with appropriate protein resolution (one- and two-dimensional gel electrophoresis (2-DE), solution isoelectric focusing (sIEF), liquid chromatography (LC)) and identification (peptide mass fingerprinting (PMF), tandem mass spectrometry (MS/MS), Western blotting) techniques for that protein subset is the key to achieving an optimal result. Sometimes, the simplest approach (e.g. Western blotting) can be highly informative (see Figure 3 below; (Jordan et al. 2004)). Similar enrichment of PSD proteins has been demonstrated in this manner (Morciano et al. 2005; Cheng et al. 2006; Trinidad et al. 2006).

When a global assessment of the total synaptosome proteome is desired, a combination of several proteomic platforms should be used, as these have been shown to be complementary rather than alternative ways of measuring protein abundances in biological systems (Wu et al. 2006). The commonly used protein identification techniques include matrix-assisted laser-desorption ionization (MALDI) based PMF and LC-MS/MS sequencing. LC-MS/MS sequencing is particularly useful for direct protein identification or for robust confirmation of protein IDs derived from PMF. 2-DE-based protein separation followed by the MS identification of differentially expressed proteins remains the most commonly used platform for the quantitation of differentially expressed proteins, particularly for a multi-group comparison. Other quantitative proteomic approaches, such as 2-D (Fluorescence) Difference Gel Electrophoresis (2D-DIGE, see below), the Isotope-Coded Affinity Tag (ICAT) reagent approach, and label free-quantitative proteomics are in continued development and improvement, aiming to address the problems associated with the traditional 2-DE gel based assays.

4. TECHNICAL CHALLENGES OF SYNAPTOSOMAL PROTEOMICS

A significant challenge for neuroscientists in studying the membranous synaptosomal proteome rests squarely in the analysis of its constituent hydrophobic and membrane-bound proteins. These include receptors, transporters, ion channels, and the molecular machinery for synaptic vesicle cycling. Hydrophobic and membrane-bound proteins are poorly resolved by traditional IEF gel technology. Similarly, they tend to resist in-gel tryptic digestion, leading to poor rates of protein identification by PMF (van Montfort et al. 2002).

Acidic PAGE systems using cationic detergents such as benzyldimethyl-*n*-hexadecylammonium chloride (16-BAC) are useful for resolving basic and membrane

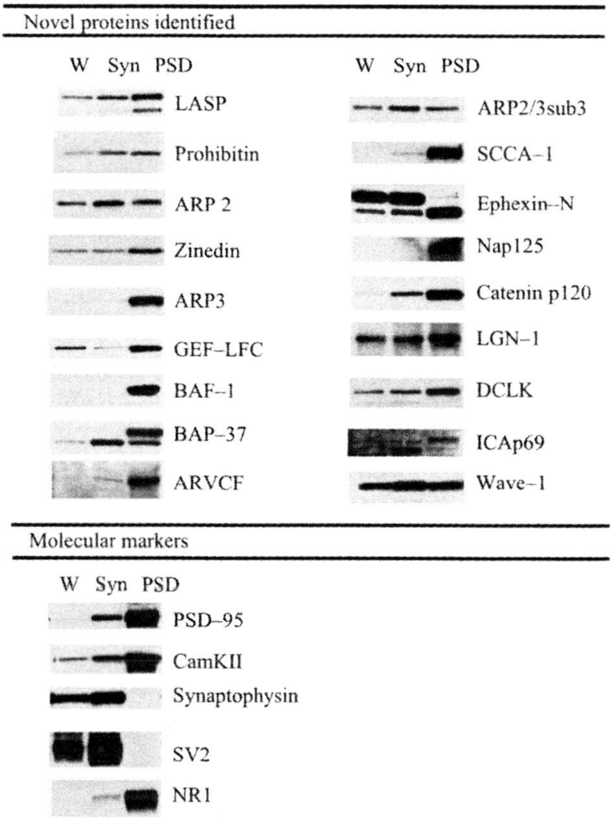

Figure 3. Western blots of several novel members of the PSD. Ten to 20 μg of PSD protein were separated by SDS-PAGE and blotted using a variety of antibodies. W, whole brain extract; Syn, synaptosomal fraction; and PSD, postsynaptic density. Known molecular markers were also blotted to demonstrate the purity of the fractions used for MS/MS. ARP, actin-related protein; BAF-1, barrier to autointegration 1; BAP-37, B cell receptor-associated protein; ARVCF, armadillo-repeat-velo-cardio-facial syndrome protein; ARP2/3sub5, actin-related protein complex 2/3 subunit 5; SCCA-1, squamous cell carcinoma antigen-1, serpin3a; NAP-125, NCK-associated protein 1; LGN-1, leucine-rich repeat LGI family, member 1; DCLK, doublecortin-like kinase/mKIAA0369; ICAp69, islet cell autoantigen protein p69. (From: Jordan et al. 2004).

proteins. Using a cationic 16-BAC polyacrylamide gel system in the first and SDS-PAGE in the second dimension, immunoaffinity purified pools of mouse brain synaptic vesicles have been analyzed successfully (Kramer 2006). Many integral and membrane associated proteins were identified from a fraction containing free synaptic vesicles and a second fraction containing synaptic vesicles plus presynaptic membranes, using either MALDI-TOF or Western Blot analysis. Some of the proteins detected included the integral synaptic vesicle proteins such as SV2A and SV2B, the zinc transporter ZnT-3, synaptotagmins I and II, synaptophysin, v-SNARE VAMP

II, t-SNARE syntaxin 1B2 and SNAP-25, vesicular neurotransmitter transporters for acetylcholine (vAChT), glutamate (vGluT1), and GABA (vGAT), the proton pumping vacuolar ATPase (V1: A1, B1, B2, C1, D, E1, E2, G2; V0: a, d1), peripheral synaptic vesicle proteins such as the synapsins Ia, Ib, IIa and IIb, alpha subunit of calcium/calmodulin dependent kinase II, proteins transiently associated with the synaptic vesicle compartment such as syntaxin binding protein 1, N-ethylamide sensitive factor (NSF), munc18, munc13, and presynaptic membrane protein N-type Ca^{2+}-channel.

In addition to the inherent difficulty of resolving membrane and basic proteins using 2-DE, gel-gel reproducibility also has been a significant challenge. 2D-DIGE technology was developed to address the gel variability issue. In 2D-DIGE analysis, two samples for comparison are labeled respectively with one of the CyDye fluors. The mixture of the labeled samples, together with a labeled standard can be resolved on a single 2-D gel (Unlu et al. 1997). Identical protein from different samples labeled with each of the fluors migrates to the same position on a 2-D gel, enabling the analysis of differences in protein abundance by comparing the fluorescent intensity of the overlaying protein spots, and subsequent identification by MS techniques. Unfortunately, this technique is only feasible for the simultaneous comparison of sample pairs. Though the application of 2D-DIGE has not yet been reported in the analysis of synaptosomal proteins, a recent study using this technique for the detection of phosphorylation pattern changes in rat cortical neuron cultures upon phosphatase treatment has been conducted (Raggiaschi et al. 2006), and its potential relevance to synaptosomal studies is clear.

Another technical challenge using the 2-DE-based approach relates to its narrow dynamic range of protein detection, a limitation that makes it almost impossible to detect and compare the expression of many very low abundance proteins, even with sample subfractionation. Though various aforementioned protein enrichment techniques are available, the application of LC-based protein separation coupled with highly sensitive mass spectrometric protein identification can address the issue of dynamic range, to a certain extent. This approach also improves the resolution and identification of proteins with extreme pI, large mass, and, in some cases, membrane associations.

For differential, relative protein quantitation, ICAT reagents have been employed in concert with tandem mass spectrometry, enabling LC-MS/MS based analysis of differentially expressed proteins from different physiological and pathological states (Gygi et al. 1999). Each ICAT reagent contains an affinity tag (biotin moiety), an isotopically labeled linker (contains heavy or light stable isotopes, such as d8-heavy or d0-light), and a chemical reactive group (such as a thiol-specific reactive group). After the free thiol groups of cysteine residues from each of the two populations of proteins are tagged with a heavy or a light ICAT reagent, same amount of proteins are combined and tryptic digested, followed by affinity enrichment of the labeled peptides using an avidin column. These peptides are then analyzed by LC-MS/MS. The ratios of signal intensities of differentially mass-tagged peptide pairs are quantified to determine the relative levels of proteins in the two samples. The sensitivity and reliability of the technique have been further improved by the generation of cleavable ICAT reagents (Hansen et al. 2003).

In a recent study, the ICAT reagent was employed for the proteomic characterization of synaptosomes isolated from mouse brain (Schrimpf et al. 2005). Though only cysteine containing proteins were analyzed, more than a thousand proteins were identified by LC-MS/MS, including proteins involved in synaptic vesicle exocytosis, synaptic vesicle recycling, postsynaptic receptors and proteins constituting the PSD, as well as a large number of soluble and membrane-bound proteins for synaptic signal transduction and metabolism. In a similar application, the effect of chronic morphine exposure on the synaptic plasma-membrane subproteome in rats was studied using ICAT technology (Prokai et al. 2005). These researchers observed a number of significant expression changes including the down-regulation of an integral membrane protein Na^+/K^+ ATPase (α-subunit) involved in regulation of the cell membrane potential that correlated to a decline in electrogenic Na^+, K^+ pumping of identical magnitude.

More recently, a study was conducted on the rat brain to compare the regional differences in the protein expression profile of PSD proteins between forebrain and cerebellum (Cheng et al. 2006). LC-MS/MS analysis of ICAT reagent-labeled peptides (including the absolute quantification, AQUA, strategy) revealed crucial molecular differences in PSD between the brain areas. Among 296 proteins identified and quantified, 43 differed significantly in abundance. These included proteins involved in cell adhesion, cytoskeleton/cellular scaffold, GTPase regulation, kinase/phosphatase regulation, membrane trafficking, mitochondria, motor function, receptors and channels, metabolism, and others. The differences in the expression of Nir2 (phosphatidyl inositol transfer protein), Septin 4, and GluRδ were further confirmed by immunostaining of sagittal sections of rat brain.

It bears repeating that all proteomic approaches have an intrinsic bias toward the detection of a certain subset or type of proteins. No single approach has the capacity for a truly comprehensive analysis of the total proteome. Therefore, it is necessary to combine several technical approaches to even approximate a global analysis. Our laboratory has analyzed the proteome of rat cortical synaptosomes using two types of proteomic approaches. The first, 2-DE separation coupled with MALDI-TOF and LC-MS/MS identification of proteins from gel excisions detected a total of 968 protein spots. The 94 most abundant protein spots were analyzed by MALDI-based PMF resulting in the identification of 85, representing 61 unique proteins. Another 96 low-abundance protein spots were analyzed by LC-MS/MS, yielding 79 identifications and representing 46 additional unique proteins. A total of 91 unique proteins thus were identified using the initial approach.

The second approach, a shotgun proteomic analysis (LC-MS/MS) of tryptic solution-digests of isolated synaptosomes, identified 209 unique proteins. When the two sets of identified proteins are compared, 46 were found to be common. Accounting for this intersection, the total number of unique proteins identified by the

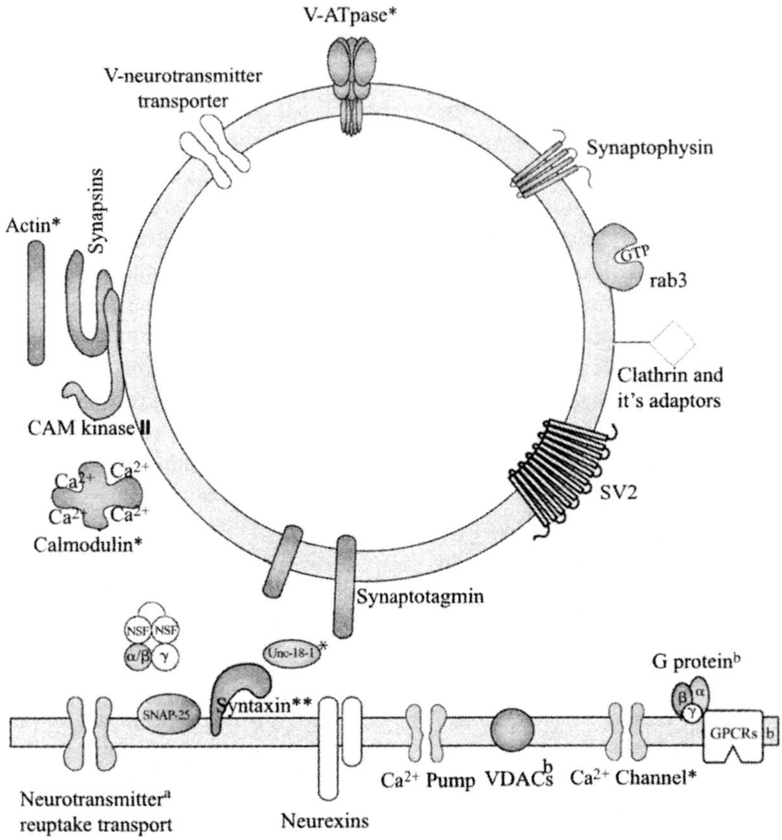

Figure 4. Diagrammatic illustration of the major pre- and postsynaptic proteins identified by 2-DE/MS and/or shotgun proteomics, and normally expected as constituent in synaptosomal preparations. Blue: the proteins that were identified by 2-DE/MS; Yellow: the proteins that were identified by shotgun proteomics; Green: the proteins that were identified by both 2-DE/MS and shotgun proteomics; and Blank (white): those major constituents expected but not identified. *proteins were identified by shotgun proteomics only after the PTM analysis; **proteins were identified as a complex with other proteins by 2-DE/MS; [a]EAA1, EAA2, and GABA transporter. [b]post-synaptic proteins. (From: Witzmann et al. 2005)

combined methods was 254. Of these 254 proteins, 61 were identified by PubMed literature search as having synapse-specific function. A total of 19 identified proteins were involved in synaptic vesicle trafficking or docking, 9 served receptor or transporter functions, 9 were involved in intra-cellular signaling cascades that affect synaptic transmission, and 24 had other synapse-specific functions. Figure 4 illustrates the major pre- and postsynaptic proteins identified by 2-DE/MS and/or shotgun proteomics in this study, a set of proteins normally expected as constituent in synaptosomal preparations (Witzmann et al. 2005).

5. DELINEATION OF THE FUNCTIONAL RELATIONSHIPS OF SYNAPTIC PROTEINS – POST-TRANSLATIONAL MODIFICATIONS

Many published studies have significantly improved our understanding of synaptic function by describing the protein constituency of synaptosomes, and the number of these documented synaptic proteins is rapidly expanding. However, more efforts are needed for an integrated functional understanding of the synapse. The analysis of the changes in the state of protein post-translational modification (PTM) is an effective approach for this purpose because PTM regulates nearly every aspect of synaptic function, including neurotransmitter release, synaptic vesicle recycling, receptor/ion channel function, and local protein synthesis (Barry et al. 2002; Lisman et al. 2002; Korolchuk et al. 2003; Purcell et al. 2003; Waltereit et al. 2003; Han et al. 2004).

PTM also plays a critical role in synaptic plasticity and memory (Purcell et al. 2003; Routtenberg et al. 2005). For instance, tyrosine kinase signal transduction cascades of Trk receptor tyrosine kinases, the Src family of non-receptor tyrosine kinases and the Eph receptor tyrosine kinases have been shown to associate with the synaptic plasticity and memory processing (Purcell et al. 2003). The main protein constituent of PSD, calcium/calmodulin-dependent protein kinase II (CaMKII) has been proposed to act as a bi-stable switch by auto phosphorylation/dephosphorylation for the long-term storage of synaptic memory (Lisman et al. 2002). The activity of CaMKII is regulated by Ca^{2+}/calmodulin, where Ca^{2+} release is controlled by NMDA receptor activation. CaMKII dictates synaptic strength, mainly by affecting the functional properties of AMPA (α-amino-3-hydroxy-5-methyl-4-isoxazole propionic acid) receptors, as well as their trafficking and anchorage to the postsynaptic membrane (Barry et al. 2002).

Clathrin-mediated endocytosis is a major vesicular transport mechanism in the neuron, which enables the internalization of plasma membrane-bound proteins, nutrients, hormones and other molecules associated with the plasma membrane into intracellular compartments. Clathrin and various adaptor and accessory proteins work in concert at different stages of clathrin coated vesicle formation and disassembly, and many of these proteins (such as clathrin light chain, AP-2, dynamin 1, synaptojanin 1, and the amphiphysins) are substrates for protein kinases (Korolchuk et al. 2003). In addition, it has been suggested that directing synaptotagmin 1 to the synaptic vesicle is dependent on the N-terminal glycosylation of this protein (Han et al. 2004).

Using immobilized metal affinity chromatography (IMAC), phosphopeptides haven been isolated and enriched from tryptically digested proteins extracted from freshly isolated human cortical synaptosomes (DeGiorgis et al. 2005). LC-MS/MS analysis identified 26 synaptic proteins with 1, 2 or 3 phosphoserine(s), including synapsin 1, syntaxin 1, SNIP, PSD-93, NCAM, GABA-B receptor, chaperone molecules, and protein kinases. In another study (Collins et al. 2005a), LC-MS/MS analysis of IMAC enriched phosphopeptides from mouse forebrain synaptosomes, identified 228 potential synaptic phosphoproteins and characterized over 350 phosphopeptides containing 331 sites of phosphorylation, including the identification of highly phosphorylated MAP1B (33 sites) and Bassoon (30 sites). Bioinformatic tools, such as Scansite (www.scansite.mit.edu/) and Net-PhosK (www.cbs.dtu.dk/services/NetPhos/), were employed to predict the likeliness of these identified proteins to be phosphorylated, the possible phosphorylation sites and the responsible kinases. Additionally, protein-protein interaction database (www.ppid.org) was utilized to analyze the protein-protein interaction of the NMDA receptor.

Fractionation of proteins by strong cation exchange (SCX) chromatography, followed by IMAC enrichment of phosphopeptides from SCX fractions, led to a comprehensive identification of phosphoproteins of PSD isolated from mouse brain using LC-MS/MS (Trinidad et al. 2006). In this study, phosphorylation site(s) were mapped to 287 proteins from a total of 1,264 unique proteins identified. This translates into a 23% phosphorylation rate, comparable to an expected 33% rate in the general proteome (Johnson et al. 2005). The 287 phosphoproteins were derived from a total of 998 unique phosphorylated peptides, and the phosphorylations were mapped to 723 unique sites. Most of these occurred on serines, to a lesser extent on threonines, and only minimally on tyrosines (Figure 5A).

Proteins within a broad range of functional categories were involved, including structural proteins, scaffolding proteins, kinases and phosphatases, membrane neurotransmitter receptors, and voltage-gated ion channels (Figure 5B; **see colour insert**). When the number of unique phosphorylation sites, was analyzed, as a function of the protein functional class, adhesion/cytoskeleton proteins appeared to carry the most modification followed by adaptor/sorting proteins, presynaptic proteins, and kinases/phosphatases (Figure 5B).

Another PMT with functional importance and complex interplay with phosphorylation in the brain is the cytosolic- and nuclear-specific *O*-linked *N*-acetylglucosamine (*O*-GlcNAc) glycosylation. This is an abundant, dynamic, and inducible carbohydrate modification of serines and threonines by *N*-acetylglucosamine that plays a role in protein regulation of cellular signaling networks, nuclear transcription, protein turnover, and the cell cycle (Slawson et al. 2006). *O*-GlcNAc transferase (OGT) and *O*-GlcNAcase, which are responsible for the addition and removal of *O*-GlcNac protein modifications, are highly expressed in the brain (Gao et al. 2001; Iyer et al. 2003).

Recently, a chemoenzymatic approach was employed for the identification of *O*-GlcNAc modified proteins from rat brain. This approach allows the transfer of a ketone-containing galactose analogue selectively to the C-4 hydroxyl of GlcNAc by an engineered α-1,4-galactosyltransferase enzyme. The ketone functionality is further reacted with an aminooxy biotin nucleophile, which enables the enrichment of *O*-GlcNAc modified peptides using avidin affinity chromatography. Using this approach, 25 unique proteins were identified from rat forebrain, including 23 newly identified and 2 previously reported proteins in *O*-GlcNAc-glycosylated form (microtubule-associated protein 2B and host cell factor) (Khidekel et al. 2004). As expected, these proteins were functionally categorized as proteins associated with gene expression, neuronal signaling, and synaptic plasticity.

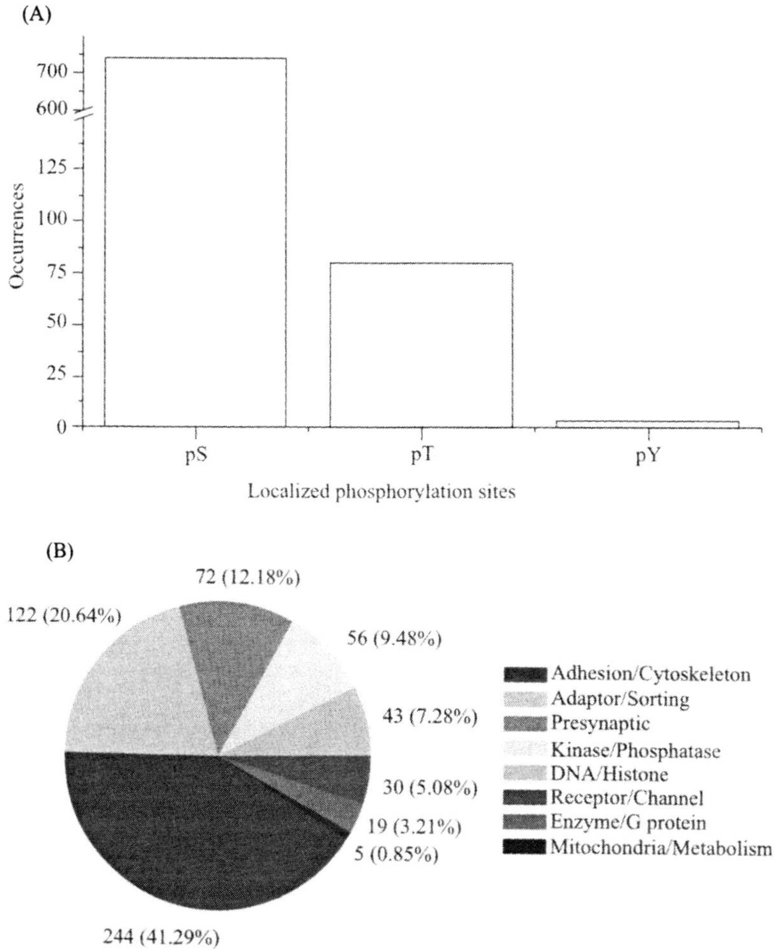

Figure 5. (A) The relative distribution of phosphoserines (pS), phosphothreonines (pT), and phosphotyrosines (pY). The majority of identified sites were phosphoserines (739) followed by phosphothreonines (79) and lastly phosphotyrosines (three). (B) The number of unique phosphorylation sites as a function of the protein functional class. Proteins involved in adhesion/cytoskeleton represent the most prevalent class followed by proteins involved in adaptor/sorting functions. (From: Trinidad et al. 2006)

Using *in situ* hybridization and immuno-labeling, Akimoto et al. (2003) found OGT transcripts and proteins to be enriched in the cortical neurons of rat cerebellum, particularly in the Purkinje cells. In nerve terminals, the OGT enzyme was observed around synaptic vesicles, in the pre- and post-synaptic terminals. The localization of immuno-labeled *O*-GlcNAc was also in general agreement with the distribution of OGT. In a related study, abundant OGT and *O*-GlcNAcase activities were detected in intact, functional synaptosomes, particularly in the soluble portion of the synaptosomal fraction (Cole et al. 2001). Using wheat germ agglutinin-based lectin weak affinity chromatography for the enrichment of *O*-GlcNAc modified peptides, 145 unique *O*-GlcNAc modified peptides were identified from a PSD preparation of mouse brain (Vosseller et al. 2006). Sixty five of these *O*-GlcNAc modified peptides were sequenced, and belonged to such functional classes as synapse assembly, synaptic vesicle trafficking, and post-synaptic signaling.

It is widely acknowledged that the efficient implementation of currently available and newly emerging bioinformatic tools has become critical for the extraction of physiological and biological principles from extremely complex proteomic data sets. This, in turn, can facilitate the elucidation of mechanisms and disease processes, and the development of effective therapeutics. Continuous improvement in proteomic technologies has facilitated a rapid expansion of identified synaptic components. However, the biological interpretation of these data has lagged behind. Systems Biology approaches (Ideker et al. 2001; Grant 2003; Jordan et al. 2006) that seek to integrate multiple levels of information (e.g. gene and protein networks) to understand how biological systems function, are needed to overcome this deficiency. Exactly such an approach has been recently applied for the analysis of NMDA receptor complex, using the protein-protein interaction network maps and annotated functions of individual components (Pocklington et al. 2006).

6. CONCLUSION

A significant goal of understanding synaptic molecular composition and function is to determine the pathogenic mechanisms through which neuropsychological disorders develop, and eventually to devise efficacious interventions, be they biological or pharmaceutical. To achieve this goal, the mapping and profiling of structural and functional synaptic protein components is merely a first step. Reliable and reproducible determination of *quantitative* or *qualitative* changes, *or both*, in these components is ultimately essential. With recent technological advances that have produced mass spectrometers with unprecedented sensitivity/resolution, LC apparatus with great flow stability and compositional accuracy, and auto-samplers with very high injection precision, our knowledge of identified synaptic components and their myriad modifications has exploded. This chapter has merely touched the surface of synaptic molecular complexity. In doing so, we hope it has illuminated the tremendous possibilities that lie ahead for neuroscientists when the powerful tools of cellular subfractionation and proteomics are skillfully combined.

REFERENCES

Akimoto, Y., Comer, F.I., Cole, R.N., Kudo, A., Kawakami, H., Hirano, H. and Hart, G.W. (2003) Localization of the O-GlcNAc transferase and O-GlcNAc-modified proteins in rat cerebellar cortex. Brain Res. *966*, 194–205.

Barry, M.F. and Ziff, E.B. (2002) Receptor trafficking and the plasticity of excitatory synapses. Curr. Opin. Neurobiol. *12*, 279–286.

Beranova-Giorgianni, S., Pabst, M.J., Russell, T.M., Giorgianni, F., Goldowitz, D. and Desiderio, D.M. (2002) Preliminary analysis of the mouse cerebellum proteome. Brain Res. Mol. Brain Res. *98*, 135–140.

Booth, R.F. and Clark, J.B. (1978) A rapid method for the preparation of relatively pure metabolically competent synaptosomes from rat brain. Biochem. J. *176*, 365–370.

Boyd-Kimball, D., Castegna, A., Sultana, R., Poon, H.F., Petroze, R., Lynn, B.C., Klein, J.B. and Butterfield, D.A. (2005) Proteomic identification of proteins oxidized by A[beta] (1–42) in synaptosomes: implications for Alzheimer's disease. Brain Res. *1044*, 206–215.

Cheng, D., Hoogenraad, C.C., Rush, J., Ramm, E., Schlager, M.A., Duong, D.M., Xu, P., Wijayawardana, S.R., Hanfelt, J., Nakagawa, T., Sheng, M. and Peng, J. (2006) Relative and absolute quantification of postsynaptic density proteome isolated from rat forebrain and cerebellum. Mol. Cell. Proteomics *5*, 1158–1170.

Cole, R.N. and Hart, G.W. (2001) Cytosolic O-glycosylation is abundant in nerve terminals. J. Neurochem. *79*, 1080–1089.

Collins, M.O. and Yu, L. et al. (2005a) Proteomic analysis of *in vivo* phosphorylated synaptic proteins. J. Biol. Chem. *280*, 5972–5982.

Collins, M.O. and Yu, L. et al. (2005b) Robust enrichment of phosphorylated species in complex mixtures by sequential protein and peptide metal-affinity chromatography and analysis by tandem mass spectrometry. Sci. STKE *2005*, l6.

Coughenour, H.D., Spaulding, R.S. and Thompson, C.M. (2004) The synaptic vesicle proteome: a comparative study in membrane protein identification. Proteomics *4*, 3141–3155.

DeGiorgis, J.A., Jaffe, H., Moreira, J.E., Carlotti, C.G., Leite, J.P., Pant, H.C. and Dosemeci, A. (2005) Phosphoproteomic Analysis of Synaptosomes from Human Cerebral Cortex. J. Proteome Res. *4*, 306–315.

Dosemeci, A., Tao-Cheng, J.H., Vinade, L. and Jaffe, H. (2006) Preparation of postsynaptic density fraction from hippocampal slices and proteomic analysis. Biochem. Biophys. Res. Commun. *339*, 687–694.

Edgar, P.F., Douglas, J.E., Knight, C., Cooper, G.J., Faull, R.L. and Kydd, R. (1999) Proteome map of the human hippocampus. Hippocampus *9*, 644–650.

Fountoulakis, M., Schuller, E., Hardmeier, R., Berndt, P. and Lubec, G. (1999) Rat brain proteins: two-dimensional protein database and variations in the expression level. Electrophoresis *20*, 3572–3579.

Gao, Y., Wells, L., Comer, F.I., Parker, G.J. and Hart, G.W. (2001) Dynamic O-glycosylation of nuclear and cytosolic proteins: cloning and characterization of a neutral, cytosolic beta-*N*-acetylglucosaminidase from human brain. J. Biol. Chem. *276*, 9838–9845.

Gauss, C., Kalkum, M., Lowe, M., Lehrach, H. and Klose, J. (1999) Analysis of the mouse proteome. (I) Brain proteins: separation by two-dimensional electrophoresis and identification by mass spectrometry and genetic variation. Electrophoresis *20*, 575–600.

Grant, S.G. (2003) Systems biology in neuroscience: bridging genes to cognition. Curr. Opin. Neurobiol. *13*, 577–582.

Grant, S.G. (2006) The synapse proteome and phosphoproteome: a new paradigm for synapse biology. Biochem. Soc. Trans. *34*, 59–63.

Gray, E.G. (1959) Axo-somatic and axo-dendritic synapses of the cerebral cortex: an electron microscope study. J. Anat. *93*, 420–433.

Gray, E.G. and Whittaker, V.P. (1962) The isolation of nerve endings from brain: an electron-microscopic study of cell fragments derived by homogenization and centrifugation. J. Anat. *96*, 79–88.

Gygi, S.P., Rist, B., Gerber, S.A., Turecek, F., Gelb, M.H. and Aebersold, R. (1999) Quantitative analysis of complex protein mixtures using isotope-coded affinity tags. Nat. Biotechnol. *17*, 994–999.

Han, W., Rhee, J.S., Maximov, A., Lao, Y., Mashimo, T., Rosenmund, C. and Sudhof, T.C. (2004) *N*-glycosylation is essential for vesicular targeting of synaptotagmin 1. Neuron *41*, 85–99.

Hansen, K.C., Schmitt-Ulms, G., Chalkley, R.J., Hirsch, J., Baldwin, M.A. and Burlingame, A.L. (2003) Mass spectrometric analysis of protein mixtures at low levels using cleavable 13C-isotope-coded affinity tag and multidimensional chromatography. Mol. Cell Proteomics *2*, 299–314.

Hansson, E. and Ronnback, L. (2003) Glial neuronal signaling in the central nervous system. FASEB J. *17*, 341–348.

Husi, H., Ward, M.A., Choudhary, J.S., Blackstock, W.P. and Grant, S.G. (2000) Proteomic analysis of NMDA receptor-adhesion protein signaling complexes. Nat. Neurosci. *3*, 661–669.

Ideker, T., Galitski, T. and Hood, L. (2001) A new approach to decoding life: systems biology. Annu. Rev. Genomics Hum. Genet. *2*, 343–372.

Iyer, S.P. and Hart, G.W. (2003) Dynamic nuclear and cytoplasmic glycosylation: enzymes of O-GlcNAc cycling. Biochemistry *42*, 2493–2499.

Jiang, C. and Schuman, E.M. (2002) Regulation and function of local protein synthesis in neuronal dendrites. Trends in Biochem. Sci. *27*, 506–513.

Johnson, S.A. and Hunter, T. (2005) Kinomics: methods for deciphering the kinome. Nat. Methods *2*, 17–25.

Jordan, B. and Ziff, E. (2006) Getting to synaptic complexes through systems biology. Genome Biol. *7*, 214.

Jordan, B.A., Fernholz, B.D., Boussac, M., Xu, C., Grigorean, G., Ziff, E.B. and Neubert, T.A. (2004) Identification and Verification of Novel Rodent Postsynaptic Density Proteins. Mol. Cell Proteomics *3*, 857–871.

Jordan, B.A., Fernholz, B.D., Neubert, T.A. and Ziff, E.B. (2006) New tricks for an old dog: proteomics of the PSD. In: The Dynamic Synapse: Molecular Methods in Ionotropic Receptor Biology. J.T. Kittler and S.J. Moss, eds. (Boca Raton: CRC/Taylor & Francis) pp. 37–55.

Kennedy, M.B. (1993) The postsynaptic density. Curr. Opin. Neurobiol. *3*, 732–737.

Khidekel, N., Ficarro, S.B., Peters, E.C. and Hsieh-Wilson, L.C. (2004) Exploring the O-GlcNAc proteome: direct identification of O-GlcNAc-modified proteins from the brain. Proc. Natl Acad. Sci. U.S.A. *101*, 13132–13137.

Kim, S.I., Voshol, H., van Oostrum, J., Hastings, T.G., Cascio, M. and Glucksman, M.J. (2004) Neuroproteomics: expression profiling of the brain's proteomes in health and disease. Neurochem. Res. *29*, 1317–1331.

Korolchuk, V. and Banting, G. (2003) Kinases in clathrin-mediated endocytosis. Biochem. Soc. Trans. *31*, 857–860.

Kramer, M.L. (2006) A new multiphasic buffer system for benzyldimethyl-*n*-hexadecylammonium chloride polyacrylamide gel electrophoresis of proteins providing efficient stacking. Electrophoresis *27*, 347–356.

Krapfenbauer, K., Fountoulakis, M. and Lubec, G. (2003) A rat brain protein expression map including cytosolic and enriched mitochondrial and microsomal fractions. Electrophoresis *24*, 1847–1870.

Langen, H., Berndt, P., Roder, D., Cairns, N., Lubec, G. and Fountoulakis, M. (1999) Two-dimensional map of human brain proteins. Electrophoresis *20*, 907–916.

Leng, A., Feldon, J. and Ferger, B. (2004) Long-term social isolation and medial prefrontal cortex: dopaminergic and cholinergic neurotransmission. Pharmacol. Biochem. Behav. *77*, 371–379.

Leski, M.L. and Steward, O. (1996) Protein synthesis within dendrites: ionic and neurotransmitter modulation of synthesis of particular polypeptides characterized by gel electrophoresis. Neurochem. Res. *21*, 681–690.

Li, K.W., Hornshaw, M.P., Van Der Schors, R.C., Watson, R., Tate, S., Casetta, B., Jimenez, C.R., Gouwenberg, Y., Gundelfinger, E.D., Smalla, K.H. and Smit, A.B. (2004) Proteomics analysis of rat brain postsynaptic density. Implications of the diverse protein functional groups for the integration of synaptic physiology. J. Biol. Chem. *279*, 987–1002.

Lisman, J., Schulman, H. and Cline, H. (2002) The molecular basis of CaMKII function in synaptic and behavioural memory. Nat. Rev. Neurosci. *3*, 175–190.

Morciano, M., Burre, J., Corvey, C., Karas, M., Zimmermann, H. and Volknandt, W. (2005) Immunoisolation of two synaptic vesicle pools from synaptosomes: a proteomics analysis. J. Neurochem. *95*, 1732–1745.

Mouledous, L., Hunt, S., Harcourt, R., Harry, J.L., Williams, K.L. and Gutstein, H.B. (2003) Proteomic analysis of immunostained, laser-capture microdissected brain samples. Electrophoresis *24*, 296–302.

Nagy, A. and Delgado-Escueta, A.V. (1984) Rapid preparation of synaptosomes from mammalian brain using nontoxic isoosmotic gradient material (Percoll). J. Neurochem. *43*, 1114–1123.

Nazarian, J., Bouri, K. and Hoffman, E.P. (2005) Intracellular expression profiling by laser capture microdissection: three novel components of the neuromuscular junction. Physiol. Genomics *21*, 70–80.

Palay, S.L. (1958) The morphology of synapses in the central nervous system. Exp. Cell Res. *14*, 275–293.

Peng, J., Kim, M.J., Cheng, D., Duong, D.M., Gygi, S.P. and Sheng, M. (2004) Semiquantitative proteomic analysis of rat forebrain postsynaptic density fractions by mass spectrometry. J. Biol. Chem. *279*, 21003–21011.

Phillips, G.R., Florens, L., Tanaka, H., Khaing, Z.Z., Fidler, L., Yates, Jr. 3rd and Colman, D.R. (2005) Proteomic comparison of two fractions derived from the transsynaptic scaffold. J. Neurosci. Res. *81*, 762–775.

Phizicky, E.M. and Fields, S. (1995) Protein-protein interactions: methods for detection and analysis. Microbiol. Rev. *59*, 94–123.

Pocklington, A.J., Armstrong, J.D. and Grant, S.G. (2006) Organization of brain complexity—synapse proteome form and function. Brief. Funct. Genomic. Proteomic. *5*, 66–73.

Prokai, L., Zharikova, A.D. and Stevens, S.M. Jr. (2005) Effect of chronic morphine exposure on the synaptic plasma-membrane subproteome of rats: a quantitative protein profiling study based on isotope-coded affinity tags and liquid chromatography/mass spectrometry. J. Mass Spectrom. *40*, 169–175.

Purcell, A.L. and Carew, T.J. (2003) Tyrosine kinases, synaptic plasticity and memory: insights from vertebrates and invertebrates. Trends Neurosci. *26*, 625–630.

Puryes, D. (2004) Neuroscience. Sunderland, Massachusetts: Sinauer Associates.

Raggiaschi, R., Lorenzetto, C., Diodato, E., Caricasole, A., Gotta, S. and Terstappen, G.C. (2006) Detection of phosphorylation patterns in rat cortical neurons by combining phosphatase treatment and DIGE technology. Proteomics *6*, 748–756.

Rao, A. and Steward, O. (1993) Evaluation of RNAs present in synaptodendrosomes: dendritic, glial, and neuronal cell body contribution. J. Neurochem. *61*, 835–844.

Routtenberg, A. and Rekart, J.L. (2005) Post-translational protein modification as the substrate for long-lasting memory. Trends Neurosci. *28*, 12–19.

Satoh, K., Takeuchi, M., Oda, Y., Deguchi-Tawarada, M., Sakamoto, Y., Matsubara, K., Nagasu, T. and Takai, Y. (2002) Identification of activity-regulated proteins in the postsynaptic density fraction. Genes Cells *7*, 187–197.

Schrimpf, S., Meskenaite, V., Brunner, E., Rutishauser, D., Walther, P., Eng, J., Aebersold, R. and Sonderegger, P. (2005) Proteomic analysis of synaptosomes using isotope-coded affinity tags and mass spectrometry. Proteomics *5*, 2531–2541.

Slawson, C., Housley, M.P. and Hart, G.W. (2006) O-GlcNAc cycling: how a single sugar post-translational modification is changing the way we think about signaling networks. J. Cell Biochem. *97*, 71–83.

Stevens, S.M. Jr., Zharikova, A.D. and Prokai, L. (2003) Proteomic analysis of the synaptic plasma membrane fraction isolated from rat forebrain. Brain Res. Mol. Brain Res. *117*, 116–128.

Steward, O. and Schuman, E.M. (2003) Compartmentalized synthesis and degradation of proteins in neurons. Neuron *40*, 347–359.

Trinidad, J.C., Specht, C.G., Thalhammer, A., Schoepfer, R. and Burlingame, A.L. (2006) Comprehensive Identification of Phosphorylation Sites in Postsynaptic Density Preparations. Mol. Cell Proteomics *5*, 914–922.

Unlu, M., Morgan, M.E. and Minden, J.S. (1997) Difference gel electrophoresis: a single gel method for detecting changes in protein extracts. Electrophoresis *18*, 2071–2077.

van Montfort, B.A., Canas, B., Duurkens, R., Godovac-Zimmermann, J. and Robillard, G.T. (2002) Improved in-gel approaches to generate peptide maps of integral membrane proteins with matrix-assisted laser desorption/ionization time-of-flight mass spectrometry. J. Mass Spectrom. *37*, 322–330

Villanueva, S. and Steward, O. (2001) Protein synthesis at the synapse: developmental changes, subcellular localization and regional distribution of polypeptides synthesized in isolated dendritic fragments. Brain Res. Mol. Brain Res. *91*, 148–153.

Vosseller, K., Trinidad, J.C., Chalkley, R.J., Specht, C.G., Thalhammer, A., Lynn, A.J., Snedecor, J.O., Guan, S., Medzihradszky, K.F., Maltby, D.A., Schoepfer, R. and Burlingame, A.L. (2006) O-Linked N-Acetylglucosamine proteomics of postsynaptic density preparations using lectin weak affinity chromatography and mass spectrometry. Mol. Cell Proteomics *5*, 923–934.

Walikonis, R.S., Jensen, O.N., Mann, M., Provance, D.W. Jr., Mercer, J.A. and Kennedy, M.B. (2000) Identification of proteins in the postsynaptic density fraction by mass spectrometry. J. Neurosci. *20*, 4069–4080.

Waltereit, R. and Weller, M. (2003) Signaling from cAMP/PKA to MAPK and synaptic plasticity. Mol. Neurobiol. *27*, 99–106.

Whittaker, V.P. and Michaelson, I.A. et al. (1964) The separation of synaptic vesicles from nerve-ending particles ("synaptosomes"). Biochem. J. *90*, 293–303.

Williams, R.W. and Herrup, K. (1988) The control of neuron number. Annu. Rev. Neurosci. *11*, 423–453.

Witzmann, F.A., Arnold, R.J., Bai, F., Hrncirova, P., Kimpel, M.W., Mechref, Y.S., McBride, W.J., Novotny, M.V., Pedrick, N.M., Ringham, H.N. and Simon, J.R. (2005) A proteomic survey of rat cerebral cortical synaptosomes. Proteomics *5*, 2177–2201.

Witzmann, F.A., Li, J., Strother, W.N., McBride, W.J., Hunter, L., Crabb, D.W., Lumeng, L. and Li, T.K. (2003) Innate differences in protein expression in the nucleus accumbens and hippocampus of inbred alcohol-preferring and -nonpreferring rats. Proteomics *3*, 1335–1344.

Wu, W.W., Wang, G., Baek, S.J. and Shen, R.F. (2006) Comparative study of three proteomic quantitative methods, DIGE, cICAT, and iTRAQ, using 2D gel- or LC-MALDI TOF/TOF. J. Proteome Res. *5*, 651–658.

Yoshimura, Y., Yamauchi, Y., Shinkawa, T., Taoka, M., Donai, H., Takahashi, N., Isobe, T. and Yamauchi, T. (2004) Molecular constituents of the postsynaptic density fraction revealed by proteomic analysis using multidimensional liquid chromatography-tandem mass spectrometry. J. Neurochem. *88*, 759–768.

Zoidl, G. and Dermietzel, R. (2002) On the search for the electrical synapse: a glimpse at the future. Cell Tissue Res. *310*, 137–142.

Corresponding author: Dr. Frank A. Witzmann (fwitzman@iupui.edu)

CHAPTER 7

PROTEOMIC ANALYSIS OF SECRETED EXOSOMES

CHRISTINE OLVER[1] and MICHEL VIDAL[2]
[1] *Colorado State University, USA*
[2] *University Montpellier II, France*

Table of Contents

E. Bertrand and M. Faupel (eds.), Subcellular Proteomics, 99–131.

Abstract: This chapter focuses on the contribution of proteomic analysis to the understanding of the process of exosome secretion and the mechanism and function of exosomes. It also describes the potential of exosome proteomic analysis to aid in the development of exosomes for therapeutic use.

1. INTRODUCTION

1.1. Defining Exosomes

Historically, the word "exosome," (beside designating the multienzyme ribonuclease complex identified in *S.cerevisiae*), has been used to define different types of vesicles released by cells (Figure 1). We feel it is critical to differentiate between the various kinds of vesicles since their mode of biogenesis could be directly related to their physiologic function and/or to the state of the productive cell.

The term "exosome" was first coined by Trams et al. (1981) to name exfoliated membrane vesicles collected from supernatants of normal or neoplastic cell lines. These surface-shed vesicles showing diameters from 40 to 1000 nm were found to contain plasma membrane 5′-nucleotidase activity. Such plasma membrane fragments possessed a high concentration of tumor antigens and were found to be shed at a higher rate by highly "metastatic" cells (Taylor et al. 1988). Since these vesicles, also called microvesicles, are formed from outward budding of the plasma membrane, we would like to suggest calling them "ectosomes," as proposed by Schifferli and coll. (Hess et al. 1999) when they isolated vesicles released by human polymorphonuclear neutrophils. By contrast, we would reserve the term exosomes, used in 1987 by R.M. Johnstone (Johnstone et al. 1987), to refer to vesicles released in the extracellular medium upon fusion of a multivesicular endosome (MVE) with the cell surface (Figure 1), as first demonstrated to occur during reticulocyte maturation (Harding et al. 1983; Pan and Johnstone 1983). However, as ectosomes and exosomes can have similar size (around 60–100 nm), regardless of the process of formation, it is difficult, if not impossible, to discriminate between the two processes without evidencing MVE-PM fusion event by electron microscopy. Finally, there appears to be molecular overlap between the two structures. It has been recently shown that components directly involved in multivesicular endosome formation are also associated with plasma membrane domains which give rise to vesicle budding from the cell surface (Taylor and Gercel-Taylor 2005; Booth et al. 2006).

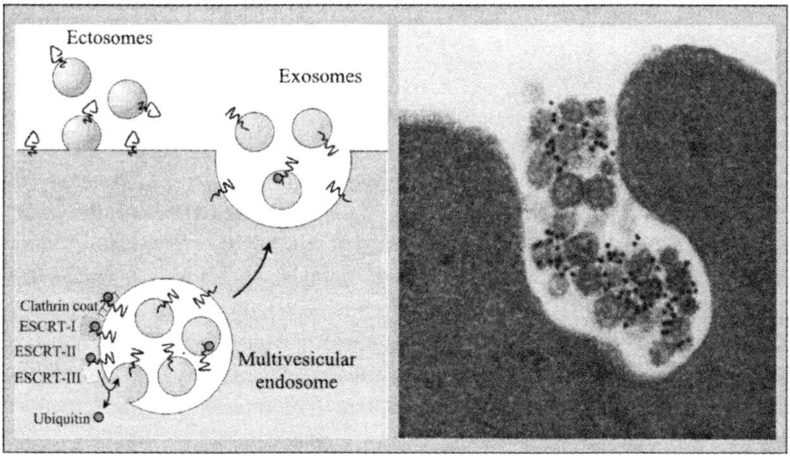

Figure 1. Exosome biogenesis. Diagram representing the budding and release of ectosomes from the plasma membrane in contrast to exosomes secreted by fusion of a multivesicular structure with the plasma membrane (left panel). Image of the fusion event leading to exosome secretion by RN cells (right panel). Colloidal gold particles label cholesterol. Bar = 200 nm. (Reproduced, with permission, from Möbius et al.: Immunoelectron Microscopic Localization of Cholesterol Using Biotinylated and Non-cytolytic Perfringolysin O. Journal of histochemistry & Cytochemistry, 50: 43–56, 2002)

1.2. A Sub-cellular Fraction Purified From the Extracellular Space

In contrast to the purification of other subcellular compartments (cf. other chapters of this volume), exosomes are isolated without cell homogenization. They are harvested by differential centrifugation from the extracellular space, thus avoiding the common problems of purification from a complex mixture and possible contamination by other organelles. However, one potential pitfall in their isolation is that fetal calf serum contains endogenous vesicles that can easily be co-purified with cell-secreted exosomes. More recently, body fluids such as plasma (Caby et al. 2005), epididymal fluid (Gatti et al. 2005), semen (Utleg et al. 2003) and urine (Pisitkun et al. 2004), or malignant effusions from cancer patients (Bard et al. 2004) were used to isolate vesicles for proteomic analysis. Although potentially very useful and instructive, the contribution of this approach should take into account the possible heterogeneity of the fraction, such as the purity of the collected vesicles (i.e. secretion by different cell types) or the process of vesicle release, as discussed in the previous section (ectosomes vs. exosomes).

1.3. Advances in Exosome Analysis Techniques: Proteomics

The proteome is technically the full complement of proteins expressed by an organism, tissue, cell or subcellular compartment at any given time and/or any given environmental condition (Aebersold and Mann 2003; Cristea et al. 2004). Proteomics, therefore, is the study and/or identification of this set or a subset of these proteins. The protein complement of the exosome would be considered a "subproteome." In reality, proteomics experiments are widely varied in their goals and techniques, and a full description of all techniques is beyond the scope of this review. Several recent reviews on the subject are available to the reader (Mann et al. 2001; Westermeier and Naven 2002; Aebersold and Mann 2003; Domon and Aebersold 2006). Suffice it to say that a proteomics experiment can be the identification of a single protein harvested from a polyacrylamide gel, or the comprehensive identification of the entire set of proteins in a whole cell lysate. The recent explosion in the field of proteomics has been made possible by the recent completion of the human genome, and the progress toward completion of the genomes of other species (http://www.ncbi.nlm.nih.gov/Genomes/), and recent advances in mass spectrometry. The study of exosomes has benefited from these techniques and the result is a long list of proteins contained within these vesicles.

1.4. Components of a Proteomics Experiment

Regardless of the specific techniques use for protein identification, a proteomics experiment can be divided into consistent steps (Domon and Aebersold 2006). The first step is the isolation of proteins from the material of choice (e.g. tissue, whole cell, subcellular organelle) and an initial separation (electrophoresis, for instance). Selected proteins are proteolytically digested (typically with trypsin) prior to identification. In some experiments, the proteins are digested before the initial separation step (liquid chromatography). The second step is mass spectrometric measurement of peptide masses. The third step involves matching measured peptide masses or fragment ions with putative amino acid sequences stored in protein or genomic databases. Appropriate statistical analyses are then used to ensure that the identifications are valid. The goals of proteomics experiments vary from the identification of one protein to a comprehensive identification of as many proteins as possible from a complex mixture of proteins. Proteomics is often comparative, where differences in protein expression in cell lysates or tissues, or in our case, exosomes derived from various cell types, in various conditions (e.g. diseased vs. normal) are measured.

2. PROTEOMIC ANALYSIS

2.1. Isolation of Vesicles

Exosomes are prepared from the supernatant of cells cultured using serum depleted of contaminating vesicles and protein aggregates by overnight ultracentrifugation. The general purification protocol of exosomes consists of successive centrifugations to eliminate cells and debris, followed by an ultracentrifugation to pellet the vesicles, as initially described (Pan and Johnstone 1984). The exosome pellet is then generally washed once in large volume of buffer and the exosomes can be further purified by sedimentation on sucrose gradient (Faure et al. 2006) and immunoisolation using derivatized Dynabeads (Wubbolts et al. 2003) before proteomic analysis. Similar protocols have been used for exosome isolation from body fluids.

2.2. Protein Separation

2.2.1. "Traditional" methods of separation

Complex mixtures of proteins are commonly separated using some form of polyacrylamide gel electrophoresis (PAGE). One dimensional (1D) PAGE separates proteins predominantly based on size, as incubation with the anionic detergent sodium dodecyl sulfate (SDS, also known as sodium lauryl sulfate) and a thiol reagent coats them with a uniform negative charge (independent of amino acid composition) (Simpson 2004). The gel is then stained with a protein stain (e.g. Coomassie Blue) so that distinct protein bands are detected. Discrete bands, or a series of segments of the gel, are then excised and digested within the gel with a proteolytic enzyme (usually trypsin) for application to a mass spectrometer. Two dimensional (2D) PAGE includes an initial separation of the protein mixture using isoelectric focusing. The second dimension separates proteins based on their mass, as in conventional 1D electrophoresis. Because there are two different and sequential modes of separation, 2DE has the potential to separate a large number of proteins into visible "spots." Immobilized pH gradient strips (Bjellqvist et al. 1993; Westermeier and Naven 2002; Simpson 2004) have made the first dimension more straightforward. These strips are prepared so that a series of stable buffers with pK values between 1 and >12 are immobilized on a plastic strip (Simpson 2004). After focusing, the strip is incubated with SDS, which denatures the proteins and coats them with a uniform negative charge. The strip is placed perpendicular to a polyacrylamide gel slab. The proteins are then visible as "spots" rather than "bands" after staining the gel with an appropriate protein stain. These spots can be excised, digested and subjected to mass spectrometry as described below. 2D PAGE has its limitations (Chevallet et al. 1998; Rabilloud et al. 1999; Low et al. 2002; Westermeier and Naven 2002). Membrane proteins are poorly soluble and are subject to interferences in 2DE because of the high quantity of lipids that they contain. Also, large or low abundant proteins, and those of a highly acidic or basic nature may be absent on a 2D gel. Other methods (1D PAGE, multidimensional chromatography) may be required to separate and identify these proteins. The studies to date that describe the proteome of exosomes have primarily used 1D separation (Table 1), thus many integral membrane proteins have been identified.

2.2.2. Liquid separation of peptides: High performance liquid chromatography

The peptide mixture that results from tryptic digestion of a protein mixture is usually fractionated prior to being submitted to mass spectrometric analysis. The goal of the separation is to reduce the complexity of the peptide mixture to simplify analysis of peptide mass peaks. The most commonly used method in exosome proteome studies is reverse phase high performance liquid chromatography (Fevrier et al. 2004; Pisitkun et al. 2004; Potolicchio et al. 2005; Segura et al. 2005; Faure et al. 2006). For this method, a mixture of peptides in aqueous solution is loaded onto a column packed with

Table 1. Methods used for exosome proteomic analysis

References	Separation		Mass analysis		Software	#
	1st	2nd	MS	MS/MS		
J. Immunol. 2001, 166:868	1D	NA	MALDI-TOF	NA	MS FIT	1
J. Cell Biol. 1999, 147:599	1D	NA	MALDI-TOF	NA	MS FIT	2
J. Immunol. 2001, 166:7309	1D	NA	MALDI-TOF	ESI-Q-TOF	MS FIT	2′
J. Biol.Chem. 2003, 278:10963	1D	NA	MALDI-TOF	ESI-Q-TOF	PPSS2	3
Gut 2003, 52:1690	1D	NA	MALDI-TOF	NA	MS FIT	4
Am. J. Pathol. 2004, 164:1807	1D	NA	MALDI-TOF	NA	Mascot	5
Proteomics 2004, 4:4019	2D	NA	MALDI-TOF	MALDI-TOF-TOF	Protein Prospector	6
Proc. Natl. Acad. Sci. 2004, 101:9683	1D	RP-HPLC		ESI-Q-TOF	Mascot	7
J. Immunol. 2005, 175:2237	1D	RP-HPLC		ESI-Q-IT	Unknown	8
Blood 2005, 106:216	1D	RP-HPLC		ESI-Q-TOF	Mascot	9
Mol. Cell. Neurosci. 2006, 31:342	1D	RP-HPLC		ESI-Q-TOF	Mascot	10
Proc. Natl. Acad. Sci. 2004, 101:13368	1D	RP-HPLC		ESI-LIT	Bioworks	11

The methods for each study are divided into the initial protein separation step, a second separation step if applicable, the type of mass analysis, and the software used for peptide identification. 1D = one dimensional polyacrylamide gel electrophoresis, 2D = two dimensional polyacrylamide gel electrophoresis, MS = mass spectrometry (peptide mass fingerprinting), MS/MS = tandem mass spectrometry, MALDI-TOF = matrix assisted laser desorption/ionization-time of flight, MS FIT = software from Protein Prospector, http://prospector.ucsf.edu/, ESI = electrospray ionization, Q-TOF = quadrupole-time of flight, PPSS2 = Protana's Proteomic Software Suite (ProtanaEngineering, Odense, Denmark), Mascot = Matrix Science, http://www.matrixscience.com/, TOF-TOF = MALDI plus TOF tandem mass spectrometry, RP-HPLC = reverse phase high performance liquid chromatography, Q-IT = quadrupole ion trap, LIT = linear ion trap, Bioworks = Thermo Electron Corporation.

a hydrophobic packing material. The proteins interact with the packing material based on their hydrophobic nature, and are eluted by gradually increasing the concentration of organic solvent.

2.3. MS Analysis

2.3.1. General overview of mass spectrometry

Mass analysis requires that peptides are ionized so that they travel according to their mass to charge ratio (*m*/z) in an electric or magnetic field (Westermeier and Naven 2002). Hence, mass spectrometers consist of an ionization source, a mass analyzer and detectors which detect the ions that have traveled through the analyzer. For most proteomics experiments, peptides are ionized either by electrospray (ESI) (Fenn et al. 1989) or by matrix assisted laser/desorption and ionization (MALDI) (Karas and Hillenkamp 1988). For ESI, an electric field at the sample exit causes a fine mist of droplets to form. Repulsion of like charges and evaporation of the droplets decreases their size while maintaining their charge. Eventually, nanometer-sized droplets, often multiply charged, are produced for entry into the mass analyzer. For MALDI, the peptides are embedded in an organic, light-absorbing matrix (e.g. α-cyano-4-hydroxy cinnamic acid or sinapinic acid) which, when activated by a laser, ionizes each peptide as it enters the gas phase. The majority of proteomics experiments involving exosomes used 1D-PAGE for initial separation of the protein mixture followed by band excision and MALDI-TOF mass spectrometry (Table 1).

2.3.2. Protein identification using mass spectrometry: MS and MS/MS

Mass spectrometry can either identify proteins by matching measured peptide masses (MS) with putative masses derived from a protein or genomic database (peptide mass fingerprinting), or can provide structural information (e.g. sequence) using tandem mass spectrometry (MS/MS). For peptide mass fingerprinting, the list of peptide masses obtained from a mass analyzer is compared with peptide masses derived from *in silico* digestion of proteins, or putative amino acid sequences, from established sequence databases (Apweiler et al. 2004). This involves the measurement of peptides produced from tryptic digestion of proteins by mass analysis of the ions produced by the ionization source. Tandem mass spectrometry requires that peptide ions of specific masses, selected in one mass analyzer, are fragmented in their peptide backbone (at variable but predictable locations) and the masses of the fragments are subsequently measured in a second mass analyzer. The spectrum of the measured fragmented ions is compared to a theoretical fragmentation spectrum derived from protein and genomic databases in order to identify the parent protein.

2.3.3. Types of mass analyzers

After ionization, the peptide mass is measured in a mass analyzer. All mass analyzers measure the mass of peptide ions by determining their travel behavior in an electric or magnetic field. Three types of mass analyzers have been used in the systematic identification of exosome proteins, the most common of which is the "time-of-flight"

(TOF) analyzer. Peptide ions enter this mass analyzer at a velocity that is proportional to their mass. The flight tube is "field-free" and thus the velocity of the ion remains constant until it reaches the detector. The time that it takes the ion to travel from the ionization source to the detector, through the flight tube, is the time of flight, and is proportional to the m/z of the ion. The second mass analyzer is the quadrupole. This consists of four metal rods which, together, generate an oscillating electric field. Only ions of a certain mass can pass through the rods to the detector at certain voltages (which can be controlled by the operator). Voltages are scanned to detect all of the ions in the sample. The third type of mass analyzer is the ion trap. This mass analyzer is similar to the quadrupole but also contains a ring electrode and two endcaps. Ions are trapped using a two-dimensional (linear ion trap) or a three-dimensional electric field. Ions of specific masses are ejected when the electric field is changed, and are then measured by the detector. All of these mass analyzers can be combined in sequence, separated by a collision cell, so that MS/MS can be performed.

2.4. "Shotgun" Methods

Shotgun proteomics refers to the global analysis of complex protein samples (e.g. whole cell lysates, tissue extracts or lysates of subcellular organelles) which are proteolytically digested, separated with (usually) multidimensional separation techniques and subjected to mass spectrometry analysis (Lin et al. 2003). Multidimensional separation techniques include sequential separations capitalizing on different properties of the peptides. For instance, strong cation exchange chromatography followed by reverse phase HPLC. Shotgun proteomics is performed on high throughput instruments such as ion trap mass spectrometers. The goal of shotgun proteomics is to increase proteomic coverage of a sample efficiently.

2.5. Search Engines, Databases and Statistical Analysis

The most commonly used search engines for correlating exosomal mass spectrometry data with protein and genomic databases are MS-FIT (www.prospector.ucsf.edu) and MASCOT (Matrix Science, www.matrixscience.com), although there are many more available on the internet. Both search engines allow for the analysis MS data by comparing experimental peptide masses to theoretical peptide masses of the proteins in the data base (peptide mass fingerprinting). MASCOT is also capable of analyzing MS/MS data by comparing experimental fragmentation ions with theoretical ones. A "score" is reported for each possible peptide match, regardless of the search engine. These scores reflect the probability that a peptide or protein match is a true positive. There are many protein databases in the public domain which can be searched by these engines. These include NCBI-nr (maintained by the National Center for Biotechnology Information, http://www.ncbi.nlm.nih.gov/entrez/query.fcgi?db=Protein) and Swiss-Prot (http://ca.expasy.org/sprot/).

3. INSIGHT INTO THE MECHANISM OF BIOGENESIS

Table 2 is a compilation of 11 proteomic studies of exosomes released by various cells in culture, compared with the data of the complete study done on exosomes collected from urine. Note that of the 22 proteins found in at least 4 of the 11 studies, 20 were also found in urine exosomes. These findings suggest that these proteins are very likely involved in the biogenesis or sorting of molecules in the vesicles. Other proteins are not systematically found in the various studies and might reflect variation between cell types, such as differential expression of cargo proteins.

3.1. Compartment of Formation

Classically, the endosomal compartment is described as the site of exosome biogenesis. Accordingly, mitochondria, ER/golgi or lysosome-associated proteins are very poorly represented in proteomic analyses of exosomes (Table 2). The endosomal origin of the secreted vesicles was suggested in the early 80s with the first observations of vesicle secretion by reticulocytes, in which the endocytosed transferrin receptor (TfR) was described to be rerouted from recycling to the plasma membrane towards cell expulsion via exosomal sorting (Harding et al. 1983). Early reports of exosome secretion by antigen presenting cells indicated that the location of formation was more likely a later pre-lysosomal compartment containing markers such as lamp-1 (Raposo et al. 1996), rather than the classical early endosome (TfR positive) compartment. This was strengthened by the fact that exosomes secreted by B lymphocytes were reported to be very poor with TfR (Raposo et al. 1996). Moreover, CD63, a marker of late endosomes, was often found associated with released vesicles using immunodetection (Escola et al. 1998; Heijnen et al. 1999; van Niel et al. 2001). CD63 has never been identified by proteomic analysis (Table 2), possibly because of poor resolution characteristics by SDS-PAGE (Escola et al. 1998).

The small rab GTPases are considered organelle markers since each rab protein regulates a distinct intracellular transport step (Zerial and McBride 2001). They are associated with the endocytic and secretion pathways and are recognized for their roles in both membrane transport and fusion. As shown in Table 2, various rab proteins were identified by proteomic analyses of exosomes secreted by different cell types. A variety of rab proteins involved in endocytic trafficking were found, suggesting the possibility that exosome biogenesis could occur from endosomal compartments differently located along the endocytic pathway. It is noteworthy that among them, rab proteins such as rab 4,5,10,11,14 are involved in intracellular vesicle translocation to, and fusion with, the plasma membrane (Ullrich et al. 1996; Larance et al. 2005; Ward et al. 2005). Similarly, the main exosomal cargo proteins, which are generally specific of the cell type referred (e.g. TfR, MHC class II molecules (MHCII) or aquaporin-2 (AQP2) associated with exosomes released by reticulocytes, secreted by APCs, or collected from urine, respectively) are proteins cycling between an intracellular compartment and the plasma membrane, potentially regulated by these GTPases (Ullrich et al. 1996; Barile et al. 2005). On the other

Table 2. Identified proteins secreted in exosomes and viruses

	1	2	3	4	5	6	7	8	9	10	U	Σ	V
Cytoplasm (endosomal trafficking)													
Alix		x				x		x	x	x	x	o	v
Annexin I		x		x	x		x		x			o	
Annexin II		x		x	x	x	x	x	x		x	o	v
Annexin III							x				x		
Annexin IV		x		x			x		x		x		
Annexin V				x	x	x	x		x		x	o	
Annexin VI	x				x	x	x		x	x	x	o	
Annexin VII		x							x		x		
Annexin XI							x		x		x		
AP-1 adaptor									x				
Arf 3,5,6							x				x		
Cdc42							x	x			x		
Clathrin heavy chain			x	x				x	x	x		o	v
Copine I									x				
Dynamin 2									x				
EH-domain containing protein 1				x							x		
H-Ras							x		x		x		
Polyubiquitin								x					
Programmed cell death 6							x				x		
Rab1A							x						
Rab2				x					x		x		
Rab4									x		x		
Rab5A or C							x		x		x		
Rab6									x		x		
Rab7		x		x			x		x		x	o	
Rab10							x		x		x		
Rab11		x					x		x		x		
Rab13,14									x		x		
Rab15									x				
RabGDI 3		x											
RabGDI α or β						x	x		x				
Ral B							x				x		
Ran						x	x						
Rap1A or B		x					x				x		
Rap2B							x		x				
Rho A									x		x		
Rho G								x	x		x		
Rho							x						
RhoGDI 1									x				
Small GTP-binding protein								x					
SNAP 23									x				
Syntaxin 7									x				
Syntaxin-binding protein 1,2 or 3						x			x				
Tsg101		x									x		v
Ubiquitin				x				x	x	x	x	o	v

Cytoplasm (soluble)													
14-3-3 ε						x	x	x	x		x		
14-3-3 ζ				x			x	x	x	x	x	o	
14-3-3 γ or δ		x		x			x		x	x	x	o	
14-3-3 β								x	x		x		
14-3-3 η		x											
2′3′cyclic-nucleot.3′phosphodiesterase					x								
Adenosylhomocysteinase							x						
Adenylyl cyclase associated protein								x	x				
Aldehyde reductase							x		x				
Aldolase A								x			x		
Aldolase reductase						x					x		
ATP citrate lyase						x			x		x		
Aspartate aminotransferase									x				
Carboxylesterase										x			
Cdc25	x												
Cyclophilin A,B or C							x	x	x	x	x	o	v
Cytosolic malate dehydrogenase						x					x		
Dipeptidylpeptidase 4									x				
Enolase 1			x	x	x	x	x	x			x	o	
eEF-1α1		x	x				x		x		x		
eEF-1αS								x					
eEF-2						x		x	x				
eIF-3ε								x					
eIF-4AI or II		x			x			x			x		
ERK2									x				
Farnesyl diphosphate synthase							x						
Fatty acid synthase									x				
Ferritin * (light or heavy chain)		x		x					x		x		
Formin-related protein (FRL)									x				
Fructose-bisphosphate aldolase			x										
GAPDH			x		x	x	x	x	x	x	x	o	v
Glucose 6 phosphate isomerase									x		x		
Glutathione S-transferase						x					x		
Grb 3-3									x				
hsp70		x	x	x	x	x	x	x	x	x	x	o	v
hsp90		x	x	x	x		x	x	x		x	o	v
Keratin 10				x		x							
L-plastin								x					
Lactate dehydrogenase						x	x				x		
Lipoprotein lipase								x					
M2 pyruvate kinase									x				
MVP (Major vault protein)									x				
NAP 22 (Neuronal tissue acidic protein)				x				x					
Peroxiredoxin							x		x		x		
Phosphogluconate deshydrogenase									x				
Phosphoglycerate kinase 1							x		x		x		
Phosphoglycerate mutase 1						x	x	x			x		

(*Continued*)

Table 2. (*Continued*)

	1	2	3	4	5	6	7	8	9	10	U	Σ	V
Phospholipase Cα									x				
Phosphoserine aminotransferase							x						
Proteasome 20S or 26S							x		x				
Protein kinase (cdc2-like)					x								
Protein phosphatase 1						x							
Pyruvate kinase			x	x	x		x	x		x	x	o	
Ribosomal protein S6 kinase	x												
Serine protease HTRA1 precursor				x									
Serine proteinase inhibitor							x						
SH2 phosphatase 1									x				
STAT-1									x				
TCP-1 (β,δ,ε,τ,ζ)									x				
TCTP									x				
Thioredoxine peroxidase II		x											
Transitional ER ATPase							x						
Transketolase									x				
Translin-assoc. factor X interact. prot.						x							
Triose phosphate isomerase						x	x						
Tyk 2									x				
Plasma membrane (integral)													
CD9 (Tetraspanin)		x		x				x	x		x	o	
CD11a (Integrin αL)									x				
CD11b (Integrin αM)		x						x	x				
CD11c (Integrin αX)									x				
CD13	x							x	x				
CD14 (GPI-protein)								x					
CD18 (Integrin β2)		x						x	x				
CD29 (Integrin β1)				x			x		x				v
CD45			x										v
CD49c (Integrin α3)				x	x		x						
CD49d (Integrin α4)			x										
CD51 (Integrin αV)							x						
CD71 (Transferrin receptor)									x				
CD81 (Tetraspanin)				x				x			x		v
CD81 partner-3 (Ig superfamily member)						x			x				
CD82 (Tetraspanin)				x									
CD98									x	x			
CD153						x							
CD166 (Ig superfamily member)									x				
CD39-like							x						
Cadherin 1 (Adhesion molecule)									x				
Chloride intracellular channel 1							x				x		
Chloride intracellular channel 4								x					
Claudin-1 (Adhesion molecule)									x				
Dectin 1,2β, 2γ (Lectin-like receptor)									x				
Fc receptor IgE								x					

Flotillin 1 (Raft assoc. protein)								x		x		
Glial glutamate transporter									x			
GP42/Basigin (Ig superfamily member)							x					
Integrin α7						x						
Invariant chain								x				
Monocarboxylate transporter							x					
MHC class I		x	x	x				x		x	o	v
MHC class II	x	x										v
Na/K ATPase		x					x	x				
Neuropilin 1 (Receptor)								x				
Nicastrin (Receptor)								x				
Pira7 (paired-Ig-like receptor A7)								x				
Plexin A1, C1								x				
Plasmalemma vesicle-assoc. prot. (Caveolae prot.)				x								
SHPS-1/ SIRP-α (Ig-like protein)								x				
Solute carrier family 3						x						
Stomatin (Raft assoc. protein)								x		x		
Tumor assoc. antigen L6 (Tetraspanin)			x									
Extracellular (secreted)												
α2 - macroglobulin *	x								x			
α - fetoprotein						x						
Albumin *	x			x	x	x				x	o	
Apolipoprotein A1	x				x							
Casein-κ *	x											
Coagulation factor X *	x											
Complement C3 *	x					x		x		x		
DEL-1 (Integrin binding protein)				x		x						
Fetuin A (α-2-HS-glycoprotein)					x							
Fibronectin precursor			x	x						x		
Galectin 3/MAC-2	x						x	x				
Hemoglobin *	x					x				x		
Insulin-like growth factor binding-5						x						
MFGE-8/lactadherin or mP47 (mouse homolog)	x		x			x	x				o	
Osteoblast specific factor-2						x						
Pigment epithelium-derived factor	x					x						
PK-120 * (not in databases…)	x											
Plasminogen *	x				x							
Tenascin precursor						x						
Thrombospondin						x		x		x		
Plasma membrane (peripheral)												
α1 - macroglobulin						x						
Advillin								x				
CAP1								x				
Catenins (Eα, β, δ1)					x			x				
c-src kinase								x				
Desmoplakin				x								

(*Continued*)

Table 2. (*Continued*)

	1	2	3	4	5	6	7	8	9	10	U	Σ	V
Ezrin (villin)					x	x			x		x		v
Fascin					x				x				
Fyn									x				
Gi protein α2 subunit		x	x				x		x		x	o	
Gi protein α3 subunit									x		x		
Gs protein α subunit							x	x	x		x		
G protein α subunit										x			
G protein β1 subunit						x	x				x		
G protein αq subunit							x				x		
G protein α11 subunit						x							
G protein α13 subunit							x						
Moesin			x	x	x	x	x	x	x		x	o	v
Phosphatidylethanolamine binding protein							x						
Radixin						x			x		x		
Talin									x				
Vinculin									x				
WD repeat containing protein						x			x				
Cytoplasm (cytoskeleton-motor)													
Actin	x	x	x	x	x	x		x	x		x	o	v
Actinβ, γ							x			x	x		
Actinin α4					x		x		x				
Arp 2/3							x		x		x		
Cofilin-1		x					x	x	x		x	o	v
Coronin									x				
Epithelial microtubule associated protein					x								
F-actin capping prot. β-subunit						x							
Gelsolin										x			
Kinesin-like protein 2					x								
IQ GAP1									x				
Myosin					x			x					
Myosin light chain							x						
Myosin heavy chain IX									x				
Profilin-1		x					x				x		
Septin/Nedd5						x							
Syntenin		x				x		x	x			o	
Tropomyosin								x					
Tubulin α or β		x	x		x		x	x	x	x	x	o	v
Vimentin							x		x				
Lysosome													
Lamp2							x						
Cathepsin B								x					
Plasma membrane (GPI-linked)													
Ceruloplasmin (GPI-protein)										x			
Nucleus													
Histones (H1,H2A,H2B,H3,H4)		x		x			x	x	x		x	o	
ER/Golgi													
FPRP 2 (Ig-like molecule)				x									
Mitochondria													
Prohibitin (Raft assoc. protein)									x				
Solute carrier 25 NT							x						

Viral proteins				
Gag polyprotein (MLV)	x			
Reverse transcriptase/pol (MLV)	x			
Others				
Putative p150		x		
Ribosomal proteins (L6, L18, S3, S18)			x	x

1-Exosomes from Mast cells	(J. Immunol. 2001, 166:868)
2-Exosomes from Dendritic cells	(J. Cell Biol. 1999, 147:599; J. Immunol. 2001, 166:7309)
3-Exosomes from B cells	(J. Biol. Chem. 2003, 278:10963)
4-Exosomes from Intestinal cells	(Gut 2003, 52:1690)
5-Exosomes from Mesothelial cells	(Am. J. Pathol. 2004, 164:1807)
6-Exosomes from Melanoma cells	(Proteomics 2004, 4:4019)
7-Exosomes from Neuroglial cells	(Proc. Natl. Acad. Sci. USA 2004, 101:9683)
8-Exosomes from Microglial cells	(J. Immunol. 2005, 175:2237)
9-Exosomes from Dendritic cells	(Blood 2005, 106:216)
10-Exosomes from Cortical neurones	(Mol. Cell Neurosci. 2006, 31:342)
U-Exosomes collected from Urine	(Proc. Natl. Acad. Sci. USA 2004, 101:13368)
V-Viral particles	(J. Virol. 2005, 79:6577)
Σ-Proteins identified at least 4 times among the 11studies.	

The proteins are categorized by subcellular origin. They are indicated in bold characters when identified in at least 4 studies on exosomes secreted by cultured cells (Σ), and underlined when also identified in viral particles (V).

Note: For exosomes collected from urine and viruses, only proteins also identified in cell-secreted exosomes (#1–10) are indicated. Asterisks indicate bovine proteins from foetal calf serum.

hand, vesicles budding directly from the cell surface are not enriched with these cycling molecules but instead can contain a large quantity of other specific proteins, as recently demonstrated with the presence of prominin-1, a surface marker of stem cells, in ectosomes released by neural progenitors and other epithelial cells (Marzesco et al. 2005). It was shown in the same work that such ectosomes could be purified from body fluids, in particular from urine. Accordingly, prominin-1 was identified in the proteomic analysis of the exosomes collected from urine (Pisitkun et al. 2004), suggesting that the "exosomal preparation" might also contain this kind of shed vesicles.

3.2. Biogenesis of Multivesicular Endosomes

Exosomes are released upon fusion of a multivesicular structure with the plasma membrane. Such multivesicular structures are generally described for endosomal carrier vesicles/late endosomes which allow sorting of transmembrane proteins for delivery into lysosomes for degradation. Proteins that are not sorted into the lumenal vesicles of late endosomes recycle back to the plasma membrane or remain on the limiting membrane, thereby escaping downregulation.

3.2.1. The ESCRT machinery

A signal for protein sorting into the lumenal vesicles has recently been discovered as conjugation of a monoubiquitin to the cytoplasmic domain of proteins

(Katzmann et al. 2002; Raiborg et al. 2003). The single ubiquitin moiety is recognized by a series of proteins that contain specific monoubiquitin-binding domains. These proteins are part of the so-called ESCRT (endosomal sorting complex required for transport) machinery organized as three protein complexes (ESCRT-I, II and III) that subsequently drive the ubiquitinated proteins to the budding site, that is, where the limiting membrane invaginates. Various proteins required for ESCRT recruitment to endosomal membranes have been found associated with exosomes by immunoblotting (Thery et al. 2001; Blanchard et al. 2002; De Gassart et al. 2003), suggesting that the same machinery could be involved in exosomal sorting. Several components related to this ESCRT sorting mechanism have been identified in the proteomic studies compiled here. Alix (ALG-2-interacting protein X), first characterized as a protein interacting with ALG-2, a calcium-binding protein necessary for cell death, is one of the proteins most often identified in the proteomic analyses (Table 2). Alix has been demonstrated to interact with Tsg101 (Tumor susceptibility gene 101) and CHMP4 (Charged multivesicular body protein 4), components of ESCRT-I and ESCRT-III complexes, respectively (Katoh et al. 2003; von Schwedler et al. 2003). Moreover, it has been shown that Bro1, the homologue of Alix in *S.cerevisiae*, functions at a late step of MVE sorting and is required for deubiquitination of cargo proteins before their enclosure within lumenal vesicles, by recruiting the ubiquitin thiolesterase Doa4 (Luhtala and Odorizzi 2004). As depicted in Figure 1, deubiquitination could be limited, leading to sorting of conjugated proteins in the lumenal vesicles. Accordingly, ubiquitinated proteins have been detected in exosomes by western blot (Buschow et al. 2005), and ubiquitin was identified in several exosome proteomic analyses as shown in Table 2. The importance of Alix and ubiquitin in the process of MVE formation is clear, and their detection in exosomes produced by various cell cultures and collected from urine point to involvement of ESCRT in the biogenesis of the secreted vesicles. In line with this, the presence of clathrin heavy chain in various proteomic analyses can be connected to the uncovering of flat clathrin coats on the surface of endosomes (Raposo et al. 2001), likely playing a role in retention of ubiquitinated membrane proteins before their inclusion in lumenal vesicles of MVE. These clathrin containing microdomains contain ubiquitin binding proteins associated in a conserved complex, the Hrs (Hepatocyte growth factor regulated tyrosine kinase substrate) complex, and acting in cooperation to sort ubiquitinated cargos upstream of ESCRT complexes (Raiborg et al. 2002; Sachse et al. 2002). Other proteins of the molecular machinery involved in MVE targeting and biogenesis have been identified in the extensive analysis carried out on urine exosomes: components of the ESCRT-I and III, and Vps4 (vacuolar protein sorting 4), an AAA-type ATPase that could have a role similar to NSF (NEM sensitive factor) involved in membrane fusion (Table 3). Tsg101, one of the ESCRT-I complex proteins discovered in urine exosomes and also identified in proteomic analysis of dendritic exosomes (Table 2), is considered as a marker of exosomes since its association with vesicles secreted by various cells has been demonstrated by immunoblotting (De Gassart et al. 2003; Fevrier et al. 2004; Segura et al. 2005; Faure et al. 2006). However, the picture might not be so clear since Tsg101 association with released vesicles has been found with cells known to produce

Table 3. Proteins involved in MVE sorting, identified in exosomes and viruses

Proteins	Exosomes	Viruses*
Ubiquitin	IE, Mgl, DC, CNeur, Ur	HIV-1, SIV, MLV, MMLV
Clathrin	BC, IE, Ngl, Mgl, DC	MLV
Tsg101	DC, Ur	HIV-1, MPMV
Vps28	Ur	HIV-1
CHMPs (1.5-2-2.5-4B-5)	Ur	
Alix	DC, Mel, Ngl, Mgl, Ur	HIV-1, SIV
Vps4B	Ur	HIV-1

Exosomes
IE: intestinal epithelial cells, Mgl: microglial cells, DC: dendritic cells, CNeur: cortical neurons, Ur: urine collected, BC: B cells, Ngl: neuroglial cells, Mel: melanoma cells.
Viruses
HIV-1: human immunodeficiency virus-1, SIV: simian immunodeficiency virus, MLV: murine leukaemia virus, MMLV: Moloney MLV, MPMV: Mason-Pfizer monkey virus.
(*data from Cantin et al. 2005).

vesicles principally through membrane shedding from the cell surface (Taylor and Gercel-Taylor 2005).

3.2.2. *Annexins and lipids*

As appears in Table 2, annexins 1, 2, 5 and 6 are very often identified in proteomic analyses of exosomes. Annexins comprise a family of calcium- and phospholipid-binding proteins, involved in membrane-related processes although their precise functions remain unclear. They have been implicated in membrane organization and traffic, especially due to their capacity to form scaffolds that might be involved in formation or stabilization of assembly sites. Annexins 1, 2 and 6 are present on the endosomal compartment and regulate various endocytic transport steps (Emans et al. 1993; Seeman et al. 1996; Grewal et al. 2000). Notably, annexins 1 and 2 have been implicated in multivesicular endosome biogenesis. Annexin 1 is a substrate for the epidermal growth factor receptor (EGFR) kinase, a property that was suggested to regulate formation of EGFR-containing multivesicular endosomes through a role in driving inward vesiculation (Futter et al. 1993). More recently, this role of annexin 1 has been confirmed, operating downstream of Hrs and the ESCRT proteins, and increasing the number of MVB intralumenal vesicles in EGF-stimulated cells (White et al. 2006). Annexin 2 can bind directly and specifically to phosphatidylinositol-4,5-bisphosphate (Hayes et al. 2004) and to cholesterol-stabilized domains (Zeuschner et al. 2001). A proposed role of annexin 2 is to organize cholesterol-rich platforms on the endosomal surface, which would be involved in detachment of nascent endosomal carrier vesicles/multivesicular bodies from early endosomes, without interfering with the inward vesiculation process (Mayran et al. 2003). In line with this, it is important to note that cholesterol-enriched domains have been isolated from exosomes secreted by reticulocytes (De Gassart et al. 2003) and B lymphocytes, (Wubbolts et al. 2003) due to their insolubility to non-ionic detergent such as Triton X-100 at 4°C, and were shown to be enriched with specific lipids (e.g. sphingomyelin and gangliosides) and proteins (e.g. flotillin). An important pool of cholesterol was found in the internal vesicles of MVE and released exosomes (Figure 1), by using a biotinylated derivative of the cholesterol-binding θ-toxin (perfringolysin O) in cryosections of cultured human lymphoblastoid cells (RN) by electron microscopy (Möbius et al. 2002). Flotillin has been found associated with exosomes released by various cells by western blot (De Gassart et al. 2003; Fevrier et al. 2004; Faure et al. 2006), which was confirmed by proteomic analysis, as shown in Table 2. Moreover, Wubbolts et al. have shown that lipid domains prepared from B lymphocyte exosomes contain tetraspanin proteins such as CD81 and CD63 (Wubbolts et al. 2003). Previously, several tetraspanins were found enriched in exosomes derived from various cells (Escola et al. 1998; Heijnen et al. 1999; Thery et al. 1999) but as we already mentioned, likely because of poor resolution of tetraspanins by SDS-PAGE, CD9 was the tetraspanin most often found in proteomic analyses (Table 2), and CD81 and CD82 were identified less commonly. It has been demonstrated that tetraspanin proteins can form large transmembrane protein networks by associating with each other as well as with other membrane proteins such as integrins and immunoglobulin superfamily proteins (Hemler 1998; Boucheix and Rubinstein 2001), and also with signaling enzymes such as protein kinase C (Zhang et al. 2001) or phosphatidylinositol-4 kinase (Carloni et al. 2004). Phosphatidylinositol-4 kinase has recently been demonstrated to be involved in the correct endocytic traffic and downregulation of activated epidermal growth factor receptor (Minogue et al. 2006). It is thus tempting to speculate that these cholesterol-enriched domains provide mechanistic frameworks important for the biogenesis of multivesicular endosomes, and that facilitation of their formation is provided by various cytosolic and/or membrane proteins. However, these lipid domains could also have a role in molecule sorting since MHC class II molecules (MHC II), demonstrated to be present in such tetraspanin-containing domains at the plasma membrane of antigen presenting cells (Hiltbold et al. 2003), have been found in lipid domains isolated from exosomes (De Gassart et al. 2003; Wubbolts et al. 2003). In the same line, some proteins anchored in the membrane through acyl chains, such as glycosylphosphatidylinositol (GPI)-anchored proteins, Gα subunits of heterotrimeric G proteins or tyrosine kinases of the src family, have a tendency to be incorporated in raft domains (Parolini et al. 1996; Harder et al. 1998) and are often found in exosomes (Rabesandratana et al. 1998; Clayton et al. 2003; De Gassart et al. 2003 and Table 2).

3.3. Virus Connection

The "Trojan exosome hypothesis" has recently been proposed with regard to the biogenesis of retroviruses (Gould et al. 2003). This hypothesis states that some retroviruses exploit the physiologic exosomal pathway for producing retroviral particles and taking advantage of exosome characteristics to increase their infectivity. We will first analyze the similarities in the biogenesis of viral and exosomal particles, and discuss later the possibility that some exosomal-derived determinants raise virus infectivity.

There are some obvious similarities between exosomes and virus particles. Both are single membrane particles of approximately 100 nm diameter, with the same topology, that is, formed by membrane budding away from the cytoplasm, released by cells in the extracellular environment and having the ability to travel and interact with other cells. There are molecular similarities, as well. Actually, it has been known for more than a decade that so-called viral "late" domains (L-domains) located in the Gag or matrix proteins are critical for virus egress from cells (Gottlinger et al. 1991; Huang et al. 1995). It has been demonstrated that these domains act to recruit the ESCRT cellular machinery necessary for viral budding and membrane fission (Freed 2002; Morita and Sundquist 2004). As shown in Table 3, several components of the ESCRT machinery and molecules involved in the process of MVE formation, identified in exosomes by mass spectroscopy, have also been found incorporated into different viruses. Moreover, recent studies have revealed that retroviruses can bud through internal cellular membrane (Raposo et al. 2002; Nydegger et al. 2003; Pelchen-Matthews et al. 2003), in addition to budding at the plasma membrane (Martin-Serrano et al. 2001). In HIV-infected macrophages, it has been observed that virus accumulates in MIIC compartments in which the final stages of MHC II maturation take place and where MHC II acquire antigenic peptides for presentation to T cells. The virus-containing compartments can then fuse directly with the cell surface in an exocytic fashion, releasing intracellular viral particles into the extracellular environment (Raposo et al. 2002). This strongly suggests that both kinds of particles make use of the same cellular machinery for their biogenesis. Interestingly, budding of equine infectious anemia virus (EIAV) has been demonstrated to be insensitive to proteasome inhibitors that decrease the cellular concentration of ubiquitin (Patnaik et al. 2002), likely because direct binding of Alix to a YPXL motif contained in the virus Gag protein could allow it to bypass the need for ubiquitination (Vincent et al. 2003). Retroviruses have thus evolved to adapt host cellular machinery for their replication, and we have shown that such direct binding of Alix to an exosomal cargo protein occurs in exosomes released by reticulocytes (Géminard et al. 2004).

Other cellular host proteins are known to be incorporated into enveloped viruses (Ott 2002; Cantin et al. 2005). As indicated in Table 2, several of these cellular proteins correspond to proteins often identified in exosomes. Interestingly, some of them have chaperone activities (e.g. hsc70, hsp90, cyclophilin A) which might suggest a role in maintaining proper protein conformation during particle biogenesis. However, although incorporation of cyclophilin A has been shown critical for HIV-1 infectivity (Yin et al. 1998), its role in assembly and disassembly of the HIV-1 life cycle has not yet been elucidated. Along the same lines, the role of hsc70 in exosome biogenesis is not known, even though this chaperone protein is considered as an exosomal marker because of its presence in almost all proteomic studies (Table 2). In reticulocyte exosomes, hsc70 binds to the cytosolic domain of TfR (Mathew et al. 1995), but this interaction does not trigger TfR sorting into exosomes (Géminard et al. 2001). Some cellular proteins found in both exosomal and viral particles are related to cytoskeleton (e.g. actin, tubulin, cofilin-1) or belong to the ERM (ezrin, radixin, moesin) family of cytoskeleton-membrane linkers. The spatial arrangement of endosomal compartments is coordinated by dynamic interactions with microtubules and actin filaments. It is well known that endosomes move bidirectionally along microtubules via minus-end (dynein) and plus-end (kinesin) directed motor proteins attached to these organelles (Hamm-Alvarez 1998). Actin polymerization/depolymerization dynamics are required for multiple distinct stages of clathrin-coated vesicle formation (Yarar et al. 2005). It was also shown that endosomes can nucleate actin assembly as comet tails, causing the vesicles to move through the cytoplasm *in vitro* (Taunton et al. 2000). More recently, Alix has been shown to regulate cellular distribution of endosomes by connecting cortical actin and these organelles (Cabezas et al. 2005). ERM proteins provide regulated linkage between membrane-associated proteins and the actin cytoskeleton (Mangeat et al. 1999). Thereby, they are involved in plasma membrane organization, having a role during the formation and stabilization of membrane ruffles. However, ERM proteins have also been found associated with endosomal compartments as a protein complex along with annexin 2, actin and α-actinin (Harder et al. 1997). Ezrin in association with an actin rich "shell" observed around clustered late endocytic organelles has been suggested to be involved in aggregation and fusion of late endosomes with lysosomes (Poupon et al. 2003). More recently, moesin has been demonstrated to be involved in translocation of AQP2 from intracellular vesicles to the cell surface (Tamma et al. 2005). Despite a very intriguing presence in exosomes and/or virions, a role of cytoskeleton proteins in the biogenesis of both particles still awaits to be demonstrated. Annexin 2 which as previously indicated is a major exosomal protein, has been recently found incorporated into HIV-1 particles due to an interaction with the Gag protein (Ryzhova et al. 2006). Moreover, since annexin 2 is not expressed in peripheral blood lymphocytes contrarily to macrophages, this led the authors to hypothesize that annexin 2 might be responsible for the differences in HIV-1 assembly sites observed between cell types (Ryzhova et al. 2006).

On the other hand, viral proteins have been identified in exosomes secreted by dendritic cells (Table 2) and might reflect the presence of endogenous retroviruses in the cell line used (Thery et al. 1999, 2001). In addition, we have expressed a chimeric protein containing the cytosolic domain of the transmembrane envelope protein of bovine leukaemia virus in K562 cells and found that the chimera was sorted into constitutively secreted exosomes (unpublished results).

All these data strongly suggest that some retroviruses utilize the same secretion pathway as exosomes, that is budding into a MVE followed by its fusion with the plasma membrane to reach the extracellular medium.

4. FUNCTIONAL ASPECTS

At this point in time, the physiologic function of exosomes is far from clear. Various roles have been suggested depending on the cell type producing the vesicles. As their mode of biogenesis likely involves ESCRT machinery, we anticipate that a function of exosomes is related to specifically sorted membrane proteins. As seen in Table 2, proteins with previously discussed functions in vesicle sorting and/or biogenesis are abundant. Interestingly, there is also a collection of proteins (more than 220) of widely variable functions, depending on cell type, that have been identified in exosome proteomic studies. The only cytosolic enzymes identified in more than four of the studies compiled here are involved in glycolysis (i.e. enolase, GAPDH and pyruvate kinase), cytosolic members of the 14-3-3 protein family, histones and extracellular proteins (i.e. albumin and lactadherin). Identification of these cytosolic proteins in exosomes by mass spectrometry is likely due to their omnipresence and substantial expression in all cell types. While albumin is probably a component introduced by culture medium, the presence of lactadherin on the surface of exosomes will be discussed later in this article.

As already mentioned, protein composition of exosomes is roughly dependent on the cell type which produces them. Consequently, the role of the secreted vesicles is often hypothesized as related to the physiologic function of the original cell type. Overall, the biological role of exosome production can be examined at the (i) cellular level, that is a down-regulation of cell components, and (ii) tissular level, that is ability to interact with bystanding cells.

4.1. Remodeling of the Membrane Surface

Generally, once secreted within exosomes, cargo proteins such as receptors (TfR, IgE-FcR), pumps (Na^+/K^+ ATPase) or channels (aquaporin-2, chloride intracellular channel) can not function normally unless they are reinserted in the cell membrane. Reticulocytes were the first cells described to release membrane proteins in association with exosomes (Harding et al. 1983; Pan and Johnstone 1983). It is known that cell remodeling occurs during the final stage of erythropoiesis, since various proteins are partially or completely lost during this maturation step (Figure 2). For instance, exosome release is concomitant with the complete loss of transferrin receptor from the surface of maturing reticulocytes, and a 50% loss of acetylcholinesterase (Johnstone et al. 1987). Exosome production is thus considered as a way to remodel the membrane of maturing red cells. This is further exemplified by the loss of integrin $\alpha 4\beta 1$ (Rieu et al. 2000) and CD55/CD59 (Rabesandratana et al. 1998) in secreted vesicles and concomitant decreased erythrocyte adhesion to fibronectin (Vuillet-Gaugler et al. 1990) and differential susceptibility to PNH (Paroxysmal nocturnal hemoglobinuria) (Ware et al. 1995), respectively. As previously indicated, heat shock proteins are often found in exosomes (Table 2). Although the role of the exosomal chaperones is not known, we do know that they are associated with unfolded proteins; proteins that might be destined for degradation. In line with this, the hsp70/hsp90-interacting protein CHIP (C-terminal of Hsp70-interacting protein) is an E3 ubiquitin ligase that promotes poly- (Jiang et al. 2001) and mono-ubiquination (Timsit et al. 2005) of proteins, playing an important function in chaperone-dependent protein quality control. Interestingly, the tumor antigen ErbB2 (or Her2/Neu), a member of the ErbB receptor tyrosine kinase family frequently found associated with exosomes derived

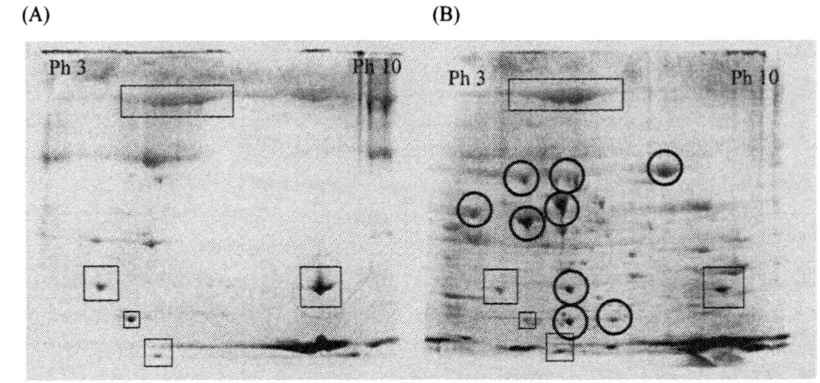

Figure 2. Loss of membrane proteins during reticulocyte maturation. Reticulocyte-poor (A) and reticulocyte-rich (B) erythrocyte membrane ghost proteins were subjected to 2DE and stained with Coomassie blue (0.5%). Selected spots identical between the protein samples are identified with black boxes. Several of the many spots that are unique to reticulocyte-rich erythrocytes are shown with the black circles. (Reproduced with permission from Prenni and Olver: Proteomics: A review and an illustration. Veterinary Clinical Pathology, in press.)

from tumours (Andre et al. 2002), has been demonstrated to be down-regulated through CHIP ubiquitination (Zhou et al. 2003).

Dendritic cells, depending on their degree of maturation, may or may not secrete the contents of the MVE. Rather than being released in the extracellular medium, intralumenal vesicles contained in MIIC compartment can fuse back with the limiting membrane of the multivesicular structure (Kleijmeer et al. 2001). The tetraspanins CD9 and CD81, which have been shown to be involved in cell fusion regulation (Rubinstein et al. 2006), could participate in this cellular event. Moreover, this process of back fusion is related to activation and maturation of dendritic cells. As a consequence, the intracellular "exosomal" pool of MHC class II molecules (>50% of the total pool in immature DC) rapidly redistributes to the plasma membrane (Kleijmeer et al. 2001). Thus in both cases, dendritic cells and red cells use "exosomes" to remodel their surface as a function of their differentiation requirements.

In the case of AQP2-containing exosomes collected from urine (Pisitkun et al. 2004), it is not known if the intralumenal vesicles of multivesicular structures are able to fuse back with the limiting membrane, but one can imagine that exosome secretion might be involved in AQP2 homeostasis in kidney. Very recently, cultured cortical neurons were shown to dispose of AMPA receptors through exosome secretion (Faure et al. 2006). Interestingly, the secretion process by the cortical neurons was enhanced by potassium-induced depolarisation and might thus be involved in receptor modulation in a physiologic context.

Various transmembrane proteins (e.g. TfR, CD44, CD171) secreted intact within exosomes have been demonstrated to be cleaved by proteases either associated with bystander cells (Johnstone 1996) or present on the exosome surface (Stoeck et al.

2006), leading to the release of soluble protein ectodomains (ectodomain shedding). Recently, the full-length tumor necrosis factor receptor (TNF-R) was found to be secreted within exosomes, whereas previously it was demonstrated to be shed in a soluble and truncated form at sites of chronic inflammation. This full-length TNF-R is thought to be an example of constitutive expression of soluble cytokine receptors (Hawari et al. 2004).

4.2. Intercellular Communication

Once released in the extracellular medium, exosomes are able to interact with cells located nearby or distantly, and thereby exosomes can be a way for carrying information from one cell to another one. Transfer of information to adjacent or distant cells could thus occur by three ways: (1) by direct contact between the exosomal and target plasma membranes; (2) by cell internalization of the vesicle; or (3) by fusion of the two membranes.

The presence of MHC I, MHC II and co-stimulatory molecules on the surface of exosomes secreted by antigen presenting cells suggests that exosomes have an immuno-modulatory role. Indeed, exosomes derived from both human and murine B lymphocytes induced antigen-specific MHC class II-restricted T cell responses *in vitro* (Raposo et al. 1996), reflecting a direct interaction between presentable peptide-MHC class II complex on the surface of exosomes and the T cell receptor. Moreover, tumor peptide-pulsed DC-derived exosomes were shown to suppress growth of established murine tumors when injected into mice, in a T cell-dependent manner (Zitvogel et al. 1998). However, the way exosomes induce antigen-specific immune responses is not yet elucidated. It seems likely that *in vivo* responses involve dendritic cells that would use MHC I- and II-peptide complexes present on exosomes to activate antigen-specific T cells (Vincent-Schneider et al. 2002; Andre et al. 2004; Segura et al. 2005). On the other hand, it has been shown that follicular dendritic cells (FDC) carry MHC II-expressing vesicles at their surface although not secreting exosomes themselves (Denzer et al. 2000). This has led to the suggestion that FDC-membrane attachment of exosomes produced by other cells (e.g. B lymphocytes) could allow FDCs to recruit specific T-helper cells.

Up to now, there is no experimental evidence to support exosome fusion with a cell membrane. However, exosomes contain a series of cell-specific transmembrane proteins, including α- and β-chains of integrins, tetraspanins, immunoglobulin-family members, adhesion molecules and receptors (Table 2), that could facilitate interaction with specific bystander cells. These determinants might also be involved in the kind of functional T-cell response (i.e. tolerance vs. priming) (Karlsson et al. 2001; Van Niel et al. 2003). For example, it has been demonstrated that vesicles secreted by immature or mature DCs differ regarding the expression of ICAM-1 (Intercellular adhesion molecule 1), and that ICAM-1 was necessary for the immunogenicity of the released exosomes. Interestingly, ICAM-1 is a major constituent of HIV-1 envelope which has been demonstrated to increase its infectivity (Paquette et al. 1998) and influence its entry route (Tardif and Tremblay 2003).

As another mechanism of exosome-plasma membrane interaction, soluble proteins bound on the surface of exosomes could favour their attraction to target cells. Lactadherin/MFGE-8 (Milk fat globule EGF-factor VIII), identified in exosomes secreted by various kind of cells (Table 2), has been shown to bind phosphatidylserine residues expressed on the surface of apoptotic cells and facilitate their *in vitro* phagocytosis by activated peritoneal macrophages (Hanayama et al. 2002). Lactadherin/MFGE-8 binds to integrins (αvβ3 and αvβ5) and its presence on the exosome surface could thus facilitate interaction of the vesicles with dendritic cells and macrophages expressing such integrins (Morelli et al. 2004).

In addition to the potent immunomodulatory functions of exosomes through antigen presentation, a role in induction of apoptosis has been reported for FasL-bearing vesicles (Andreola et al. 2002). It was suggested that these vesicles released by tumor cells could be involved in tumor evasion by inducing apoptosis of T lymphocytes, as observed in the peripheral circulation of patients with cancer (Kim et al. 2005). As previously indicated, several GPI-anchored proteins have been found to be associated with exosomes, likely brought about through their association with lipid rafts. In agreement with this, the GPI-anchored prion protein (PrP) is contained in exosomes released by cortical neurones (Faure et al. 2006). Interestingly, it has been shown that exosomes containing the abnormally folded prion protein (PrPsc) are infectious, potentially contributing to prion propagation (Fevrier et al. 2004). The molecular basis of exosome interaction with target cells leading PrPc transformation into the PrPsc form is not known. Various mechanisms could be anticipated, such as trans-interaction between the two membranes, but also cis-interaction after transfer of GPI-anchored PrPsc from the exosomal to the plasma membrane. As a matter of fact, in contrast to transmembrane proteins stably inserted into membranes, GPI-linked proteins associate via their acyl chains to the outer leaflet of membrane and as such, have been shown to transfer to target plasma membrane both *in vitro* (Medof et al. 1985) and *in vivo* (Kooyman et al. 1995).

5. THERAPEUTIC APPLICATIONS

5.1. Tumor Cells Release Exosomes

A very promising application of exosomes is their use in immunotherapy of cancer. For this, Zitvogel and coll. established that mouse tumor cell lines release exosomes containing tumor antigens in the extracellular medium, and that these antigens could be transferred to dendritic cells to induce potent T-cell-dependent antitumor effects on established mouse tumors (Wolfers et al. 2001). Very interestingly, exosomes collected from the supernatant of various tumor cell lines were shown to cross-protect against establishment of a specific tumor, reflecting the presence of shared tumor antigens within exosomes. Moreover, malignant effusions from patients were shown to accumulate high numbers of exosome-like vesicles bearing tumor antigens (Her2/Neu, Mart1/Melan-A, TRP, gp100) representing a natural and novel source of tumor rejection antigens (Andre et al. 2002). These antigen-containing exosomes are efficiently taken up and presented by autologous dendritic cells to cytotoxic T-lymphocytes *in vitro*. More recently, purification of exosomes and their proteomic analysis have been carried out from pleural effusions of patients with various type of cancer (Bard et al. 2004). Some proteins identified in this study (e.g. MHC I and II, cytoskeletal proteins, heat shock proteins) were reported in exosomes from other origins as we already mentioned. However, other proteins that might be related to malignancy were also found: B-cell translocation gene 1 (BTG1), a member of the anti-proliferative gene family that regulates cell growth and differentiation (Iwai et al. 2004); basement membrane-chondroitin sulphate proteoglycan (bamacan) protein shown to cause transformation when overexpressed (Ghiselli and Iozzo 2000), thrombospondin-2 involved in the regulation of cell proliferation, adhesion and migration of various cells (Streit et al. 1999), and pigment epithelium-derived factor (PEDF) promoting neuronal differentiation and survival, as well as inhibiting angiogenesis (Rege et al. 2005).

5.2. Biomarkers

Exosomes have been isolated from various biological fluids (Utleg et al. 2003; Bard et al. 2004; Pisitkun et al. 2004; Caby et al. 2005; Gatti et al. 2005) and thus might be used as a source of specific markers. The membrane vesicles are easily separated from abundant soluble proteins contained in biological fluids (e.g. albumin in plasma) favoring detection of potent exosomal biomarkers. In line with the paragraph above, identification of protein and peptide antigens on tumor cells lead to use them as biomarkers in disease detection and/or monitoring. For example, BTG-1 has been suggested as a biomarker for monitoring the remission of acute myeloid leukaemia (Cho et al. 2004). A longstanding example of a circulating biomarker is the soluble transferrin receptor in human serum which provides a valuable clinical index of tissue iron needs (Kohgo et al. 1986) and erythropoiesis (Huebers et al. 1990; Cooper et al. 2005). As erythroid cells possess most of the body's transferrin receptors, the plasma level of "soluble TfR" is a very good biomarker of erythroid diseases such as hyperplastic erythropoiesis (e.g. increased level in β-thalassemia) or hypoplastic erythropoiesis (e.g. decreased level in aplastic anemia) (Huebers et al. 1990).

More recently, urine has been used as a novel source for isolating exosomes (Pisitkun et al. 2004). The non-invasive way of fluid collection allows purification of exosomes in sufficient amounts for detecting potent biomarkers of renal diseases. Several proteins (e.g. aquaporin-2, polycystin-1, podocin) reported to be associated with some kidney diseases have been identified in exosomes (Pisitkun et al. 2004). Potentially, the level of these potential biomarkers in exosomes from patients could be compared with those of biomarkers in control exosomes.

6. CONCLUSION AND OUTLOOK

Improvements in the methods and strategies to identify membrane proteins provide better characterization of organelles in terms of functioning in the intracellular environment (Yates et al. 2005). Although clinical studies using exosomes are already underway (Escudier et al. 2005; Morse et al. 2005), the picture of their physiological role is still fragmented. Proteomic techniques should allow a complete description of the protein composition of exosomes, and thereby giving clues to the molecular basis of their biogenesis and a better understanding of their function. This might also provide new possibilities for tailoring vesicles to exhibit defined activities (Delcayre et al. 2005). Finally, their secretion in various fluids might provide the opportunity to detect specific diseases in the early development phase.

REFERENCES

Aebersold, R. and Mann, M. (2003) Mass spectrometry-based proteomics. Nature *422*, 198–207.

Andre, F., Chaput, N., Schartz, N.E., Flament, C., Aubert, N., Bernard, J., Lemonnier, F., Raposo, G., Escudier, B., Hsu, D.H., Tursz, T., Amigorena, S., Angevin, E. and Zitvogel, L. (2004) Exosomes as potent cell-free peptide-based vaccine. I. Dendritic cell-derived exosomes transfer functional MHC class I/peptide complexes to dendritic cells. J. Immunol. *172*, 2126–2136.

Andre, F., Schartz, N.E.C., Movassagh, M., Flament, C., Pautier, P., Morice, P., Pomel, C., Lhomme, C., Escudier, B., Le Chevalier, T., Tursz, T., Amigorena, S., Raposo, G., Angevin, E. and Zitvogel, L. (2002) Malignant effusions and immunogenic tumour-derived exosomes. Lancet *360*, 295–305.

Andreola, G., Rivoltini, L., Castelli, C., Huber, V., Perego, P., Deho, P., Squarcina, P., Accornero, P., Lozupone, F., Lugini, L., Stringaro, A., Molinari, A., Arancia, G., Gentile, M., Parmiani, G. and Fais, S. (2002) Induction of lymphocyte apoptosis by tumor cell secretion of FasL-bearing microvesicles. J. Exp. Med. *195*, 1303–1316.

Apweiler, R., Bairoch, A. and Wu, C.H. (2004) Protein sequence databases. Curr. Opin. Chem. Biol. *8*, 76–80.

Bard, M.P., Hegmans, J.P., Hemmes, A., Luider, T.M., Willemsen, R., Severijnen, L.A., van Meerbeeck, J.P., Burgers, S.A., Hoogsteden, H.C. and Lambrecht, B.N. (2004) Proteomic analysis of exosomes isolated from human malignant pleural effusions. Am. J. Respir. Cell Mol. Biol. *31*, 114–121.

Barile, M., Pisitkun, T., Yu, M.J., Chou, C.L., Verbalis, M.J., Shen, R.F. and Knepper, M.A. (2005) Large scale protein identification in intracellular aquaporin-2 vesicles from renal inner medullary collecting duct. Mol. Cell Proteomics *4*, 1095–1106.

Bjellqvist, B., Sanchez, J.C., Pasquali, C., Ravier, F., Paquet, N., Frutiger, S., Hughes, G.J. and Hochstrasser, D. (1993) Micropreparative two-dimensional electrophoresis allowing the separation of samples containing milligram amounts of proteins. Electrophoresis *14*, 1375–1378.

Blanchard, N., Lankar, D., Faure, F., Regnault, A., Dumont, C., Raposo, G. and Hivroz, C. (2002) TCR activation of human T cells induces the production of exosomes bearing the TCR/CD3/zeta complex. J. immunol. *168*, 3235–3241.

Booth, A.M., Fang, Y., Fallon, J.K., Yang, J.M., Hildreth, J.E. and Gould, S.J. (2006) Exosomes and HIV Gag bud from endosome-like domains of the T cell plasma membrane. J. Cell Biol. *172*, 923–935.

Buschow, S.I., Liefhebber, J.M., Wubbolts, R. and Stoorvogel, W. (2005) Exosomes contain ubiquitinated proteins. Blood Cells Mol. Dis. *35*, 398–403.

Cabezas, A., Bache, K.G., Brech, A. and Stenmark, H. (2005) Alix regulates cortical actin and the spatial distribution of endosomes. J. Cell Sci. *118*, 2625–2635.

Caby, M.P., Lankar, D., Vincendeau-Scherrer, C., Raposo, G. and Bonnerot, C. (2005) Exosomal-like vesicles are present in human blood plasma. Int. Immunol. *17*, 879–887.

Cantin, R., Methot, S. and Tremblay, M.J. (2005) Plunder and stowaways: incorporation of cellular proteins by enveloped viruses. J. Virol. *79*, 6577–6587.

Carloni, V., Mazzocca, A. and Ravichandran, K.S. (2004) Tetraspanin CD81 is linked to ERK/MAPKinase signaling by Shc in liver tumor cells. Oncogene *23*, 1566–1574.

Chevallet, M., Santoni, V., Poinas, A., Rouquie, D., Fuchs, A., Kieffer, S., Rossignol, M., Lunardi, J., Garin, J. and Rabilloud, T. (1998) New zwitterionic detergents improve the analysis of membrane proteins by two-dimensional electrophoresis. Electrophoresis *19*, 1901–1909.

Cho, J.W., Kim, J.J., Park, S.G., Lee do, H., Lee, S.C., Kim, H.J., Park, B.C. and Cho, S. (2004) Identification of B-cell translocation gene 1 as a biomarker for monitoring the remission of acute myeloid leukemia. Proteomics *4*, 3456–3463.

Clayton, A., Harris, C.L., Court, J., Mason, M.D. and Morgan, B.P. (2003) Antigen-presenting cell exosomes are protected from complement-mediated lysis by expression of CD55 and CD59. Eur. J. Immunol. *33*, 522–531.

Cooper, C., Sears, W. and Bienzle, D. (2005) Reticulocyte changes after experimental anemia and erythropoietin treatment of horses. J. Appl. Physiol. *99*, 915–921.

Cristea, I.M., Gaskell, S.J. and Whetton, A.D. (2004) Proteomics techniques and their application to hematology. Blood *103*, 3624–3634.

De Gassart, A., Géminard, C., Février, B., Raposo, G. and Vidal, M. (2003) Lipid raft-associated proteins sorting in exosomes. Blood *102*, 4336–4344.

Delcayre, A., Estelles, A., Sperinde, J., Roulon, T., Paz, P., Aguilar, B., Villanueva, J., Khine, S. and Le Pecq, J.B. (2005) Exosome Display technology: applications to the development of new diagnostics and therapeutics. Blood Cells Mol. Dis. *35*, 158–168.

Denzer, K., van Eijk, M., Kleijmeer, M.J., Jakobson, E., de Groot, C. and Geuze, H.J. (2000) Follicular dendritic cells carry MHC class II-expressing microvesicles at their surface. J. Immunol. *165*, 1259–1265.

Domon, B. and Aebersold, R. (2006) Mass spectrometry and protein analysis. Science *312*, 212–217.

Emans, N., Gorvel, J.P., Walter, C., Gerke, V., Kellner, R., Griffiths, G. and Gruenberg, J. (1993) Annexin II is a major component of fusogenic endosomal vesicles. J. Cell Biol. *120*, 1357–1369.

Escola, J.M., Kleijmeer, M.J., Stoorvogel, W., Griffith, J.M., Yoshie, O. and Geuze, H.J. (1998) Selective enrichment of tetraspan proteins on the internal vesicles of multivesicular endosomes and on exosomes secreted by human B-lymphocytes. J. Biol. Chem. *273*, 20121–20127.

Escudier, B., Dorval, T., Chaput, N., Andre, F., Caby, M.P., Novault, S., Flament, C., Leboulaire, C., Borg, C., Amigorena, S., Boccaccio, C., Bonnerot, C., Dhellin, O., Movassagh, M., Piperno, S., Robert, C., Serra, V., Valente, N., Le Pecq, J.B., Spatz, A., Lantz, O., Tursz, T., Angevin, E. and Zitvogel, L. (2005) Vaccination of metastatic melanoma patients with autologous dendritic cell (DC) derived-exosomes: results of the first phase I clinical trial. J. Transl. Med. *3*, 10.

Faure, J., Lachenal, G., Court, M., Hirrlinger, J., Chatellard-Causse, C., Blot, B., Grange, J., Schoehn, G., Goldberg, Y., Boyer, V., Kirchhoff, F., Raposo, G., Garin, J. and Sadoul, R. (2006) Exosomes are released by cultured cortical neurones. Mol. Cell Neurosci. *31*, 642–648.

Fenn, J.B., Mann, M., Meng, C.K., Wong, S.F. and Whitehouse, C.M. (1989) Electrospray ionization for mass spectrometry of large biomolecules. Science *246*, 64–71.

Fevrier, B., Vilette, D., Archer, F., Loew, D., Faigle, W., Vidal, M., Laude, H. and Raposo, G. (2004) Cells release prions in association with exosomes. Proc. Natl Acad. Sci. U.S.A. *101*, 9683–9688.

Freed, E.O. (2002) Viral late domains. J. Virol. *76*, 4679–4687.

Futter, C.E., Felder, S., Schlessinger, J., Ullrich, A. and Hopkins, C.R. (1993) Annexin I is phosphorylated in the multivesicular body during the processing of the epidermal growth factor receptor. J. Cell Biol. *120*, 77–83.

Gatti, J.L., Metayer, S., Belghazi, M., Dacheux, F. and Dacheux, J.L. (2005) Identification, proteomic profiling, and origin of ram epididymal fluid exosome-like vesicles. Biol. Reprod. *72*, 1452–1465.

Géminard, C., de Gassart, A., Blanc, L. and Vidal, M. (2004) Degradation of AP2 during reticulocyte maturation enhances binding of hsc70 and Alix to a common site on TfR for sorting into exosomes. Traffic *5*, 183–195.

Géminard, C., Nault, F., Johnstone, R. and Vidal, M. (2001) Characteristics of the interaction between hsc70 and the transferrin receptor in exosomes released during reticulocyte maturation. J. Biol. Chem. *276*, 9910–9916.

Ghiselli, G. and Iozzo, R.V. (2000) Overexpression of bamacan/SMC3 causes transformation. J. Biol. Chem. *275*, 20235–20238.

Gottlinger, H.G., Dorfman, T., Sodroski, J.G. and Haseltine, W.A. (1991) Effect of mutations affecting the p6 gag protein on human immunodeficiency virus particle release. Proc. Natl Acad. Sci. U.S.A. *88*, 3195–3199.

Gould, S.J., Booth, A.M. and Hildreth, J.E. (2003) The Trojan exosome hypothesis. Proc. Natl Acad. Sci. U.S.A. *100*, 10592–10597.

Grewal, T., Heeren, J., Mewawala, D., Schnitgerhans, T., Wendt, D., Salomon, G., Enrich, C., Beisiegel, U. and Jackle, S. (2000) Annexin VI stimulates endocytosis and is involved in the trafficking of low density lipoprotein to the prelysosomal compartment. J. Biol. Chem. *275*, 33806–33813.

Hamm-Alvarez, S.F. (1998) Molecular motors and their role in membrane traffic. Adv. Drug Deliv. Rev. *29*, 229–242.

Hanayama, R., Tanaka, M., Miwa, K., Shinohara, A., Iwamatsu, A. and Nagata, S. (2002) Identification of a factor that links apoptotic cells to phagocytes. Nature *417*, 182–187.

Harder, T., Kellner, R., Parton, R.G. and Gruenberg, J. (1997) Specific release of membrane-bound annexin II and cortical cytoskeletal elements by sequestration of membrane cholesterol. Mol. Biol. Cell *8*, 533–545.

Harder, T., Scheiffele, P., Verkade, P. and Simons, K. (1998) Lipid domain structure of the plasma membrane revealed by patching of membrane components. J. Cell Biol. *141*, 929–942.

Harding, C., Heuser, J. and Stahl, P. (1983) Receptor-mediated endocytosis of transferrin and recycling of the transferrin receptor in rat reticulocytes. J. Cell Biol. *97*, 329–339.

Hawari, F.I., Rouhani, F.N., Cui, X., Yu, Z.X., Buckley, C., Kaler, M. and Levine, S.J. (2004) Release of full-length 55-kDa TNF receptor 1 in exosome-like vesicles: a mechanism for generation of soluble cytokine receptors. Proc. Natl Acad. Sci. U.S.A. *101*, 1297–1302.

Hayes, M.J., Merrifield, C.J., Shao, D., Ayala-Sanmartin, J., Schorey, C.D., Levine, T.P., Proust, J., Curran, J., Bailly, M. and Moss, S.E. (2004) Annexin 2 binding to phosphatidylinositol 4,5-bisphosphate on endocytic vesicles is regulated by the stress response pathway. J. Biol. Chem. *279*, 14157–14164.

Heijnen, H.F., Schiel, A.E., Fijnheer, R., Geuze, H.J. and Sixma, J.J. (1999) Activated platelets release two types of membrane vesicles: microvesicles by surface shedding and exosomes derived from exocytosis of multivesicular bodies and alpha-granules. Blood *94*, 3791–3794.

Hemler, M.E. (1998) Integrin associated proteins. Curr. Opin. Cell Biol. *10*, 578–585.

Hess, C., Sadallah, S., Hefti, A., Landmann, R. and Schifferli, J.A. (1999) Ectosomes released by human neutrophils are specialized functional units. J. Immunol. *163*, 4564–4573.

Hiltbold, E.M., Poloso, N.J. and Roche, P.A. (2003) MHC class II-peptide complexes and APC lipid rafts accumulate at the immunological synapse. J. Immunol. *170*, 1329–1338.

Huang, M., Orenstein, J.M., Martin, M.A. and Freed, E.O. (1995) p6Gag is required for particle production from full-length human immunodeficiency virus type 1 molecular clones expressing protease. J. Virol. *69*, 6810–6818.

Huebers, H.A., Beguin, Y., Pootrakul, P., Einspahr, D. and Finch, C.A. (1990) Intact transferrin receptors in human plasma and their relation to erythropoiesis. Blood *75*, 102–107.

Iwai, K., Hirata, K., Ishida, T., Takeuchi, S., Hirase, T., Rikitake, Y., Kojima, Y., Inoue, N., Kawashima, S. and Yokoyama, M. (2004) An anti-proliferative gene BTG1 regulates angiogenesis in vitro. Biochem. Biophys. Res. Commun. *316*, 628–635.

Jiang, J., Ballinger, C.A., Wu, Y., Dai, Q., Cyr, D.M., Hohfeld, J. and Patterson, C. (2001) CHIP is a U-box-dependent E3 ubiquitin ligase: identification of Hsc70 as a target for ubiquitylation. J. Biol. Chem. *276*, 42938–42944.

Johnstone, R.M. (1996) Cleavage of the transferrin receptor by human granulocytes: preferential proteolysis of the exosome-bound TfR. J. Cell Physiol. *168*, 333–345.

Johnstone, R.M., Adam, M., Hammond, J.R., Orr, L. and Turbide, C. (1987) Vesicle formation during reticulocyte maturation. Association of plasma membrane activities with released vesicles (exosomes) J. Biol. Chem. *262*, 9412–9420.

Karas, M. and Hillenkamp, F. (1988) Laser desorption ionization of proteins with molecular masses exceeding 10,000 daltons. Anal. Chem. *60*, 2299–2301.

Karlsson, M., Lundin, S., Dahlgren, U., Kahu, H., Pettersson, I. and Telemo, E. (2001) "Tolerosomes" are produced by intestinal epithelial cells. Eur. J. Immunol. *31*, 2892–2900.

Katoh, K., Shibata, H., Suzuki, H., Nara, A., Ishidoh, K., Kominami, E., Yoshimori, T. and Maki, M. (2003) The ALG-2-interacting protein Alix associates with CHMP4b, a human homologue of yeast Snf7 that is involved in multivesicular body sorting. J. Biol. Chem. *278*, 39104–39113.

Katzmann, D.J., Odorizzi, G. and Emr, S.D. (2002) Receptor downregulation and multivesicular-body sorting. Nat. Rev. Mol. Cell Biol. *3*, 893–905.

Kim, J.W., Wieckowski, E., Taylor, D.D., Reichert, T.E., Watkins, S. and Whiteside, T.L. (2005) Fas ligand-positive membranous vesicles isolated from sera of patients with oral cancer induce apoptosis of activated T lymphocytes. Clin. Cancer Res. *11*, 1010–1020.

Kleijmeer, M., Ramm, G., Schuurhuis, D., Griffith, J., Rescigno, M., Ricciardi-Castagnoli, P., Rudensky, A.Y., Ossendorp, F., Melief, C.J., Stoorvogel, W. and Geuze, H.J. (2001) Reorganization of multivesicular bodies regulates MHC class II antigen presentation by dendritic cells. J. Cell Biol. *155*, 53–63.

Kohgo, Y., Nishisato, T., Kondo, H., Tsushima, N., Niitsu, Y. and Urushizaki, I. (1986) Circulating transferrin receptor in human serum. Br. J. Haematol. *64*, 277–281.

Kooyman, D.L., Byrne, G.W., McClellan, S., Nielsen, D., Tone, M., Waldmann, H., Coffman, T.M., McCurry, K.R., Platt, J.L. and Logan, J.S. (1995) *In vivo* transfer of GPI-linked complement restriction factors from erythrocytes to the endothelium. Science *269*, 89–92.

Larance, M., Ramm, G., Stockli, J., van Dam, E.M., Winata, S., Wasinger, V., Simpson, F., Graham, M., Junutula,. J.R., Guilhaus, M. and James, D.E. (2005) Characterization of the role of the Rab GTPase-activating protein AS160 in insulin-regulated GLUT4 trafficking. J. Biol. Chem. *280*, 37803–37813.

Lin, D., Tabb, D.L. and Yates, J.R. 3rd (2003) Large-scale protein identification using mass spectrometry. Biochim. Biophys. Acta. *1646*, 1–10.

Low, T.Y., Seow, T.K. and Chung, M.C. (2002) Separation of human erythrocyte membrane associated proteins with one-dimensional and two-dimensional gel electrophoresis followed by identification with matrix-assisted laser desorption/ionization-time of flight mass spectrometry. Proteomics *2*, 1229–1239.

Luhtala, N. and Odorizzi, G. (2004) Bro1 coordinates deubiquitination in the multivesicular body pathway by recruiting Doa4 to endosomes. J. Cell Biol. *166*, 717–729.

Mangeat, P., Roy, C. and Martin, M. (1999) ERM proteins in cell adhesion and membrane dynamics. Trends Cell Biol. *9*, 187–192.

Mann, M., Hendrickson, R.C. and Pandey, A. (2001) Analysis of proteins and proteomes by mass spectrometry. Annu. Rev. Biochem. *70*, 437–473.

Martin-Serrano, J., Zang, T. and Bieniasz, P.D. (2001) HIV-1 and Ebola virus encode small peptide motifs that recruit Tsg101 to sites of particle assembly to facilitate egress. Nat. Med. *7*, 1313–1319.

Marzesco, A.M., Janich, P., Wilsch-Brauninger, M., Dubreuil, V., Langenfeld, K., Corbeil, D. and Huttner, W.B. (2005) Release of extracellular membrane particles carrying the stem cell marker prominin-1 (CD133) from neural progenitors and other epithelial cells. J. Cell Sci. *118*, 2849–2858.

Mathew, A., Bell, A. and Johnstone, R.M. (1995) Hsp 70 is closely associated with the transferrin receptor in exosomes from maturing reticulocytes. Biochem. J. *308*, 823–830.

Mayran, N., Parton, R.G. and Gruenberg, J. (2003) Annexin II regulates multivesicular endosome biogenesis in the degradation pathway of animal cells. Embo. J. *22*, 3242–3253.

Medof, M.E., Kinoshita, T., Silber, R. and Nussenzweig, V. (1985) Amelioration of lytic abnormalities of paroxysmal nocturnal hemoglobinuria with decay-accelerating factor. Proc. Natl Acad. Sci. U.S.A. *82*, 2980–2984.

Minogue, S., Waugh, M.G., De Matteis, M.A., Stephens, D.J., Berditchevski, F. and Hsuan, J.J. (2006) Phosphatidylinositol 4-kinase is required for endosomal trafficking and degradation of the EGF receptor. J. Cell. Sci. *119*, 571–581.

Möbius, W., Ohno-Iwashita, Y., van Donselaar, E.G., Oorschot, V.M., Shimada, Y., Fujimoto, T., Heijnen, H.F., Geuze, H.J. and Slot, J.W. (2002) Immunoelectron microscopic localization of cholesterol using biotinylated and non-cytolytic perfringolysin O. J. Histochem. Cytochem. *50*, 43–55.

Morelli, A.E., Larregina, A.T., Shufesky, W.J., Sullivan, M.L., Stolz, D.B., Papworth, G.D., Zahorchak, A.F., Logar, A.J., Wang, Z., Watkins, S.C., Falo, L.D. Jr. and Thomson, A.W. (2004) Endocytosis, Intracellular Sorting and Processing of Exosomes by Dendritic Cells. Blood *104*, 3257–3266.

Morita, E. and Sundquist, W.I. (2004) Retrovirus budding. Annu. Rev. Cell. Dev. Biol. *20*, 395–425.

Morse, M.A., Garst, J., Osada, T., Khan, S., Hobeika, A., Clay, T.M., Valente, N., Shreeniwas, R., Sutton, M.A., Delcayre, A., Hsu, D.H., Le Pecq, J.B. and Lyerly, H.K. (2005) A phase I study of dexosome immunotherapy in patients with advanced non-small cell lung cancer. J. Transl. Med. *3*(9), 1–8.

Nydegger. S., Foti, M., Derdowski, A., Spearman, P. and Thali, M. (2003) HIV-1 egress is gated through late endosomal membranes. Traffic *4*, 902–910.

Ott, D.E. (2002) Potential roles of cellular proteins in HIV-1. Rev. Med. Virol. *12*, 359–374.

Pan, B.T. and Johnstone, R. (1984) Selective externalization of the transferrin receptor by sheep reticulocytes *in vitro*. Response to ligands and inhibitors of endocytosis. J. Biol. Chem. *259*, 9776–9782.

Pan, B.T. and Johnstone, R.M. (1983) Fate of the transferrin receptor during maturation of sheep reticulocytes *in vitro*: selective externalization of the receptor. Cell *33*, 967–977.

Paquette, J.S., Fortin, J.F., Blanchard, L. and Tremblay, M.J. (1998) Level of ICAM-1 surface expression on virus producer cells influences both the amount of virion-bound host ICAM-1 and human immunodeficiency virus type 1 infectivity. J. Virol. *72*, 9329–9336.

Parolini, I., Sargiacomo, M., Lisanti, M.P. and Peschle, C. (1996) Signal transduction and glycophosphatidylinositol-linked proteins (LYN, LCK, CD4, CD45, G proteins and CD55) selectively localize in triton-insoluble plasma membrane domains of human leukemic cell lines and normal granulocytes. Blood *87*, 3783–3794.

Patnaik, A., Chau, V., Li, F., Montelaro, R.C. and Wills, J.W. (2002) Budding of equine infectious anemia virus is insensitive to proteasome inhibitors. J. Virol. *76*, 2641–2647.

Pelchen-Matthews, A., Kramer, B. and Marsh, M. (2003) Infectious HIV-1 assembles in late endosomes in primary macrophages. J. Cell Biol. *162*, 443–455.

Pisitkun, T., Shen, R.F. and Knepper, M.A. (2004) Identification and proteomic profiling of exosomes in human urine. Proc. Natl Acad. Sci. U.S.A. *101*, 13368–13373.

Potolicchio, I., Carven, G.J., Xu, X., Stipp, C., Riese, R.J., Stern, L.J. and Santambrogio, L. (2005) Proteomic analysis of microglia-derived exosomes: metabolic role of the aminopeptidase CD13 in neuropeptide catabolism. J. Immunol. *175*, 2237–2243.

Poupon, V., Stewart, A., Gray, S.R., Piper, R.C. and Luzio, J.P. (2003) The role of mVps18p in clustering, fusion, and intracellular localization of late endocytic organelles. Mol. Biol. Cell *14*, 4015–4027.

Rabesandratana, H., Toutant, J.P., Reggio, H. and Vidal, M. (1998) Decay-accelerating factor (CD55) and membrane inhibitor of reactive lysis (CD59) are released within exosomes during *in vitro* maturation of reticulocyte. Blood *91*, 2573–2580.

Rabilloud, T., Blisnick, T., Heller, M., Luche, S., Aebersold, R., Lunardi, J. and Braun-Breton, C. (1999) Analysis of membrane proteins by two-dimensional electrophoresis: comparison of the proteins extracted from normal or Plasmodium falciparum-infected erythrocyte ghosts. Electrophoresis *20*, 3603–3610.

Raiborg, C,. Bache, K.G., Gillooly, D.J., Madshus, I.H., Stang, E. and Stenmark, H. (2002) Hrs sorts ubiquitinated proteins into clathrin-coated microdomains of early endosomes. Nat. Cell Biol. *4*, 394–398.

Raiborg, C., Rusten, T.E. and Stenmark, H. (2003) Protein sorting into multivesicular endosomes. Curr. Opin. Cell Biol. *15*, 446–455.

Raposo, G., Moore, M., Innes, D., Leijendekker, R., Leigh-Brown, A., Benaroch, P. and Geuze, H. (2002) Human macrophages accumulate HIV-1 particles in MHC II compartments. Traffic *3*, 718–729.

Raposo, G., Nijman, H.W., Stoorvogel, W., Leidendekker, R., Harding, C.V., Meleif, C.J.M. and Geuze, H. (1996) B lymphocytes secrete antigen-presenting vesicles. J. Exp. Med. *183*, 1161–1172.

Raposo, G., Tenza, D., Murphy, D.M., Berson, J.F. and Marks, M.S. (2001) Distinct protein sorting and localization to premelanosomes, melanosomes, and lysosomes in pigmented melanocytic cells. J. Cell Biol. *152*, 809–824.

Rege, T.A., Fears, C.Y. and Gladson, C.L. (2005) Endogenous inhibitors of angiogenesis in malignant gliomas: nature's antiangiogenic therapy. Neuro-oncol. *7*, 106–121.

Rieu, S., Géminard, C., Rabesandratana, H., Sainte-Marie, J. and Vidal, M. (2000) Exosomes released during reticulocyte maturation bind to fibronectin via integrin $\alpha 4\beta 1$. Eur. J. Biochem. *267*, 583–590.

Rubinstein, E. (2001) Tetraspanins. Cell Mol. Life Sci. *58*, 1189–1205.

Rubinstein, E., Ziyyat, A., Wolf, J.P., Le Naour, F. and Boucheix, C. (2006) The molecular players of sperm-egg fusion in mammals. Semin. Cell Dev. Biol. *17*, 254–263.

Ryzhova, E.V., Vos, R.M., Albright, A.V., Harrist, A.V., Harvey, T. and Gonzalez-Scarano, F. (2006) Annexin 2: a novel human immunodeficiency virus type 1 Gag binding protein involved in replication in monocyte-derived macrophages. J. Virol. *80*, 2694–2704.

Sachse, M., Urbe, S., Oorschot, V., Strous, G.J. and Klumperman, J. (2002) Bilayered clathrin coats on endosomal vacuoles are involved in protein sorting toward lysosomes. Mol. Biol. Cell *13*, 1313–1328.

Seeman, J., Weber, K., Osborn, M., Parton, R.G. and Gerke, V. (1996) The association of annexin I with early endosomes is regulated by Ca^{2+} and requires an intact N-terminal domain. Mol. Biol. Cell *7*, 1359–1374.

Segura, E., Nicco, C., Lombard, B., Veron, P., Raposo, G., Batteux, F., Amigorena, S. and Thery, C. (2005) ICAM-1 on exosomes from mature dendritic cells is critical for efficient naive T cell priming. Blood *106*, 216–223.

Simpson, R. (2004) Strategies for purifying proteins. In: Purifying Proteins for Proteomics: A Laboratory Manual. R. Simpson, ed. (Cold Spring Harbor, New York: Cold Spring Harbor Laboratory Press), pp. 17–40.

Stoeck, A., Keller, S., Riedle, S., Sanderson, M.P., Runz, S., Le Naour, F., Gutwein, P., Ludwig, A., Rubinstein, E. and Altevogt, P. (2006) A role for exosomes in the constitutive and stimulus-induced ectodomain cleavage of L1 and CD44. Biochem. J. *393*, 609–618.

Streit, M., Riccardi, L., Velasco, P., Brown, L.F., Hawighorst, T., Bornstein, P. and Detmar, M. (1999) Thrombospondin-2: a potent endogenous inhibitor of tumor growth and angiogenesis. Proc. Natl Acad. Sci. U.S.A. *96*, 14888–14893.

Tamma, G., Klussmann, E., Oehlke, J., Krause, E., Rosenthal, W., Svelto, M. and Valenti, G. (2005) Actin remodeling requires ERM function to facilitate AQP2 apical targeting. J. Cell Sci. *118*, 3623–3630.

Tardif, M.R. and Tremblay, M.J. (2003) Presence of host ICAM-1 in human immunodeficiency virus type 1 virions increases productive infection of CD4+ T lymphocytes by favoring cytosolic delivery of viral material. J. Virol. *77*, 12299–12309.

Taunton, J., Rowning, B.A., Coughlin, M.L., Wu, M., Moon, R.T., Mitchison, T.J. and Larabell, C.A. (2000) Actin-dependent propulsion of endosomes and lysosomes by recruitment of N-WASP. J. Cell Biol. *148*, 519–530.

Taylor, D.D. and Gercel-Taylor, C. (2005) Tumour-derived exosomes and their role in cancer-associated T-cell signalling defects. Br. J. Cancer *92*, 305–311.

Taylor, D.D., Taylor, C.G., Jiang, C.G. and Black, P.H. (1988) Characterization of plasma membrane shedding from murine melanoma cells. Int. J. Cancer *41*, 629–635.

Thery, C., Boussac, M., Veron, P., Ricciardi-Castagnoli, P., Raposo, G., Garin, J. and Amigorena, S. (2001) Proteomic analysis of dendritic cell-derived exosomes: a secreted subcellular compartment distinct from apoptotic vesicles. J. Immunol. *166*, 7309–7318.

Thery, C., Regnault, A., Garin, J., Wolfers, J., Zitvogel, L., Ricciardi-Castagnoli, P., Raposo, G. and Amigorena, S. (1999) Molecular characterization of dendritic cell-derived exosomes: selective accumulation of the heat shock protein hsc73. J. Cell Biol. *147*, 599–610.

Timsit, Y.E., Miller, S.L., Mohney, R.P. and O'Bryan, J.P. (2005) The U-box ligase carboxyl-terminus of Hsc 70-interacting protein ubiquitylates Epsin. Biochem. Biophys. Res. Commun. *328*, 550–559.

Trams, E.G., Lauter, C.J., Salem, N. Jr. and Heine, U. (1981) Exfoliation of membrane ecto-enzymes in the form of micro-vesicles. Biochim. Biophys. Acta *645*, 63–70.

Ullrich, O., Reinsch, S., Urbé, S., Zerial, M. and Parton, R.G. (1996) Rab11 regulates recycling through the pericentriolar recycling endosome. J. Cell Biol. *135*, 913–924.

Utleg, A.G., Yi, E.C., Xie, T., Shannon, P., White, J.T., Goodlett, D.R., Hood, L. and Lin, B. (2003) Proteomic analysis of human prostasomes. Prostate *56*, 150–161.

van Niel, G., Mallegol, J., Bevilacqua, C., Candalh, C., Brugiere, S., Tomaskovic-Crook, E., Heath, J.K., Cerf-Bensussan, N. and Heyman, M. (2003) Intestinal epithelial exosomes carry MHC class II/peptides able to inform the immune system in mice. Gut *52*, 1690–1697.

van Niel, G., Raposo, G., Candalh, C., Boussac, M., Hershberg, R., Cerf-Bensussan, N. and Heyman, M. (2001) Intestinal epithelial cells secrete exosome-like vesicles. Gastroenterology *121*, 337–349.

Vincent-Schneider, H., Stumptner-Cuvette, P., Lankar D, Pain S., Raposo, G., Benaroch, P. and Bonnerot, C. (2002) Exosomes bearing HLA-DR1 molecules need dendritic cells to efficiently stimulate specific T cells. Int. Immunol. *14*, 713–722.

Vincent, O., Rainbow, L., Tilburn, J., Arst, H.N. Jr. and Penalva, M.A. (2003) YPXL/I is a protein interaction motif recognized by aspergillus PalA and its human homologue, AIP1/Alix. Mol. Cell Biol. *23*, 1647–1655.

von Schwedler, U.K., Stuchell, M., Muller, B., Ward, D.M., Chung, H.Y., Morita, E., Wang, H.E., Davis, T., He, G.P., Cimbora, D.M., Scott, A., Krausslich, H.G., Kaplan, J., Morham, S.G. and Sundquist, W.I. (2003) The protein network of HIV budding. Cell *114*, 701–713.

Vuillet-Gaugler, M.H., Breton-Gorius, J., Vainchencker, W., Guichard, J., Leroy, C., Tchernia, G. and Coulombel, L. (1990) Loss of attachment to fibronectin with terminal human erythroid differentiation. Blood *76*, 865–873.

Ward, E.S., Martinez, C., Vaccaro, C., Zhou, J., Tang, Q. and Ober, R.J. (2005) From sorting endosomes to exocytosis: association of Rab4 and Rab11 GTPases with the Fc receptor, FcRn, during recycling. Mol. Biol. Cell *16*, 2028–2038.

Ware, R.E., Rosse, W.F. and Hall, S.E. (1995) Immunophenotypic analysis of reticulocytes in paroxysmal nocturnal hemoglobinuria. Blood *86*, 1586–1589.

Westermeier, R. and Naven, T. (2002) Proteomics in practice: a laboratory manual of proteome analysis. Weinheim: Wiley-VCH Verlag-GmbH.

White, I.J., Bailey, L.M., Aghakhani, M.R., Moss, S.E. and Futter, C.E. (2006) EGF stimulates annexin 1-dependent inward vesiculation in a multivesicular endosome subpopulation. EMBO J. *25*, 1–12.

Wolfers, J., Lozier, A., Raposo, G., Regnault, A., Thery, C., Masurier, C., Flament, C., Pouzieux, S., Faure, F., Tursz, T., Angevin, E., Amigorena, S. and Zitvogel, L. (2001) Tumor-derived exosomes are a source of shared tumor rejection antigens for CTL cross-priming. Nat. Med. *7*, 297–303.

Wubbolts, R.W., Leckie, R.S., Veenhuizen, P.T., Schwartzmann, G., Moebius, W., Hoernschemeyer, J., Slot, J.W., Geuze, H.J. and Stoorvogel, W. (2003) Proteomic and biochemical analyses of human B cell-derived exosomes: potential implications for their function and multivesicular body formation. J. Biol. Chem. *7*, 10963–10972.

Yarar, D., Waterman-Storer, C.M. and Schmid, S.L. (2005) A dynamic actin cytoskeleton functions at multiple stages of clathrin-mediated endocytosis. Mol. Biol. Cell *16*, 964–975.

Yates, J.R., 3rd, Gilchrist. A., Howell, K.E. and Bergeron, J.J. (2005) Proteomics of organelles and large cellular structures. Nat. Rev. Mol. Cell Biol. *6*, 702–714.

Yin, L., Braaten, D. and Luban, J. (1998) Human immunodeficiency virus type 1 replication is modulated by host cyclophilin A expression levels. J. Virol. *72*, 6430–6436.

Zerial, M. and McBride, H. (2001) Rab proteins as membrane organizers. Nat. Rev. Mol. Cell Biol. *2*, 107–117.

Zeuschner, D., Stoorvogel, W. and Gerke, V. (2001) Association of annexin 2 with recycling endosomes requires either calcium- or cholesterol-stabilized membrane domains. Eur. J. Cell Biol. *80*, 499–507.
Zhang, X.A., Bontrager, A.L. and Hemler, M.E. (2001) Transmembrane-4 superfamily proteins associate with activated protein kinase C (PKC) and link PKC to specific beta(1) integrins. J. Biol. Chem. *276*, 25005–25013.
Zhou, P., Fernandes, N., Dodge, I.L., Reddi, A.L., Rao, N., Safran, H., DiPetrillo, T.A., Wazer, D.E., Band, V. and Band, H. (2003) ErbB2 degradation mediated by the co-chaperone protein CHIP. J. Biol. Chem. *278*, 13829–13837.
Zitvogel, L., Regnault, A., Lozier, A., Wolfers, J., Flament, C., Tenza, D., Ricciardi-Castagnoli, P., Raposo, G. and Amigorena, S. (1998) Eradication of established murine tumors using a novel cell-free vaccine: dendritic cell-derived exosomes. Nat. Med. *4*, 594–600.

Corresponding authors: Dr. Christine Olver (christine.olver@colostate.edu) and Dr. Michel Vidal (mvidal@univ-montp2.fr)

SECTION 3

CHARACTERIZATION OF SUPRAMOLECULAR PROTEIN COMPLEXES

CHAPTER 8

FROM PROTEIN–PROTEIN COMPLEXES TO INTERACTOMICS

VINCENT COLLURA AND GUILLAUME BOISSY
Hybrigenics S.A., France

Table of Contents

E. Bertrand and M. Faupel (eds.), Subcellular Proteomics, 135–183.

Abstract: Protein–protein interactions (or PPIs) are key elements for the normal functioning of a living cell. A large description of the protein interactomics field is given in this review where different aspects will be discussed. We first give an introduction of the different large scale experimental approaches from yeast two-hybrid to mass spectrometry used to discover PPIs and build protein interaction maps. Single PPI validation techniques such as co-immunoprecipitation or fluorescence methods are then presented as they are more and more integrated in global PPI discovery strategy. Data from different experimental sets are compared and an assessment of the different large scale technologies is presented. Bioinformatics tools can also predict with a good accuracy PPIs *in silico*, PPIs databases are now numerous and topological analysis has led to interesting insights into the nature of network connection. Finally, PPI, as an association of two proteins, has been structurally characterized for many protein complexes and is largely discussed throughout existing examples. The results obtained so far already provide the biologist with a large set of structured data from which knowledge on pathways and associated protein function can be extracted.

1. INTRODUCTION

1.1. Interactomics

Interactomics is the study of physical molecular interactions at the cell or organ level and may thus be considered in its protein–protein interaction version as a part of the functional proteomics. Compared to genomics (genes identification and regulation), interactomics is in its infancy. Indeed, genomics has provided the scientific community with almost 250 completed genome-sequencing projects scanning a wide range of eukaryotes and prokaryotes species. To date, only fractions of interactomes have been collected and are far from completed for the examined species including *Homo sapiens* and a whole set of model organisms (*Saccharomyces cerevisiae* yeast, *Drosophila* fly, *Caenorhabditis elegans* worm, *Helicobacter pylori* and *Escherichia coli* bacteria, T7 bacteriophage and HIV virus; see Chapter 2 for detailed listing) with, remarkably, no plant representatives. One of the interactomics challenges comes from the difficulty to deal simultaneously with several space and time dimensions. First, disparate interactions occur depending on organism, organ or tissue. They may also display specific or altered pattern depending on whether they occur in differentiated or undifferentiated tissues and in healthy or diseased cells. Second, many of them are transient, likely to be correlated with regulation purposes, and adopt sub-location preferences with respect to their function.

Reaching the same degree of comprehensiveness encountered for genomics or proteomics could sound quite unrealistic but this would not account for the huge amount of on going and planed collaborative international efforts dedicated to this task. Regarding time lines, the next decade should be crucial. However, past achievements support some optimism: who would have bet 10 years ago that the complete Human genome would have been released in 2001 and that present sequencing devices allow to perform the same task in a few weeks only?

1.2. Interaction Networks as Part of Cell Pathways

Protein–protein interactions produced experimentally can be consolidated into larger functional units focused on a specific cellular role. These units are distributed in several categories like metabolic pathways, for example the one that governs galactose metabolism. In these very loose pathways in terms of protein networking, proteins are enzymes joined by small molecules. These networks often involve low-level assembled proteins (more often homo-oligomeric) and have been widely studied last decades as part of enzymology. Signaling transduction pathways are another branch where an events cascade occurs to transfer a signal triggered by the binding of a ligand to the cytoplasmic membrane target up to cytoplasmic or nuclear location eliciting a precise and controlled response. Interactions are often transient, that is assembly undergoes changes in oligomeric state involving subunits that can assemble and disassemble consecutively to environment changes, allosteric mechanism or effector binding. We can thus mention the kinase post-translational modifications which facilitate interactions between targeted molecule and other proteins as illustrated in Figure 1A by the human signalling pathway (Smad system) regulated by members of the transforming growth factor β (TGFβ). Moreover, scaffold proteins may facilitate signal transduction by coupling molecules when, used as a backbone for multiprotein assembly, they recruit specific molecules simultaneously through a unique domain (Vondriska et al. 2004). Remy and Michnick (2001) have demonstrated that such pathway can be extensively screened to concomitantly reveal new interaction partners, their pharmacological profiling and cellular location. Lastly, genes regulation networks (DNA transcription, replication) complete the family of protein interaction networks by providing many analyzed patterns downstream of signalling pathways. As a matter of fact, transcription machinery is orchestrated by a plethora of transcription factors, activators and suppressors whose complexity in terms of components interaction is still to be understood.

The analysis of protein networks in signalling pathways will allow us to progress towards a better understanding on how a perturbation consecutive to genetic or environmental causes may induce pathological phenotypes and may assist to find curative solutions targeted against an individual component (Pellegrini et al. 2004).

1.3. Diversity of Protein–Protein Complexes

Compared to other protein assemblies, supramolecular complexes keep similar features excepted that the higher number of implicated components often requires specifically designed strategies to reduce the complexity level before reincorporating all the accumulated hybrid data issued from smaller experiments. The division of complexes into smallest workable units prior to be convoluted as a more understandable entity is well suited for overall structure determination as discussed by Sali et al. (2003). Several examples of supramolecular complexes are found in cellular trafficking linked to Golgi membrane transport (Oka et al. 2005) or the exocyst involved in secretory vesicles trafficking to plasma membrane as illustrated in Figure 1B.

A good starting point to get a picture of complexes resides in one quote from former authors (Sali et al. 2003): "Just as words must be assembled into sentences, paragraphs, chapters and books to make sense, vital cellular functions are performed by structured ensembles of proteins (i.e. complexes)." Bestiary of complexes shows so much diversity that it would be an illusion to try to itemize it: some may be regarded as molecular machines (Alberts 1998), others participate in various regulatory processes as transduction signal networks. Nooren & Thornton (2003a) have classified complexes into obligate (i.e. protomers do not form stable structure on their own *in vivo*) or non-obligate (e.g. receptor-ligand or enzyme-inhibitor for which pairing partners are initially not co-localized requiring therefore to maintain stability as single). Oligomeric state and macromolecular complex lifetime, depend on the control of multifactorial parameters lying with both subunits concentration (e.g. co-localization in time and space, local concentrations that require special consideration for anchored transmembrane proteins) and binding affinity of the complex (e.g. effectors like Mn^{2+}

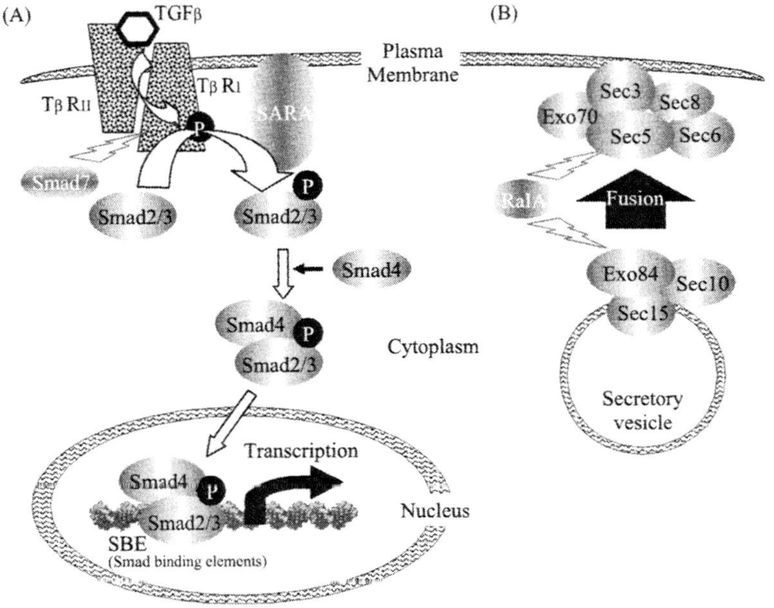

Figure 1. (A) The TGFβ signaling pathway: the Smad signaling system is regulated by members of the transforming growth factor β (TGFβ) that trigger a signal transduction through cell compartments, from extracellular location to nucleus, mediated by a cascading network of transient interactions and complexes activation (e.g. phosphorylation). Upon ligand binding, type II serine/threonine kinase receptor (TβRII) phosphorylates and activates type I receptor (TβRI) leading to the phosphorylation of receptor-activated Smad2 and Smad3 (R-Smads), localized beforehand to the membrane by SARA. They translocate in the nucleus with their common partner Smad4 where they recruit other transcription factors to regulate target gene expression. All along this signal transduction pathway, effectors can regulate downstream biological effects for example Smad7 or LAPTm5 (Colland et al. 2004). (B) Example of a supramolecular assembly: the exocyst is a 734 kDa multisubunit molecular machine that mediates the targeting and tethering of post-Golgi secretory vesicles for subsequent membrane fusion. This octomeric complex is required for polarized exocytosis encountered in eukaryotic cells that is crucial for cell-growth, intercellular communication and cell polarity establishment. Exocytosis therefore mediates secretory events (e.g. hormones, neurotransmitters) and targets the incorporation of membrane components (e.g. proteins, lipids) (Hsu et al. 2004). Exocyst is a well-conserved dynamic complex comprising eight proteins: Sec3, Sec5, Sec6, Sec8, Sec10, Sec15, Exo70 and Exo84 which the relative positions have to be further investigated. It is likewise assembled from subcomplexes located both on vesicles and plasma membrane (Camonis and White 2005). Ral GTPases promote this assembly through dual subunit interactions: Sec5 and Exo84 have therefore been identified as RalA/B effectors. Other small G-proteins can associate with exocyst subunits thus participating to its dynamic regulation.

or GTP, surrounding ions and macromolecules, pH, temperature, covalent modifications). Following the same path, the Thornton group has dissected several features characteristic of transient interactions (Nooren and Thornton 2003b). Allosteric conformational changes (Goh et al. 2004) can also be promoted by large assemblies providing cooperative crosstalk between domains and components to fine-regulate or optimize cellular mechanisms. Because self-association of proteins often ends in dimers and higher oligomeric structures, they have been studied at length providing the biologist with many disparate and original discovered functions (Marianayagam et al. 2004). Thus, geometry features observed on the four core histones (H2A-H2B)$_2$ and H3-H4 dimers located in the nucleosome explain the affinity for binding double-strand DNA. However, does the ability to determine whether proteins form a dimer or not still constitute a delicate technical issue? The G Protein-Coupled Receptors (GPCRs) examples attest that it remains quite questionable, although, in this special case, it accounts for the important heterogeneity of function and behaviour encountered across this family (Milligan 2004).

Oligomeric states of similarly folded subunits can be very disparate in architecture through evolution and have implications in terms of functions as it has been well exemplified with the hemoglobin family (Royer et al. 2005). The gain of modularity expected by switching from a single domain to multidomains proteins and then to supramolecular assemblies, the fact that components of stable complexes are more conserved than transient ones as well as the fact that essential proteins tend to be subunit of complexes have been discussed from the point of view of evolution (Pereira–Leal et al. 2006; Bornberg-Bauer et al. 2005).

The above distinction between pathways and supramolecular complexes has however to be toned down since the assembly of the majority of complex components is up-regulated by the transient complexes involved in pathways.

1.4. Protein–Protein Interaction Special Focus

Looking for protein–protein interactions (PPIs) contributes to proteomics by providing some uncharacterized protein with functional assignment (Wojcik and Hamburger 2003). Mayer and Hieter (2000) have first proposed the application of this "guilt-by-association" rule by attempting to infer functional annotation from well-known interacting partners. This is now largely accepted as a standard strategy based for example on cellular role or sub-cellular localization annotations. It is thereby legitimate to examine these new components as putative original targets.

Switching from interactome mapping to functional and dynamical interactome modelling is not obvious. The data interpretation is often blurred by the quite high level of false negatives/positives or the interactome partial coverage. In all cases, further steps of validation are needed (see Chapter 5 for details) as well as the incorporation of functional genomics such as expression profiling, gene knock-outs or RNA interference in a wide integrative analysis of biological systems (Vidal 2005).

PPIs are so widely dispersed in any biological process that they appear of particular interest as drug targets (Toogood 2002) considering the narrowed space of targets commonly used by pharmaceutical industries.

In this review we will give a snapshot of current available collections of PPIs produced by large-scale experimental approaches. An assessment of their reproducibility, their quality and the level of false negatives/positives will be discussed. Current effort to build proteins or domains interaction maps consolidated into integrated databases as well as algorithms to infer, compare, enrich and check raw or refined data will be presented. A magnification from protein network to individual component materialized by complex will end the presentation. Short description of methodologies deployed to validate the composition and relationships between subunits will be given, notably those allowing to derive domains and interface features that describe properties of the assembly. We will show how structural biology data have revolutionized insights about supramolecular complexes and we will explore current ways envisaged to modulate interactions, particularly from the point of view of drug discovery and diagnostic tests.

2. LARGE SCALE DETERMINATION OF PROTEIN INTERACTION MAPS

In the last decade, large data sets of PPIs have been made available to the scientific community. This interest for PPIs has been driven by the need to perform an in depth functional characterization of the large number of genomes sequenced in the past 10 years. Despite the fact that numerous genomes have been sequenced comprehensively, many of them including the human genome are not fully annotated. The part of functional proteomics dedicated to the characterization of molecular interactions, and most particularly PPI, can provide the biologist with key information on regulation, networks and pathways in cells. Proteic complexes identification and fine protein–protein partnering characterization have been addressed by *in vitro* or *in vivo* methodologies carrying out different nature of read-out: genetic selection (e.g. two-hybrid and phage-display assays), biochemical (e.g. GST pull-down, far western, co-immunoprecipitation [coIP], chips, chemical cross-linking, density gradient centrifugation, affinity/size-exclusion separations), or spectroscopic (e.g. Fluorescence Resonance Energy Transfer [FRET], mass spectroscopy analyses of affinity purified macro-complexes). If many of them constitute the core of the classical methods well-suited to describe any one-by-one association, others have been elected to perform large scale production because of their ability to be partly or wholly industrialized while preserving both efficacy and quality. We will report progress achieved by these medium- or high-throughput applications and will discuss about their respective advantages and drawbacks in a comparative manner. We will more particularly describe two-hybrid methods in different model organisms (fly, yeast and worm) and in human that have until now undoubtedly proposed to the biologist the most populated proteome-wide interaction maps.

2.1. Yeast Two-Hybrid Screening of Fragments Libraries

Within this part, we will describe quite exhaustively our platform dedicated to the generation of PPIs by yeast two-hybrid (Y2H) screening as first described by Rain et al. (2001) that has allowed to build a protein interaction map (PIM) covering 46.6% of *Helicobacter pylori* proteome. Additionally, we will take the opportunity of this specific use case to emphasize the delicate and challenging issue that consists in successfully transforming raw data to refined accurate interactions. We will also address some features common to protein network building such as bioinformatics pipe analysis designed for an industrial process.

2.1.1. The two-hybrid system in yeast

The original two-hybrid system, first implemented by Fields and Song in 1989, was designed to detect protein interactions through the transcriptional activation of one or several reporter genes. It was based on the key-role of an eukaryotic transcription activator composed of a domain that specifically binds to DNA sequences (the DNA-Binding domain, *DB*) and a domain able to recruit the transcription machinery (the Activation Domain, *AD*). Both domains constitute two structurally distinctive components that can be dissociated leading to the loss of the whole activity. It has been shown that a weak linkage of these domains, possibly mediated by a non-covalent bridge, is sufficient to recover an active protein ready to promote the activation of the RNA polymerase at the corresponding promoter locations and consequently give rise to the transcription of a reference bacterial reporter gene. The two-hybrid approach consists in grafting protein into DB and AD (to form the bait and the prey respectively) whose potential interaction will be tested by a read-out adapted to the reporter gene. Many "variations on a same theme" have since been described (Vidal and Legrain 1999).

2.1.2. Principles and experimental workflow of fragment library screening

This variation of the yeast two-hybrid system constitutes one of the many yeast two-hybrid procedures. In this assay, a protein of interest (the bait) is fused with the DNA-binding domain (*DB*) within a plasmid to be transfected into a competent haploid yeast strain. Putative counterpart polypeptides, which are randomly generated fragments prepared from genomic or cDNA materials, are fused to the activation domain (*AD*) to build the library of preys. The *in vivo* genetic selection of this assay is driven both by an efficient mating strategy developed for a saturating coverage of the library and by submitting resulting diploid cells to a selective pressure sufficiently stringent to discriminate the diploids for which interaction occurs. Both mating and saturation have brought significant improvements compared to the original Y2H library screening method (Luban et al. 1993) notably by leading to a statistical treatment of the results. Positive clones have to be analyzed to precisely locate and identify the region of the open reading frame (ORF) matching the grafted prey fragments. To perform this task, clones are submitted to industrialized and robotized molecular biology procedures. Our Y2H platform is supported by a customized in-house LIMS (Laboratory Information Management System, see Figure 2).

2.1.3. Bioinformatics designed to perform high content results analysis

Once positive clones have been detected, prey fragments are amplified by PCR, their $5'$ and $3'$ chromatograms are analyzed. The resulting sequences are processed

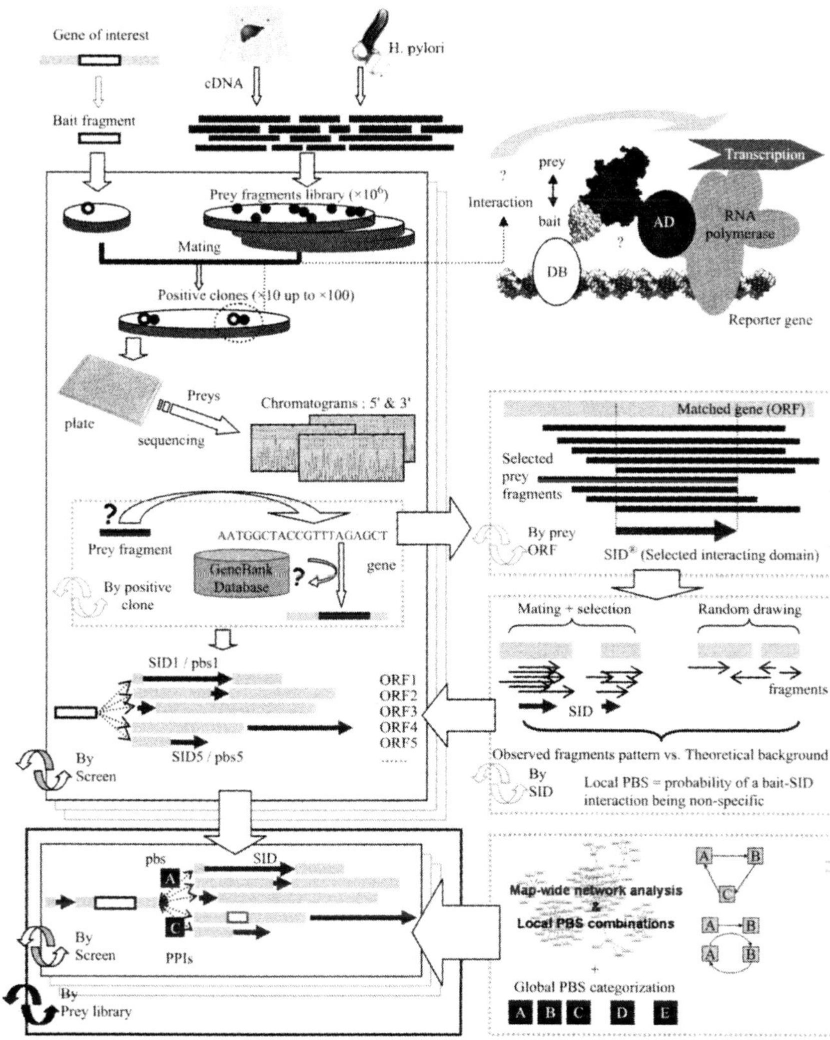

Figure 2. Details of the experimental Y2H screening and production flow-chart supported by Hybrigenics's LIMS platform (automate interfacing, samples and plates tracking, projects management, user-restriction to data access, planning and reporting tools, screen results analysis algorithms) to guarantee a high degree of completeness and total quality control.

by a proprietary algorithm whose flowchart is summarized in Figure 2. The prey sequences are assembled in order to take advantage of the overlapping similarity between fragments sharing a same interacting region. For identification purpose, the resulting contigs are searched with BLASTn against the nt database (Genbank, NCBI). A tailored heuristic is used to assign the best match by taking into account, among other things, the gene annotations incorporated into hit entries. This compels for example to make of a human gene the preferential choice from a batch of very close mammalian hits when the library originates from a human tissue. The identified prey fragments are then collected according to their attributed protein match to build a final list of distinct interacting partners. Taking advantage of the multiplicity of aggregated prey fragments, every interaction is analyzed, characterized and coupled to a single interacting domain (or SID®) computed as the minimal overlapping sequence matching all of them. The biological relevance of each interaction is assessed by computing a predictive score referred to as the PBS (Formstecher et al. 2005, see supplemental method). This screen-level local Predictive Biological Score is a statistical indicator representing the probability of a bait-SID interaction to be non-specific. This parameter has the dimension and the significance of an E-value as computed by BLAST. Encompassing the whole set of interactions, this score is of special interest by providing the researcher with adapted tools to sort, filter and detect the most attractive or worst set of interactions among the pool of putative partners revealed by the screen analysis.

2.1.4. Libraries considerations

Library approaches offer the ability to explore a great variety of time- or spatial-dependent interacting contexts: although genomic libraries are more suitable for compact genomes (e.g. prokaryotes), cDNA libraries fit better for genomes with intron/exon structure (e.g. metazoans) by providing scientists with appropriate tools to study tissues characteristics, differential PPIs patterns linked to a transient state (e.g. differentiated/undifferentiated tissues) as well as interactions correlated to sickness (healthy/diseased tissues).

Genomic libraries complexity is estimated from the combination ofboth the number of independent clones and the genome size allowing a sound calibration. They are made in such a manner that a coding sequence is screened by tens of fragments allowing each nucleotide of the genome to be covered about 50 times (Legrain et al. 2001). Conversely the diversity of cDNA libraries depends on the original biological sample where the cDNA quantity reflects the endogenous expression level of each corresponding gene. Redundancy and complexity are required to guarantee that a meaningful fraction of fragments encodes genuine interaction domains since the large majority of them are not biological chimera consecutive to meaningless location, improper orientation or shifted reading frame. Typical complexities are around 10 millions independent fragments. Routine screening conditions are made to guarantee that every interaction is evaluated up to five times.

2.1.5. The false negatives/positives issues

Intrinsic limitations of Y2H combined with its high-throughput application make unavoidable the production of a substantial amount of false positives and false negatives which has made mandatory the constitution of a range of suitable solutions to obtain high quality PIM (Legrain and Selig 2000).

False negatives issues are mainly addressed by reaching exhaustiveness through saturation of screening conditions applied to highly complex libraries coupled with the identification of all the positive clones. Selective pressure is fine tuned for individual screen in order to reach a homogeneous selectivity and decrease the background level. This participates to both reproducibility and results normalization promoting an easiness of screens comparison.

Additional sources of false negatives originate from the expression of artefactual and heterologous fusion proteins in yeast such as misfolding, unstability, inactivity, mislocalization, and toxicity. For example, to correct some of these defaults, bait can be carefully designed by taking into consideration the protein domain topology (e.g. transmembrane segments) and also by removing the non-nuclear localization signals. Moreover, some specific dominant mutant alleles with defined biological/biochemical properties stabilizing or promoting protein interactions (e.g. mutations that convert small GTPases to constitutively activated GTP-bound forms) have to be deployed to get successful screening experiments thus suggesting that template protocols had to be repeated for cognate or similar genes. If the design is manually performed for optimizing the bait, for preys, analogous results are obtained by the intrinsic nature of the libraries which consist in multiple fragments partially or completely covering protein domains. Some post-translational modifications of bait/prey proteins may be required to promote recognition and binding. An illustration of such endogenous enzymes activity in yeast was demonstrated with the finding of the Bromo domain known to bind lysines once acetylated (Formstecher et al. 2005). An alternative solution consists in introducing into the yeast the missing enzyme (Tirode et al. 1997).

Conversely, both experimental adjustments and bioinformatics tracking tools are needed to tackle the false positives issue. Some false positives are artefacts coming from direct interference with the Y2H system by triggering the transcription by themselves. There are composed of bait interacting directly with DNA binding region (e.g. transcription factors) and preys with a natural affinity for the DNA binding domain. Other false positives can be distributed in two families: the highly connected proteins which main functional property is to bind multiple partners (e.g. chaperone HSP90) and the "sticky" proteins coming from the unnatural biochemical behaviours of some prey fragments competent to bind many other proteins in an unspecific manner.

These properties would nearly be impossible to unearth without incorporating data coming from other screens performed with the same library and for which such a repeated event is encountered sufficiently enough to be detected by the bioinformatics analysis process. For this reason, a strategy has been implemented to enrich and refine the raw results obtained from an individual screen. It consists in widening the analysis context to the whole PIM built from screens sharing the same library. By applying adapted heuristics, several bait-SID interactions are merged to give single protein–protein interactions (PPI) at the PIM-level: the global connectivity of the map is thus analyzed to flag highly connected prey polypeptides and extract the precise domain responsible for the coding sequence "stickiness." The PIM is also searched for small connectivity patterns linking three proteins together (3-cycles). When a cycle is found, it increases the PBS values because it likely represents a biologically relevant network. In the end, SIDs are refined by clusters making and each interaction is associated with a global improved PBS tabulated into three categories encoded by a single character from A (best) to C with two additional categories associated to peculiar cases: D (unique bait-SID interaction defined by a singleton prey fragment) and E (highly connected or sticky domain).

2.1.6. Results and PIMs connectivity

Helicobacter pylori PIM is the largest bacterial one obtained up to now (Rain et al. 2001). More than one billion bait/prey pair interactions were tested from 261 selected ORFs. Consolidated results gave a total of 1524 interactions: 1280 displayed a good relevance (PBS range from A to C), included 62 homo-oligomeric interactions and accounted for an average connectivity of 3.36 partners per connected protein.

Another example of a partial protein interaction map produced with the technological platform described above is the fruitfly PIM (Formstecher et al. 2005). More than 2300 PPIs were identified from 102 baits chosen as orthologous of human cancer-related and/or signalling proteins, of which 710 are of high confidence (PBS A to C). For this set of relevant interactions, mean connectivity is 2.22 partners per connected proteins (7 for baits).

2.2. Other Y2H Applications

2.2.1. Y2H matrix approach

The Y2H matrix method diverges from the library screen Y2H mainly by the preparation of the preys and the mating procedure. For the one-by-one matrix approach which is labour intensive but allows theoretically a full coverage of the proteome under study, a protein array of AD-preys is prepared in yeast cells, mated individually with a single DB-bait and finally selected by the reporter gene activation. In the case of the high-throughput matrix procedure, the mating is performed in bulk where one given DB-bait plasmid is mated with a pool of cells transformed with AD-prey plasmids but is known to increase the level of false negatives since each predefined full length protein is tested only once. Nevertheless, the bias induced by the protein abundance in the case of the cDNA library screen approach is fully eliminated as the set of studied genes is comprehensively tested on an individual basis. However, the problem of misfolded DB-bait or AD-prey evoked with the library screen Y2H is probably more acute with the matrix approach as the protein coding genes are integrated in their full length. This approach, first tested on the drosophila cell cycle regulators (Finley and Brent 1994) and the T7 phage proteome (Bartel et al. 1996), has been used by Ito et al. (2001) and Uetz et al. (2000) providing the first yeast protein interaction maps. The *Drosophila* interactome by Giot et al. (2003) was carried out by combining data using both the matrix and the non saturating library screen approach.

2.2.2. *Special treatment of membrane proteins*

A subtle approach was designed to address more specifically the identification of membrane protein partners. Johnsson and Varshavsky (1994) found that the native ubiquitin can be split into a N-terminal (Nub) and a C-terminal (Cub) half and still retain a basic affinity for each other as they reassemble spontaneously to form quasi-native ubiquitin. The basic idea behind the split-ubiquitin assay system is to fuse a reporter protein to Cub which, when cleaved off by Ubiquitin Specific Proteases (USPs), is released. If a point mutation in Nub (NubG) is introduced it abolishes the affinity of Nub for Cub and if NubG and Cub are fused to two proteins X and Y respectively that restore the NubG and Cub association, the gene reporter will be released. This elegant principle has been extended to build a high-throughput Y2H membrane-based assay by converting the re-association event between NubG and Cub (respectively associated to membrane proteins X and Y) into a transcriptional output that can be detected easily (Stagljar et al. 1998; Stagljar and Fields 2002). This technique has been recently adapted for prey library screening (Thaminy et al. 2003), has also been applied to a large scale identification screening of yeast integral membrane protein interactions (131 of high confidence found; Miller et al. 2005) and to the construction of a human GPCR interactome which found several interesting GPCR interacting proteins (Bürkle et al. 2005).

2.3. Protein Complexes Analysis by Mass Spectrometry

Protein identification by mass spectrometry (MS) has emerged as a very powerful and very sensitive tool to detect and identify protein aliquots (Mann et al. 2001). 2D gel is the traditional tool to perform protein separation and still constitutes the main support for analyzing complex protein mixtures. The PPIs analysis requires a different approach to preserve the physical integrity of protein complexes. The combined development of affinity purification protocols, epitopes tagging, high throughput purification and MS analysis of protein complexes has thus made possible the mapping of full proteome especially for yeast. Ho et al. (2002) used a specific strategy (HMS-PCI) for deciphering a part of the yeast interactome. A vector encoding a protein (bait) linked to a peptide tag for which specific antibodies are available is constructed. The yeast cells are then transformed with the vector and localization and interaction are let to occur. After an immuno-purification step for capturing the FLAG tag linked to the bait complexes and a separation by gel electrophoresis, liquid chromatography ESI-MS/MS (Electrospray ionization tandem mass spectrometry) is performed in a high throughput manner for identifying the proteins. Starting from an initial set of 725 baits including kinases, phosphatases, regulatory subunits and proteins implicated in DNA damage response, they identified 1578 interacting proteins involved in 3617 interactions. In summary, the main feature of this method is to use overexpressed proteins in plasmid enabling the detection of interaction even for low abundant protein.

The affinity purification method also called TAP-TAG (Rigaut et al. 1999; Puig et al. 2001) was used by Gavin et al. (2002) in a slightly different strategy from the Ho group in order to analyze the yeast protein interaction network. The procedure consists in fusing a set of baits (from the yeast ORF proteome) by homologous recombination in yeast cells with a sequence tag encoding a calmodulin binding protein, a TEV cleavage site and the protein A. The baits are then extracted from cells and purified by two sequential steps where the tagged bait, potentially associated with other proteins to form a complex, is recovered from cell extracts on an IgG matrix. In the next step TEV protease is added to release the bound complex which is then incubated with calmodulin coated beads in presence of calcium to perform the last purification stage by removing the contaminants as well as the remaining TEV protease. Finally the purified protein assemblies are separated on a 1D Gel, analyzed by MALDI-TOF Mass Spectrometry and identified by database search algorithms. Based on a 25% coverage of the yeast proteome, Gavin group first identified 232 distinct protein complexes. A recent update of this proteome interaction map was carried out by the same group (Gavin et al. 2006) providing the biological community with a very comprehensive interactome of the yeast covering 60% of the proteome and identifying 491 complexes of which 257 were found novel.

While the procedure used by Ho group was based on inducible overexpression, the Gavin procedure enables through homologous recombination the expression of each tagged protein at their natural endogenous level making the whole analysis system as close as possible to an *in vivo* situation. For essential genes, the Gavin methodology demonstrates also the functionality of the fused tagged proteins that is the tag does not perturb the gene expression and function.

2.4. Protein Chips

The principle of protein arrays is to immobilize a set of different proteins onto a solid support. Protein microarrays have demonstrated their usefulness for the detection of protein–protein interaction, enzyme-substrate assays, protein-DNA, protein-oligosaccharide and protein-drug interaction (see Templin et al. 2003 for a review). The pioneering work of MacBeath and Schreiber (2000) showed that it was possible to study PPIs, identify protein substrates (kinases) and protein targets of small molecules (DIG/mouse monoclonal antibody, biotin/steptavidin and FKBP12). Zhu et al. (2001) were the first to present a genome-wide interaction analysis of 5800 yeast ORFs focusing on phosphoinositide-binding and calmodulin-binding proteins. The calmodulin probed 39 partners among which six were known calmodulin-binding partners. A recent application (Jones et al. 2006) has focused on a comprehensive analysis of SH2/PTN- mediated interactions with ErbB receptors where 43 previously known interactions were found and 116 new partners of this receptor family were uncovered. One important feature is the ability of these protein microarrays to deliver quantitative estimate of the binding affinity constant enabling to build a system-level view of EGFR, ErB2 and ErbB3 at different affinity threshold. Recently, an in depth analysis of PDZ domain has been carried out by the same group (Stiffler et al. 2006) revealing a quantitative PDZ interaction domain network with a very low rate of false positives and false negatives (14%). Thirty-nine new

PPIs were discovered of which 12 were confirmed by co-expression analysis. The main advantage of the protein chips is to provide quantitative estimation of the PPIs strength compared to the binary type of information provided by Y2H and co-affinity purification coupled to Mass Spectrometry procedures.

2.5. Phage, Ribosome and mRNA Display

In the phage display system, the proteins of interest are fused to a subunit of a viral capsid and the bait protein selection is performed *in vitro*. Phages that show affinity are then amplified and sequenced to identify the interacting partner. This method is often employed to screen small peptides against bait protein. Tong et al. (2002) combined yeast two hybrid and phage display to build a map of yeast SH3 domains interacting partners finding 59 high confidence interactions. mRNA display is an alternative method as illustrated by Hammond et al. (2001) where several known and unknown Bcl-X_L interactors have been found. Ribosome display, which is a technical variation of the mRNA display method, has been recently applied to the identification of regions of the Nogo receptor involved in ligand binding (Schimmele and Plückthun 2005). This latter study is a good illustration on how to uncover interacting domain. To conclude, these "display" methods are not only used for interaction discovery but also for optimizing interaction which can be very useful for designing new types of protein molecules (see review Lipovsek and Plückthun 2004).

2.6. Discussion

As already discussed specifically for the Y2H in Chapter 2, large scale approaches suffer from the existence of artificial interaction (i.e. false positives) or undetected true interaction (i.e. the false negatives). Perhaps should we add one more category to distinguish true *in vivo* interaction from true *in vitro* interaction.

Figures presented in this paragraph, taken altogether, without understanding the experimental conditions of the different methodologies could led the reader to draw severe conclusions on the validity and the usefulness of large scale approaches.We think it is necessary to look in details into the different experimental systems and give a short summary of the overall differences and features in the methodologies described so far. Furthermore, many interactomes studied so far at the exception of the latest delivery of the Gavin group do not provide a full coverage of the proteome under scrutiny. The interactome undersampling makes difficult to compare different data set especially across species. TAP-TAG methodology is very reproducible (70%) compared to HMS-PCI (20%) as described by Dziembowski and Seraphin (2004) but HMS-PCI is probably better at picking up low abundant protein interaction.

Y2H screening is probably more sensitive for detecting transient interactions whereas TAP-TAG is more efficient at detecting connections between proteins in stable complexes although the washing step during the elution process may fail to identify complexes which components are linked by weak interactions. On one hand Y2H gives comprehensive data on binary interactions, on the other hand TAP-TAG identifies a set of related proteins in a complex without details on physical interactions that connect them. As a matter of fact, both methods are complementary (Lebreton et al. 2006).

Discrepancies between the matrix approach and library screen results for Y2H stress more the method differences rather than their sensitivity to detect PPI. The matrix approach has the advantage of overcoming the cDNA library normalization problem but does not cancel the problems related to full length ORFs and its consequences in terms of artificial interaction. The library screen method enables the use of partial optimized bait to avoid this problem, allows a statistical treatment of the Y2H screen which finally estimates an interaction confidence score (see above) and identifies interaction domain. The two methods are complementary and the resulting maps hit different part of the interactome space.

The quality assessment, the reproducibility estimation and the validation process are all important indicators when PIMs originating from different experiment conditions are compared.

How many interactions share two sets of the same organism? It has been shown for the first two Y2H experiments performed on the yeast proteome that the interactions common to both sets were estimated at 15–20% (Ito et al. 2001). A comparison between the data sets coming from TAP-TAG (6285 connections detected; Gavin et al. 2002) and HMS-PCI (9005 connections detected) has revealed a similar trend between the two data sets: 1728 interactions are common to both sets which means that 19.2% of HMS-PCI match with 27.5% of TAP-TAG (von Mering et al. 2002). The same study estimated the level of data accuracy if all interactions are counted with respect to a reference set of known complexes showing a 12.5% accuracy for the TAP-TAG methodology, 2% for HMS-PCI and 3.7% for the two Y2H assays (Ito and Uetz data set).

The comparison of the Y2H *Drosophila* PIMs from Giot et al. (2003) and Formstecher et al. (2005) confirmed that less than 10% of the interactions from the Giot data set can be found in the Formstecher data set. Stanyon et al. (2004), taking the same method than Giot but just using a different bait/prey construct (Gal4 vs. LexA), applied on cell cycle regulators, found 1814 PPIs involving 488 proteins of which only 28 were found to overlap with the Giot data set.

Sprinzak et al. (2003) found by a rigorous computation that the overall reliability of high throughput Y2H assay was about 50% (based on Ito and Uetz yeast proteome Y2H screening). The quality of an interaction was assessed with the potential co-localization or a possible common cellular role of the two proteins. This was integrated into a formula which determined the ratio of true positives.

In a brighter outlook, a map of *C. elegans* has been produced (Li et al. 2004) and 65% of the interactions were retested positive using an independent co-affinity purification assay. On a smaller subset of the *C. elegans* proteome, the TGF-β pathway (Tewari et al. 2004), 90% of the interactions retested by co-affinity purification assays confirmed the Y2H assay. Colland et al. (2004), on the Smad signalling pathway, proved the overall reliability of the Y2H assay, not only retrieving known actors of this pathway but also finding new protagonists confirmed by RNAi genetic perturbation method and overexpression study. These two studies showed that large scale approaches combined with more focused introspection method could unravel new targets and proposed functional role for unknown proteins. It would be very interesting to compare the newer and comprehensive data set from Gavin et al. (2006) with the other available PPIs obtained from the same proteome.

Rual et al. (2005) built a human PIM covering 10% of the total search space. They also controlled representative samples of interactions using an *in vivo* pull down assay (affinity purification with glutathione S-transferase, GST) and confirmed 78% of the Y2H interactions found. Interestingly, this ratio is about 62% in the case of literature curated interactions underlining that this type of data set coming from the scientific literature should not be taken as the reference set *per se* without considering the inherent bias included in any curation process. In the same period of time, Stelzl et al. (2005) produced a Human interaction map identifying 3186 novel interactions. Independent co-immunoprecipitation and pull down assays confirmed on an interaction subset 2/3 of them supporting the overall quality of this human PIM.

It is therefore very encouraging to observe the evolution of the large scale approaches from an intensive use of a single technique to the combination of methods to obtain the most accurate map. It is now strongly and repeatedly suggested to test the same interaction with two different assays rather than repeating the same procedure many times. Indeed interactions revealed by more than one method have been shown to be the most reliable.

3. *IN SILICO* PREDICTION AND INFERENCE OF PIMs

3.1. Genomic-Based Predictive Methods

It is perhaps not necessary to emphasize that bioinformatics in its widest acceptation i.e. algorithms and information technology (graphics visualization, relational database), given the combinatorial aspect of the PPI has been utilized intensively as a companion to the many different PPI discovery methods. We will give some examples and flavour of the different use of bioinformatics tools to predict, analyse and/or validate PPI.

The prediction methods rely on the notion that the genes coding for potentially interacting proteins tend to be associated with each other on genomes (Figure 3). Two genes can either form a single multi-domain polypeptide in one organism or be functional as independent protein domains in a different organism. Where it occurs, it is named a gene fusion event from which it can be deduced that the two single protein domains potentially interact with each other (Marcotte et al. 1999). Close to this method is the one that searches for the conservation of gene neighbourhood among species that can also be an indicator of potentially interacting proteins (Dandekar et al. 1998). The method of phylogenetic profiles (Pellegrini et al. 1999) can compare and detect the absence or presence of genes in complete genomes. If a pair of genes is consistently present in a set of different genomes, one can deduce that at least they both have a related functional role and perhaps a direct physical interaction

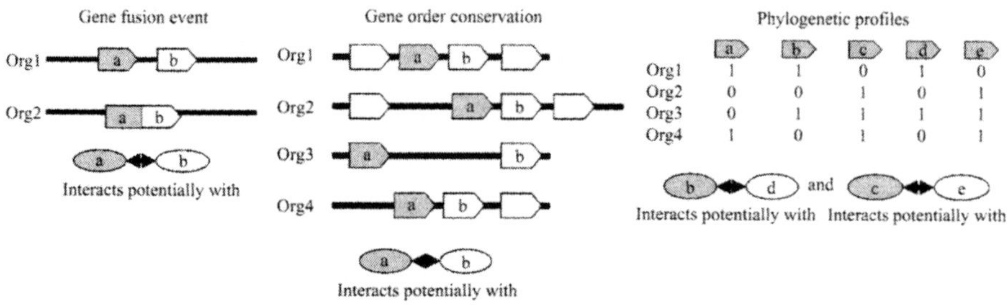

Figure 3. In silico methods for predicting protein interactions adapted from Pazos and Valencia (2002) and Huynen et al. (2003) with Org = organism.

although it should be not automatically implied. On *Mycoplasma genitalium* (480 genes), Huynen et al. (2000) have estimated to 50% of this genome the degree of coverage that can be drawn from the three genome context prediction methods. The gene fusion events, the gene-order conservation and the phylogenetic methods have demonstrated for this prokaryote genome a high rate of reliability in predicting PPIs with a success rate of 78, 80 and 63% respectively. It is important to note that the type of interaction detected by these methods should not be taken in its strictest sense that is direct physical interaction but rather in general terms where two proteins can either present a direct interaction, belong to the same macromolecular complex, pathway or be part of the same biological process. Another method (*in silico* two-hybrid; Pazos and Valencia 2002) compares the degree of similarity of correlated mutations between two proteins and has shown to be capable of extracting information on PPI keeping in mind that this heuristic needs a high quality multiple sequence alignment to derive meaningful mutation correlations.

We can also mention the mirror tree method (Pazos and Valencia 2001) where the entire structure (branches and lengths) of phylogenetic trees of a pair of proteins is compared, their degree of similarity will be considered as a serious indication of two proteins that have co-evolved in time and constitutes possibly two interacting proteins. Applied to *E.coli*, 2742 potential interactions were found. Recent improvements in the so-called mirror tree method have been made by taking into account the fact that an interaction is not always one to one but should be regarded as protein families interacting potentially with each other (Ramani and Marcotte 2003). Sato et al. (2005) reduced drastically the number of false positives prediction in *E.coli* by excluding the information about the phylogenetic relationships. Finally Pazos et al. (2005), in a parallel approach, obtained improved predictions still on *E.coli* by incorporating to the core method described previously information given by the canonical tree of life in order to correct the inherent background similarity effect between two protein families which tree is constructed on the same set of source organisms. Additionnally, they also address the problem of the non-standard evolutionary event (horizontal gene transfer).

An important point to notice is the favourable conjunction of both the availability of affordable powerful computer and the arrival of fully annotated genomes from many different organisms and species, both have allowed to carry out comparative genomics and is at the source of the aforementioned methods. One can expect with newer completed genomes that these PPIs prediction methods will become more accurate and comprehensive. These approaches providing very reliable predictions already feed many PPI databases that will be discussed in the next chapter.

3.2. Cross-Organisms Inferred PIMs

The human interactome is of course the interactome of prime interest for the pharmaceutical industry. In the first yeast PIM performed by TAP-TAG mass spectrometry (Gavin et al. 2002), it was shown that the analysis of yeast complexes could be propagated to the prediction of their human counterparts. Walhout et al. (2000) were the first to propose an interolog concept (based on sequence similarity and clustering) which was subsequently verified by yeast two-hybrid experiments (Matthews et al. 2001) on *C.elegans* where it was found that 16 to 31% of interologs could be detected between yeast and *C.elegans*. Li et al. (2004) used this procedure to expand the *C.elegans* protein interactions data set using again the yeast PIM.

A related approach based on known genome interaction map used an interacting domain profile pair which clusters protein domains by sequence and connectivity similarities and builds a map of *Escherichia coli* with a 12% prediction rate (Wojcik et al. 2002) inferred from *Helicobacter pylori*. Small conserved subnetworks were uncovered by comparing the yeast and the *H.pylori* interactome (Kelley et al. 2003) using PATHBLAST, a method specifically devised for this task. Based on yeast, fly and worm proteomes, the same group (Sharan et al. 2005) performed a protein network comparison and found 71 network regions that were conserved and predicted 2609 new protein interactions (176 for yeast, 1139 for worm, 1294 for the fruitfly) with a 50% accuracy. These methods can be very useful to derive good quality protein network at low cost from different known organisms PIM in order to identify the parts of the proteome that are organism specific and therefore should be studied experimentally.

3.3. Literature and Data Mining

The picture will not be complete if we did not describe a simple but efficient manner to find interactions: text mining. Although these techniques are not as firmly based as the genome context analysis seen above, they enable the fast and automated extraction from online abstracts and play a substantial role in PPIs databases (see below). Where do the literature PPIs data come from? Simply from the accumulation of studies (not particularly focused on interaction discovery) carried out over the years using a large variety of interaction discovery technologies from traditional genetic analyses and co-immunoprecipitation to pull down or synthetic lethals methodologies. This source of PPIs is considered to be of good overall quality although one has to be meticulous when it comes to integrate these data in specialized PPI databases. To be more specific, text mining heuristics do not completely eliminate the tedious process of manual curation and annotation but it certainly helps to speed up the process. Language speech and semantic analysis are a vast area of research in computer sciences but the algorithms used have been simplified and restricted to address the specific problem of PPI extraction. Blaschke et al. (1999) proposed an automated process to extract PPI with assumptions such as pre-specified protein names and a limited set of verbs to guide the search analysis. Ono et al. (2001) employed a protein name dictionary and obtained precision of 95% and recall of 85% tested on yeast and *E.coli*. Marcotte et al. (2001) with a bayesian approach and a list of 80 discriminating words (complex, interactions) can sort through abstract from Medline with a fairly good overall accuracy. This process has been used to expand the interactions in DIP (see Chapter 4). More recently, the same group (Ramani et al. 2005) has combined experimental PPIs data set with a set of literature extracted PPIs to build a framework for validating future large-scale human protein interaction assays. It is also worth mentioning the work of Albert et al. (2003) which used an automatic extraction procedure to find nuclear receptor interacting partners and compared them favourably to experimental data performed with Y2H. It was followed by a Y2H validation study and they uncovered 64 new potential interacting partners of nuclear receptors among 377 high quality interaction pairs (Albers et al. 2005).

4. SPECIALIZED REPOSITORIES AND ASSOCIATED TOOLS: RESOURCES DEDICATED TO LARGE-SCALE COLLABORATIVE PROJECTS

4.1. Public and Private PPIs Databases

The need to store the wealth of data produced by large scale biological projects and to provide a robust framework for analyzing and studying PPIs is obvious. Nearly all the projects described in Chapters 2 and 3, whether produced by experimental methods, *in silico* predictive techniques or extracted from the scientific literature, have their data collapsed and structured in databases (see Table 1).

4.2. Interaction Domains Annotations

By integrating information from database such as Interpro and Gene Ontology classification, the subcellular location and the destination of such or such protein taking part in an interaction can be obtained. This is particularly interesting to assess the quality of large scale screening data set originating from different experimental conditions. The Gene Ontology (GO) consortium has arisen from the common need expressed by the genomic and proteomic community to build a common gene annotation framework. It was first based on three genome projects, FlyBase, SGD

Table 1. Description of protein–protein interaction databases either publicly accessible or on a license/fee-for-service basis

Name	Url	Access	Comments	References
MIPS	mips.gsf.de/services/ppi	No restriction	Comprehensive coverage for the yeast and mammalian interactome	Pagel et al. (2005)
BIND	www.bind.ca/	Licence for commercial user	General interaction data set archiving biomolecular interaction, complex and pathway information	Bader et al. (2003)
DIP	dip.doe-mbi.ucla.edu/	Licence for commercial user	Interactome built from combining diverse sources of PPIs (experimental, *in silico* prediction, literature-based)	Salwinski et al. (2004)
MINT	mint.bio.uniroma2.it/	No restriction	Database focused on PPIs mined from the literature	Zanzoni et al. (2002)
STRING	string.embl.de	No restriction	PPIs derived from genomic context analysis, experiments, coexpression and literature-based (cover 179 species)	von Mering et al. (2005)
HPRD	www.hprd.org/	Licence for commercial user	Human proteome annotation database including 33,000 manually curated Human protein interaction	Peri et al. (2003)
Intact	www.ebi.ac.uk/intact/	No restriction	EBI protein interaction data	Hermjakob et al. (2004)

(*Continued*)

Table 1. (Continued)

Name	Url	Access	Comments	References
The Grid	www.thebiogrid.org/	No restriction	Protein and genetic interactions (fly, worm, yeast and human)	Stark et al. (2006)
RIKEN	fantom21.gsc.riken.go.jp/PPI/	No restriction	Mouse protein–protein interactions extracted from experimental source and combined with BIND, DIP and literature data	Suzuki et al. (2003)
OPHID	ophid.utoronto.ca/ophid/	No restriction	Specialized in human protein interaction Combined data from BIND, HPRD, MINT and predictions from model organism (Yeast, Worm, Fruitfly), PSI compliant	Brown and Jurisica (2005)
iHOP	www.ihop-net.org/UniPub/iHOP/	No restriction	Gene network for navigating the literature linking information from phenotype, pathologies and gene function	Hoffmann and Valencia (2004)
HPID	wilab.inha.ac.kr/hpid/	No restriction	Human protein interaction database	Han et al. (2004a)
Hybrigenics	pim.hybrigenics.com	Free access for selected protein maps	*Helicobacter pylori*, TGF-β and fly protein interaction map and HIV litterature	Rain et al. (2001) Colland et al. (2004) Formstecher et al. (2005)
Prolexys	www.ariadnegenomics.com/products/prolexys/	Licence required	Prolexys HyNet Human interactome	–

(yeast) and MGD (Mouse Genome Database) and was successively extended to other databases and species. In a nutshell, GO describes three ontologies: the biological process, the cellular components and the molecular function. To search the GO database, the site amigo (www.godatabase.org) is the most uptodate. Additionnally, many tools have been made on the basis of the Gene Ontology guidelines (see site www.geneontology.org for a comprehensive list of tools associated to GO), an interesting by-product is the mapping of the Interpro database with the Gene Ontology database (interpro2go www.geneontology.org/external2go/interpro2go). The Interpro database is a large and very comprehensive compendium of annotated protein domains of all species. Once a sequence is scanned with iprscan, it can be associated to the corresponding Gene Ontology with the interpro2go mapping. Among the valuable information we can extract is the "cellular components" ontology which can be very useful to enrich PPI annotation (by upgrading the value of the interaction when found co-localized, all the same for interaction involving proteins participating to the same biological process). If the information concerning the cellular localization is not contained in the databases it can be computed by a subcellular prediction software such as psort (Nakai and Horton 1999) which will compute with a fairly good accuracy an estimation of the probability for a protein to be in such or such cellular compartment.

As mentioned in Chapter 2 for the description of our Y2H library screening procedure, what we obtained after a bioinformatics analysis is the location of the interacting domain within the bait or the prey protein. In reality, we do not measure interaction between full length proteins but between domains within proteins which is a closer representation of an *in vivo* interaction where only some key fragments of each protein interact with one another in space. These domains were called Selected Interacting Domain or SID. It is therefore clear that bioinformatics sequence analysis that compare SIDs obtained from Y2H experiments with Interpro domains can lead to very interesting observation regarding the domain keen to interact and can help to discriminate more specifically true and false positives.

4.3. Dedicated Tools for Topology Analysis and Visualization

Displaying map on a user-friendly graphical interface is absolutely necessary if one hopes to study, compare or analyze even small protein networks. Softwares fabricated for this goal are now numerous. It has become the natural way to display large PIM from full genome. Common functionalities in these softwares allow the biologist to navigate through large set of interaction data, searching neighbours, zooming or having a map panoramic view. The main and common idea behind the map display among the different graphical interfaces is to describe the many interactions between proteins as a graph that is nodes (proteins) connected by edges (interactions). This basic description enables the general representation of any interaction data sets such as pathway (signalling, transduction, metabolic) and macromolecular interactions (protein–protein, DNA-protein). Figure 4 **(see colour insert)** shows an example of

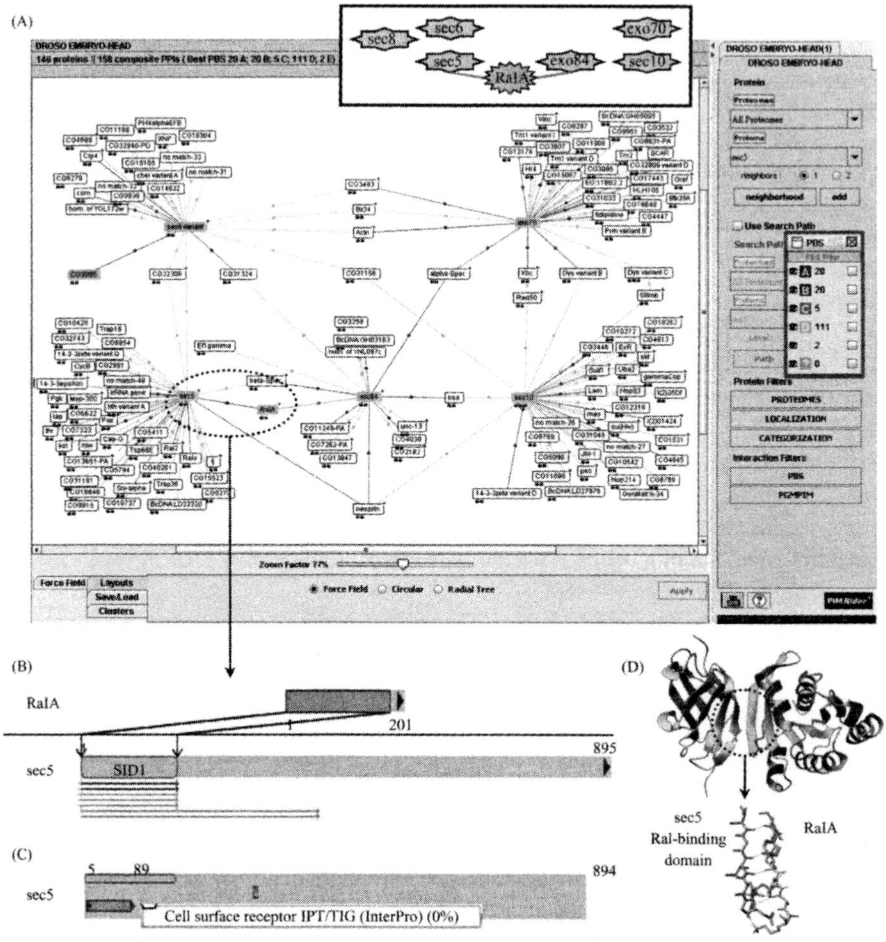

Figure 4. Typical dissection of a supramolecular complex (the exocyst, see Figure 1) carried out with PIMRider platform tools (pim.hybrigenics.com). (A) *PIM viewer* tool is used to display the partial PIM arising from the yeast two-hybrid high-throughput screening of *Drosophila* (Formstecher et al. 2005) focused on some components of Sec6/8 secretory machine (i.e. exocyst): Sec5, RalA, Exo84, Sec10, Exo70, Sec6, Sec8 (CG2095) are orange-boxed. The resulting graph (proteins are nodes, interactions are edges, baits have typical radiating-hub display) may be used to browse and navigate through a pathway or assembly subunits. Interaction scores follow a coloring scheme from red (PBS A) to pink (PBS E) transcribed on PPI linkage and may be used to filter and decrease PIM densification. (B) Screen copy of the *Interaction Viewer* tool showing the highly confident (PBS A) PPI between RalA and sec5 and the SID (Selected Interacting Domain) resulting from common part of experimental prey fragments (arrows). (C) Snapshot of the *Domain Viewer* tool that allows to map sec5 SID to IPT, a Ral-binding domain annoted within InterPro database. (D) X-ray crystallographic structure of the complex between the 9-183 fragment of RalA and the 120 residues N-terminal fragment of Sec5, that adopts an original immunoglobulin-like β-sandwich fold (Fukai et al. 2003; pdb entry 1UAD). These authors have well-depicted the hydrogen bonds networks formed at the junction of the intermolecular antiparallel β-sheet formed by sec5 β1 and RalA β2. 3D pictures performed with TRIPOS 3D-Explorer.

how such a tool (PIMRider®) can be used to dissect Y2H screening results on the exocyst complex already presented in Figure 1.

A good introduction to the concepts underlying the graphical display of large interaction data sets can be found in Uetz et al. (2002). An inventory list of visualization tools for protein network and for any set of interacting data is given in Table 2.

Interpretation of network topology has also naturally arisen from the availability of large scale PIM of different organisms. From network analysis, it may be derived crucial information on the potential impact of choosing a particular protein as a drug target, on the robustness of a protein network against mutation or external conditions, on the right strategy to obtain a wider interactome coverage, on how a protein network has evolved over time and finally on how the network is organized. How can a protein network be best described? As a random model, a scale-free model or a hierarchical model? Barabasi and Oltvai (2004) in their review gave a comprehensive description of the network theory applied to the understanding of the cell organization. Protein network represented by available data appears to be best represented by the scale-free model where most proteins participate in few interactions whereas a few proteins have many interaction partners (hubs). However a recent study (Han et al. 2005) has shown that if, on the partial networks provided by Y2H experiments scale-free is a good approximation to represent the map topology, it cannot be inferred that it will be the case for full interactome. The human protein network recently built by Rual et al. (2005) follows a power-law distribution which is the characteristic of a scale-free network. In this study, it is interesting to mention that Rual group used the MCODE graph clustering algorithm (Bader and Hogue 2003) which allows the detection of densely interconnected regions such as molecular complexes.

Han et al. (2004b) described the dynamical aspect of the modular organization and proposed the concept of "date hubs" (connecting different biological processes) which introduced the notion of time and spatial location. Gavin et al. (2006) describe the modular organization of the typical macromolecular complex composed of cores representing functional units (proteins highly co-expressed and co-localized) and modules which can be found acting in many different complexes. This dynamical organization enables the cell to perform a wide variety of tasks in time and space with a limited set of proteins. Indeed, the next step will be to add to the 2D topology network representation a third (spatial sub-cellular location) and a fourth temporal dimension to have a true image of the network topology dynamics in a cell (Bork and Serrano 2005). Topology analysis is undoubtedly useful but it will take some years before having access to larger and higher confidence interaction data sets which are a mandatory prerequisite to build topology hypotheses closer and closer to reality.

4.4. Promote Data Exchange and Comparisons by Implementing Standards

At this stage of the review, we have certainly convinced the reader that the wealth of data and the natural inclination to cross compare large sets of data obtained from

Table 2. List of protein–protein interaction viewers

Name	Url	Access	Comments	References
PimRider	pim.hybrigenics.com	Free access and licence required	Functional proteomics software platform, dedicated to the exploration of Protein Interaction Maps (PIM). It integrates four viewers and is GO compliant. See Figure 4	Rain et al. (2001)
PimWalker	pim.hybrigenics.com/pimwalker	Freeware	Light platform for the visualization of protein network (PSI compliant)	Meil et al. (2005)
I ogma™(GenoLink)	www.genostar.com/	Licence required	Powerful graphical browser which can display and organize large sets of heterogeneous data into large network of entities. Very useful querying functionalities for network exploration and new relationships inference	Durand et al. (2006)
Medusa	www.bork.embl.de/medusa	Freeware	General interaction tool for graph analysis, contains interesting features such as representation of different subcellular compartment	Hooper and Bork (2005)

Visant	visant.bu.edu/	Freeware	Integrative visual data-mining of multi-scale Bio-Network/Pathways (GO compliant)	Hu et al. (2004)
ProViz	cbi.labri.fr/eng/proviz.htm	Freeware	Protein–protein interaction graphs visualisation tool	Iragne et al. (2005)
CytoScape	www.cytoscape.org/	Freeware	Platform for visualizing and integrating molecular interactions including mRNA expression profile (GO compliant)	Shannon et al. (2003)
Osprey	biodata.mshri.on.ca/osprey/	Freeware	Platform for visualizing complex molecular interactions (GO compliant)	Breitkreutz et al. (2003)
Cnplot	csb.stanford.edu/~nbatada/VCN.html	Freeware	Visualization of global connectivity within pre-clustered network, interesting features with nodes and edges outline proportional to their number	Batada (2004)
Pajek	vlado.fmf.uni-lj.si/pub/networks/pajek/default.htm	Licence required	Package for general and large network analysis and visualization	See references in the web site

different techniques and experimental conditions had created the need to adopt a community standard data model for the representation and exchange of protein interaction data. This data model (Hermjakob et al. 2004) and the on-going regular improvements have been developed by the Proteomics Standards Initiatives (PSI) consortium which is a work group of the Human Proteome Organization (HUPO). This data model is fully integrated in our Y2H production platform as one of the output format and used as an input data model for the light PPI platform derived from the PimRider, PimWalker™(Meil et al. 2005). Most of the databases described above also support the PSI exchange format standard.

4.5. On-going and Planed Collaborative Projects

Many large proteomics projects are under way (www.hupo.org), only one interactome, *Saccharomyces cerevisiae*, can be said to have been thoroughly covered in terms of protein interactions. Other interactomes such as fly, *C.elegans* and of course *Homo sapiens* are far from being completed although comparative and predictive interactomics help to increase the number of reliable interactions. It is therefore natural to expect a human interactome project on a per basis in terms of resource to the human genome sequencing effort. The human interactome is the biology next big challenge and the first sketch of the project is in sight as attested by the article published by Marc Vidal (2006). It will be a highly collaborative project as it will not be limited to the protein network (e.g. mRNA co-expression metabolic pathways) and very certainly will involve a persistent integrative effort where an important role will be played to collect and structure all the data. We hope we have shown in these lines that bioinformatics tools and the use of well designed databases are necessary to provide the biologist with a robust framework onto which one can carry on analysis and build new hypothesis. The pharmaceutical industry will of course be very interested by the panoramic view a completed human interactome could provide but it will take years before obtaining a full picture of this interactome. In the meantime, small scale interactome on disease targets and pathways could be as informative and useful as it has been shown by the studies on TNF-α (Bouwmeester et al. 2004) and TGF-β (Colland et al. 2004).

5. INSIDE MOLECULAR COMPLEXES: HIGHLIGHTING AND MODULATING PPI FEATURES

The determination of PPIs in large numbers suggests putative new targets for which the interaction has to be validated by independent method prior to go further. Drawn up hypotheses need to undergo an additional validation step to test their biological relevance. A battery of conventional approaches enriched with more recent additional techniques can be used for assessing an interaction and providing a more quantitative description of the interface.

5.1. Biochemical and Physico-Chemical Validation of Protein Complexes

5.1.1. Available technologies: a quick survey

Recent reviews have focused on establishing a census report of available biochemical and biophysical methods to get a better understanding of PPI physico-chemical features. Pichler (2005) brought an overview of cutting-edge methodologies to identify and characterize PPIs (e.g. solid-phase detection like surface plasmon resonance (SPR)) while enlightening a few emerging techniques dedicated to single-complex detection as AFM (atomic force microscopy, see this book for further details). In the same way, numerous spin-off methods have been derived from NMR to fine study subunits organization within macromolecular assemblies. They proved to be powerful tools for scrutinizing weak and transient complexes (Bonvin et al. 2005). When overhauser effects (NOEs) are coupled to residual dipolar coupling (RDCs), analysis of chemical shifts perturbation allows to extract the relative orientations of docked components and scrutinize the interface topography. The precise determination of the oligomeric state allows to build a fitting model to be matched with collected experimental data in order to extrapolate quantitative information as stoichiometric ratio, binding affinity and kinetics rate constants (Lakey and Raggett 1998; Schreiber 2002). For example, isothermal titration calorimetry (ITC) can compute these parameters from a single experiment data set on the outcome of curve-fitting procedure. Finally, supramolecular assemblies may call for methods from structural biology to access special features as overall architecture and shapes or to draw the delimitation of subunits interfaces (Sali et al. 2003).

Lastly, we have to note that integral membrane proteins have been shown to have their own way to assemble, as illustrated with secondary transport proteins that drive some translocation reactions (Veenhoff et al. 2002). These authors list methods that specifically apply to membrane proteins whether studied in detergent solution or anchored to membrane. Because they are outside the scope of this review, these experimental approaches will only be enumerated into Table 3 at the exception of a very classical method (co-IP) and a cutting-edge technology (FRET/BRET).

5.1.2. CoIP: example of a powerful conventional approach

The principle of co-immunoprecipitation (coIP) is simple as protein X physiologically associates stably with protein Y then Y may co-precipitate with an antibody against X. It requires a prior knowledge of the protein of interest, a reasonable quantity of starting material and antibodies of high affinity (Séraphin 2002). One of the difficulties is to work under conditions that preserve the PPI as many wash steps are utilized during the procedure. Another pitfall to overcome is to reduce the background noise of non specific binding. The cell lysis disrupts the compartmentalization and two proteins can possibly associate when the different cell barriers are broken and therefore create an artificial interaction. The specificity and the quality of the antibody must be controlled as well. The rest of the co-IP procedure consists in identifying the complex for which mass spectrometry has become the tool of choice. Improvement in this technique has

Table 3. Summary of technologies dedicated to biochemical, physico-chemical and structural characterization of protein complexes v/V (*in vitro/in vivo*); M/HTS (Medium/High Throughput Screening)

Technique	Scope	PPI Ident.	Cell. Loc.	Affinity (K_d)	Specificities	References
Classical methods *(may be combined)*						
Site-directed mutagenesis	V	–	–	+	Interface characterisation	
Co-immunoprecipitation (coIP)	v/V - HTS	++	–	–		See Chapter 5
GST pull down	v - MTS	++	–	–		
Analytical centrifugation *Density gradient centrifugation*	v	++	–	+	Molecular Weight complex stoichiometry	
Gel filtration	v	+	–	++	Oligomeric state	
Chemical cross-linking	v/(V)	++	–	–	Specific protein–protein contact points	Trakselis et al. (2005)
(Far) Western blotting	v	++	–	+		
Large-scale identification						
Tandem affinity chromatography coupled to Mass Spectrometry (MS) *TAP-TAG, HMS-PCI*	v - HTS	+++	+	–	Supramolecular complexes	See Chapter 2
Yeast two-hybrid	V - HTS	+++	–	+	No purification needs interacting domains oligomeric state	See Chapter 2
Protein arrays (chips)	v - HTS	++	+	+		See Chapter 2

Phage, ribosome & mRNA display	v/V - HTS	++	–	–		See Chapter 2
Characterization of complex attributes						
Protein fragment complementation	v/V	+++		–		
Ubiquitin-based Y2H	MTS				Membrane proteins	See Chapter 2.2
DHFR PCA			++			Michnick (2001)
Fluorescence assay (BiFC)			+++			Kerppola (2006)
Circular dichroism spectroscopy	v	–	–	++		Lakey et al. (1998)
Enzymatic & chemical footprinting	v	+			Map protein sites involved in PPI	Heyduk et al. (1994)
Solid-phase detection	v - MTS	+	–	+++		Karlsson (2004)
Surface plasmon resonance (SPR)						
Isothermal titration calorimetry (ITC)	v	–	–	+++	Stoichiometry pH, salt, effects changes in ΔG, ΔH, ΔS	Lakey et al. (1998)
Fluorescence spectroscopy (FS)	v/V					
FRET/BRET	MTS	+	+++	+		See Chapter 5
Fluorescence Polarisation				++		Neduva et al. (2005)
coupled with confocal spectroscopy			+++		Conformational fluctuations	

(*Continued*)

Table 3. (*Continued*)

Technique	Scope	PPI Ident.	Cell. Loc.	Affinity $(K)_d$	Specificities	References
Single molecule detection						
AFM	v	+			Localisation	Piehler (2005)
Fluorescence (cross-)correlation spectroscopy	v/V		+++		Dynamics of interactions	Piehler (2005)
Supramolecular complexes						
3D features						
NMR	v	+	−	+		Bonvin et al. (2005)
NOE/TROSY/residual dipolar coupling (RDC)					Docking	
Chemical shift perturbations (CSP)					Relative orientations	
(Cross-)Saturation transfer (SAT)					Residues in proximity	
Electron microscopy	v	+	++	−	Images of complexes	
Cryo-electron tomography			+++		Cellular interactome	Nickell et al. (2006)
Immuno-electron microscopy		++	+++		Sub-cellular localisation	
x-ray crystallography	v	−	−	+	Interaction interfaces	

been brought by the use of immunomatrix which results in decreasing the level of contaminating light and heavy antibody chains (Qoronfleh et al. 2003). The range of the possibility of this technique may be greatly widened by initiatives such as the development of a human atlas of mono-specific antibodies. Nilsson et al. (2005) have shown that specific antibodies can be generated in a high throughput manner.

5.1.3. FRET/BRET technologies for in vitro *and* in vivo *characterization*

The discovery of auto-fluorescent proteins (Green, Yellow, and Cyan Fluorescent Protein) has ignited the development of novel techniques for studying gene function and their respective location. Among them, fluorescence resonance energy transfer (FRET) enables to detect PPI between fluorescent tags on interacting proteins. It is a non-radiative, dipole-dipole coupling process that transfers energy from an excited fluorophore to an acceptor fluorophore within close proximity (60Å). The GFP or its variants are genetically fused to the proteins of interest. The full power of this technique for mapping PPIs is when FRET is coupled to microscopy technology (Kenworthy 2001) and its development to fluorescence life-time imaging (FLIM, Clayton et al. 2002). Many PPIs have been uncovered by FRET or derived technology (FLIM, Flash-FRET) such as the pre-association of calmodulin with voltage-gated Ca^{2+} channels (Erickson et al. 2001), the interaction between Bcl-2 and Bax (Mahajan et al. 1998), between Pit1 and Ets-1 transcription factor (Day, 1998), between CXCR4 and the protein Kinase C (Peter et al. 2005) and Nox proteins with p22phox (Ambasta et al. 2004) just to name a few applications of the FRET technology.

The bioluminescence resonance energy transfer technology (BRET), very similar in principle to the FRET method, differs by using a bioluminescent donor instead of a fluorophore. The main advantage of BRET over FRET is to avoid photobleaching, high background of cellular auto-fluorescence and does not require exogenous illumination which can bias the interaction measurements. BRET seems also to be more amenable to high throughput screening. Concerning PPI, the most interesting studies have been performed on G-protein coupled receptor oligomerization (Angers et al. 2000; Germain-Desprez et al. 2003; Galès et al. 2005; Milligan et al. 2005).

Although it is still required to overexpress the proteins under study, the ability to measure directly the potential interaction between two proteins is a clear advantage of FRET and BRET. It preserves the physiological conditions under which the partners interact in the living cell which is not always the case of co-IP. It is still a technical and demanding technique and to date has not been used for proteome-scale studies comparable to TAP-TAG or Y2H.

5.1.4. Cross-experiments validation approaches

Successful and exhaustive understanding of complexes is generally achieved through the accumulation of disparate data coming from approaches that have to be scrupulously chosen based on expected easiness, level of accuracy, background noise, specificity, constraints as labelling needs or *in vitro/in vivo* carrying out.

Fortunately, affordable instruments (e.g. BIAcore systems for SPR implementation) or the close vicinity of dedicated facilities have made this aggregative trend more accessible for current investigation (Wilkinson 2004). As an illustration, Reichmann et al. (2005) have monitored the importance of key residues on the complex stability formed between TEM1-βlactamase and the β-lactamase inhibitor protein. For this they have measured change in free energy by SPR after mutations in conjunction with the analysis of mutational effects on the binding site structure as resolved by x-ray crystallography. Eggers et al. (2001) went further to dissect the heterotetrameric complex between ecotin, a serine protease inhibitor, and enzymes for which it displays pan-specificity and more especially they focused on the role of its dimerization in broad inhibitor specificity. They combined gel filtration analysis (oligomeric state), sedimentation equilibrium analysis (molecular weight), FRET monitoring of subunits exchange, fluorescence titration as well as measures of kinetics and dissociation constants involving a wide range of ecotin mutants for trypsin or chymotrypsin. Finally, Houtman et al. (2005) employed the hematopoietic-specific adapter protein (LAT) as a model to exemplify how signaling complexes can be characterized using a wide range of techniques such as confocal microscopy (visualization of protein co-localization), FRET microscopy (PPI quantification in cell), EM (high-resolution images of PPI localization), ITC (thermodynamics constants, PPI affinity and stoichiometry) and analytical ultracentrifugation (multiprotein complexes characterization).

5.2. Structural Outlook on Protein–Protein Interaction

Availability of large amount of structural data has opened new investigating ways in the field of PPIs by providing details at the atomic level. The foundation of a unique repository, the Protein Data Bank (PDB, see www.rcsb.org), at the earliest hours of structural biology, has greatly favoured the constitution of an exhaustive and homogeneous collection of biological macromolecules. Usually, the data format complies with experimental methods but this is not always biologically relevant. As an illustration, x-ray crystallographic data depict the content of the asymmetric unit that is the smallest part that is needed to reconstitute the entire cell and therefore the crystal by applying rotational and translational symmetries encoded into space groups. For example, the 2-fold axis inner to some homodimeric complex that allows to switch from a chain to its counterpart is very often indivisible of a crystallographic transformation axis resulting in a asymmetric unit filled with a single monomer. As a consequence, specialised server like Protein Quaternary Structure, PQS, hosted at EBI (Henrick and Thornton 1998; see pqs.ebi.ac.uk) offers to generate the biologically functional unit that acknowledges for the functional realness and quaternary structure of complexes. Dutta et al. (2005) have given a snapshot of the PDB: about 28000 structures of which 430 entries represent biological functional units greater than 500kDa. Examples of these macromolecular complexes can be found with icosahedral viruses (first published by 1984), ribosome complexes (2000), proteasome (1995) and chaperonin complex GroEL-GroES (1997).

5.2.1. Interface local structural environment

Many structural biology teams have demonstrated for the last two decades a great interest in the protein–protein interfaces. By performing a scrupulous dissection of recognition sites in terms of geometry (e.g. size, shape, complementarity), chemical nature (e.g. chemical groups, amino acids, hydrophobicity, electrostatics interactions or hydrogen bonds) crossed with sequence survey and 3D available data, these teams have brought new insights into interface stability and specificity.

Chakrabarti and Janin (2002) have proposed that recognition sites can be represented as a single patch or a collection of a few downsized patches. Accessibility to solvent measurements have allowed to specify the intimate pattern of standard-sized 1200–2000 Å^2 quite flat patch that may be viewed as a core and a rim, both filled with an equivalent number of amino acid residues, typically 12 by subunit. However, due to the differential distribution of their respective atoms regarding solvent accessibility and burying, they contribute differently to the overall patch surface, respectively 72 and 28%. Another dissonance was revealed by the amino acid composition: whereas the rim resembles the remainder of the protein surface, the core displays an excess of aliphatic and aromatic residues and is depleted in charged residues excepted Arg. These figures match well with those published by another group using radically distinct method based on binding energy changes (ΔG_d) after mutations. They corroborated the high propensity of Trp/Tyr to be part of the core while Ser/Thr are under-represented. Observed patterns of hydratation are quite heterogeneous (Rodier et al. 2005): on one hand the average value is around 10–11 water molecules per 1000 Å^2, on the other hand it can be found counterexamples as the completely dry interface of the dimeric protein kinase C (PKC) interacting domain. Water molecules and Arg-Asp/Glu salt bridges (about 2.2 per homodimer interface) contribute to the reinforcement of the assembly stability. It is worth noting that, contrary to an intuitive feeling, main chains constitute a well-populated pool of contributors for promoting interactions: side chain/side chain H-bonds weigh for only 38% encountered hydrogen bonds.

Another interesting point concerns the relative limited number of required recognition-patch residues compared to the admitted size of globular interacting domain. This is to be put in correlation with the distribution of these residues along discontinuous segments of the polypeptidic chain as well as the fact these interaction domains fit generally well with structural- or fold-related domains. Some idiosyncrasy has been spotted for homodimeric protein compared to other complexes (Bahadur et al. 2003). Their subunit interfaces are about twice as large as in complexes. This information should be coupled with the observation of an increase of larger patches as well as a hydrophibicity strengthening very close to the protein interior. This reflects quite well their differential mode of assembly: whereas complexes are mostly transient and are assembled once completely folded, therefore requiring the recognition site surface to be solvent compatible, homo-dimerization often occurs very soon and this permanent pairing is favoured by large hydrophobic patches. The same authors have pointed out remarkable disparities between specific and unspecific interactions that are devoid of biological relevance as encountered inside protein crystals to promote the crystalline packing (Bahadur et al. 2004). Non-specific interfaces are mainly characterised by a slack-off packing correlated with a debased shape complementary of the contacting protein surfaces. Additionally, it has been observed downsized recognition site filled with less fully buried atoms and hydrophobic residues but marked by an enhancement of water molecules representation.

Among panel of other encountered strategies related to interfaces scrutiny, Kim and Ison (2005) inquired how a domain family deals with multiple partnerships. They have catalogued about 4000 distinct interfaces as collections of points of contact between face-to-face domain family pairs mapped from SCOP classification (Andreeva et al. 2004). They have thoroughly investigated interaction specificity, interface diversity and sequence conservation. Resulting data set collected 1719 domain families among which 62% interact with a single family and 38% have partners from multiple families. If only 31% of families have been shown to interact with a unique face, interacting domain pairs are mediated on average through 1.2-1.9 distinct interfaces sometimes engaged in different orientations. Both, SH2 and SH3 domains illustrate nicely this fact by displaying at least three binding interfaces. They noticed that faces not only choose for 90% of them their other partner face but are also 94% selective inasmuch as a face determines its interaction partner as well as its binding orientation. It is proposed that the development of distinct faces for each partner may have provided each domain family with evolutionary linked modularity tools to prevent competitions between them.

5.2.2. Survey and classification of structural domain–domain interfaces

Several databases are publicly available across dedicated web-applications in order to investigate domain–domain interfaces in proteins of known structure. SCOPPI (Winter et al. 2006) is filled with about 8400 interface types obtained from a selective procedure. SCOPPI viewers allow to browse data, to consult multiple alignments, to access all complexes hosting an interface type and to filter by accessible surface area. Moreover, additional information are given concerning if domains reside on the same or a separate chain that is if the interaction is inter- or intra-molecular. A similar strategy was used to build PSIMAP/PSIbase (Gong et al. 2005). We can quote other interesting databases such as PIBASE (Davis and Salì 2005) and 3did (Stein et al. 2005), the latest being more focused on domain-domain interaction networks.

All interactions are not mediated by pairs of globular domains: many, required for cell signalling or trafficking, involve the binding of a protein domain to short linear motifs laying in disordered regions and matching particular sequence pattern, typically 3-8 residues long. As an illustration, SH3 and Clathrin domains bind respectively to PxxP and LDxL motifs. SH3-binding motifs may also be extended with extra residues (e.g. RxPxxP, PxxPxK) and can exhibit additional modulated recognition specificities towards SH3-embedded proteins. As a consequence it can bind to very different partners (Neduva et al. 2005).

5.2.3. Methods Combination dedicated to macromolecular complexes reconstruction

By mixing appropriate information coming from high- or low-resolution methods, hybrid procedures have already brought satisfying response to different structural issues (Russell et al. 2004). Indeed, even if x-ray crystallography does not have any theoretical size limitations, very large complexes continue to be seldom solved due to purification or crystallization bottlenecks as well as intrinsic exacerbated flexibility features. Conversely, cryo-electron microscopy (EM) and its derived methods are well-suited to catch a glimpse of the overall structure of very large complex architectures up to 7–10 Å resolution.

Approximate atomic model can be derived by fitting the atomic-resolution structures obtained independently for each subunit onto low-resolution cryo-EM model matching the complete assembly. Repositioning is hand-made or obtained through computational docking that usually requires a two-stage procedure where realistic positioning solutions are generated first and in a second step the best of them are refined using computed score to take into account particularities of the modelled interfaces. Dengue virus or tailed bacteriophages T4 reconstructions have been recently reviewed by Rossmann and collaborators (2005). In the case where some subunits would not have been provided with an atomic-level structure some circuitous alternatives have been imagined. Based on the widely accepted notion that proteins are likely to be folded and interact in the same way that close homologous with which they share a sequence similarity greater than 30–40% identity (Aloy et al. 2003), it is therefore possible to get a valuable model of a protein polypeptide by inferring its structure from comparative structure modelling that is homology modelling or even threading of far sequence-related proteins suspected to adopt a similar fold.

In a very analogous way, Aloy et al. (2006) have underlined the benefit of structural data together with performing the complete reconstitution of the fibroblast-growth-factor (FGF) signalling pathway. Thus, they plotted the usual static snapshot modelled by shaped pieces and arrows versus the corresponding dynamic view between interacting proteins displayed as 3D structures while highlighting key interaction domains and residues.

5.3. Acting on Interactions

The key-role of interacting protein domains in a wide range of cellular functions has sparked off a growing and lively interest in new tools to identify and validate original targets suitable for drug development (Legrain and Strosberg 2002).

Interacting domains can serve as dominant negative mutants, as modulators acting on individual proteins or as potentially interfering elements with some protein complexes existing in a diseased cell in order to recover a normal state. Further considerations make interactions themselves a groundbreaking target material that can be searched for novel agonists or antagonists. Recent excitement about past and present successes of therapeutic antibodies has demonstrated that the concept of interactions as extracellular drug targets is sound. In the past, some cytotoxic drugs (e.g. anti-tubulin) appeared *a posteriori* to be active as intracellular interaction inhibitors. The directed design of new intracellular inhibitors is too vast a topic to be comprehensively addressed in this review: we have made the choice to exemplify current strategies by three examples describing innovative directions.

5.3.1. Identify and engineer key residues governing interaction

Protein–protein interaction can dissociate through mutation on one partner and the wild-type protein expressed by cells for *in vivo* assay may interfere with such interaction-defective allele leading to unworkable results by compromising the phenotypic alteration for which the mutant was designed. Reverse two-hybrid system (Vidal and Legrain 1999) proposes an elegant solution to partly bypass these drawbacks by reducing the risk to generate unstable or misfolded proteins especially when no structural data are available to conduct a rational design of expected mutations. In opposition to forward Y2H described in Chapter 2, the genetic selection takes advantage of the interaction impairment. Emiliani et al. (2005) have recently used loss of affinity screening to unravel some interesting key point-mutations of HIV-1 integrase linked to defective viral replication. Based on the outcome (interaction of HIV-1 integrase with the transcriptional co-activator LEDGF/p75) of a classical Y2H screening, authors have carried out a library of integrase mutants (predominantly with single or no mutation) screened against LEDGF/p75 interaction domain used as the bait. They isolate two point-mutants of interest keeping the wild-type enzymatic activity but for which interaction does not occur preventing the virus to replicate. Further validation have conducted to underline the role of LEDGF/p75 in anchoring integrase to chromatin before the enzyme catalyses the chromosome inclusion of viral nucleotidic material. It is an illustration of two aspects already discussed in this review: the role and importance of co-localisation of proteins in the assembly of functional complexes and the therapeutical potential of interactions since it can be postulated that a compound mimicking the effect of mutations could finally be a drug with anti-viral activity.

5.3.2. Use of an interacting domain as a modulator

A strategy to efficiently and rapidly design interaction inhibitors is to check the inhibition efficacy of SID domains as determined by Y2H identification process described in Chapter 2. Colland et al. (2001), based on the *Helicobacter pylori* interaction map, first identified HP1122 as the anti-σ^{28} factor controlling the transcription of FlaA which is known as a major component of the *H.Pylori* flagellar filament. Overproduction of 29 C-terminal residues detected experimentally as the minimal subpart of HP1122 promoting interaction with the σ^{28} factor (i.e. the SID), succeeds in down-regulating the flagella synthesis therefore reducing the mobility of the *H. pylori*, considered as a key feature of its virulence.

5.3.3. Pharmacological modulation of protein–protein interactions

Up to a few years, PPIs inhibition was preferentially targeted by antibodies, dominant negatives proteins or medium-sized peptides. Difficulties met to design small-molecules inhibitors in this particular field can be explained by the inherent nature of interfaces (i.e. often flat and depleted of pockets or crevices) and also by the lack of natural inhibitors in opposition to enzymes for which competitive antagonists of all sorts are widely available and can be used as scaffold to design new inhibitors. Although this quest seems to be more challenging, successful examples are nevertheless on the increase: thus, following a phage-display screening on a small molecules library, Guo et al. (2000) designed a compound that restores the disrupted binding affinity of the engineered complex between a human growth hormone (hGH) and its receptor. Modified and upgraded two-hybrid systems for example Reverse two-hybrid system (Vidal and Endoh 1999) and Dual Bait two-hybrid system (Serebriiskii et al. 2002) have provided the biologist with alternative solutions to identify and qualify new inhibitors. Arkin and Wells (2004) showed how the current PPI inhibitors discovery strategy takes advantage of the knowledge of stoichiometry features, 3D structure of the binding site and QSAR analysis. Notably, several laboratories have recently reported low Kd (100 nM–10 μM) BCL inhibitors using high throughput screening or HSQC (^{1}H-^{15}N heteronuclear single quantum correlation) nuclear magnetic resonance (NMR). The compounds were designed against BCL2 and BCL-X_L, known anti-apoptotic proteins whose function is regulated by the binding of pro-apoptotic factors such as Bak and Bad notably characterised by an α-helice (the BH3 region). Recently, Oltersdorf et al. (2005) have built a Bad-derived BH3 peptide that has been shown to bind a large hydrophobic groove of BCL-X_L. This group performed "SAR by NMR" high throughput screening for compound identification followed by structure-based refinement for improving both potency and selectivity. They succeeded in designing a lead compound which prevents BCL-2 family to interact. Its usefulness for the lymphoma treatment was further validated by the observation of xenograft tumour regression. These encouraging results legitimate the hypothesis that these lead compounds may constitute the first intracellular PPIs inhibitors.

6. CONCLUSION

The long but not exhaustive list of methods and results presented in this review leads us to think that the study of pathway, signalling and trafficking through large or medium scale approach are just in its infancy. System biology can be described as the study of the cellular network and cell components organization. In this regard, PIMs already contribute significantly to its development. PIMs are noisy and contain a significant level of false positives and negatives. However the combination of introspective methods to discover and validate new interactions has certainly helped in the past three years to improve the general quality of large scale approaches. It will also be required to continue the development of robust bioinformatics tools and algorithms to organize, sort, filter and mine the data. The Protein Standard Initiative has paved the way for sharing such data at the community level and systematic use of it should be recommended for the scientific community to benefit from all existing data. In the near future, we can also expect to bring even more insights into the protein interaction field from structural genomics project. Crossing systematically structural data with interactomics data set will open up new perspectives and increase dramatically the level of details that are needed by the scientific community. System biology is evidently not only linked to protein interactions maps, it is also necessary to integrate other data coming from various functional genomics and proteomics projects such as transcriptomics and high-troughput phenotypic analyses.

However, even with sparse data based on focused aspect of the interactome, it is possible to derive very useful organizational information from the protein interaction maps. The human interactome has already been started to be analyzed and certainly will prove its interest as long as it is kept in mind that there is not a single human proteome but many sub-proteome depending on cells origin and conditions. Next decade should be marked by an increasing interest in PPI as a major provider of therapeutic targets or diagnostic assays. Before, we will previously have to get a better understanding of the PPIs role as key element of cell processes (Joyce et al. 2006). We think it is more than necessary to continue the patient construction and validation of protein interactions maps to set the background knowledge in order for the biologist to zoom in or out at will, to answer fundamental biological questions.

ACKNOWLEDGEMENTS

We thank Dr J.-C. Rain, Dr E. Formstecher, Dr L. Daviet, Dr F. Colland and Dr R. Delansorne for their useful comments and fruitful discussions on the manuscript.

REFERENCES

Albers, M., Kranz, H., Kober, I., Kaiser, C., Klink, M., Suckow, J., Kern, R. and Koegl, M. (2005) Automated yeast two-hybrid screening for nuclear receptor-interacting proteins. Mol. Cell Proteomics *4*, 205–213.

Albert, S., Gaudan, S., Knigge, H., Raetsch, A., Delgado, A., Huhse, B., Kirsch, H., Albers, M., Rebholz-Schuhmann, D. and Koegl, M. (2003) Computer-assisted generation of a protein-interaction database for nuclear receptors. Mol. Endocrinol. *17*, 1555–1567.

Alberts, B. (1998) The cell as a collection of protein machines: preparing the next generation of molecular biologists. Cell *92*, 291–294.

Aloy, P., Ceulemans, H., Stark, A. and Russell. R.B. (2003) The relationship between sequence and interaction divergence in proteins. J. Mol. Biol. *332*, 989–998.

Aloy, P. and Russell, R.B. (2006) Structural systems biology: modelling protein interactions. Nat. Rev. Mol. Cell Biol. *7*, 188–197.

Ambasta, R.K., Kumar, P., Griendling, K.K., Schmidt, H.H., Busse, R. and Brandes, R.P. (2004) Direct interaction of the novel Nox proteins with p22phox is required for the formation of a functionally active NADPH oxidase. J. Biol. Chem. *279*, 45935–45941.

Andreeva, A., Howorth, D., Brenner, S.E., Hubbard, T.J., Chothia, C. and Murzin, A.G. (2004) SCOP database in 2004: refinements integrate structure and sequence family data. Nucleic Acids Res. *32*, D226–D229.

Angers, S., Salahpour, A., Joly, E., Hilairet, S., Chelsky, D., Dennis, M. and Bouvier, M. (2000) Detection of beta 2-adrenergic receptor dimerization in living cells using bioluminescence resonance energy transfer (BRET). Proc. Natl Acad. Sci. U.S.A. *97*, 3684–3689.

Arkin, M.R. and Wells, J.A. (2004) Small-molecule inhibitors of protein–protein interactions: progressing towards the dream. Nat. Rev. Drug Discov. *3*, 301–317.

Bader, G.D. and Hogue, C.W. (2003) An automated method for finding molecular complexes in large protein interaction networks. BMC Bioinformatics *4*, 1–27.

Bahadur, R.P., Chakrabarti, P., Rodier, F. and Janin, J. (2003) Dissecting subunit interfaces in homodimeric proteins. Proteins *53*, 708–719.

Bahadur, R.P., Chakrabarti, P., Rodier, F. and Janin, J. (2004) A dissection of specific and non-specific protein–protein interfaces. J. Mol. Biol. *336*, 943–955.

Barabasi, A.L. and Oltvai, Z.N. (2004) Network biology: understanding the cell's functional organization. Nat. Rev. Genet. *5*, 101–113.

Bartel, P.L., Roecklein, J.A., SenGupta, D. and Fields, S. (1996) A protein linkage map of *Escherichia coli* bacteriophage T7. Nat. Genet. *12*, 72–77.

Batada, N.N. (2004) CNplot: visualizing pre-clustered networks. Bioinformatics *20*, 1455–1456.

Blaschke, C., Andrade, M.A. Ouzounis, C. and Valencia, A. (1999) Automatic extraction of biological information from scientific text: protein-protein interactions. Proc. Int. Conf. Intell. Syst. Mol. Biol., 60–67.

Bonvin, A.M., Boelens, R. and Kaptein, R. (2005) NMR analysis of protein interactions. Curr. Opin. Chem. Biol. *9*, 501–508.

Bork, P. and Serrano, L. (2005) Towards cellular systems in 4D. Cell 121, 507–509.

Bornberg-Bauer, E., Beaussart, F., Kummerfeld, S.K., Teichmann, S.A. and Weiner, J. 3rd (2005) The evolution of domain arrangements in proteins and interaction networks. Cell Mol. Life Sci. *62*, 435–445.

Bouwmeester, T., Bauch, A., Ruffner, H., Angrand, P.O., Bergamini, G., Croughton, K. Cruciat, C., Eberhard, D., Gagneur, J., Ghidelli, S., Hopf, C., Huhse, B., Mangano, R., Michon, A.M., Schirle, M., Schlegl, J., Schwab, M., Stein, M.A., Bauer, A., Casari, G., Drewes, G., Gavin, A.C., Jackson, D.B., Joberty, G., Neubauer, G., Rick, J., Kuster, B. and Superti-Furga, G. (2004) A physical and functional map of the human TNF-alpha/NF-kappa B signal transduction pathway. Nat. Cell Biol. *6*, 97–105.

Breitkreutz, B.J., Stark, C. and Tyers, M. (2003) Osprey: a network visualization system. Genome Biol. *4*, R22.

Brown, K.R. and Jurisica, I. (2005) Online predicted human interaction database. Bioinformatics *21*, 2076–2082.

Bürkle, L., Dinkel, M., Auerbach, D. and Stagljar, I. (2005) A GPCR interactome: a comprehensive membrane protein interaction map of Human G-protein coupled receptors using the membrane-based yeast two-hybrid approach. Mol. Cell. Proteomics *4* (Suppl. 1), S45.

Camonis, J.H. and White, M.A. (2005) Ral GTPases: corrupting the exocyst in cancer cells. Trends Cell Biol. *15*, 327–332.

Chakrabarti, P. and Janin, J. (2002) Dissecting protein-protein recognition sites. Proteins *47*, 334–343.

Clayton, A.H., Hanley, Q.S., Arndt Jovin, D.J., Subramaniam, V. and Jovin, T.M. (2002) Dynamic fluorescence anisotropy imaging microscopy in the frequency domain (rFLIM). Biophys. J. *83*, 1631–1649.

Colland, F., Jacq, X., Trouplin, V., Mougin, C., Groizeleau, C., Hamburger, A., Meil, A., Wojcik, J., Legrain, P. and Gauthier, J.M. (2004) Functional proteomics mapping of a human signaling pathway. Genome Res. *14*, 1324–1332.

Colland, F., Rain, J.C., Gounon, P., Labigne, A., Legrain, P. and De Reuse, H. (2001) Identification of the Helicobacter pylori anti-sigma28 factor. Mol. Microbiol. *41*, 477–487.

Dandekar, T., Snel, B., Huynen, M. and Bork, P. (1998) Conservation of gene order: a fingerprint of proteins that physically interact. Trends Biochem. Sci. *23*, 324–328.

Davis, F.P. and Sali, A. (2005) PIBASE: a comprehensive database of structurally defined protein interfaces. Bioinformatics *21*, 1901–1907.

Day, R.N. (1998) Visualization of Pit-1 transcription factor interactions in the living cell nucleus by fluorescence resonance energy transfer microscopy. Mol. Endocrinol. *12*, 1410–1419.

Durand, P., Labarre, L., Meil, A., Divol, J.L., Vandenbrouck, Y., Viari, A. and Wojcik, J. (2006) GenoLink: a graph-based querying and browsing system for investigating the function of genes and proteins. BMC Bioinformatics *7*, 21.

Dutta, S. and Berman, H.M. (2005) Large macromolecular complexes in the Protein Data Bank: a status report. Structure *13*, 381–388.

Dziembowski, A. and Seraphin, B. (2004) Recent developments in the analysis of protein complexes. FEBS Lett. 556, 1–6.

Eggers, C.T., Wang, S.X., Fletterick, R.J. and Craik, C.S. (2001) The role of ecotin dimerization in protease inhibition. J. Mol. Biol. *308*, 975–991.

Emiliani, S., Mousnier, A., Busschots, K., Maroun, M., Van Maele, B., Tempe, D., Vandekerckhove, L., Moisant, F., Ben-Slama, L., Witvrouw, M., Christ, F., Rain, J.C., Dargemont, C., Debyser, Z. and Benarous, R. (2005) Integrase mutants defective for interaction with LEDGF/p75 are impaired in chromosome tethering and HIV-1 replication. J. Biol. Chem. *280*, 25517–25523.

Erickson, M.G., Alseikhan, B.A., Peterson, B.Z. and Yue, D.T. (2001) Preassociation of calmodulin with voltage-gated Ca(2+) channels revealed by FRET in single living cells. Neuron. *31*, 973–985.

Fields, S. and Song, O. (1989) A novel genetic system to detect protein-protein interactions. Nature *340*, 245–246.

Finley, R.L. Jr. and Brent, R. (1994) Interaction mating reveals binary and ternary connections between Drosophila cell cycle regulators. Proc. Natl Acad. Sci. U.S.A. *91*, 12980–12984.

Formstecher, E., Aresta, S., Collura, V., Hamburger, A., Meil, A., Trehin, A., Reverdy, C., Betin, V., Maire, S., Brun, C., Jacq, B., Arpin, M., Bellaiche, Y., Bellusci, S., Benaroch, P., Bornens, M., Chanet, R., Chavrier, P., Delattre, O., Doye, V., Fehon R., Faye, G., Galli, T., Girault, J.A., Goud, B., de Gunzburg, J., Johannes, L., Junier, M.P. Mirouse, V., Mukherjee, A., Papadopoulo, D., Perez, F., Plessis, A., Rosse, C., Saule, S., Stoppa-Lyonnet, D., Vincent, A., White, M., Legrain, P., Wojcik, J., Camonis, J. and Daviet. L. (2005) Protein interaction mapping: a Drosophila case study. Genome Res. *15*, 376–384.

Fukai, S., Matern, H.T., Jagath, J.R., Scheller, R.H. and Brunger, A.T. (2003) Structural basis of the interaction between RalA and Sec5, a subunit of the sec6/8 complex. EMBO J. *22*, 3267–3278.

Gales, C., Rebois, R.V., Hogue, M., Trieu, P., Breit, A., Hebert, T.E. and Bouvier, M. (2005) Real-time monitoring of receptor and G-protein interactions in living cells. Nat. Methods *2*, 177–184.

Gavin, A.C., Aloy, P., Grandi, P., Krause, R., Boesche, M., Marzioch, M., Rau, C., Jensen, L.J., Bastuck, S., Dumpelfeld, B., Edelmann, A., Heurtier, M.A., Hoffman, V., Hoefert, C., Klein, K., Hudak, M., Michon, A.M., Schelder, M., Schirle, M., Remor, M., Rudi, T., Hooper, S., Bauer, A., Bouwmeester, T., Casari, G., Drewes, G., Neubauer, G., Rick, J.M., Kuster, B., Bork, P., Russell, R.B. and Superti-Furga, G. (2006) Proteome survey reveals modularity of the yeast cell machinery. Nature *440*, 631–636.

Gavin, A.C., Bosche, M., Krause, R., Grandi, P., Marzioch, M., Bauer, A., Schultz, J., Rick, J.M., Michon, A.M., Cruciat, C.M., Remor, M., Hofert, C., Schelder, M., Brajenovic, M., Ruffner, H., Merino, A., Klein, K., Hudak, M., Dickson, D., Rudi, T., Gnau, V., Bauch, A., Bastuck, S., Huhse, B., Leutwein, C., Heurtier, M.A., Copley, R.R., Edelmann, A., Querfurth, E., Rybin, V., Drewes, G., Raida, M., Bouwmeester, T., Bork, P., Seraphin, B., Kuster, B., Neubauer, G. and Superti-Furga, G. (2002) Functional organization of the yeast proteome by systematic analysis of protein complexes. Nature *415*, 141–147.

Germain-Desprez, D., Bazinet, M., Bouvier, M. and Aubry, M. (2003) Oligomerization of transcriptional intermediary factor 1 regulators and interaction with ZNF74 nuclear matrix protein revealed by bioluminescence resonance energy transfer in living cells. J. Biol. Chem. *278*, 22367–22373.

Giot, L., Bader, J.S., Brouwer, C., Chaudhuri, A., Kuang, B., Li, Y., Hao, Y.L., Ooi, C.E., Godwin, B., Vitols, E., Vijayadamodar, G., Pochart, P., Machineni, H., Welsh, M., Kong, Y., Zerhusen, B., Malcolm, R., Varrone, Z., Collis, A., Minto, M., Burgess, S., McDaniel, L., Stimpson, E., Spriggs, F., Williams, J., Neurath, K., Ioime, N., Agee, M. Voss, E. Furtak, K. Renzulli, R. Aanensen, N. Carrolla, S. Bickelhaupt, E. Lazovatsky, Y. DaSilva, A. Zhong, J. Stanyon, C.A. Finley, R.L. Jr., White,

K.P., Braverman, M., Jarvie, T., Gold, S., Leach, M., Knight, J., Shimkets, R.A., McKenna, M.P., Chant, J. and Rothberg, J.M. (2003) A protein interaction map of Drosophila melanogaster. Science *302*, 1727–1736.

Goh, C.S., Milburn, D. and Gerstein, M. (2004) Conformational changes associated with protein-protein interactions. Curr. Opin. Struct. Biol. *14*, 104–109.

Gong, S., Yoon, G., Jang, I., Bolser, D., Dafas, P., Schroeder, M., Choi, H., Cho, Y., Han, K., Lee, S., Choi, H., Lappe, M., Holm, L., Kim, S., Oh, D. and Bhak, J. (2005) PSIbase: a database of Protein Structural Interactome map (PSIMAP). Bioinformatics *21*, 2541–2543.

Guo, Z., Zhou, D. and Schultz, P.G. (2000) Designing small-molecule switches for protein–protein interactions. Science *288*, 2042–2045.

Hammond, P.W., Alpin, J. Rise, C.E., Wright, M. and Kreider, B.L. (2001) *In vitro* selection and characterization of Bcl-X(L)-binding proteins from a mix of tissue-specific mRNA display libraries. J. Biol. Chem. 276, 20898–20906.

Han, J.D., Bertin, N., Hao, T., Goldberg, D.S., Berriz, G.F., Zhang, L.V., Dupuy, D., Walhout, A.J., Cusick, M.E., Roth, F.P. and Vidal, M. (2004b) Evidence for dynamically organized modularity in the yeast protein-protein interaction network. Nature *430*, 88–93.

Han, J.D., Dupuy, D., Bertin, N., Cusick, M.E. and Vidal, M. (2005) Effect of sampling on topology predictions of protein-protein interaction networks. Nat. Biotechnol. *23*, 839–844.

Han, K., Park, B., Kim, H., Hong, J. and Park, J. (2004a) HPID: the human protein interaction database. Bioinformatics *20*, 2466–2470.

Henrick, K. and Thornton, J.M. (1998) PQS: a protein quaternary structure file server. Trends Biochem. Sci. *23*, 358–361.

Hermjakob, H., Montecchi-Palazzi, L., Bader, G., Wojcik, J., Salwinski, L., Ceol, A., Moore, S., Orchard, S., Sarkans, U., von Mering, C., Roechert, B., Poux, S., Jung, E., Mersch, H., Kersey, P., Lappe, M., Li, Y., Zeng, R., Rana, D., Nikolski, M., Husi, H., Brun, C., Shanker, K., Grant, S.G., Sander, C., Bork, P., Zhu, W., Pandey, A., Brazma, A., Jacq, B., Vidal, M., Sherman, D., Legrain, P., Cesareni, G., Xenarios, I., Eisenberg, D., Steipe, B., Hogue, C. and Apweiler, R. (2004) The HUPO PSI's molecular interaction format – a community standard for the representation of protein interaction data. Nat. Biotechnol. *22*, 177–183.

Heyduk, E. and Heyduk, T. (1994) Mapping protein domains involved in macromolecular interactions: a novel protein footprinting approach. Biochemistry *33*, 9643–9650.

Ho, Y., Gruhler, A., Heilbut, A., Bader, G.D., Moore, L., Adams, S.L., Millar, A., Taylor, P., Bennett, K., Boutilier, K., Yang, L., Wolting, C., Donaldson, I., Schandorff, S., Shewnarane, J., Vo, M., Taggart, J., Goudreault, M., Muskat, B., Alfarano, C., Dewar, D., Lin, Z., Michalickova, K., Willems, A.R., Sassi, H., Nielsen, P.A., Rasmussen, K.J., Andersen, J.R., Johansen, L.E., Hansen, L.H., Jespersen, H., Podtelejnikov, A., Nielsen, E., Crawford, J., Poulsen, V., Sorensen, B.D., Matthiesen, J., Hendrickson, R.C., Gleeson, F., Pawson, T., Moran, M.F., Durocher, D., Mann, M., Hogue, C.W., Figeys, D. and Tyers, M. (2002) Systematic identification of protein complexes in Saccharomyces cerevisiae by mass spectrometry. Nature *415*, 180–183.

Hoffmann, R. and Valencia, A. (2004) A gene network for navigating the literature. Nat. Genet. *36*, 664.

Hooper, S.D. and Bork, P. (2005) Medusa: a simple tool for interaction graph analysis. Bioinformatics *21*, 4432–4433.

Houtman, J.C., Barda-Saad, M. and Samelson, L.E. (2005) Examining multiprotein signaling complexes from all angles. FEBS J. *272*, 5426–5435.

Hsu, S.C., TerBush, D., Abraham, M. and Guo, W. (2004) The exocyst complex in polarized exocytosis. Int. Rev. Cytol. *233*, 243–265.

Hu, Z., Mellor, J., Wu, J. and DeLisi, C. (2004) VisANT: an online visualization and analysis tool for biological interaction data. BMC Bioinformatics *5*, 17.

Huynen, M., Snel, B., Lathe, W. 3rd and Bork, P. (2000) Predicting protein function by genomic context: quantitative evaluation and qualitative inferences. Genome Res. *10*, 1204–1210.

Huynen, M.A., Snel, B., von Mering, C. and Bork, P. (2003) Function prediction and protein networks. Curr. Opin. Cell. Biol. *15*, 191–198.

Iragne, F., Nikolski, M., Mathieu, B., Auber, D. and Sherman, D. (2005) ProViz: protein interaction visualization and exploration. Bioinformatics *21*, 272–274.

Ito, T., Chiba, T., Ozawa, R., Yoshida, M., Hattori, M. and Sakaki, Y. (2001) A comprehensive two-hybrid analysis to explore the yeast protein interactome. Proc. Natl Acad. Sci. U.S.A. *98*, 4569–4574.

Johnsson, N. and Varshavsky, A. (1994) Split ubiquitin as a sensor of protein interactions *in vivo*. Proc. Natl Acad. Sci. U.S.A. *91*, 10340–10344.

Jones, R.B., Gordus, A., Krall, J.A. and MacBeath, G. (2006) A quantitative protein interaction network for the ErbB receptors using protein microarrays. Nature *439*, 168–174.

Joyce, A.R. and Palsson, B.O. (2006) The model organism as a system: integrating 'omics' data sets. Nat. Rev. Mol. Cell Biol. *7*, 198–210.

Karlsson, R. (2004) SPR for molecular interaction analysis: a review of emerging application areas. J. Mol. Recognit. *17*, 151–161.

Kelley, B.P., Sharan, R., Karp, R.M., Sittler, T., Root, D.E., Stockwell, B.R. and Ideker, T. (2003) Conserved pathways within bacteria and yeast as revealed by global protein network alignment. Proc. Natl Acad. Sci. U.S.A. *100*, 11394–11399.

Kenworthy, A.K. (2001) Imaging protein-protein interactions using fluorescence resonance energy transfer microscopy. Methods *24*, 289–296.

Kerppola, T.K. (2006) Visualization of molecular interactions by fluorescence complementation. Nat. Rev. Mol. Cell Biol. *7*, 449–456.

Kim, W.K. and Ison, J.C. (2005) Survey of the geometric association of domain-domain interfaces. Proteins *61*, 1075–1088.

Lakey, J.H. and Raggett, E.M. (1998) Measuring protein-protein interactions. Curr. Opin. Struct. Biol. *8*, 119–123.

Lebreton, A., Saveanu, C., Decourty, L., Rain, J.C., Jacquier, A. and Fromont-Racine, M. (2006) A functional network involved in the recycling of nucleocytoplasmic pre-60S factors. J. Cell Biol. *173*, 349–360.

Legrain, P. and Selig, L. (2000) Genome-wide protein interaction maps using two-hybrid systems. FEBS Lett. *480*, 32–36.

Legrain, P. and Strosberg, D. (2002) Protein interaction domain mapping for the selection of validated targets and lead compounds in the anti-infectious area. Curr. Pharm. Des. *8*, 1189–1198.

Legrain, P., Wojcik, J. and Gauthier, J.M. (2001) Protein–protein interaction maps: a lead towards cellular functions. Trends Genet. *17*, 346–352.

Li, S., Armstrong, C.M., Bertin, N., Ge, H., Milstein, S., Boxem, M., Vidalain, P.O., Han, J.D., Chesneau, A., Hao, T., Goldberg, D.S., Li, N., Martinez, M., Rual, J.F., Lamesch, P., Xu, L., Tewari, M., Wong, S.L., Zhang, L.V., Berriz, G.F., Jacotot, L., Vaglio, P., Reboul, J., Hirozane-Kishikawa, T., Li, Q., Gabel, H.W., Elewa, A., Baumgartner, B., Rose, D.J., Yu, H., Bosak, S., Sequerra, R., Fraser, A., Mango, S.E., Saxton, W.M., Strome, S., Van Den Heuvel, S., Piano, F., Vandenhaute, J., Sardet, C., Gerstein, M., Doucette-Stamm, L., Gunsalus, K.C., Harper, J.W., Cusick, M.E., Roth, F.P., Hill, D.E. and Vidal, M. (2004) A map of the interactome network of the metazoan *C.Elegans*. Science *303*, 540–543.

Lipovsek, D. and Pluckthun, A. (2004) *In-vitro* protein evolution by ribosome display and mRNA display. J. Immunol. Methods *290*, 51–67.

Luban, J., Bossolt, K.L., Franke, E.K., Kalpana, G.V. and Goff, S.P. (1993) Human immunodeficiency virus type 1 Gag protein binds to cyclophilins A and B. Cell *73*, 1067–1078.

MacBeath, G. and Schreiber, S.L. (2000) Printing proteins as microarrays for high-throughput function determination. Science *289*, 1760–1763.

Mahajan, N.P., Linder, K., Berry, G., Gordon, G.W., Heim, R. and Herman, B. (1998) Bcl-2 and Bax interactions in mitochondria probed with green fluorescent protein and fluorescence resonance energy transfer. Nat. Biotechnol. *16*, 547–552.

Mann, M., Hendrickson, R.C. and Pandey, A. (2001) Analysis of proteins and proteomes by mass spectrometry. Annu. Rev. Biochem. *70*, 437–473.

Marcotte, E.M., Pellegrini, M., Ng, H.L., Rice, D.W., Yeates, T.O. and Eisenberg, D. (1999) Detecting protein function and protein-protein interactions from genome sequences. Science *285*, 751–753.

Marcotte, E.M., Xenarios, I. and Eisenberg, D. (2001) Mining literature for protein-protein interactions. Bioinformatics *17*, 359–363.

Marianayagam, N.J., Sunde, M. and Matthews, J.M. (2004) The power of two: protein dimerization in biology. Trends Biochem. Sci. *29*, 618–625.

Matthews, L.R., Vaglio, P., Reboul, J., Ge, H., Davis, B.P., Garrels, J., Vincent, S. and Vidal, M. (2001) Identification of potential interaction networks using sequence-based searches for conserved protein-protein interactions or "interologs." Genome Res. *11*, 2120–2126.

Mayer, M.L. and Hieter, P. (2000) Protein networks-built by association. Nat. Biotechnol. 18, 1242–1243.

Meil, A., Durand, P. and Wojcik, J. (2005) PIMWalker: visualising protein interaction networks using the HUPO PSI molecular interaction format. Appl. Bioinformatics *4*, 137–139.

Michnick, S.W. (2001) Exploring protein interactions by interaction-induced folding of proteins from complementary peptide fragments. Curr. Opin. Struct. Biol. *11*, 472–477.

Miller, J.P., Lo, R.S., Ben-Hur, A., Desmarais, C., Stagljar, I., Noble, W.S. and Fields, S. (2005) Large-scale identification of yeast integral membrane protein interactions. Proc. Natl Acad. Sci. U.S.A. *102*, 12123–12128.

Milligan, G. (2004) G protein-coupled receptor dimerization: function and ligand pharmacology. Mol. Pharmacol. *66*, 1–7.

Milligan, G., Wilson, S. and Lopez-Gimenez, J.F. (2005) The specificity and molecular basis of alpha1-adrenoceptor and CXCR chemokine receptor dimerization. J. Mol. Neurosci. *26*, 161–168.

Nakai, K. and Horton, P. (1999) PSORT: a program for detecting sorting signals in proteins and predicting their subcellular localization. Trends Biochem. Sci. *24*, 34–36.

Neduva, V., Linding, R., Su-Angrand, I., Stark, A., de Masi, F., Gibson, T.J., Lewis, J., Serrano, L. and Russell, R.B. (2005) Systematic discovery of new recognition peptides mediating protein interaction networks. PLoS Biol. *3*, e405.

Nickell, S., Kofler, C., Leis, A.P. and Baumeister, W. (2006) A visual approach to proteomics. Nat. Rev. Mol. Cell Biol. *7*, 225–230.

Nilsson, P., Paavilainen, L., Larsson, K., Odling, J., Sundberg, M., Andersson, A.C., Kampf, C., Persson, A., Al-Khalili Szigyarto, C., Ottosson, J., Bjorling, E., Hober, S., Wernerus, H., Wester, K., Ponten, F. and Uhlen, M. (2005) Towards a human proteome atlas: high-throughput generation of mono-specific antibodies for tissue profiling. Proteomics *5*, 4327–4337.

Nooren, I.M. and Thornton, J.M. (2003a) Diversity of protein-protein interactions. EMBO J. *22*, 3486–3492.

Nooren, I.M. and Thornton. J.M. (2003b) Structural characterisation and functional significance of transient protein-protein interactions. J. Mol. Biol. *325*, 991–1018.

Oka, T. and Krieger, M. (2005) Multi-component protein complexes and Golgi membrane trafficking. J. Biochem. (Tokyo) *137*, 109–114.

Oltersdorf, T., Elmore, S.W., Shoemaker, A.R., Armstrong, R.C., Augeri, D.J., Belli, B.A., Bruncko, M., Deckwerth, T.L., Dinges, J., Hajduk, P.J., Joseph, M.K., Kitada, S., Korsmeyer, S.J., Kunzer, A.R., Letai, A., Li, C., Mitten, M.J., Nettesheim, D.G., Ng, S., Nimmer, P.M., O'Connor, J.M., Oleksijew, A., Petros, A.M., Reed, J.C., Shen, W., Tahir, S.K., Thompson, C.B., Tomaselli, K.J., Wang, B., Wendt, M.D., Zhang, H., Fesik, S.W. and Rosenberg, S.H. (2005) An inhibitor of Bcl-2 family proteins induces regression of solid tumours. Nature *435*, 677–681.

Ono, T., Hishigaki, H., Tanigami, A. and Takagi, T. (2001) Automated extraction of information on protein-protein interactions from the biological literature. Bioinformatics *17*, 155–161.

Pagel, P., Kovac, S., Oesterheld, M., Brauner, B., Dunger-Kaltenbach, I., Frishman, G., Montrone, C., Mark, P., Stumpflen, V., Mewes, H.W., Ruepp, A. and Frishman, D. (2005) The MIPS mammalian protein-protein interaction database. Bioinformatics *21*, 832–834.

Pazos, F., Ranea, J.A., Juan, D. and Sternberg, M.J. (2005) Assessing protein co-evolution in the context of the tree of life assists in the prediction of the interactome. J. Mol. Biol. *352*, 1002–1015.

Pazos, F. and Valencia, A. (2001) Similarity of phylogenetic trees as indicator of protein-protein interaction. Protein Eng. *14*, 609–614.

Pazos, F. and Valencia, A. (2002) *In silico* two-hybrid system for the selection of physically interacting protein pairs. Proteins *47*, 219–227.

Pellegrini, M., Haynor, D. and Johnson, J.M. (2004) Protein interaction networks. Expert Rev Proteomics *1*, 239–249.

Pellegrini, M., Marcotte, E.M., Thompson, M.J., Eisenberg, D. and Yeates, T.O. (1999) Assigning protein functions by comparative genome analysis: protein phylogenetic profiles. Proc. Natl Acad. Sci. U.S.A. *96*, 4285–4288.

Pereira-Leal, J.B., Levy, E.D. and Teichmann, S.A. (2006) The origins and evolution of functional modules: lessons from protein complexes. Philos. Trans. R. Soc. Lond., B, Biol. Sci. *361*, 507–517.

Peri, S., Navarro, J.D., Amanchy, R., Kristiansen, T.Z., Jonnalagadda, C.K., Surendranath, V., Niranjan, V., Muthusamy, B., Gandhi, T.K., Gronborg, M., Ibarrola, N., Deshpande, N., Shanker, K., Shivashankar, H.N., Rashmi, B.P., Ramya, M.A., Zhao, Z., Chandrika, K.N., Padma, N., Harsha, H.C., Yatish, A.J., Kavitha, M.P., Menezes, M., Choudhury, D.R., Suresh, S., Ghosh, N., Saravana, R., Chandran, S., Krishna, S., Joy, M., Anand, S.K., Madavan, V., Joseph, A., Wong, G.W., Schiemann, W.P., Constantinescu, S.N., Huang, L., Khosravi-Far, R., Steen, H., Tewari, M., Ghaffari, S., Blobe, G.C., Dang, C.V., Garcia, J.G., Pevsner, J., Jensen, O.N., Roepstorff, P., Deshpande, K.S., Chinnaiyan, A.M., Hamosh, A., Chakravarti, A. and Pandey, A. (2003) Development of human protein reference database as an initial platform for approaching systems biology in humans. Genome. Res. *13*, 2363–2371.

Peter, M., Ameer-Beg, S.M., Hughes, M.K., Keppler, M.D., Prag, S., Marsh, M., Vojnovic, B. and Ng, T. (2005) Multiphoton-FLIM quantification of the EGFP-mRFP1 FRET pair for localization of membrane receptor-kinase interactions. Biophys. J. *88*, 1224–1237.

Piehler, J. (2005) New methodologies for measuring protein interactions *in vivo* and *in vitro*. Curr. Opin. Struct. Biol. *15*, 4–14.

Puig, O., Caspary, F., Rigaut, G., Rutz, B., Bouveret, E., Bragado-Nilsson, E., Wilm, M. and Seraphin, B. (2001) The tandem affinity purification (TAP) method: a general procedure of protein complex purification. Methods *24*, 218–229.

Qoronfleh, M.W., Ren, L., Emery, D., Perr, M. and Kaboord, B. (2003) Use of immunomatrix Methods to improve protein-protein interaction detection. J. Biomed. Biotechnol. *2003*, 291–298.

Rain, J.C., Selig, L., De Reuse, H., Battaglia, V., Reverdy, C., Simon, S., Lenzen, G., Petel, F., Wojcik, J., Schachter, V., Chemama, Y., Labigne, A. and Legrain, P. (2001) The protein-protein interaction map of Helicobacter pylori. Nature *409*, 211–215.

Ramani, A.K., Bunescu, R.C., Mooney, R.J. and Marcotte, E.M. (2005) Consolidating the set of known human protein-protein interactions in preparation for large-scale mapping of the human interactome. Genome Biol. *6*, R40.

Ramani, A.K. and Marcotte, E.M. (2003) Exploiting the co-evolution of interacting proteins to discover interaction specificity. J. Mol. Biol. *327*, 273–284.

Reichmann, D., Rahat, O., Albeck, S., Meged, R., Dym, O. and Schreiber, G. (2005) The modular architecture of protein-protein binding interfaces. Proc. Natl Acad. Sci. U.S.A. *102*, 57–62.

Remy, I. and Michnick, S.W. (2001) Visualization of biochemical networks in living cells. Proc. Natl Acad. Sci. U.S.A. 98, 7678–7683.

Rigaut, G., Shevchenko, A., Rutz, B., Wilm, M., Mann, M. and Seraphin, B. (1999) A generic protein purification method for protein complex characterization and proteome exploration. Nat. Biotechnol. *17*, 1030–1032.

Rodier, F., Bahadur, R.P., Chakrabarti, P. and Janin, J. (2005) Hydration of protein-protein interfaces. Proteins *60*, 36–45.

Rossmann, M.G., Morais, M.C., Leiman, P.G. and Zhang, W. (2005) Combining X-ray crystallography and electron microscopy. Structure *13*, 355–362.

Royer, W.E. Jr, Zhu, H., Gorr, T.A., Flores, J.F. and Knapp, J.E. (2005) Allosteric hemoglobin assembly: diversity and similarity. J. Biol. Chem. *280*, 27477–27480.

Rual, J.F., Venkatesan, K., Hao, T., Hirozane-Kishikawa, T., Dricot, A., Li, N., Berriz, G.F., Gibbons, F.D., Dreze, M., Ayivi-Guedehoussou, N., Klitgord, N., Simon, C., Boxem, M., Milstein, S., Rosenberg, J., Goldberg, D.S., Zhang, L.V., Wong, S.L., Franklin, G., Li, S., Albala, J.S., Lim, J., Fraughton, C., Llamosas, E., Cevik, S., Bex, C., Lamesch, P., Sikorski, R.S., Vandenhaute, J., Zoghbi, H.Y., Smolyar, A., Bosak, S., Sequerra, R., Doucette-Stamm, L., Cusick, M.E., Hill, D.E., Roth, F.P. and Vidal, M. (2005) Towards a proteome-scale map of the human protein-protein interaction network. Nature *437*, 1173–1178.

Russell, R.B., Alber, F., Aloy, P., Davis, F.P., Korkin, D., Pichaud, M., Topf, M. and Sali, A. (2004) A structural perspective on protein-protein interactions. Curr. Opin. Struct. Biol. *14*, 313–324.

Sali, A., Glaeser, R., Earnest, T. and Baumeisterm, W. (2003) From words to literature in structural proteomics. Nature *422*, 216–225.

Salwinski, L., Miller, C.S., Smith, A.J., Pettit, F.K., Bowie, J.U. and Eisenberg, D. (2004) The Database of interacting proteins: 2004 update. Nucleic Acids Res. *32*, D449–D451.

Sato, T., Yamanishi, Y., Kanehisa, M. and Toh, H. (2005) The inference of protein-protein interactions by co-evolutionary analysis is improved by excluding the information about the phylogenetic relationships. Bioinformatics *21*, 3482–3489.

Schimmele, B. and Pluckthun, A. (2005) Identification of a functional epitope of the Nogo receptor by a combinatorial approach using ribosome display. J. Mol. Biol. *352*, 229–241.

Schreiber, G. (2002) Kinetic studies of protein-protein interactions. Curr. Opin. Struct. Biol. *12*, 41–47.

Séraphin, B. (2002) Identification of transiently interacting proteins and of stable protein complexes. In: Advances in Protein Chemistry. J. Janin and S.J. Wodak, eds. (U.S.A.: Elsevier Science), vol. 61, pp. 99–117.

Serebriiskii, I.G., Mitina, O., Pugacheva, E.N., Benevolenskaya, E., Kotova, E., Toby, G.G., Khazak, V., Kaelin, W.G., Chernoff, J. and Golemis, E.A. (2002) Detection of peptides, proteins, and drugs that selectively interact with protein targets. Genome Res. 12, 1785–1791.

Shannon, P., Markiel, A., Ozier, O., Baliga, N.S., Wang, J.T., Ramage, D., Amin, N., Schwikowski, B. and Ideker, T. (2003) Cytoscape: a software environment for integrated models of biomolecular interaction networks. Genome Res. *13*, 2498–2504.

Sharan, R., Suthram, S., Kelley, R.M., Kuhn, T., McCuine, S., Uetz, P., Sittler, T., Karp, R.M. and Ideker. T. (2005) Conserved patterns of protein interaction in multiple species. Proc. Natl Acad. Sci. U.S.A. *102*, 1974–1979.

Sprinzak, E., Sattath, S. and Margalit, H. (2003) How reliable are experimental protein-protein interaction data? J. Mol. Biol. *327*, 919–923.

Stagljar, I. and Fields, S. (2002) Analysis of membrane protein interactions using yeast-based technologies. Trends. Biochem. Sci. *27*, 559–563.

Stagljar, I., Korostensky, C., Johnsson, N. and te Heesen, S. (1998) A genetic system based on split-ubiquitin for the analysis of interactions between membrane proteins *in vivo*. Proc. Natl Acad. Sci. U.S.A. *95*, 5187–5192.

Stanyon, C.A., Liu, G., Mangiola, B.A., Patel, N., Giot, L., Kuang, B., Zhang, H., Zhong, J. and Finley, Jr. R.L. (2004) A Drosophila protein-interaction map centered on cell-cycle regulators. Genome Biol. *5*, R96.

Stark, C., Breitkreutz, B.J., Reguly, T., Boucher, L., Breitkreutz, A. and Tyers, M. (2006) BioGRID: a general repository for interaction datasets. Nucleic Acids Res. *34*, D535–D539.

Stein, A., Russell, R.B. and Aloy, P. (2005) 3did: interacting protein domains of known three-dimensional structure. Nucleic Acids Res. *33*, D413–D417.

Stelzl, U., Worm, U., Lalowski, M., Haenig, C., Brembeck, F.H., Goehler, H., Stroedicke, M., Zenkner, M., Schoenherr, A., Koeppen, S., Timm, J., Mintzlaff, S., Abraham, C., Bock, N., Kietzmann, S., Goedde, A., Toksoz, E., Droege, A., Krobitsch, S., Korn, B., Birchmeier, W., Lehrach, H. and Wanker, E.E. (2005) A human protein-protein interaction network: a resource for annotating the proteome. Cell *122*, 957–968.

Stiffler, M.A., Grantcharova, V.P., Sevecka, M. and Macbeath, G. (2006) Uncovering quantitative protein interaction networks for mouse PDZ domains using protein microarrays. J. Am. Chem. Soc. *128*, 5913–5922.

Suzuki, H., Saito, R., Kanamori, M., Kai, C., Schonbach, C., Nagashima, T., Hosaka, J. and Hayashizaki, Y. (2003) The mammalian protein-protein interaction database and its viewing system that is linked to the main FANTOM2 viewer. Genome Res. *13*, 1534–1541.

Templin, M.F., Stoll, D., Schwenk, J.M., Potz, O., Kramer, S. and Joos, T.O. (2003) Protein microarrays: promising tools for proteomic research. Proteomics *3*, 2155–2166.

Tewari, M., Hu, P.J., Ahn, J.S., Ayivi-Guedehoussou, N., Vidalain, P.O., Li, S., Milstein, S., Armstrong, C.M., Boxem, M., Butler, M.D., Busiguina, S., Rual, J.F., Ibarrola, N., Chaklos, S.T., Bertin, N., Vaglio, P., Edgley, M.L., King, K.V., Albert, P.S., Vandenhaute, J., Pandey, A., Riddle, D.L., Ruvkun, G. and Vidal, M. (2004) Systematic interactome mapping and genetic perturbation analysis of a *C.Elegans* TGF-beta signaling network. Mol. Cell *13*, 469–482.

Thaminy, S., Auerbach, D., Arnoldo, A. and Stagljar, I. (2003) Identification of novel ErbB3-interacting factors using the split-ubiquitin membrane yeast two-hybrid system. Genome Res. *13*, 1744–1753.

Tirode, F., Malaguti, C., Romero, F., Attar, R., Camonis, J. and Egly, J.M. (1997) A conditionally expressed third partner stabilizes or prevents the formation of a transcriptional activator in a three-hybrid system. J. Biol. Chem. *272*, 22995–22999.

Tong, A.H., Drees, B., Nardelli, G., Bader, G.D., Brannetti, B., Castagnoli, L., Evangelista, M., Ferracuti, S., Nelson, B., Paoluzi, S., Quondam, M., Zucconi, A., Hogue C.W., Fields, S., Boone, C. and Cesareni, G. (2002) A combined experimental and computational strategy to define protein interaction networks for peptide recognition modules. Science *295*, 321–324.

Toogood, P.L. (2002) Inhibition of protein-protein association by small molecules: approaches and progress. J. Med. Chem. *45*, 1543–1558.

Trakselis, M.A., Alley, S.C. and Ishmael, F.T. (2005) Identification and mapping of protein-protein interactions by a combination of cross-linking, cleavage, and proteomics. Bioconjug. Chem. *16*, 741–750.

Uetz, P., Giot, L., Cagney, G., Mansfield, T.A., Judson, R.S., Knight, J.R., Lockshon, D., Narayan, V., Srinivasan, M., Pochart, P., Qureshi-Emili, A., Li, Y., Godwin, B., Conover, D., Kalbfleisch, T., Vijayadamodar, G., Yang, M., Johnston, M., Fields, S. and Rothberg, J.M. (2000) A comprehensive analysis of protein-protein interactions in *Saccharomyces cerevisiae*. Nature *403*, 623–627.

Uetz, P., Ideker, T. and Schwikowski, B. (2002) Visualization an integration of protein-protein interactions. In: Protein-protein interactions. E. Golemis, ed. (New York: CSHL Press), pp. 623–646.

Veenhoff, L.M., Heuberger, E.H. and Poolman, B. (2002) Quaternary structure and function of transport proteins. Trends Biochem. Sci. *27*, 242–249.

Vidal, M. (2005) Interactome modeling. FEBS Lett. 579, 1834–1838.

Vidal, M. (2006) Time for a human interactome project? The Scientist *20*, 46–51.

Vidal, M. and Endoh, H. (1999) Prospects for drug screening using the reverse two-hybrid system. Trends Biotechnol. *17*, 374–381.

Vidal, M. and Legrain, P. (1999) Yeast forward and reverse 'n'-hybrid systems. Nucleic Acids Res. *27*, 919–929.

von Mering, C., Jensen, L.J., Snel, B., Hooper, S.D., Krupp, M., Foglierini, M., Jouffre, N., Huynen, M.A. and Bork, P. (2005) STRING: known and predicted protein-protein associations, integrated and transferred across organisms. Nucleic Acids Res. *33*, D433–D437.

von Mering, C., Krause, R., Snel, B., Cornell, M., Oliver, S.G., Fields, S. and Bork, P. (2002) Comparative assessment of large-scale data sets of protein–protein interactions. Nature *417*, 399–403.

Vondriska, T.M., Pass, J.M. and Ping, P. (2004) Scaffold proteins and assembly of multiprotein signaling complexes. J. Mol. Cell Cardiol. *37*, 391–397.

Walhout, A.J., Sordella, R., Lu, X., Hartley, J.L., Temple, G.F., Brasch, M.A., Thierry-Mieg, N. and Vidal, M. (2000) Protein interaction mapping in *C.Elegans* using proteins involved in vulval development. Science *287*, 116–122.

Wilkinson, K.D. (2004) Quantitative analysis of protein-protein interactions. Methods Mol. Biol. *261*, 15–32.
Winter, C., Henschel, A., Kim, W.K. and Schroeder, M. (2006) SCOPPI: a structural classification of protein-protein interfaces. Nucleic Acids Res. *34*, D310–D314.
Wojcik, J., Boneca, I.G. and Legrain, P. (2002) Prediction, assessment and validation of protein interaction maps in bacteria. J. Mol. Biol. *323*, 763–770.
Wojcik, J. and Hamburger, A. (2003) Proteomic Informatics. In: Bioinformatics for Geneticists. M.R. Barnes and I.C. Gray, eds. (Paris, France: John Wiley & Sons, Ltd), pp. 345–371.
Zanzoni, A., Montecchi-Palazzi, L., Quondam, M., Ausiello, G., Helmer-Citterich, M. and Cesareni, G. (2002) MINT: a Molecular INTeraction database. FEBS Lett. *513*, 135–140.
Zhu, H., Bilgin, M., Bangham, R., Hall, D., Casamayor, A., Bertone, P., Lan, N., Jansen, R., Bidlingmaier, S., Houfek, T., Mitchell, T., Miller, P., Dean, R.A., Gerstein, M. and Snyder, M. (2001) Global analysis of protein activities using proteome chips. Science *293*, 2101–2105.

Corresponding authors: Dr. Vincent Collura (vcollura@hybrigenics.com) and Dr. Guillaume Boissy (gboissy@hybrigenics.com)

CHAPTER 9

SUPRAMOLECULAR SIGNALLING COMPLEXES IN THE NERVOUS SYSTEM

M.O. COLLINS and S.G.N. GRANT
The Wellcome Trust Sanger Institute, United Kingdom

Table of Contents

E. Bertrand and M. Faupel (eds.), Subcellular Proteomics, 185–207.

Abstract: It is now apparent that multiprotein signalling complexes or "signalling machines" are responsible for orchestrating many complex signalling pathways in the cell. The synapse is a sub-cellular specialisation which transmits and converts patterns of electrical activity into cellular memory. This processing of electrical information is mediated by the protein components of the synapse. The organisation of synaptic proteins has been investigated over the last number of years using proteomic methods and with the application of bioinformatics; a landscape of modular protein complexes at the synapse is emerging. Many share a common organisation centred on a receptor/channel, a protein scaffold, (in which the signalling molecules are localised) and membrane to cytoskeleton interactions. The use of PDZ-domain based protein scaffolds is a particularly common feature in the construction of neuronal protein complexes and the differential presence of these proteins in complexes can have functional consequences. Here we overview current proteomic methodologies for the analysis of multiprotein complexes. In addition, we describe the characterisation of a number of multiprotein complexes associated with ion channels (NMDAR, P2X7 and Kir2) and GPCRs (5-HT2A/5-HT2C, D2 and mGluR5) and discuss common their common components and organisation.

1. INTRODUCTION

The field of proteomics aims to characterise proteins at many levels of complexity in a biological system. This ranges from identification of the complement of protein expressed in that system to how they are modified by posttranslational events. Comprehensive proteome analysis also seeks to elucidate protein sub-cellular distribution and protein-protein interactions, which integrate proteins into functional molecular complexes that ultimately perform cellular functions. Protein-protein interactions provide the basis for the macromolecular organisation of cells and the architecture of these interactions, for example connectivity of proteins in a network, reflect the assembly of proteins into complexes.

The composition of multiprotein complexes or organelles is variable between cell types, tissues, regions of tissues and between species. In addition, the content of these complexes is likely to be highly dynamic and temporal and thus such complexes can exist in many different forms in a cell at a single time point. Classical plasma membrane associated complexes generally contain receptor or channel protein(s) with an associated matrix of scaffolding proteins which may be linked to the cytoskeleton and other subcellular organelles. Within this scaffold there exists signalling proteins which mediate the transmission of incoming signals from the membrane to the interior, e.g. signaling to the nucleus to regulate gene expression. The types of protein interactions in protein complexes range from stable structural protein domain interactions such as PDZ domain-based interactions to highly transient interactions mediated by phosphorylation or other posttranslational events.

The capture of multiprotein complexes in sufficient quantity and purity whilst maintaining their integrity has been and still remains quite a challenge. There are many approaches to study multiprotein complexes and here we overview methods for the isolation of tagged or native complexes from brain and how the components of these complexes can be characterised. The application of these approaches will be reviewed to highlight the diversity of proteomic applications in neuroscience.

2. METHODS FOR ISOLATION OF PROTEIN COMPLEXES

2.1. Biochemical Isolation

The approach employed to biochemically isolate a protein complex depends heavily of the abundance of the complex in the starting material. Abundant complexes can be isolated by a physical fractionation scheme, which exploits physical characteristics such as the density or ionic charge of the complex (Dynlacht et al. 1991). However, many proteins and the multiprotein complexes that they form are of low abundance and such physical fractionation schemes are not suitable. In these cases, generation of highly enriched samples containing the target molecule can be achieved by employing various affinity chromatography steps. These can range from ligand affinity (drugs, substrates or co-factors) to the more widely used approach of immunoprecipitation, involving specific antibodies to the molecule of interest.

2.1.1. Immunoprecipitation/Immunoaffinity Chromatography

The use of antibodies for the precipitation or isolation of proteins and protein complexes has been the most common approach for many years. Antibodies raised to whole recombinant proteins or short peptides can be generated that exhibit high affinity binding and specificity to the target protein. In immunoprecipitation, the antibody is incubated with the solubilised protein extract and then the antibody-target protein complex with associated proteins, is precipitated using beads or resin conjugated with Protein A or Protein G. Immunoaffinity chromatography involves coupling the antibody to a resin for example, via amine groups to N-hydroxy-succimide (NHS)-activated Sepharose. The antibody-coupled resin is incubated with the solubilised protein extract and the protein complex is captured directly (Figure 1A). Extensive washing of the resin in both of these antibody-based isolation methods is necessary to reduce non-specific binding. Elution of the complex can be achieved by a variety of methods differing in their efficiency and specificity. In a standard experiment, the washed resin with attached protein complex is boiled in SDS-PAGE loading buffer that contains a high concentration of SDS and a reducing agent to interrupt the antibody-antigen interaction. An aliquot of the supernatant is run on a 1D SDS-PAGE gel and protein-containing bands, as visualised by colloidal coomassie staining, can be excised for downstream analysis by mass spectrometry or for western blotting. A major drawback in using antibody-based approaches for proteomic analysis is the introduction of the IgG itself, which interferes with the analysis by overshadowing proteins of interest that co-migrate during the gel-separation. This can be greatly reduced by using alternative elution strategies such as competitive elution with the immunising peptide. This approach is much cleaner with dramatically less non-specific binding proteins, however complete elution of target protein from the resin is difficult to achieve. An alternative elution method involves raising or lowering the pH, thereby disturbing the antibody-antigen interaction. This method is still cleaner than boiling elution but some prominent non-specific protein interactions are still evident.

2.1.2. Fusion-protein pull-down

The two most commonly used fusion proteins for protein complex purifications are GST (Glutathione-S-transferase) and His (>6 Histidine residues) tagged. In this approach, the tagged protein is expressed in *Escherichia coli* as a recombinant fusion protein and subsequently immobilized on a resin. The GST-tagged protein is purified using a Glutathione sepharose resin whereas the His-tagged protein is purified by metal-affinity chromatography using for example, a nickel-sepharose resin (Figure 1B). These tagged proteins are incubated with a solubilised protein extract and interacting proteins are co-purified. The main advantages of such fusion-proteins are that it is useful for detecting weak interactions and for low abundance proteins. However, the drawback for such an approach is that the complex must form *in vitro*; post-translational modification of the protein may be necessary for protein complex formation and competition with the pre-assembled complex may occur.

2.1.3. Epitope-tagging

This method combines the use of antibodies and the expression of fusion proteins to deliver an approach which can be applied in a very generic way. The protein of interest is expressed in a cell-line with an epitope-tag which is recognised by an antibody (Figure 1C). Many different cDNA's can be fused to the same tag allowing purification of many different complexes in parallel using the same strategy. Many

Figure 1. Methods for isolation of multiprotein complexes. These methods can be broadly divided into two categories; one which is independent of the starting material such as immunopurification and ligand affinity purification and those which are based on the introduction of a tag into the target protein in the biological sample, for example Fusion protein, Epitope tag and TAP-tag methods. Immunopurification of protein complexes is achieved by the production of antibodies with high affinity to the target protein and isolation of the complex by incubation of the immobilised antibody with a protein extract (Panel A). Peptide affinity purification exploits the affinity of the target protein to a short immobilised peptide for example PDZ domain-containing protein PSD-95 binding to the PDZ motif contained in the last six amino acids (SIESDV) of NR2B (Panel D). The other methods all require engineering of a tag into the gene of the target protein and expression of this tagged protein in a cell line or model organism. The tag can be a short peptide sequence which can be isolated using an antibody specific to that sequence (Flag-tag, Panel C) or a peptide sequence which can be captured by metal-affinity chromatography (His-tag, Panel B). All of these methods suffer from some degree of non-specific binding which can be problematic. A tandem affinity approach reduces the amount of contaminating non-specific binding by using sequential purifications based on for example Protein A/IgG interactions and CBP (Calmodulin binding protein)/Calmodulin interactions (Panel E). The complex is eluted from the first purification by cleavage at a TEV protease site between the tags and from the second purification using a chelating buffer to disturb the calmodulin interaction which is calcium-dependent.

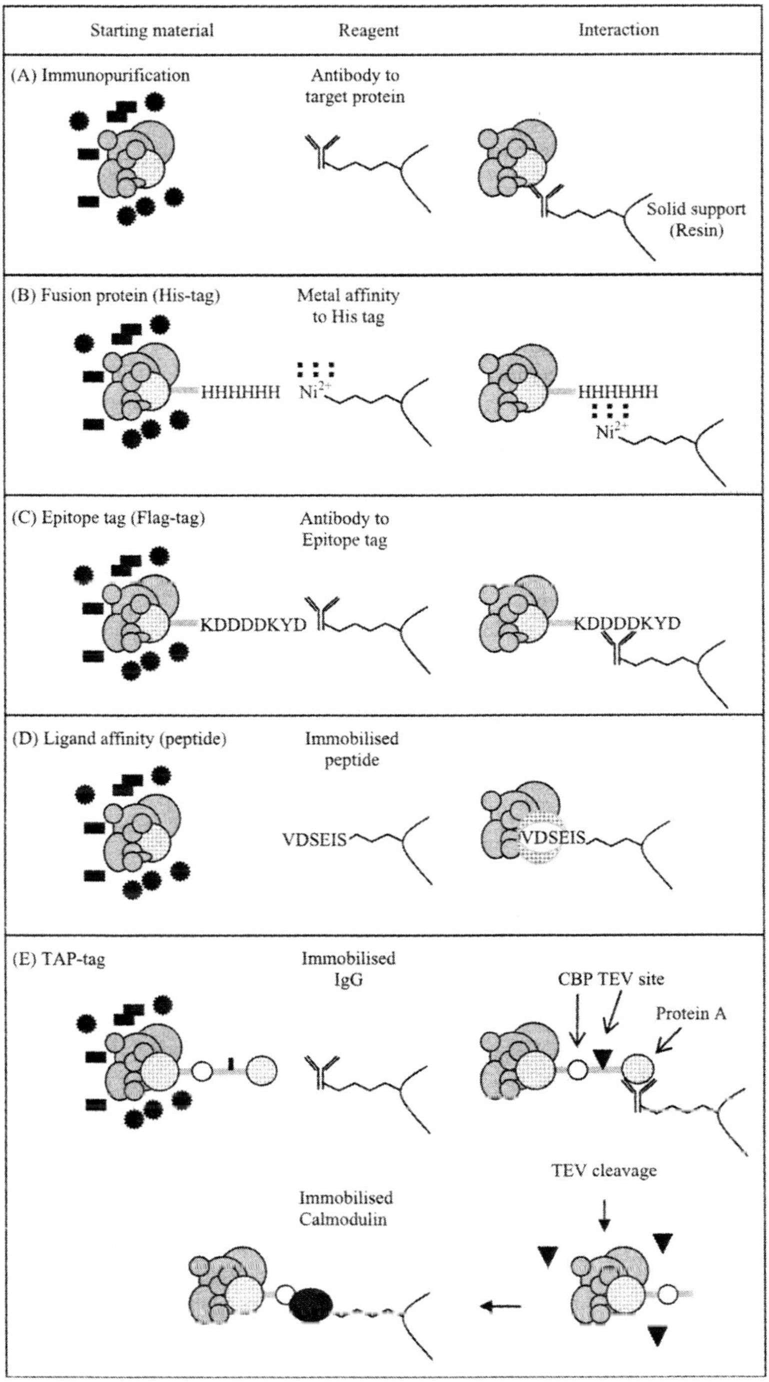
Starting material
Reagent
Interaction
(A) Immunopurification
Antibody to
target protein
Solid support
(Resin)
(B) Fusion protein (His-tag)
Metal affinity
to His tag
HHHHHH
Ni2+
HHHHHH
Ni2+
(C) Epitope tag (Flag-tag)
Antibody to
Epitope tag
KDDDDKYD
KDDDDKYD
(D) Ligand affinity (peptide)
Immobilised
peptide
VDSEIS
VDSEIS
(E) TAP-tag
Immobilised
IgG
CBP TEV site
Protein A
TEV cleavage
Immobilised
Calmodulin

such tags exist (e.g. Flag, Myc) and high affinity antibodies to these tags are commercially available. This approach has the advantage that large numbers of complexes can be purified in a consistent way; however problems concerning levels of expression of the tagged-protein and interference of the tag with protein interactions are recognised caveats.

2.1.4. Ligand affinity purification

An alternative to antibody-based or tagging-based methods to isolate protein complexes is to utilise the affinity of a ligand to capture the target protein and associated proteins. Membrane receptors and channels and many classes of enzymes are quite amenable to this kind of approach as many already have well characterised ligand binding pockets. Analogs of their natural ligands have been developed and are used as drugs to modulate the activity and therefore can be used as bait for affinity purification. Single proteins and (Furuichi et al. 2000) or protein complexes (Husi and Grant, 2001) can be isolated by immobilisation of the ligand on a suitable matrix, incubation with a protein extract, and after extensive washing the captured protein(s) can be eluted. The success of such an approach depends of the availability of functional groups or "handles" with which the drug can be immobilised and also upon the specificity of the drug in question.

There are many discrete protein domains which bind to defined linear peptide sequences in their interacting proteins. Often these kinds of interactions are central to the structure of multiprotein complexes and can be exploited to capture these complexes biochemically (Husi and Grant 2001), (Collins et al. 2006), (Becamel et al. 2004). The peptide sequence which forms the binding site for the interacting protein can be synthesised and immobilised readily using the N-terminal amino group (Figure 1D). As these types of interactions are mediated by such short peptide sequences, a synthesised and immobilised version of the sequence is adequate to capture the *in vivo* interaction. As the peptide has to compete with an interaction in a pre-assembled complex in the protein extract, a longer incubation time is usually required for satisfactory capture. Elution can be performed by boiling of the resin, pH elution or by competitive elution with the free synthesised peptide. As for peptide elution of antibody-antigen interactions, peptide elution in this case is also specific and therefore a cleaner elution is achieved. However, long incubation times with the peptide are necessary to increase the efficiency of elution.

2.1.5. Tandem affinity purification

Purification of protein complexes in a high throughput manner requires a method which is generally applicable to any protein and must be reproducible, specific and efficient. Tandem affinity purification (TAP) was developed in the late 1990s and has proved to be very useful for large-scale analysis of protein complexes (Gavin et al. 2002). In this approach, two tags engineered into the protein are used for sequential purification of the complex. The TAP tagged protein is usually stably expressed in a cell line along with the endogenous protein or can be targeted into the genome of an organism by homologous recombination to replace the endogenous gene. The protein of interest is engineered in frame with an N-terminal or a C-terminal tag; usually both are used for the purification to check for disruption of protein interactions caused by the tag. The tag module is composed of two different affinity tags separated by a tobacco etch virus protease (TEV) cleavage site (Figure 1E). The first purification exploits the protein A tag which is captured efficiently on IgG Sepharose. Then the tag is cleaved at the protease cleave site, thereby releasing the rest of the tag module and protein complex. Next the second tag, which can be a calmodulin binding peptide (CBP) (Gavin et al. 2002), a 6 Histidine residue repeat (Rubio et al. 2005), hemagglutinin (HA) peptide or a myc repeat (Rubio et al. 2005) is captured by calmodulin sepharose, metal chelate sepharose, by antibodies to HA and finally by antibodies to myc, respectively (see Cheeseman and Desai 2005 for review). The final elution of the protein complex is achieved by using EGTA (a calcium chelator) in the case of the CBP or by imidazole in the case of the His-tag. These two elutions are efficient and are free from contaminating IgG. The main advantages of using such two-step purification are that less stringent washing conditions can be used to allow capture of less stable complexes and that non-specific protein binding is low. In addition, because the same purification strategy can be applied to many complexes, the resultant data is directly comparable, a point which is quite important in reciprocal or reverse tagging for protein complex network expansion.

2.2. Identification of Components

2.2.1. Western blotting

Identification of isolated proteins can be readily achieved using western blotting or mass spectrometry. Western blotting (also known as immunoblotting) utilises specific antibodies to detect a protein that has been transferred from a gel onto a membrane. A major advantage of western blotting is its sensitivity and ability to detect amounts of protein beyond the range of current sequencing-based or mass spectrometry methodologies. In fact, the range of protein abundance which was detected by western blotting in yeast was from 50 to more than 10^6 molecules per cell (Ghaemmaghami et al. 2003) which is at least an order of magnitude more sensitive than current LC-MS/MS analyses of complex samples (de Godoy et al. 2005).

It is obvious that this approach is biased by the assumption that a given molecule might be present, and cannot be used to identify unknown or unsuspected proteins. However, when there is a large amount of evidence concerning the putative involvement of molecules that influence many target proteins and such proteins pose ideal targets for western blotting approaches. Additionally, changes in protein levels, as a result of modulation or modification of a receptor's environment, can easily be visualised using specific antibodies against the proteins under investigation. A disadvantage of this method is its labour intensive nature and thus difficulty in performing on a large scale.

2.2.2. Proteomic mass spectrometry

Mass spectrometry (MS) permits the identification of proteins by the determination of the exact mass of peptides, and the fragmentation of these peptides to determine the amino acid sequence. Protein samples, such as purified organelles or protein complexes are usually separated by their relative molecular weight on SDS-PAGE gels. The protein bands are visualized by dyes such as Coomassie brilliant blue and then are excised from the gel. The protein-containing gel slices are washed and incubated with proteolytic enzymes such as trypsin and the resultant peptides extracted from the gel-slices.

A mass spectrometer consists of an ion source, a mass analyzer that measures the mass to charge ratio (m/z) of peptides and a detector that counts the numbers of ions at each m/z value. Electrospray ionization (ESI) mass spectrometers use a heated capillary needle through which the peptide solution elutes and are subsequently ionized and introduced into the mass spectrometer. This type of ion source is readily compatible with liquid chromatographic (LC) separations resulting in an LC-MS/MS setup in which thousands of peptides maybe identified in a single experiment. In a typical LC-MS/MS experiment a peptide mixture is separated by analytical reversed-phase (RP) chromatography in a HPLC (High pressure liquid chromatography) system and is introduced into the mass spectrometer via the ESI source (Figure 2). Multiply

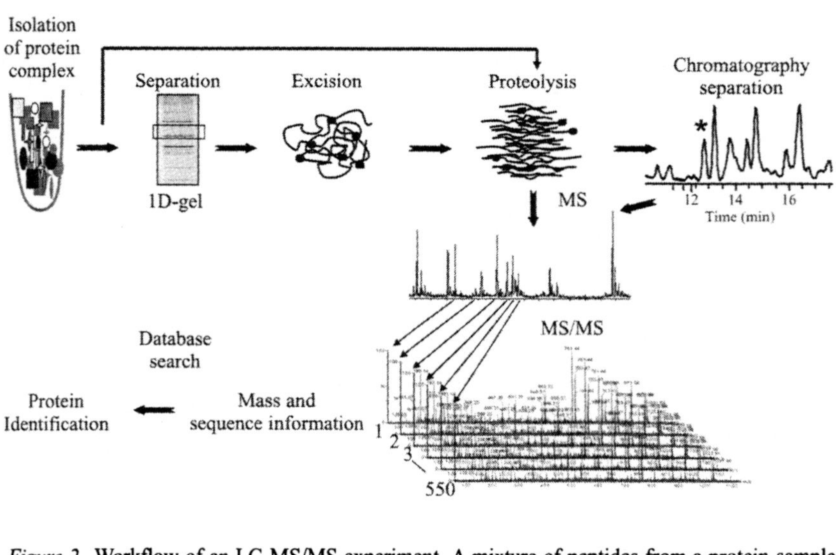

Figure 2. Workflow of an LC-MS/MS experiment. A mixture of peptides from a protein sample digest is separated by reversed-phase chromatography on a nano-flow HPLC. The peptides elute from the RP column and are ionized by an electrospray source. In the first stage of mass spectrometry, m/z values and charge states for each precursor ion are determined and the most abundant precursor ions are selected for analysis in the second stage. The ions are then fragmented with by collision-induced dissociation (CID) a gas to produce fragment ions which are detected. Using the mass (from MS-1) and sequence information (from MS-2) protein sequence databases are searched to provide peptide identifications and protein matches.

Table 1. Multiprotein complexes in the nervous system. Summary of the numbers of identified components and methods used to characterise three ion channel and 4 GPCR multiprotein complexes

Complex	No. # of proteins	Methods	Reference
Ion Channel			
NMDA receptor	186	Peptide-affinity chromatography/ Immunoprecipitation	(Husi et al. 2000)/ (Collins et al. 2006)
P2X7 channel	11	Immunoprecipitation	(Kim et al. 2001)
Kir2.x channel	11	GST-tagging	(Leonoudakis et al. 2004)
GPCR			
5-HT2A receptor	7	GST pull-down/Peptide-affinity chromatography	(Becamel et al. 2002)
5-HT2C receptor	17	Peptide-affinity chromatography	(Becamel et al. 2002)
D2 receptor	39	Various/many single protein studies	(Kabbani et al. 2005)
mGluR5 receptor	76	Immunoprecipitation	(Farr et al. 2004)

protonated peptides give rise to the first mass spectrum (MS1) and then the computer generates a prioritised list of the most intense precursor ions detected that are subsequently subjected to fragmentation by energetic collision with gas (collision induced dissociation, (CID)) in the collision cell. This MS/MS spectrum contains data relating to the amino acid composition of the peptides and the pattern of fragmentation that occurred. This process of tandem mass spectrometry provides m/z values of peptides and fragmentation data for each peptide. This data is then used by sequence database searching programs such as MASCOT (Perkins et al. 1999) or SEQUEST (Yates et al. 1995) and provides probability-based peptide matches and therefore protein identifications.

3. MULTIPROTEIN COMPLEXES IN THE NERVOUS SYSTEM

Many of the methods described in the previous section have been applied to the study of multiprotein complexes in the nervous system. The first proteomic study of a neurotransmitter receptor complex in 2000 (Husi et al. 2000) has been followed by similar studies of important classes of brain receptors and channels (Table 1). As shown in Table 1, complexes associated with ion channels such as members of the ionotropic class of glutamate receptors and GPCRs such as the metabotropic class of glutamate receptors have been described. Here we review a number of such complexes and discuss how their components relate to nervous system biology.

3.1. Ion Channel Complexes

3.1.1. NMDA receptor-adhesion protein signalling complex

The first proteomic analysis of a multiprotein complex relevant to the brain involved the purification and identification of the molecular constituents of the NMDA receptor-adhesion protein signalling complex (NRC). Using an antibody based approach targeting the NR1 subunit of the NMDA receptor, the NRC was shown to comprise 77 proteins organized into receptor, adaptor, signalling, cytoskeletal and novel proteins, of which 30 are implicated from binding studies and another 19 participate in NMDAR signalling (Husi et al. 2000).

NMDAR and metabotropic glutamate receptor subtypes are linked to cadherins and L1 cell-adhesion molecules in complexes lacking AMPA receptors (Husi et al. 2000). These neurotransmitter-adhesion receptor complexes are bound to kinases, phosphatases, GTPase-activating proteins and Ras with effectors including MAPK pathway components (Husi et al. 2000). A striking feature of the composition is that there appears to be "modules" or sets of signalling proteins that are known to comprise key components of signal transduction pathways that can be distinctly regulated. For example, all of the molecules necessary to induce the phosphorylation of mitogen-activated protein kinase (MAPK) following NMDAR stimulation are present in the NRC (including CaMKII, SynGAP, Ras, MEK and ERK). These modules may allow the NMDAR and mGluR to integrate signals within the complex and then couple to down stream cellular effector mechanisms, such as trafficking of AMPA receptors and cytoskeletal changes that mediate structural and physiological plasticity. Furthermore, at least 18 NRC constituents are regulated by synaptic activity indicating that the composition of the complex is dynamic. In the hippocampus, these activity-dependent genes are known to undergo specific temporal changes following the induction of plasticity.

3.1.2. NMDAR/MAGUK-associated signalling complexes

This NRC dataset was further expanded using a combination of peptide affinity chromatography and mass spectrometry based identification of new components of the complex. Purification of MASCs (MAGUK (membrane-associated putative guanylate kinase)-associated signalling complexes) was achieved by peptide affinity chromatography using a hexapeptide corresponding to the C-terminus of the NR2B subunit that binds MAGUK proteins (including PSD-95). Purified complexes were separated by SDS-PAGE gel electrophoresis, bands excised, digested, and analysed by mass spectrometry. Western blotting of peptide-purified complexes for candidate proteins was also performed. These complexes are similar in composition to NMDA receptor complexes and MAGUK complexes isolated with antibodies (Husi and Grant 2001; Husi et al. 2000). This approach extended the NRC dataset to comprise 186 proteins which constitute MASC complexes (Collins et al. 2006). Out of the previously identified 100 proteins identified in the NRC by analysis of anti-NR1 immunopurifications, 84 of these were also found by the peptide approach. In addition, 86 new proteins were identified by the peptide approach, many of which were in the molecular weight range obscured by the contaminating antibody chains in the immunopurification strategy. The composition and organisation of MASC has been investigated in a number of ways from the use of protein interaction data to construct molecular networks (Pocklington et al. 2006) to the analysis of domains and other protein sequence-based qualities of the components of this complex. The presence of Interpro (Mulder et al.

2005) domains in MASC proteins was investigated using a hierarchical clustering (Figure 3). Proteins in this MASC dataset that contained Interpro domains that were present more than once, and were present in more than one protein were clustered and the resultant clusters were annotated according to their protein classes. MASC protein domains cluster into 14 discrete sets of proteins (Figure 3) reflecting many of the structural and functional aspects known about this set of proteins. This complex is centred on interactions between glutamate receptors and PDZ-containing scaffolders (Table 2). This provides a framework in which a host of kinases, phosphatases and signalling molecules transduce information to gene expression regulatory pathways and to cytoskeletal and transport networks. The MASC dataset represents a comprehensive account of NMDAR-MAGUK complexes identified by large-scale western blotting and mass spectrometry based analyses of both antibody and peptide affinity-based purifications. In order to gain an overall view of the postsynaptic proteome and to validate the components of MASC complexes, a systematic analysis of purified postsynaptic densities (in which these complexes are embedded) was performed. PSD proteins were separated on an SDS-PAGE gel and gel slices were analysed by LC-MS/MS (Collins et al. 2006). 7402 peptides were identified

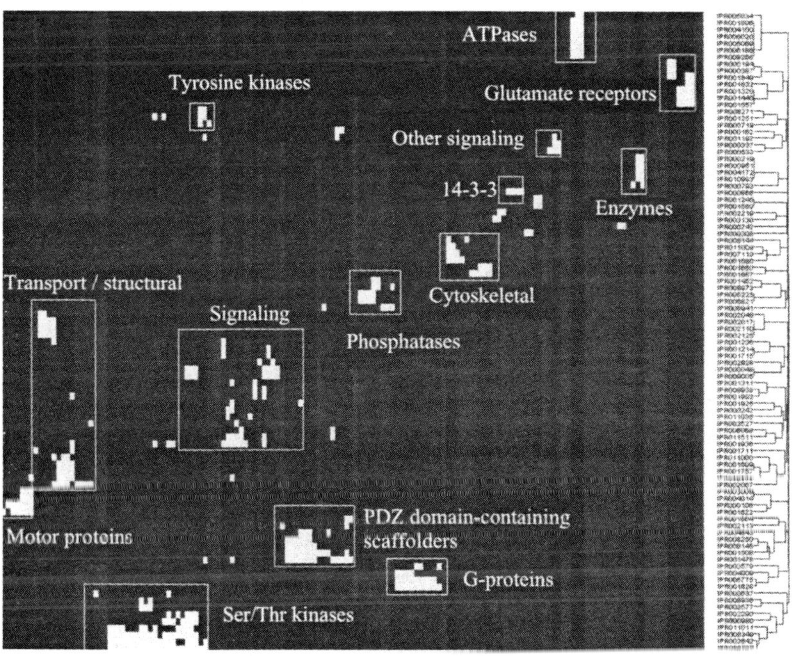

Figure 3. Hierarchical domain clustering of MASC proteins. Interpro domains in MASC proteins occurring more than once in the MASC dataset were clustered. Discrete clusters (grey boxes) were annotated with the protein class of which the majority proteins belonged to. Clusters are formed from co-occurrence of two or more domains characteristic of a protein class

Table 2. Distribution of PDZ domain-containing proteins in neuronal multiprotein complexes. As MAGUK proteins and other PDZ domain-containing proteins provide important scaffolding functions in protein complexes their distribution was analysed. The most widely used PDZ domain-containing proteins are DLGH3 and SAP97. The differential use of these proteins in complexes containing subtypes of 5-HT receptor highlights the heterogeneity and variability in complexes even in closely related receptors

UniProt	Name	MASC	mGluR5	Kir2.2(B)	Kir2.2(H)	P2X7	5HT-2a	5HT2c	D2	# times
O88910	DLGH3	X		X		X	X	X		5
Q62402	SAP97	X		X	X		X	X		5
Q62108	PSD-95	X		X			X	X		4
O70589	CASK			X	X			X		3
P70175	SAP102	X		X				X		3
Q9Z251	VELI1	X		X	X					3
O88952	VELI3			X	X		X			3
Q63622	CHAPSYN-110	X		X						2
Q9WV34	DLGH2	X		X						2
Q9Z216	HOMER1	X	X							2
Q9WU13	SHANK1	X	X							2
Q02410	MINT-1			X				X		2
Q9WVQ1	MAGI2						X	X		2
Q925T6	GRIP		X							1
P39447	ZO-1	X								1
Q61234	SYNTROPHIN				X					1
Q63ZW7	CIPP						X			1
O35274	PPP1R9B								X	1
Q9Z0G0	RGS19IP1								X	1

corresponding to 620 non-non-redundant protein identifications, with an average of 16.7 proteins identified per gel slice (Collins et al. 2006). This dataset was merged with six other published proteomic studies of the PSD (Walikonis et al. 2000; Satoh et al. 2002; Jordan et al. 2004; Li et al. 2004; Peng et al. 2004; Yoshimura et al. 2004) and 119 individual papers reporting PSD localised proteins. This combined Total PSD dataset contains 1124 proteins and the distribution of protein identifications in the Total PSD dataset show that the majority of proteins were detected only once (58%) and that 198 (18%) proteins were detected twice. In order to define a set of higher confidence proteins, those that were identified two or more times (466 proteins) were grouped and termed the "Consensus PSD" (cPSD) (Figure 4).

It is important to note that the majority of the components of MASC complexes have been validated as PSD proteins in this way and the major overlap between these complexes occur in the cPSD (Figure 4), supporting the notion that the cPSD is an important subset of the Total PSD dataset. This is consistent with a postsynaptic organization where the MASC signalling complex is connected to multiple cell biological effector mechanisms organized into their respective complexes. In addition to MASC and AMPA complexes, components of other complexes such as cell adhesion, growth factor, cytoskeletal, transport and ribosomal complexes were found in the PSP.

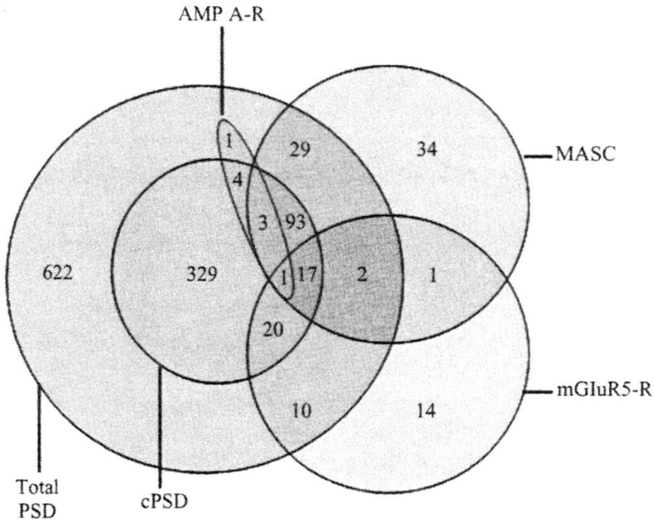

Figure 4. Multiprotein complexes in the postsynaptic density. Venn diagram illustrating the overlap of three glutamate receptor complexes with the Total and cPSD datasets. It can be seen that the majority of overlap between components of these complexes and the PSD occurs in the high confidence cPSD dataset. Proteins detected in these multiprotein complexes, which were not found in any of the PSD datasets, are generally of low abundance that are enriched in immuno-purifications of complexes compared to whole PSD analyses.

Functional annotation of each component protein in a multiprotein complex from the literature can be very useful for determining shared functions within a complex and ultimately functions and pathways specific to the protein complex in question. Using such an approach the importance of the MASC complex was highlighted by the fact that mutations or polymorphisms in 47 MASC genes are associated with 183 human disorders, of which 54 are nervous system disorders (Grant et al. 2005). In addition, 43 MASC proteins have been reported to be important in synaptic plasticity and 40 have been associated with rodent behaviour (Grant et al. 2005). Combining such phenotypic data with protein interaction data allows investigation of the distribution of functions within a network. A protein interaction network with 650 protein–protein interactions for 281 PSP proteins is shown in Figure 5 (**see colour insert**). The individual proteins in this network are represented by filled circles (nodes) and the interactions are represented by connecting lines (edges). An in-house database of mouse targeted mutations, in which a change in LTP or LTD was observed for a given gene knockout was used (see databases on synaptic plasticity and behaviour of mouse knockouts on www.genes2cogntion.org). PSP genes for which such mutant phenotypic data was available are coloured in green in Figure 5. 19 MASC genes, 29 cPSD and 8 Total PSD (not in MASC or cPSD) display changes in LTP/LTD when mutated. The size of nodes displayed in this network is proportional to the number of protein interactors or connectivity of that node. It can be seen that many of the highly connected nodes are found in the MASC complex and that nodes with a phenotype in LTD/LTP are the mostly highly connected and many are in the MASC complex. This is in agreement with the notion that synaptic plasticity is relatively robust and that perturbation with more highly connected proteins in a network is more likely to produce a phenotype than less connected proteins (Jeong et al. 2001; Pocklington et al. 2006).

Clearly, this kind of phenotypic evidence is very useful and especially when combined with diverse data sources such as gene expression and protein interaction information. Using such information one can assess the heterogeneity of the complex by for example, looking at co-expression of the protein components in brain regions or in developmental time courses in culture (Valor et al. 2007). Furthermore, functional subsets of the complex can be highlighted by using protein interaction network tools, for example such analysis of the MASC protein interaction network resulted in the identification of modularity in the complex where sub-complexes or modules were predicted to perform distinct functions that is there was distributed functionality within the complex as a whole (Pocklington et al. 2006).

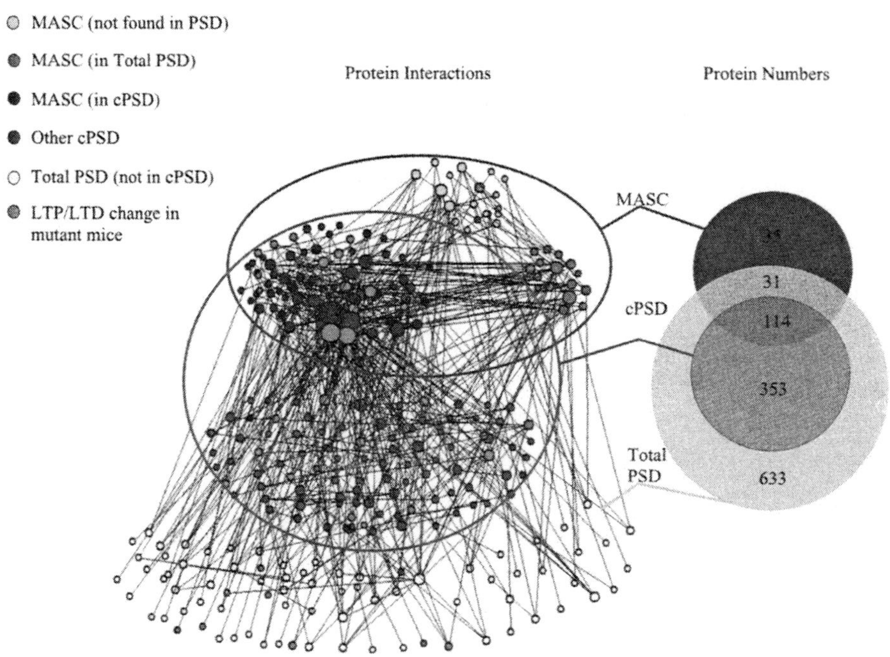

Figure 5. Network illustration of protein interactions and knockout mouse phenotype in PSP proteins. 650 protein–protein interactions for 281 PSP proteins are shown. These PSP interactions are divided into MASC (Red circle), cPSD (Blue circle) and nodes outside the circles are other components of the Total PSD. Also, a Venn diagram of the total numbers of proteins in the MASC, cPSD and Total PSD is shown. Proteins in this network for which an electrophysiological phenotype (change in LTP/LTD) has been reported in the literature are indicated by green nodes. The size of a node is proportional to its connectivity (number of interactors) and it can be seen that the largest nodes with an electrophysiological phenotype are in MASC.

3.1.3. P2X7 receptor complex

P2X receptors are ATP-gated ion channels which are exclusively found in vertebrates. There are seven subtypes of P2X receptors which may be found as heteromers or homomers composed of between three and six of these subunits (Nicke et al. 1998). The $P2X_7$ subtype form functional homomeric channels and have functions in distinct neuronal, astroglial, and microglial cells as well as in cells in non nervous tissues. The exact functions of $P2X_7$ receptors in the brain are diverse but most involve sensing of intracellular ATP levels. When such ATP levels are low, there is evidence to suggest that it can cause $P2X_7$ receptors to directly affect synaptic neurotransmission, when high ATP levels occur, widespread activation of $P2X_7$ receptors stimulate many pathological cascades and finally under persistent activation these receptors, cell death signals are generated (see Sperlagh et al. 2006 for review).

A small complex of proteins associated with the $P2X_7$ receptor has been characterised in HEK293 cells (Kim et al. 2001). The complex was immunoprecipitated with an anti-$P2X_7$ antibody from the membrane fraction of HEK293 cells stably expressing the rat form of the $P2X_7$ receptor. The components were separated by SDS-PAGE electrophoresis and prominent protein containing (as assessed by Coomassie blue staining) bands were excised, digested and the resultant peptides extracted and analysed by MALDI-TOF mass spectrometry. Eleven proteins were identified, including structural/cytoskeletal proteins such as alpha-actinin and MAGuK P55 subfamily member 3 and signalling proteins which included PI4K and RPTPbeta. The presence of a member of the MAGUK family of scaffolding proteins is reminiscent of the NMDA receptor complex indicating perhaps a common mode of protein complex organisation. The association of the complex with the cytoskeleton is expected as activation of the receptor in HEK293 cells results in cytoskeletal rearrangements which result in membrane blebbing (Virginio et al. 1999).

The presence of the tyrosine phosphatase RPTPbeta prompted investigation of tyrosine phosphorylation of the receptor and it was found that the $P2X_7$ receptor was functionally modulated by tyrosine phosphorylation and the exact site of phosphorylation was determined by mutational analysis (Kim et al. 2001). PI4K is involved in the generation of the second messenger precursor PIP_2 and its presence in the complex would confer the ability of the receptor to signal downstream through inositol triphosphate and diacylglycerol. Although this complex was characterised in a non neuronal cell type, the components identified so far hint to the signalling capacity of the receptor. A similar analysis of the complex from neuronal tissue and comparison with complexes associated with the other P2X sub-types is warranted and would be very useful in defining the potential functional similarities as well as differences in P2X signalling complexes.

3.1.4. Kir2 channel complex

Potassium channels are key regulators of cell excitability in the brain and also in other electrically active tissues such as the heart. They control cell membrane potential and aspects of action potential waveform frequency and shape. Potassium channels consist of pore-forming alpha subunits of which, there are at least 70 different types and are often associated with auxiliary subunits. The activity, trafficking and expression of potassium channels are regulated by interactions with such auxiliary proteins and identification of associated proteins may aid elucidation of channel function.

One subclass of potassium channels, the strong inward rectifier potassium channel (Kir2.x), has been characterised by proteomic analysis (Leonoudakis et al. 2004). Three of these Kir2.x channels contain PDZ interaction motif's at their N-terminus that provides a binding site to scaffolding proteins which contain PDZ domains. A C-terminal GST-Kir2.2 fusion protein was used to affinity-purify interacting proteins from brain and heart tissue. Using mass spectrometry, immunoblotting and N-terminal microsequencing 11 proteins were found to be in a complex with Kir2.2 in brain tissue. Eight members of the MAGUK family were found in the complex, seven of which are also in the MASC complex interacting with the C-terminal PDZ binding site of the NR2 subunit (Table 1). The composition of Kir2 channel complexes in cardiac tissue was also investigated and in addition to many members of the MAGUK family, components of the dystrophin-associated protein complex were identified. These additional components presumably reflect additional Kir2 channel functions in cardiac cells where cytoskeletal and extracellular links are particularly important and alterations in dystrophin can lead to muscular dystrophy (Albrecht and Froehner, 2002). The most likely function of these groups of proteins associated with Kir2.2 channels is to regulate the targeting and trafficking of the channel to discrete subcellular locations as well as scaffolding and stabilisation of the channel in the plasma membrane.

3.2. G-protein Coupled Receptor Complexes

The G-protein coupled receptor (GPCR) family is the largest receptor superfamily in mammalian genomes (Kabbani et al. 2005) and its members are involved in diverse functions from neurotransmission to chemotaxis and many other important physiological processes. This family of receptors is the target for over 50% of all medicines and characterisation of associated proteins in GPCR complexes may aid the discovery of novel modes of pharmaceutical modulation of GPCR-associated functions.

3.2.1. 5-HT-2A/C receptor complexes

All 5-HT receptors (except for 5-HT3, which is a ligand-gated channel) are members of the GPCR family and many psychoactive drugs have been discovered to target the 5-HT-2 sub-types. The 5-HT_{2A} and 5-HT_{2C} subtypes are quite similar in terms of amino acid sequence and pharmacological profile and both contain a canonical type 1 PDZ ligand ((S/T)Xϕ, where ϕ is a hydrophobic residue, SCV and SSV, respectively). In order to investigate whether these differences in the PDZ binding motif in the C-terminal of these two receptors recruited different sets of proteins, a differential proteomic approach was employed. (Becamel et al. 2002, 2004). Synthetic peptides corresponding to the last 14 amino acids of each receptor were coupled to a resin and peptide-affinity chromatography was performed using brain extract. The co-purifying proteins were identified by a combination of two-dimensional electrophoresis and MALDI mass spectrometry and by western blotting. In total, 7 proteins were found to be associated with the 5-HT_{2A} subtype and 17 proteins with the 5-HT_{2C} subtype of the receptor. Both of these sets of proteins were rich in PDZ domain-containing proteins with 9 such proteins being present in one or other of these complexes (Table 2). The presence as well as the quantity of these PDZ domain-containing proteins varied between these two purifications. This differential recruitment of scaffolding proteins due to a single amino acid change in the consensus PDZ motif is striking and illustrates the specificity inherent in PDZ domain interactions (Nourry et al. 2003).

The 5-HT_{2C} receptor preferentially interacted with PSD-95, SAP102 and MPP-3 when compared with the 5-HT_{2A} receptor. However, the most striking difference between these complexes is the presence of the tripartite Veli3-CASK-Mint scaffolding sub-complex in the 5-HT_{2C} but not the 5-HT_{2A} complex. Electron microscopy revealed that the 5-HT_{2C} receptor is highly concentrated at pre- and post-synaptic thickenings, consistent with the notion that the 5-HT_{2C} receptor is associated with protein networks that are important for its synaptic localisation and its coupling to the signalling machinery (Becamel et al. 2002, 2004). The 5-HT_{2A} receptor on the other hand, which does not interact with this scaffolding complex, is localised to mainly the cytosol of dendritic shafts. This differential localisation of these 2 sub-types of 5-HT_2 receptor correlate well with the differential binding of PDZ domain-containing scaffolding proteins determined by the single amino acid difference in the -1 position of their C-termini.

3.2.2. D2 Dopamine receptor complex

The dopamine receptor is another relatively well characterised and pharmacologically relevant GPCR, which appears to exist in a discrete receptor signalling complex (Kabbani et al. 2005). Dopamine receptors are involved in many functions, from motor control to memory formation and they mediate these functions through regulation of ion channels, sodium pumps and ion exchangers (Kabbani et al. 2005). There are five subtypes of dopamine receptor with differing affinities for dopamine, specificity for coupling to G-proteins and post-translational processing. A systematic analysis of the components of dopamine receptor complexes has not been carried out to date but collection of data from individual published studies reveals that there are at least 39 proteins which directly interact or transiently associate with the D2 type of dopamine receptor (Kabbani et al., 2005). These interacting proteins range from the usual GPCR-associated G-proteins to scaffolding/trafficking proteins and some signalling proteins. Two PDZ domain-containing proteins are associated with the D2 receptor (Table 2). GIPC (GAIP (RGS19) interacting protein C-terminus) binds to the carboxyl-terminal tail of the D2 receptor, thereby recruiting RGS19 which regulates downstream G-protein signalling. The PP1 phosphatase regulatory subunit PPP1R9B (Neurabin II) interacts directly with the third intracellular loop of the D2R by a mechanism which is independent of its PDZ domain and actin-binding domains (Smith et al. 1999). This interaction may mediate linkage of the receptor to the cytoskeleton via actin and also bring the phosphatase PP1 to the vicinity of the receptor to regulate its state of phosphorylation.

However, the most striking feature about D2 receptors is their direct interaction with many receptors and ion channels. It is generally accepted that GPCRs exist as dimmers or as part of larger oligomeric complexes formed by other types of receptors. Signalling complexes composed of receptors for different neurotransmitters might result in more efficient processing of incoming signals than if they were physically uncoupled. D2 receptors can associate with SST5 Somatostatin receptors (Rocheville et al. 2000), CB1 Cannabinoid receptors (Kearn et al. 2005) and A_{2A} Adenosine receptors (Ciruela et al. 2004). D2 receptors also interact with many ion channels including NMDA (Fiorentini and Missale, 2004), AMPA (Fukata et al. 2005) and GABA receptors (Seamans and Yang, 2004). It is highly unlikely that the dopamine receptor exists in a supramolecular complex with all of these channels and receptors but rather in many different specialised smaller complexes. Elucidation of the components of such a spectrum of complexes would be very interesting and might allow resolution of the diverse functions of the dopamine receptor and dopamine receptor/other receptor or channel complex combinations, at a molecular level.

3.2.3. Metabotropic glutamate receptor 5 complex

Metabotropic glutamate receptors, which are GPCRs and thus couple to G-proteins, modulate synaptic transmission and cell excitability. There are three sub-groups of receptor based on their G-protein coupling preference. mGluR5 is a member of group I mGluRs which couple to the Gq signalling pathway and prior to any proteomic investigation it was well known to be involved in many signalling pathways and to interact with other receptors. The mGluR5 receptor complex was isolated using an immunoprecipitation approach combined with MS-based analysis of a one-dimensional gel separation of the eluted protein and also by direct tryptic digestion of the antibody-conjugated mGluR5 complex. Seventy-six proteins were identified in mGluR5 complexes with varying degrees of confidence based on the presence of each protein in multiple experiments and not in control experiments (Farr et al. 2004). This analysis confirmed the association of the mGluR5 receptor with an intracellular calcium cascade with identification of the IP3 (Inositol 1,4,5-triphosphate) receptor, Homer and Phospholipase C-beta-1. Activation of mGluR5 results in the hydrolysis of membrane phosphatidylinositol bisphosphate (PIP2) to diacylglycerol (DAG), which activates PKC, and inositol triphosphate (IP3), which in turn activates the IP3 receptor to release intracellular calcium (Kawabata et al. 1998).

Interestingly, the IP3 receptor binds to a signalling protein called IRBIT (membrane associated IP3 (inositol triphosphate receptor) binding protein)), in an IP3 sensitive manner (Ando et al. 2003). A phosphoproteomic study of the synapse discovered phosphorylation sites in mGluR5, PLC-β and IRBIT (membrane associated IP3 (inositol triphosphate receptor) binding protein)) indicating potentially novel mechanisms of signalling through the mGluR5 receptor(Collins et al. 2005). The N-terminal region of IRBIT responsible for the interaction with the IP3 receptor was found to contain three novel phosphorylation sites, and phosphorylation in this region had been postulated to regulate this interaction (Ando et al. 2003).

mGluR5, G-proteins and PLC-β form a multiprotein complex with the IP3 receptor and IRBIT via the scaffolding properties of Homer (Tu et al. 1998). This complex seems to facilitate a signalling pathway from mGluR5 to modulation of intracellular calcium, which is known to regulate many intercellular signalling activities. Homer and Shank are prominent PDZ domain-containing proteins at the synapse (Table 2) and they seem to provide a scaffold (possibly by homo-multimerisation) between the plasma membrane bound, mGluR5 and endoplasmic membrane bound IP3 receptor (Sala et al. 2005).

mGluR5 has an important role in many NMDAR associated synaptic functions. Co-activation of mGluR5 and NMDA receptors is required for potentiation of excitatory synaptic transmission in hippocampal neurons (Kotecha et al. 2003) and this process requires IP3R-mediated release of intracellular calcium and activation of PKC for maintenance of potentiation. There is biochemical evidence to suggest that these receptors are physically associated; mGluR5 is a component of the NMDA receptor complex (Husi et al. 2000), and indeed NR2A (an NMDA receptor subunit) is present in the mGluR5 receptor complex (Farr et al. 2004). In fact, 21 proteins are shared by both the NMDAR and the mGluR5 receptor complexes (Figure 4) raising the possibility that these two receptors may co-exist in a signalling complex.

4. EMERGING THEMES IN PROTEOMIC STUDIES OF BRAIN COMPLEXES

What do these multiprotein complexes have in common? The most striking attribute of neuronal multiprotein complexes is their content of scaffolding proteins. The family of proteins which contain PDZ domains are likely to be crucial to the structure and function of many of these complexes. Nineteen PDZ domain containing proteins are found in the eight brain complexes listed in Table 2. Some of these proteins are very common with proteins such as SAP97 and DLGH3 being present in five complexes while others such as GRIP and CIPP are much more specific. Although it is interesting to compare these complexes to look for commonalities that may explain shared structure or function it is likely that the molecular differences may be more informative about the specific functions of these complexes. The association of discrete sets of proteins such as G-proteins, specific kinases and their adaptors (e.g. PKA/AKAP complexes) or other signalling modules are good indicators of what a multiprotein complex actually does. As more data concerning the components of brain complexes accumulates, extraction of what is common and what is specific to each complex will be possible. Systematic characterisation of large numbers of complexes especially within a subclass of complexes such as neurotransmitter receptor complexes, would allow identification of complex specific proteins. Such an approach might accelerate functional validation of complexes by highlighting the most likely candidates for inferring multiprotein complex function.

It is interesting to note that the size of brain protein complexes varies from less than 10 proteins to nearly 200. It is unlikely that the 186 proteins contained in NRC/MASC complexes exist as a single complex but rather this dataset is an average of a number of smaller variant complexes. Protein interaction network analyses suggest that complexes do not exist as discrete units but are linked together and share common modules. The idea of modules or sub-complexes which are used in a generic way in complexes is intriguing, indicating yet another level of molecular architecture. In addition, the physical association of complexes which are functionally linked, such as that described for the NMDAR and mGluR5 points to the fact that signalling pathways between these receptors may be physically embedded in the components of the respective complexes and not remote from each other.

5. CONCLUSIONS

In the last 6 years the concept that multiprotein complexes are employed extensively in cells has been driven by the application of many proteomic technologies, and in particular, mass spectrometry. Systematic mapping of protein complexes in yeast has revealed its molecular architecture in unprecedented detail. Characterisation of protein complexes in the nervous system has been achieved for a relatively small number of channels and receptors using mainly labour intensive approaches. The future of mapping and characterisation of protein complexes lies in the use of generic strategies such as epitope tagging or TAP tagging. So far these are the only technologies that are realistically scaleable for whole proteome analysis of protein complexes. Together with recent advances in high-throughput and sensitive mass spectrometry platforms, characterisation of protein complexes on a proteomic levels is becoming more feasible. Systematic charting of the landscape of protein complexes in higher organisms would certainly be a valuable exercise and will most certainly mark the start of a new era in cellular and molecular biology.

ACKNOWLEDGEMENT

MOC and SGNG are supported by the Wellcome Trust Sanger Institute and the Wellcome Trust Genes to Cognition Programme. Thanks to Jyoti Choudhary for Figure 2.

REFERENCES

Albrecht, D.E. and Froehner, S.C. (2002) Syntrophins and dystrobrevins: defining the dystrophin scaffold at synapses. Neurosignals *11*, 123–129.

Ando, H., Mizutani, A., Matsu-ura, T. and Mikoshiba, K. (2003) IRBIT, a novel inositol 1,4,5-trisphosphate (IP3) receptor-binding protein, is released from the IP3 receptor upon IP3 binding to the receptor. J. Biol. Chem. *273*, 10602–10612.

Becamel, C., Alonso, G., Galeotti, N., Demey, E., Jouin, P., Ullmer, C., Dumuis, A., Bockaert, J. and Marin, P. (2002) Synaptic multiprotein complexes associated with 5-HT(2C) receptors: a proteomic approach. EMBO J. *21*, 332–2342.

Becamel, C., Gavarini, S., Chanrion, B., Alonso, G., Galeotti, N., Dumuis, A., Bockaert, J. and Marin, P. (2004) The serotonin 5-HT2A and 5-HT2C receptors interact with specific sets of PDZ proteins. J. Biol. Chem. *279*, 20257–20266.

Cheeseman, I.M. and Desai, A. (2005) A combined approach for the localization and tandem affinity purification of protein complexes from metazoans. Sci. STKE *2005*, 11.

Ciruela, F., Burgueno, J., Casado, V., Canals, M., Marcellino, D., Goldberg, S.R., Bader, M., Fuxe, K., Agnati, L.F., Lluis, C., Franco, R., Ferre, S. and Woods, A.S. (2004) Combining mass spectrometry and pull-down techniques for the study of receptor heteromerization. Direct epitope-epitope electrostatic interactions between adenosine A2A and dopamine D2 receptors. Anal. Chem. *76*, 5354–5363.

Collins, M.O., Husi, H., Yu, L., Brandon, J.M., Anderson, C.N., Blackstock, W.P., Choudhary, J.S. and Grant, S.G. (2006) Molecular characterization and comparison of the components and multiprotein complexes in the postsynaptic proteome. J. Neurochem. *97*, 16–23.

Collins, M.O., Yu, L., Coba, M.P., Husi, H., Campuzano, I., Blackstock, W.P., Choudhary, J.S. and Grant, S.G. (2005) Proteomic analysis of *in vivo* phosphorylated synaptic proteins. J. Biol. Chem. *280*, 5972–5982.

de Godoy, L.M., Olsen, J.V., de Souza, G.A., Li, G., Mortensen, P. and Mann, M. (2005) Status of complete proteome analysis by mass spectrometry: SILAC labeled yeast as a model system. Genome Biol. *7*, R50.

Dynlacht, B.D., Hoey, T. and Tjian, R. (1991) Isolation of coactivators associated with the TATA-binding protein that mediate transcriptional activation. Cell *66*, 563–576.

Farr, C.D., Gafken, P.R., Norbeck, A.D., Doneanu, C.E., Stapels, M.D., Barofsky, D.F., Minami, M. and Saugstad J.A. (2004) Proteomic analysis of native metabotropic glutamate receptor 5 protein complexes reveals novel molecular constituents. J. Neurochem. *91*, 438–450.

Fiorentini, C. and Missale, C. (2004) Oligomeric assembly of dopamine D1 and glutamate NMDA receptors: molecular mechanisms and functional implications. Biochem. Soc. Trans. *32*, 1025–1028.

Fukata, Y., Tzingounis, A.V., Trinidad, J.C., Fukata, M., Burlingame, A.L., Nicoll, R.A. and Bredt, D.S. (2005) Molecular constituents of neuronal AMPA receptors. J. Cell Biol. *169*, 399–404.

Furuichi, H., Yamashita, K., Okada, M., Toyoshima, T., Hata, Y., Suzuki, S., Itano, T., Shishibori, T., Tokumitsu, H. and Kobayashi, R. (2000) Identification of tranilast-binding protein as 36-kDa microfibril-associated glycoprotein by drug affinity chromatography, and its localization in human skin. Biochem. Biophys. Res. Commun. *270*, 1002–1008.

Gavin, A.C., Bosche, M., Krause, R., Grandi, P., Marzioch, M., Bauer, A., Schultz, J., Rick, J.M., Michon, A.M., Cruciat, C.M., Remor, M., Hofert, C., Schelder, M., Brajenovic, M., Ruffner, H., Merino, A., Klein, K., Hudak, M., Dickson, D., Rudi, T., Gnau, V., Bauch, A., Bastuck, S., Huhse, B., Leutwein, C., Heurtier, M.A., Copley, R.R., Edelmann, A., Querfurth, E., Rybin, V., Drewes, G., Raida, M., Bouwmeester, T., Bork, P., Seraphin, B., Kuster, B., Neubauer, G. and Superti-Furga, G. (2002) Functional organization of the yeast proteome by systematic analysis of protein complexes. Nature *415*, 141–147.

Ghaemmaghami, S., Huh, W.K., Bower, K., Howson, R.W., Belle A., Dephoure, N., O'Shea, E.K. and Weissman, J.S. (2003) Global analysis of protein expression in yeast. Nature *425*, 737–741.

Grant, S.G., Marshall, M.C., Page, K.L., Cumiskey, M.A. and Armstrong, J.D. (2005) Synapse proteomics of multiprotein complexes: en route from genes to nervous system diseases. Hum. Mol. Genet. *14*, R225–R234.

Husi H. and Grant, S.G. (2001) Isolation of 2000-kDa complexes of N-methyl-D-aspartate receptor and postsynaptic density 95 from mouse brain. J. Neurochem. *77*, 281–291.

Husi H., Ward, M.A., Choudhary, J.S., Blackstock. W.P and Grant, S.G. (2000) Proteomic analysis of NMDA receptor-adhesion protein signaling complexes. Nat. Neurosci. *3*, 661–669.

Jeong H., Mason, S.P., Barabasi, A.L. and Oltvai Z.N. (2001) Lethality and centrality in protein networks. Nature *411*, 41–42.

Jordan, B.A., Fernholz, B.D., Boussac, M., Xu, C., Grigorean, G., Ziff, E.B. and Neubert, T.A. (2004) Identification and verification of novel rodent postsynaptic density proteins. Mol. Cell Proteomics *3*, 857–871.

Kabbani, N., Hannan, M.A., Levenson, R. (2005) Unraveling the Dopamine Receptor Signalplex by DRIPs and DRAPs. Current Proteomics *2*, 209–223.

Kawabata, S., Kohara, A., Tsutsumi, R., Itahana, H., Hayashibe, S., Yamaguchi, T. and Okada, M. (1998) Diversity of calcium signaling by metabotropic glutamate receptors. J. Biol. Chem. *273*, 17381–17385.

Kearn, C.S., Blake-Palmer, K., Daniel, E., Mackie, K. and Glass, M. (2005) Concurrent stimulation of cannabinoid CB1 and dopamine D2 receptors enhances heterodimer formation: a mechanism for receptor cross-talk? Mol. Pharmacol. *67*, 1697–1704.

Kim, M., Jiang, L.H., Wilson, H.L., North, R.A. and Surprenant, A. (2001) Proteomic and functional evidence for a P2X7 receptor signalling complex. EMBO J. *20*, 6347–6358.

Kotecha, S.A., Jackson, M.F., Al-Mahrouki, A., Roder, J.C., Orser, B.A. and MacDonald, J.F. (2003) Co-stimulation of mGluR5 and N-methyl-D-aspartate receptors is required for potentiation of excitatory synaptic transmission in hippocampal neurons. J. Biol. Chem. *278*, 27742–27749.

Leonoudakis, D., Conti, L.R., Anderson, S., Radeke, C.M., McGuire, L.M., Adams, M.E., Froehner, S.C., Yates, J.R., 3rd and Vandenberg, C.A. (2004) Protein trafficking and anchoring complexes revealed by proteomic analysis of inward rectifier potassium channel (Kir2.x)-associated proteins. J. Biol. Chem. *279*, 22331–22346.

Li, K.W., Hornshaw, M.P., Van Der Schors, R.C., Watson, R., Tate, S., Casetta, B., Jimenez, C.R., Gouwenberg, Y., Gundelfinger, E.D., Smalla, K.H. and Smit, A.B. (2004) Proteomics analysis of rat brain postsynaptic density. Implications of the diverse protein functional groups for the integration of synaptic physiology. J. Biol. Chem. *279*, 987–1002.

Mulder, N.J., Apweiler, R., Attwood, T.K., Bairoch, A., Bateman, A., Binns, D., Bradley, P., Bork, P., Bucher, P., Cerutti, L., Copley, R., Courcelle, E., Das, U., Durbin, R., Fleischmann, W., Gough, J., Haft, D., Harte, N., Hulo, N., Kahn, D., Kanapin, A., Krestyaninova, M., Lonsdale, D., Lopez, R., Letunic, I., Madera, M., Maslen, J., McDowall, J., Mitchell, A., Nikolskaya, A.N., Orchard, S., Pagni, M., Ponting, C.P., Quevillon, E., Selengut, J., Sigrist, C.J., Silventoinen, V., Studholme, D.J., Vaughan, R. and Wu, C.H. (2005) InterPro, progress and status in 2005. Nucleic Acids Res. *33*, D201–D205.

Nicke, A., Baumert, H.G., Rettinger, J., Eichele, A., Lambrecht, G., Mutschler, E. and Schmalzing, G. (1998) P2X1 and P2X3 receptors form stable trimers: a novel structural motif of ligand-gated ion channels. EMBO J. *17*, 3016–3028.

Nourry, C, Grant, S.G. and Borg, J.P. (2003) PDZ domain proteins: plug and play! Sci. STKE *2003*, RE7.

Peng, J., Kim, M.J., Cheng, D., Duong, D.M., Gygi, S.P. and Sheng, M. (2004) Semiquantitative proteomic analysis of rat forebrain postsynaptic density fractions by mass spectrometry. J. Biol. Chem. *279*, 21003–21011.

Perkinsm, D.N., Pappin, D.J., Creasy, D.M. and Cottrell, J.S. (1999) Probability-based protein identification by searching sequence databases using mass spectrometry data. Electrophoresis *20*, 3551–3567.

Pocklington, A.J., Cumiskey, M., Armstrong, J.D. and Grant, S.G. (2006) The proteomes of neurotransmitter receptor complexes form modular networks with distributed functionality underlying plasticity and behaviour. Mol. Syst. Biol. *2*, 0023.

Rocheville, M., Lange, D.C., Kumar, U., Patel, S.C., Patel, R.C. and Patel, Y.C. (2000) Receptors for dopamine and somatostatin: formation of hetero-oligomers with enhanced functional activity. Science *288*, 154–157.

Rubio, V., Shen, Y., Saijo, Y., Liu, Y., Gusmaroli, G., Dinesh-Kumar, S.P. and Deng, X.W. (2005) An alternative tandem affinity purification strategy applied to Arabidopsis protein complex isolation. Plant J. *41*, 767–778.

Sala, C., Roussignol, G., Meldolesi, J. and Fagni, L. (2005) Key role of the postsynaptic density scaffold proteins Shank and Homer in the functional architecture of Ca2+ homeostasis at dendritic spines in hippocampal neurons. J. Neurosci. *25*, 4587–4592.

Satoh, K., Takeuchi, M., Oda, Y., Deguchi-Tawarada, M., Sakamoto, Y., Matsubara, K., Nagasu, T. and Takai, Y. (2002) Identification of activity-regulated proteins in the postsynaptic density fraction. Genes Cells. *7*, 187–197.

Seamans, J.K. and Yang, C.R. (2004) The principal features and mechanisms of dopamine modulation in the prefrontal cortex. Prog. Neurobiol. *74*, 1–58.

Smith, F.D., Oxford, G.S. and Milgram, S.L. (1999) Association of the D2 dopamine receptor third cytoplasmic loop with spinophilin, a protein phosphatase-1-interacting protein. J. Biol. Chem. *274*, 19894–19900.

Sperlagh, B., Vizi, E.S., Wirkner, K. and Illes, P. (2006) P2X7 receptors in the nervous system. Prog. Neurobiol. *78*, 327–346.

Tu, J.C., Xiao, B., Yuan, J.P., Lanahan, A.A., Leoffert, K., Li, M., Linden, D.J. and Worley, P.F. (1998) Homer binds a novel proline-rich motif and links group 1 metabotropic glutamate receptors with IP3 receptors. Neuron. *21*(4), 717–726.

Valor, L.M., Charlesworth, P., Humphreys, L., Anderson, C.N. and Grant, S.G. (2007) Network activity-independent coordinated gene expression program for synapse assembly. Proc. Natl. Acad. Sci. U.S.A. *104*(11), 4658–4663.

Virginio, C., MacKenzie, A., North, R.A. and Surprenant, A. (1999) Kinetics of cell lysis, dye uptake and permeability changes in cells expressing the rat P2X7 receptor. J. Physiol. *519*, 335–346.

Walikonis, R.S., Jensen, O.N., Mann, M., Provance, D.W. Jr., Mercer, J.A. and Kennedy, M.B. (2000) Identification of proteins in the postsynaptic density fraction by mass spectrometry. J. Neurosci. *20*, 4069–4080.

Yates, J.R. 3rd, Eng, J.K., McCormack, A.L. and Schieltz, D. (1995) Method to correlate tandem mass spectra of modified peptides to amino acid sequences in the protein database. Anal. Chem. *67*, 1426–1436.

Yoshimura, Y., Yamauchi, Y., Shinkawa, T., Taoka, M., Donai, H., Takahashi, N., Isobe, T. and Yamauchi, T. (2004) Molecular constituents of the postsynaptic density fraction revealed by proteomic analysis using multidimensional liquid chromatography-tandem mass spectrometry. J. Neurochem. *88*, 759–768.

Corresponding authors: Dr. M.O. Collins (moc@sanger.ac.uk) and Dr. S.G.N. Grant (sg3@sanger.ac.uk)

CHAPTER 10

PROTEIN NETWORKS AND COMPLEXES IN PHOTORECEPTOR CILIA

RONALD ROEPMAN[1] and UWE WOLFRUM[2]
[1] *Radboud University Nijmegen Medical Centre*
[1] *Nijmegen Centre for Molecular Life Sciences, Nijmegen, The Netherlands*
[2] *Institute of Zoology, Johannes Gutenberg University of Mainz, Germany*

Table of Contents

E. Bertrand and M. Faupel (eds.), Subcellular Proteomics, 209–235.

Abstract: Vertebrate photoreceptor cells are ciliated sensory cells specialized for single photon detection. The photoreceptor outer segment corresponds to the ciliary shaft of a prototypic cilium. In the outer segment compartment, the ciliary membrane is highly modified into membranous disks which are enveloped by the plasma membrane in rod cells. At these outer segment disks, the visual transduction cascade – a prototypical G-protein coupled receptor transduction pathway is arranged. The light sensitive outer segments are linked by the socalled connecting cilium with the inner segment, the photoreceptor compartment which contains all organelles necessary for cell metabolism. The connecting cilium correlates with the transition zone, the short junction between the basal body and the axoneme of a prototypic cilium. The connecting cilium and the calycal processes, including the periciliary ridge complex, as well as the basal body complex are in close functional association with each other. In the latter ciliary compartments, the export and import from/into the outer segment of the photoreceptor cell are controlled and regulated. In all subciliary compartments, proteins are arranged in functional multiprotein complexes. In the outer segment, signaling components are arranged into complexes which provide specificity and speed for the signaling and serve in adaptation. Centrin-G-protein complexes may regulate the light driven translocation of the visual G-protein transducin through the connecting cilium. Intraflagellar transport (IFT) complexes may serve in intersegmental exchange of molecules. The import/export of molecules is thought to be regulated by proteins arranged in networks at the basal body complex. Proteins of the interactome related to the human Usher syndrome are localized in the connecting cilium and may participate in the ciliary transport, but are also arranged at interfaces between the inner segment and the connecting cilium where they probably control the cargo handover between the transport systems of the inner segment and these of the cilium. Furthermore, USH protein complexes may further provide mechanical stabilization to membrane specializations of the calycal processes and the connecting cilium. The protein complex in which the retinitis pigmentosa GTPase regulator (RPGR) participates in the ciliary compartments also plays a key role in the function and maintenance of photoreceptor cells. It further associates through the presumed scaffolding protein RPGRIP1 with the nephrocystin protein network. Although many of these proteins have been also found in prototypic cilia or primary cilia, the arrangements of the proteins in complexes can be specific for vertebrate photoreceptor cells. Defects of proteins in these complexes lead to photoreceptor cell death and retinal degeneration, underlying syndromic and non-syndromic blindness.

1. INTRODUCTION

The visual process is initiated by the detection of a light signal by photoreceptor cells in the outer retina of the vertebrate eye. Photoreceptor cells absorb light (photons) and convert it into an electric neuronal signal. Vertebrate cone and rod photoreceptor cells are highly specialized, polarized neurons, which consist of morphologically and functionally distinct cellular compartments (Figure 1). The light sensitive photoreceptor outer segment is linked with an inner segment by a small intracellular bridge, the so-called connecting cilium, through which all intracellular intersegmental exchanges occur (Besharse and Horst 1990). The inner segment compartment contains all organelles typical for the metabolism of a eukaryotic cell. It continues into the perikaryon and the synaptic region where electrical signals are transmitted from photoreceptor cells to horizontal and bipolar cells of the inner neuronal retina.

In both types of photoreceptor cells, the outer segment contains all components of the visual transduction cascade which is one of the best studied examples of

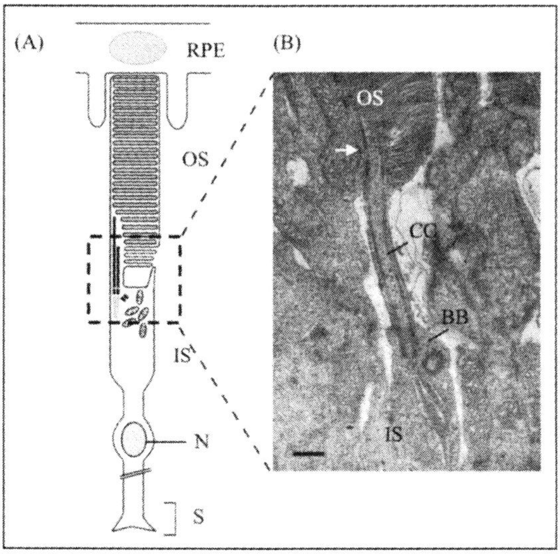

Figure 1. Structure of a ciliated photoreceptor cell in vertebrates. (A) Scheme of a rod photoreceptor cell. (B) Transmission electronmicroscopy image of a part of a mouse rod photoreceptor cell. The apical extensions of cells of the retinal pigment epithelium (RPE) evolve the tips of photoreceptors light-sensitive outer segments (OS). The OS is linked via a connecting cilium (CC) to an inner segment (IS) which bears the basal body complex (BB) in its apical region. Synaptic terminals (S) link the photoreceptor cell and the 2nd-order neurons, bipolar and horizontal cells. N = nucleus; in B, arrow point to axonemal microtubules projecting into the OS. Bar in B = 0.2 μm

G-protein mediated signal transduction cascades (Pugh and Lamb 2000; Okada et al. 2001). In rods, the cascade is arranged separate from the plasma membrane at hundreds of stacked membrane disks. Photoexcitation of the visual pigment rhodopsin (Rh*) activates a heterotrimeric G-protein (the visual G-protein transducin, composed of an α-subunit bearing the guanine nucleotide binding site and an undissociable βγ-complex) cascade, leading to cyclic GMP (cGMP) hydrolysis in the cytoplasm and closing of cGMP-gated (CNG) channels in the plasma membrane (see details reviewed in: Molday and Kaupp 2000; Kaupp and Seifert, 2002; Arshavsky et al. 2002). For termination of the visual cascade, Rh* is phosphorylated by the rhodopsin kinase which allows subsequent binding of arrestin molecules to P-Rh* inhibiting further R*-transducin interaction. CNG channels-closing leads to a decrease of the free Ca^{2+} concentration and an activation of Ca^{2+}-dependent proteins in the outer segment. In turn these proteins activate guanylate cyclases (GC) and lead to a delayed restoration of the cGMP concentration (Nakatani et al. 2002).

The photoreceptor outer segment membranous discs are continually renewed throughout lifetime. Newly synthesized disk membranes are added at the base of the outer segment by the expansion of the plasma membrane (Steinberg et al. 1980) or by incorporation of vesicular structures into nascent disc membranes (Uskura and Obata 1995), whereas disk packages at the distal outer segment tip are phagocytosed by the cells of the retinal pigment epithelium (Young 1967), which is juxtaposed to the apical rod photoreceptor outer segment (Figure 1A). The abundant membrane turnover of the photoreceptor outer segment implicates an efficient and massive vectorial transport of all disk components from the site of biogenesis, the ER and Golgi apparatus in the photoreceptor inner segment, to the base of the outer segment, the site of disk neogenesis. In addition to these unidirectional constitutive translocations of outer segment molecules, massive light dependent bidirectional movements of visual signal cascade proteins between the inner and outer segment are in the focus of current research (e.g. Pulvermüller et al. 2002; Sokolov et al. 2002; Gießl et al. 2004; Strissel et al. 2006). The massive reciprocal translocation of arrestin and transducin is thought to contribute to the long range light adaptation of rod photoreceptor cells (Sokolov et al. 2002). In any case, the intracellular exchange of molecules between the inner segment and the outer segment is forced to occur through the connecting cilium of the photoreceptor cell.

2. STRUCTURE AND FUNCTION OF THE PHOTORECEPTOR CILIUM AND THEIR RELATION TO THE PROTOTYPIC CILIA

In general, cilia are widespread finger-like cell appendages. The structure of a prototypic motile cilium is characterized as follows: the ciliary shaft originates from a basal body complex in the apical cytoplasm beneath the plasma membrane (Figure 2A).

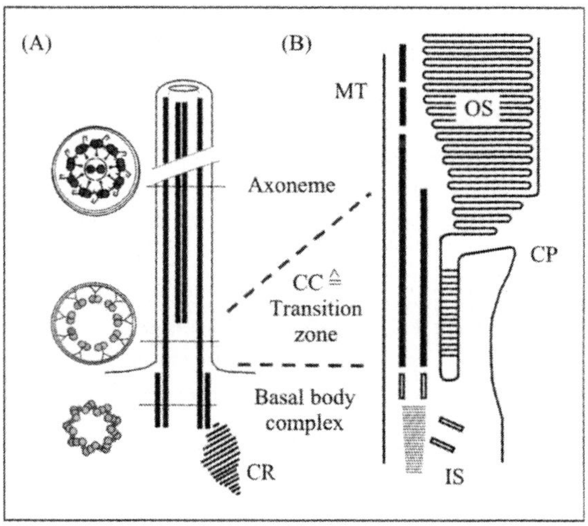

Figure 2. Schematic representations of a prototypic cilium and the photoreceptor cilium in comparison. (A) Scheme of a prototypic cilium, in longitudinal extension and cross sections through subciliary compartments: axoneme ($9 \times 2 + 2$ microtubule arrangement), transition zone ($9 \times 2 + 0$ microtubule arrangement) and centriole ($9 \times 3 + 0$ microtubule arrangement) of the basal body. (B) Scheme of the "ciliary part" of a rod photoreceptor cell. Axonemal microtubules (MT) project into the outer segment (OS). The OS is linked via the connecting cilium (CC) to the inner segment (IS). The CC corresponds to the transition zone of a prototypic cilium. The basal body complex (BB) is localized in the apical region of the IS. The calycal process (CP) of the IS is linked by extracellular fibers with the membrane of the CC.

The centriolar triplet arrangement of microtubules in the basal body is converted in the transition zone ($9 \times 2 + 0$ microtubular array) into the axoneme $9 \times 2 + 2$ (0) microtubular array. Cilia terminate at their tip in an end cap structure where axonemal microtubules are anchored at the apical ciliary membrane. Although, the principle composition of cilia is highly conserved throughout the eukaryote evolution, they show broad diversity in a single multicellular organism. Based on their axonemal microtubule array and their motile attributes, cilia are divided by current definition into four subtypes: (i) motile $9 \times 2 + 2$ cilia (e.g. motile tracheal cilia), (ii) motile $9 \times 2 + 0$ cilia (e.g. monocilia at the embryonic node), (iii) immotile $9 \times 2 + 2$ cilia (e.g. kinocilia in the inner ear), and (iv) immotile $9 \times 2 + 0$ cilia (e.g. renal cilia). Although, we can currently presume that almost all cilia have sensory functions (Singla and Reiter 2006; Scholey and Anderson 2006), animals possess specialized sensory cilia highly tuned for the perception of a single sensory modality, for example vertebrate and invertebrate olfactory cells and the photoreceptor cells of the vertebrate retina. In the latter case, the entire outer segment compartment can be considered as the highly modified distal part of an immotile cilium (Röhlich 1975; Besharse and Horst 1990).

As in prototypic cilia, the arrangement of the ciliary compartments in photoreceptor cells is not only structural but also functional: from the basal body region beneath the highly specialized apical inner segment membrane, long striated ciliary rootlets project through the inner segment into the cell body and can terminate even in the synaptic region terminals (Spira and Milman 1979). Ciliary rootlets are composed of fibers of static rootletin polymers which provide mechanical support for anchoring the basal body complex within the cytoplasm (Yang and Li 2006). The basal body region acts as the major microtubule organizing center (MTOC) in most ciliated cells and vertebrate photoreceptor cells (Troutt et al. 1990; Muresan et al. 1993; Wolfrum and Salisbury 1998). Microtubules nucleate at the MTOC and project with its fast growing plus-end into the cell. But, there is growing evidence that this region also controls the handover of cargos from the minus-end-directed microtubule-based transport through the inner segment transport (or the cell body of ciliated epithelial cells) to the molecular translocation machinery within the cilium (Sung and Tai 2000).

The connecting cilium, often regarded as the photoreceptor cilium, actually correlates with a short part of a prototypic cilium, the so-called transition zone (Röhlich 1975; Besharse and Horst 1990; Schmitt and Wolfrum 2001). In prototypic cilia, the transition zone is the short ciliary segment at the basal body-axoneme junction. It may serve as control gate at which the exchange of ciliary proteins between the cytoplasm and the ciliary compartment is controlled (Fliegauf and Omran 2006). In mammalian photoreceptor cells, the transition zone is extended to the connecting cilium (from $\sim$0.2 μm in a prototypical cilium to $\sim$1 μm in longitudinal extension) and appears to play an important role in photoreceptor organization and function in development and maintenance (Röhlich 1975; Besharse and Horst 1990; Schmitt and Wolfrum 2001). As the transition zone in prototypic cilia, the connecting cilium has a $9 \times 2 + 0$ microtubule configuration and bears a unique transmembrane assemblage: Y-shaped cross-linkers form a stable connection between cell surface glycoconjugates in the ciliary plasma membrane and the underlying microtubule cytoskeleton (Horst et al.

1987, 1990; Besharse and Horst 1990). At the base of the connecting cilium, the apical membrane of the photoreceptor inner segment is specialized as a so-called periciliary ridge complex (Papermaster et al. 1985; Papermaster 2002). Over the extension of the periciliary ridge complex, the inner segment membrane is linked by extracellular fibers with the membrane of the connecting cilium (Figure 2, Besharse and Horst 1990). At this membrane specialization, transport vesicles of the inner segment dock and hand their load (opsin and other outer segment components) over to the ciliary transport machinery (Deretic 2004). Additional specializations of the apical inner segment membrane are calycal processes which are microvilli-like extensions containing a prominent actin cytoskeleton (Pagh-Roehl et al. 1992). In their projection parallel to the outer segment they may support the outer segment against mechanical forces. In photoreceptor cells, the ciliary shaft of a prototypic cilium is extremely modified to the outer segment. In the outer segment of rod cells, thousands of membrane disks are stacked containing the visual signalling cascade (see: Chapter 1 Introduction). The axonemal cytoskeleton of the outer segment looses the stereotypical $9 \times 2 + 0$ arrangement and is reduced to a small number of "axonemal" microtubules which continue from the connecting cilium and project through cytoplasmic compartments of the outer segment, in some species for up to 80% of its length (Kaplan et al. 1987; Liu et al. 2002).

3. PROTEIN COMPLEXES IN PHOTORECEPTOR CILIA AND THEIR FUNCTIONS

Recent proteomic analysis indicated that the cilia of mammalian photoreceptor cells are significantly more complex than other eukaryotic cilium (Liu et al. 2006). Over 1200 different polypeptides have been identified by the quantitative analysis of the proteome of photoreceptor outer segment compared with the proteome of the axoneme/ciliary fraction of mouse photoreceptor cells. This data set contains all previously identified protein components of the photoreceptor cilium (e.g. Schmitt and Wolfrum 2001). An understanding how the identified proteins function in their native environment of the diverse compartments of the photoreceptor cilia will require increased knowledge of their molecular interaction and networking. Insights into the organization and composition of diverse protein complexes may also provide novel information on how functional modules of the cell, recently proposed by Hofmann et al. 2006, are connected.

3.1. Centrin-G-protein Complexes may Regulate Light-dependent G-protein Translocation through the Lumen of the Connecting Cilium

Ca^{2+}-activated centrins form complexes with the visual heterotrimeric G-protein, transducin, in the ciliary apparatus of photoreceptor cells (Pulvermüller et al. 2002; Wolfrum et al. 2002; Gießl et al. 2004a, b, 2006). The visual heterotrimeric G-protein transducin ($G_t$holo) is composed of a un-dissociable $G_t\beta\gamma$-dimer and the $G_t\alpha$-subunit which acts as the mediator and amplifier of the visual transduction cascade in the outer segment (see above Introduction Chapter 1 and Arshavsky et al. 2002). In rod photoreceptor cells, transducin light dependently shuttles between the inner and the outer segment: in the dark, transducin is localized in the outer segment whereas after light adaptation ~80% of the entire amount of transducin protein is found in the inner segment. This bidirectional intersegmental exchange of transducin through the lumen of connecting cilium is thought to be regulated by the formation of reversal centrin/transducin complexes (Pulvermüller et al. 2002; Wolfrum et al. 2002; Gießl et al. 2004a, b, 2006).

Centrins are members of a highly conserved subfamily of the EF-hand superfamily of Ca^{2+}-binding proteins commonly associated with centrioles of centrosome-related structures (Salisbury 1995; Schiebel and Bornens 1995; Wolfrum et al. 2002; Gießl et al. 2004b). In photoreceptor cells, centrins are also prominent components of the ciliary apparatus where the four centrin isoforms are differentially localized at the basal body and in the lumen of the connecting cilium (Gießl et al. 2004a, 2006). Centrin isoforms 1, 2 and 3 are localized in the lumen of the connecting cilium, centrin isoforms 2 and 3 are also present at the centrioles of the basal body complex whereas centrin 4 is restricted to the basal body. Mammalian centrins are activated by binding of two Ca^{2+} ions to EF-hands III and IV located in the C-terminal half of the molecules (Thompson et al. 2006; Park et al. 2006). In contrast, the two EF-hands in the N-terminal half do not bind, or bind Ca^{2+} ions with significant lower affinity (Park et al. 2006). N-terminus is the most diverse region among mammalian centrins and mediates protein-protein interactions for self assembly or for binding of partner proteins (Park et al 2006; Yang et al. 2006). C-terminal Ca^{2+}-binding probably induces conformational changes in the N-terminus of centrins necessary for oligomerization and protein binding.

Applying a combinative set of biochemical and biophysical protein-protein interaction assays we have demonstrated that all centrin isoforms can interact with the visual heterotrimeric G_t-protein in a Ca^{2+}-dependent manner (Pulvermüller et al. 2002; Wolfrum et al. 2002; Gießl et al. 2004a, b, 2006). All centrin isoforms interact with the $G_t$holo complex, the undissociable $G_t\beta\gamma$-dimer and the isolated $G_t\beta$-subunit, but not with $G_t\alpha$ alone. Nevertheless, centrin isoform 3 has a significant lower affinity to G_t compared to the other three centrin isoforms (Gießl et al. 2004a). Furthermore, centrin 3 interacts as a monomer while the other centrin isoform bind in form of oligomeres to $G_t\beta$ (Gießl et al. 2004a). Recent microtubule binding assays revealed binding of centrins to microtubules which suggests that centrins and their complexes with transducin are anchored to the inner surface of the ciliary microtubules of the photoreceptor connecting cilium (Ph. Trojan and U. Wolfrum unpublished).

In summary, an increase of the ciliary Ca^{2+}-concentration should induce oligomerization of centrin 1 and 2 and binding of these oligomeres to $G_t$holo complexes or to $G_t\beta\gamma$ dimer on their way through the connecting cilium. However, the protein complexes of centrins with G_t are not only regulated by Ca^{2+}, but also by phosphorylation. Recently, we observed that in mammalian retinas, centrin isoforms 1 and 2 are phosphorylated by the casein protein kinase CK2 in a light dependent manner

(Wolfrum et al. 2006; Trojan et al., in prep.). The residues phosphorylated in the dark (amino acids T138 (Cen1) and T137 (Cen2)) are specifically dephosphorylated by protein phosphatase PP2Cβ (Thissen et al. 2006, in prep., Wolfrum et al. 2006; Trojan et al. in prep.). The phosphorylations of centrin 1 and 2 drastically reduce the affinity of both iosforms to G_t. Since both enzymes, CK2 and PP2Cβ, are also found in the ciliary apparatus of photoreceptor cells (Hollander et al. 1999; Thissen et al. 2006; in prep.), it is worst to speculate that they temporarily bind to centrins and/or centrin/G_t-complexes during their enzymatic activity. Moreover, initial blot overlays indicated numerous more centrin binding proteins in the retina which may also contribute to the assembly and regulation of centrin/Gt-complexes in the photoreceptor connecting cilium.

Based on our results there is evidence that the Ca^{2+}-dependent assembly of centrin 1 and 2/G-protein complexes regulates transducin movement through the connecting cilium (Wolfrum et al. 2002; Gießl et al. 2006). Although, the source of the light modulated changes in free Ca^{2+} remains to be solved (Gießl et al. 2006), light-induced Ca^{2+} changes in the connecting cilium should induce binding of the high affine centrin isoforms 1 and 2 to G_t. The assembly into centrin/G_t complexes, together with Ca^{2+}-induced centrin oligomerization may form a barrier for further intersegmental exchanges of transducin (Figure 3) (Wolfrum et al. 2002; Gießl et al. 2006). This mechanism resembles a novel aspect of translocation regulations of signaling proteins

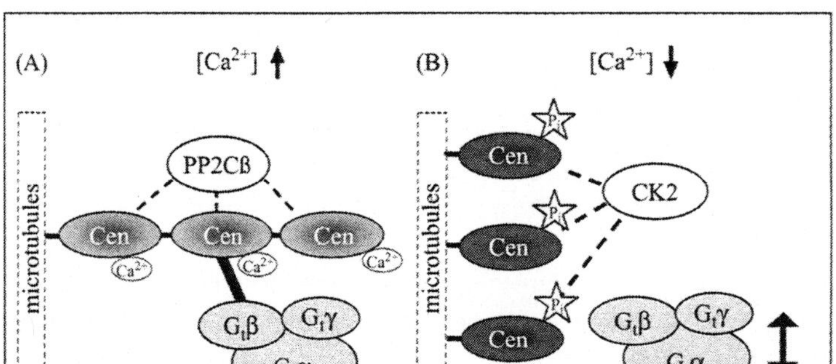

Figure 3. Model for centrin-G-protein complex assembly in the connecting cilium of photoreceptor cell. Schematic representations of a part of the inner lumen of the photoreceptor connecting cilium. Centrins (~ centrin isoforms 1 and 2) (Cen) are physically linked to the inner surface of the microtubule of the connecting cilium(CC). (A) Scenario at high free Ca^{2+} concentrations in CC: Cen are specifically dephosphorylated by protein phosphatase PP2Cβ. Ca^{2+}-binding to Cen induces Cen oligomerization and increases affinity of the $G_t\beta$-subunit of the visual heterotrimeric G-protein transducin (G_ta-$G_t\beta\gamma$). This may result in trapping G-protein molecules in the connecting cilium and G-protein diffusion is inhibited (Barrier hypothesis, Wolfrum et al. 2002). (B) Scenario at low free Ca^{2+} concentrations in CC: Cen are specifically phosphorylated by protein kinase CK2. Cen-P decreases affinity of G-protein to Cen. *Arrow* indicates that free diffusion of G-protein is possible. (for references, please see text)

in sensory cells, as well as a potential link between molecular trafficking and signal transduction in general.

3.2. Intraflagellar Transport Complexes in Mammalian Photoreceptor Cells

Intraflagellar transport (IFT) is an evolutionally conserved mechanism required for the assembly and maintenance of all eukaryotic cilia and flagella. IFT is a bidirectional transport system which moves non-membrane bound particles from basal body out to the tip of the cilium, and then returns them back to the cell body (Piperno and Mead 1997; reviewed in Rosenbaum and Witman 2002). A ~16 S IFT particle fraction, originally described in the green algae *Chlamydomonas* consists of at least 16 proteins that occur in two protein complexes, the complex A composed of four relatively high molecular weight proteins (M_r ~120–150 kDa) and the complex B which contains proteins of mostly lower molecular weight (M_r below 100 kDa) (Piperno and Mead 1997; Cole et al. 1998; reviewed in Rosenbaum and Witman 2002). IFT complexes are assembled near the basal body, moved along the axoneme by the heterotrimeric kinesin-II in an anterograde direction to the ciliary tip and back to the basal body region by the "axonemal" cytoplasmic dynein containing the dch2/1b heavy chain (Kozminski et al. 1995; Pazour et al. 1999; Pedersen et al. 2005). The IFT-system is thought to be associated with the transport of cargos, proteins and ciliary precursors essential for the assembly and maintenance of the axonemal structures, for example, α/β tubulin and axonemal dynein components (Qin et al. 2004). Furthermore, signaling pathways in *Chlamydomonas* gametes are IFT-dependent (Cole et al. 1998; Pan et al. 2005; Pan and Snell 2003; Wang et al. 2006).

The IFT concept was extended by analysis of mutations in IFT-proteins in *C. elegans* and mice (e.g. Orozco et al. 1999; Qin et al. 2001). A hypermorphic mutation in IFT88 leads to both polycystic kidney disease and retinal degeneration due to photoreceptor outer segment abnormalities in mice (Pazour et al. 2000, 2002). The knowledge on the IFT-system in mammalian photoreceptor cells mainly relies on studies by Joe Besharse and colleagues (Pazour et al. 2002; Baker et al. 2003; Besharse et al. 2003a, b). In retinal photoreceptor cells, all IFT proteins were identified so far, for which it has been searched for (Joe Besharse, personal communication). Available data on the IFT protein complex in vertebrate photoreceptor cells are summarized in Figure 4. In photoreceptor cells, IFT20 and IFT57 directly interact through coiled coil domains present in each protein within an IFT complex that also contains IFT88 and IFT52. IFT20 also connect this complex to the heterotrimeric kinesin-II by physical interaction with the kinesin-II motor subunit KIF3B necessary for the anterograde axonemal transport (Baker et al. 2003). Later assembly is regulated by ATP-hydrolysis. A search for IFT88 binding partners revealed direct binding of MRJ (mammalian relative to DnaJ), a molecular chaperone of the DnaJ/Hsp40 family (Li et al. 2004a). Further interaction of MRJ with the retinal guanylate cyclase E (GC-1) suggests that MRJ serves as a cargo linker and GC is one of the cargos for IFT in photoreceptor cells. Mis-localization of opsin and arrestin in mice with a

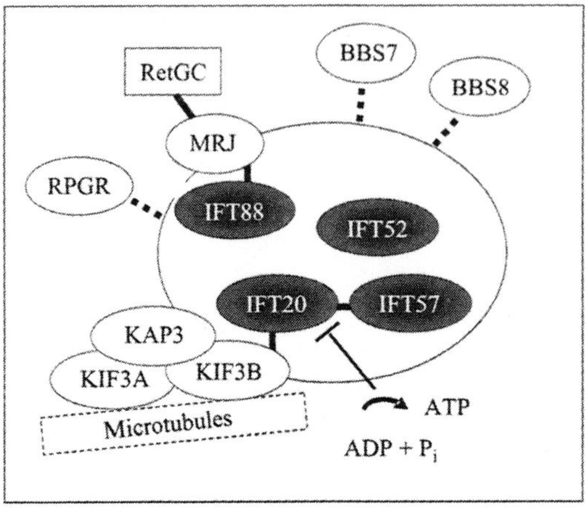

Figure 4. Model of intraflagellar transport (IFT) complex in photoreceptor cells. In the IFT complex found in photoreceptor cells IFT20 binds directly to IFT57 and to the KIF3B subunit of heterotrimeric kinesin-II, composed of KIF3B, KIF3A and KAP3. Latter interaction is reguated by ATP hydrolyses. IFT52 and IFT88 were also identified in photoreceptor IFT complexes. IFT88 interacts directly with the molecular chaperone MRJ which acts as a cargo receptor for photoreceptor specific guanylate cyclase RetGC. Association of BBS7 and BBS8 with mammalian IFT complex has been reported. (for references, please see text)

photoreceptor-specific knockout of the gene for kinesin-II motor subunit KIF3A suggested that opsin and arrestin might also be cargos for kinesin-II driven IFT (Marszalek et al. 2000). However, a direct link of the IFT complexes to these cargos has not been established and alternative motor complexes are described in photoreceptor cilia (see Section 3.3.1.1).

Expression of GFP-IFT proteins in transgenic *Xenopi* and zebrafish demonstrated targeting of IFT proteins to the cilia of photoreceptor cells (Besharse et al. 2003b). Although immunofluorescence data on IFT protein localizations in retinal photoreceptor cells indicate that IFT proteins and associated components are concentrated in the basal body region and in the connecting cilium (Pazour et al. 2000; Baker et al. 2004), they may function as bi-directional transport carriers along the axonemal microtubules in the outer segment (see Chapter 3.4). Knowledge of the molecular and spatial organization of IFT complexes, the regulation of cargo loading and unloading, and motor protein regulation will certainly elucidate the function of the IFT system in photoreceptor cells.

3.3. Protein Complexes of the Photoreceptor Cilium Involved in Human Genetic Disease

A group of inherited retinal degenerations called retinitis pigmentosa (RP) are a common cause of blindness. RP affects 1 out of every 3500 people worldwide (Berson

1993). RP is genetically heterogeneous, with 33 known different gene loci for non-syndromic RP and RP is also one part of several syndromes in which other organs are affected in addition to the retina. In the majority of the retina-related syndromes, ciliary dysfunctions are thought to be the common ground for pathogeneses of these disorders.

3.3.1. Usher syndrome (USH) "interactome" and USH protein complexes in photoreceptor cilia.

Usher syndrome (USH) is the most frequent cause of combined deaf-blindness in man. It is clinically and genetically heterogeneous and at least 11 chromosomal loci are assigned to three clinical USH types, namely *USH1B-G, USH2A,C,D, USH3A* (Gerber et al. 2006; Kremer et al. 2006; Reiners et al. 2006; Ebermann et al. 2007). The proteins encoded by the identified USH genes are members of protein classes with very different functions. Myosin VIIa (USH1B) is a motor protein, harmonin (USH1C), SANS (scaffold protein containing ankyrin repeats and SAM domain) (USH1G), and whirlin (USH2D) are scaffold proteins, cadherin 23 (USH1D) and protocadherin 15 (USH1F) are cell adhesion molecules and USH2A (usherin) and VLGR1 (very large G-coupled protein receptor) (USH2C) are transmembrane proteins with very large extracellular domains. The protein encoded by the UHS3A gene, clarin-1, is a member of the vertebrate-specific clarin family of four-transmembrane-domain proteins.

Molecular analysis of the diverse USH1 and USH2 proteins in the inner ear and retina revealed integration of USH proteins in a USH protein network or "interactome" (reviewed in Reiners et al. 2006; Kremer et al. 2006). In this network, the USH1 and USH2 proteins are thought to be assembled in a multiprotein scaffold, with a central role for the PDZ domain containing protein homologues harmonin and whirlin and the microtubule associated protein SANS (Reiners et al. 2005b, 2006; Kremer et al. 2006; van Wijk et al. 2006; Märker et al. in prep.). Although the colocalization of all USH1 and USH2 proteins at the photoreceptor synapse suggests a synaptic localization of the USH interactome in retinal photoreceptor cells, a subset of USH proteins is also present at the ciliary apparatus of photoreceptor cells indicating a USH network there (Kremer et al. 2006; Reiners et al. 2006).

3.3.1.1. Myosin VIIa (USH1B) molecular motor complex participates in ciliary transport of opsin The USH1B protein myosin VIIa is concentrated at the ciliary membrane of the photoreceptor connecting cilium (Liu et al. 1997; Wolfrum and Schmitt 2000). Immunoelectron microscopic analysis of myosin VIIa deficient shaker-1 mouse retinas showed an accumulation of opsin in the ciliary plasma membrane of photoreceptor cells (Liu et al. 1999; Wolfrum and Schmitt 2000). This excessive concentration of opsin of shaker-1 mice suggests that myosin VIIa is normally responsible for the opsin transport through the connecting cilium. Actin filaments extending beneath the ciliary membrane make myosin VIIa-based transport of opsin within the ciliary plasma membrane feasible (Wolfrum and Schmitt 2000). FERM (4.1-ezrin-redixin-mesoin) domains present as a tandem repeat in the

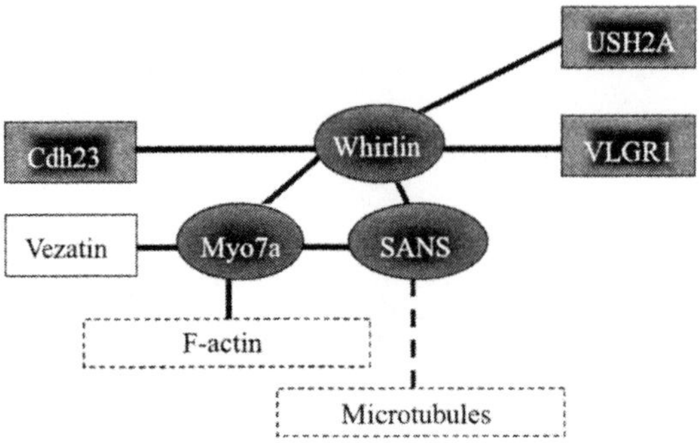

Figure 5. Usher protein complex in photoreceptor cilia and associated compartments. The three USH2 related transmembrane proteins USH2A, and VLGR1b (USH2C) bind via their C-terminal PBM to PDZ1 of whirlin, whereas cadherin 23 (Cdh23) (USH1D) interacts via its C-terminal PBM with PDZ2 of the scaffold protein whirlin (USH2D). Whirlin interacts via PDZ domains with SANS (USH1G). The scaffold protein SANS forms homomers via its central domain. The central domain of SANS is also suitable to interact with the MyTH-FERM domains in the tail of myosin VIIa (Myo7a) (USH1B). In the connecting cilium, myosin VIIa also interacts with the transmembrane protein vezatin. The present USH protein network is directly connected to the actin cytoskeleton through the actin-based molecular motor myosin VIIa dimmers and an association with microtubules is mediated by SANS. Rectangles indicate membrane proteins; dark gray fillings indicate "Usher proteins." (for references, please see text)

myosin VIIa tail has been shown to interact with lipids or proteins of the plasma membrane (Chishti et al. 1998). In the connecting cilium, transmembrane proteins, cadherin 23 (USH1D) and vezatin, are localized (Lillo et al. 2005; Wolfrum, unpublished data) which directly interact with the myosin VIIa FERM2 domain (Küssel-Andermann et al. 2000) and may anchor the molecular motor at the ciliary membrane. However, myosin VIIa is also linked to several other proteins of the USH interactome which were described for the connecting cilium. Myosin VIIa binds to the scaffold protein SANS (USH1G) and its interaction with whirlin provides a link to the USH2 proteins USH2A and VLGR1b (Reiners et al. 2006; van Wijk et al. 2006). In conclusion, these proteins may assemble into a multiprotein complex supporting the transport role of myosin VIIa in the connecting cilium (Figure 5). Thus, this myosin VIIa motor complex is a plausible alternative transport mechanism to the IFT system based on kinesin-II/dynein1b and microtubules described in Chapter 3.2.

3.3.1.2. Usher protein complexes connect the photoreceptor cilium with the apical compartments of the inner segment As described above, several USH proteins are localized in the connecting cilia associated with the membrane. Nevertheless, a prominent assembly of USH proteins is also present in the inner segment of photoreceptor cells associated with the ciliary apparatus. In the calycal processes of the apical region of the inner segment, in the absence of harmonin, the scaffold protein whirlin is localized and interacts via its PDZ domains with the PDZ-binding motifs (PBM) in the cytoplasmic tails of the USH2 proteins USH2A and VLGR1b, as well as with USH1 proteins protocadherin 15 and SANS (Figure 5) (Märker and Wolfrum, unpublished; Kremer et al. 2006). In mammals, the transmembrane cell adhesion protein protocadherin 15 is concentrated in the plasma membrane of the distal extensions of the calycal process where it faces the newly formed disk membranes of the outer segment base (Reiners et al. 2005a). Here, protocadherin 15 may associate with the photoreceptor specific cadherin prCAD (Rattner et al. 2001, 2004) stabilizing the labile newly formed disks (Reiners et al. 2005a).

In contrast, USH2A, VLGR1b and SANS are associated with the periciliary ridge complex which is thought to be the docking side for cargo loaded post-Golgi vesicles (Papermaster 2002). In mammals, this specialized domain extends over the plasma membrane of the proximal part of the calycal process which is connected via extracellular fibrous links to the plasma membrane of the connecting cilium. Recently, analogous to the fibrous links that connect stereovilli (= stereocilia) in mechanosensory hair cells, these fibers were identified to be composed of the long extracellular parts of cadherin 23 and VLGR1b (McGee et al. 2006). There is evidence that the extracellular domain of USH2A may also participate in the formation of these fibrous links (Liu et al. 2007). While, in the extracellular space between the membranes of the inner segment and the connecting cilium, homo- or/and heteromeric binding of the extracellular domains of USH2A and VLGR1b are developed, their short cytoplasmic tails are anchored by whirlin in the cytoplasm. So far, this USH protein complex at the periciliary ridge is completed by SANS which directly interacts with whirlin and may provide the molecular bridge to microtubules (van Wijk et al. 2006; Adato et al. 2005; Märker et al. in prep.). The present USH protein complex (Figure 5) should provide mechanical support to the membrane junction between the inner segment and the connecting cilium and probably also participate in the control of vesicle docking (reviewed by Deretic 2004) and cargo handover in the region of the periciliary ridge.

Immunoelectron microscopy revealed localization of the molecular components of the periciliary ridge USH protein complex in the region of the basal body of the photoreceptor connecting cilium (van Wijk et al. 2006; Märker et al. in prep.). Here at the ciliary basis, cytoplasmic splice variants of USH2A and VLGR1b together with whirlin and SANS, but also cadherin 23, might be enrolled in the MTOC function of the basal body (microtubule nucleation), but more plausible is a role of the USH protein complex proteins in the control of the import and export of molecules into or from the connecting cilium or they may participate in cargo handover. Unfortunately, little is known about the "cross talk" of the proteins of the USH complex with other proteins and protein complexes of the basal body complex reviewed in the other chapters of the present contribution.

Defects of one component of these USH complexes may cause dysfunction of the entire protein complex and induce sensorineuronal degeneration found as symptoms in USH patients.

3.3.2. The RPGR/RPGRIP1 protein network

3.3.2.1. RPGR and RPGRIP1 The X-linked gene *RPGR* (retinitis pigmentosa GTPase regulator) is mutated in patients with retinitis pigmentosa type 3 (RP3) (Roepman et al. 1996a, b Meindl et al. 1996). All missense mutations in *RPGR* have been identified in RP3 patients, in the region that encodes the ubiquitously expressed N-terminal 440 amino acids of RPGR. This domain shows significant homology to the regulator of chromosome condensation (RCC1), a guanine nucleotide exchange factor (GEF) for the small GTPase Ran (Roepman et al. 1996b; Meindl et al. 1996). Though as of yet no exchange activity on any GTPase has been described for the RCC1 homologous domain of RPGR, it binds to PDEδ (Linari et al. 1999), and to the C-terminus of RPGR interacting protein 1 (RPGRIP1) (Boylan and Wright 2000; Roepman et al. 2000a, b). The splice variant $RPGR^{ORF15}$, that is upregulated in rod and cone photoreceptors, harbours a mutational hot-spot (Vervoort et al. 2000), and mutations in this exon can give rise to a variety of retinal phenotypes (reviewed in Ferreira 2005). In addition, some mutations in this exon were found to cause RP in combination with impaired hearing and sinorespiratory infections (Van Dorp et al. 1992; Zito et al. 2003; Iannaccone et al. 2003, 2004) and RP with primary ciliary dyskinesia (PCD) (Moore et al. 2006). The latter conditions indicate a disrupted cilia function, and localization of the RPGR protein to the transitional zone of airway epithelial motile cilia supports this hypothesis (Hong et al. 2003).

In photoreceptors, RPGR localizes to the axoneme and basal bodies of connecting cilia (Hong et al. 2003; Khanna et al. 2005), where it was suggested to be involved in a correct localization of opsins in the outer segments. Localization to other subcellular sites have been described, but the exact nature of that is under debate (reviewed in Ferreira 2005). Multiple proteins were suggested to be associated with the protein complex in which the $RPGR^{ORF15}$ variant participates, including SMC (structural maintenance of chromosomes) proteins 1 and 3, IFT88, KIF3A, p150Glued, and p50-dynamitin (Khanna et al. 2005), and nucleophosmin (Shu et al. 2005). RPGR was shown to be anchored to the connecting cilium by RPGRIP1, which was suggested to be a structural component of the ciliary axoneme based on its resistance to detergent extraction (Hong et al. 2001). Immunoelectron microscopy further indicates RPGRIP1 localization external to the profiles of the microtubule doublets of the connecting cilium. These results could indicate that RPGRIP1 is a component of the microtubule-membrane cross linkers, Y-shaped structures projecting from each microtubule doublet at the junction between the A and B tubules to the adjacent plasma membrane (see Chapter 2, Besharse and Horst 1990). A general role of RPGRIP1 as a scaffold protein at these sites was therefore suggested (Hong et al. 2001).

The gene encoding the RPGR interacting protein 1 (RPGRIP1) harbors mutations that can lead to Leber congenital amaurosis (LCA), a genetically heterogeneous recessive disorder that is regarded to be the earliest and most severe form of all retinal dystrophies (Cremers et al. 2002). Absence of RPGRIP1 in $RPGRIP^{-/-}$ mutant mice leads to a reduced ERG amplitude and response sensitivity of both rod and cones, and a defect in outer segment disk formation, indicating a role in disk morphogenesis (Zhao et al. 2003). Similar to RPGR, the exact nature of subcellular localization of RPGRIP1 at other retinal sites is under debate (reviewed in Ferreira 2005). The conserved C2 domain of RPGRIP1, that appears to be encoded by a spliced variant that is expressed in a pan-retinal rather than photoreceptor-restricted fashion, strongly interacts with nephrocystin-4 (Roepman et al. 2005). Mutations in the gene encoding nephrocystin-4 (*NPHP4*) are associated with nephronophthisis type 4 (NPHP4) and Senior-Løken syndrome (SLSN) (Mollet et al. 2002; Otto et al. 2002).

3.3.2.2. The NPHP connection Nephronophthisis (NPHP) is an autosomal recessive cystic kidney disease, and the most frequent monogenic cause for end-stage renal failure in children and young adults (Smith and Graham 1945). SLSN is characterized by nephronophthisis in combination with retinal degeneration, either progressive (retinitis pigmentosa, RP) or congenital (LCA) (Olbrich et al. 2003). Nephrocystin-4 has been localized to kidney primary cilia of renal epithelial cells (Mollet et al. 2005), and interacts with the cilia-localized nephrocystin-1 protein (also called nephrocystin), involved in NPHP1 (Mollet et al. 2002). Furthermore, it is a conserved member of the flagellar apparatus and basal body proteome (Li et al. 2004b). Although nephrocystin-4 shows a panretinal localization, including the connecting cilia (Roepman et al. 2005), the interaction with RPGRIP1 strongly suggests an important functional role of this protein complex in the disease pathology of RP/LCA and NPHP/SLSN. This is emphasized by identification that the ciliary protein nephrocystin-5 (interacting with calmodulin) is involved in SLSN and exists in a complex with RPGR (Otto et al. 2005).

Within the *NPHP* gene family (*NPHP1-6*), there is a variable association with other phenotypes besides renal cyst formation and retinal degeneration, such as congenital oculomotor apraxia (COGAN syndrome) (Mollet et al. 2002), and a complex brain stem malformation and associated brain features (Joubert syndrome (Joubert et al. 1969). Recently, homozygous protein truncating mutations in the gene encoding a new centrosomal protein, CEP-290, were found to be associated with Joubert syndrome (Sayer et al. 2006; Valente et al. 2006), while a specific splice variant of the cognate gene (either homozygous or in combination with a second deleterious mutation on the other allele) is the most frequent cause of LCA known to date (den Hollander et al. 2006). Interestingly, this protein was also found to exist in a complex with RPGR (Chang et al. 2006; Sayer et al. 2006). Nephrocystin-4, inversin (nephrocystin-2) and nephrocystin-3, but not nephrocystin-5 interact with nephrocystin-1, and nephrocystin-3 has been shown to colocalize with inversin in the primary cilia of renal tubular epithelial cells (Otto et al. 2003), similar to what previously was found for the proteins associated with autosomal dominant and recessive polycystic kidney disease, polycystin-1, polycystin-2, polaris, cystin, and polyductin (reviewed in Hildebrandt and Otto 2005). This seems to point towards a unifying pathogenic mechanism, with a central role for the cilia. By combining the recent findings regarding the RPGR-RPGRIP1-nephrocystin-4 interaction with the reports of protein complexes identified by affinity purifications and immunoprecipitations, a putative interaction network arises in the connecting cilium of photoreceptor cells (Figure 6). Future analysis of these components will reveal the actual interplay of the different members of this network, and their subcellular sites of interaction.

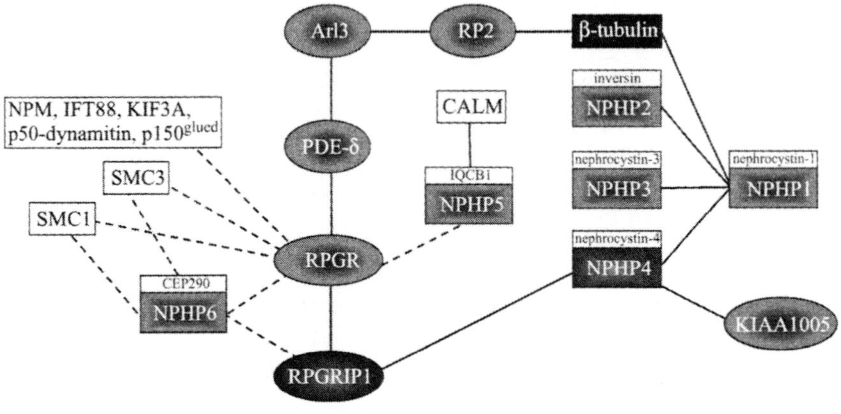

Figure 6. The RPGR-nephrocystin protein network. Putative core oculo-renal protein network in the cilia, interconnected in the retina by the RPGRIP1–nephrocystin-4 interaction. The connections are based on reported protein-protein interactions (for description and references, please see text). Direct interactions are shown by solid lines, interactions that were identified by immunoprecipitations, as part of a protein complex, are shown by dotted lines.

3.3.3. Bardet-Biedl Syndrome Proteins in ciliary protein complexes

The Bardet Biedl Syndrome (BBS) is a rare polygenetic and pleiotropic disorder associated with basal body and ciliary defects (Beales 2005). Patients with this multifaceted disease can suffer from a large number of symptoms, including retinal (rod-cone) degeneration, obesity, cystic kidneys, learning disabilities, hearing loss and anosmia. A full description of the BBS symptoms is shown in several competent overviews (Green et al. 1989; Beales et al. 1999; Moore et al. 2005). In addition to their common role in basal body and ciliary function, the 8 identified BBS proteins are involved in establishing planar cell polarity, modulation of intraflagellar transport (IFT) and lipid homeostasis, and in the regulation of intracellular trafficking and centrosomal functions. While BBS4 may serve as an adaptor protein for IFT and cilia function, BBS7 and BBS8 are certainly associated with IFT complexes and participate in particle assembly (reviewed in Chapter 3.2, Figure 4). These findings point towards a common function of BBS proteins in mediation and regulation of microtubule-based intracellular transport processes, as reviewed in Blacque and Leroux 2006.

3.4. Protein Complexes in the Light Sensitive Outer Segment of Photoreceptor Cells – The Search for the Vertebrate Photoreceptor Transducisomes

The outer segment of vertebrate photoreceptor cells can be considered as the highly modified distal part of an immotile cilium (Röhlich 1975; Besharse and Horst 1990) and contains only a small number of "axonemal" microtubules (see Introduction). The RP1 protein is specifically associated with these microtubules and is required for correct stacking of disk membranes into organized outer segments (Liu et al. 2002). Mutations in the *RP1* gene are a common cause of dominant retinitis pigmentosa (reviewed in Achenbach et al. 2004). In addition to this microtubule associated protein (MAP), proteins of the visual signal transduction cascade are also associated with the outer segment cytoskeleton (Figure 7). The guanylate cyclase (RetGC-1) was found to bind to actin filaments as well as to microtubules in photoreceptor outer segments (Hallett et al. 1996; Schrem et al. 1999). RetGC was also identified as a cargo of IFT complexes (see Chapter 3.2, Figure 5). Dimers of RetGC form a functional complex with GC-associated proteins GCAPs which regulate the GC enzymatic activity in Ca^{2+}-dependent manner. Peptide affinity chromatographies indicated binding RetGC-1/GCAP to a much bigger protein complex localized at the rim region of outer segment disks (Körschen et al. 1999). In the latter study, glutamic-acid-rich proteins (GARPs) were identified as multivalent proteins which exist as two soluble forms and as a large cytoplasmic tail of the B1-subunit of the

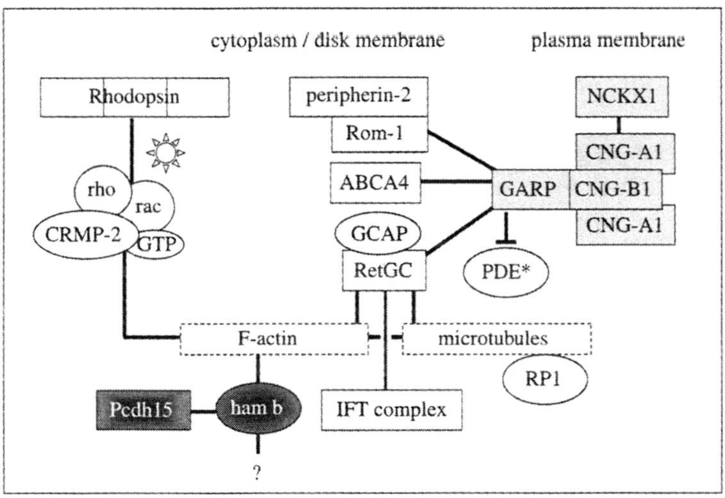

Figure 7. Schematic illustration of protein complexes in the photoreceptor outer segment (OS). The plasma membrane complex composed of the exchanger NCKX1 and the visual CNG channel (mediated by CNG-A1 subunit) interacts through the GARP domain of the CNG-B1 subunit with heteromers of peripherin-2/Rom-1 and may furthermore bind ABCR4, guanylate cyclase RetGC and activated phosphodiesterase (PDE*) at the disk membrane rim. RetGC is regulated by Ca^{2+}-binding GCAPs and serves as cargo for IFT complexes. RetGC also provides an association of disk rim complexes to actin filaments as well as to microtubules. These cytoskeletal elements bridge the latter complex to rhodopsin-associated Rac-GTP-bound protein complexes, recently identified. These multiprotein complexes contain among other components the small G-proteins rho and rac, in the GTP-bound form, and the CRMP-2 protein. The assembly of the rhodopsin-associated complexes is controlled by light ("sun"). The actin-binding scaffold protein harmonin b (hamb) (USH1C splice variant) is exclusively expressed in OS and binds via its PDZ2 domain Pcdh15 (USH1F). The microtubule associated protein RP1 specifically binds to axonemal microtubules in the OS. Rectangles indicate membrane proteins; light grey fillings indicate plasma membrane components, dark gray fillings indicate "Usher proteins." (for references, please see text)

visual cGMP-gated (CNG) channel, and interact with central players of the visual cGMP signaling pathway, RetGC-1 and phosphdiesterase (PDE), and with the ATP-binding cassette receptor (ABCR) (Figure 7). Since GARPs powerfully inhibit PDE activity, the "GARP complex" may constitute an adaptational signaling system, that inactivates active PDE molecules diffused to the disk rim and down regulates the high cGMP turnover under rod saturation conditions during daylight (Körschen et al. 1999). Nevertheless, co-immunoprecipitation experiments demonstrate a GARP containing protein complex of different composition (Poetsch et al. 2001). In the identified arrangement of outer segment membrane proteins, the visual CNG channel form a complex with the Na/Ca-K exchanger (NCKX1) via its A1-subunit in the plasma membrane. Furthermore, the CNG channel interacts through the GARP-domain of the B1-subunit with peripherin-2-ROM-1 oligomers which are localized in the rim of the outer segment disk (Schwarzer et al. 2000; Poetsch et al. 2001; Kang et al. 2003; Molday 2004). These interactions of the visual channel with peripherin-2-ROM-1 may guarantee that a significant portion of the CNG channels is situated next to a disk rim so that it can promptly response to the signaling cascade arranged at the disk membranes. A combination of the two models of the outer segment membrane protein complexes are shown in Figure 7.

The disk rim complexes are most probably associated with the outer segment cytoskeleton (Hallet et al. 1996; Körschen et al. 1999; Kajimura et al. 2000). In invertebrate photoreceptors, the PDZ protein INAD clusters the components of the visual signal transduction cascade into a signal complex associated with the rhabdomeric actin cytoskeleton (Montell 1999). In vertebrates, the scaffold protein harmonin b, a long splice variant of the USH1C gene, exhibits all features of a potent actin-binding protein (Boeda et al. 2002) and was furthermore identified to be exclusively expressed in photoreceptor outer segments (Reiners et al. 2003). Harmonin b might associate with the disk rim complexes through its interaction with the actin cytoskeleton providing its PDZ domains as attractive binding sites for the assembly of further proteins into outer segment protein complexes (Figure 7). The identification of outer segment proteins interacting with the PDZ domains of harmonin will broaden our knowledge on the organization of signal complexes in vertebrate photoreceptor cells.

In conclusion, the assembly of signaling molecules into supramolecular complexes, known as transducisomes, provides specificity, sensitivity and speed in intracellular signaling cascades (Zucker and Ranganathan 1999). A recent study provides evidence for an interdependence of visual perception with signalling networks involved in the organization of proper cellular structure. The Ueffing lab identified several candidates as novel rhodopsin interaction partners (Swiatek-De Lange et al. 2006). These candidates include small GTPases from the Rho and Rab families as well as CRMP-2 (collapsing response mediator protein 2) as a novel small GTPase-binding protein in photoreceptor cells. RhoA and Rac1 regulate protein transport and structural organisation of cells, while the CRMP family is one of main regulators of polarity development in neuronal cells. The identified Rac-GTP-bound protein complexes associate with the cytoskeleton and exhibit light and dark regulated dynamics. These complexes interact with light-activated rhodopsin (R*) which might be organized in form of dimers or higher order oligomers in the disk membrane of photoreceptor outer segments (Fotiadis et al. 2003, 2006). A possible link to the scaffold protein harmonin b is currently investigated. Although, there is good evidence for an involvement of Rac-containing protein complexes in the *de novo* synthesis of membranous disks at the base of the outer segment (Deretic et al. 1995; Deretic 2004), specific subcellular localization of the complexes in the outer segment will provide more information on the function of the light modulated outer segment complex. Nevertheless, these supramolecular complexes may also serve as a link between the transduction module (Hofmann et al. 2006) and the module of signaling networks involved in structural integrity and cell polarity in photoreceptor cells.

4. CONCLUDING REMARKS

Vertebrate photoreceptor cells are ciliated sensory cells modified for single photon detection. In all of the modified subciliary compartments of the photoreceptor cells, proteins are arranged in functional supramolecular complexes. Although, many of these proteins are also found in prototypic cilia or primary cilia, the arrangements of the proteins in complexes can be specific for vertebrate photoreceptor cells. These protein complexes may serve in functional modules specific for photoreceptor cells, for example, for visual signal transduction, adaptation, ciliary signaling, cell polarity and integrity or in specific modules of molecular translocations. The current knowledge on protein complexes in photoreceptor cilia is certainly only the peak of the iceberg and very fragmentary. In particular, the interconnections between the different protein networks and complexes and their side branches seems to be patchy. Future analysis of the molecular interactome in photoreceptor cilia will certainly expand the insights into photoreceptor function/dysfunction, in health and disease.

ACKNOWLEDGEMENTS

Authors thank Philipp Trojan and Dr. Andreas Gießl for contributing Figure 3, Heleen Arts for contributing Figure 6, and Dr. Kerstin Nagel-Wolfrum for critical comments on the manuscript. This work is supported by DFG, FAUN, ProRetina, Forschung contra Blindheit – Initative UsherSyndrom e.V. (FcB) (UW) and by European Commision IP 'EVI-GenoRet' (LSHG-CT-2005-512036), Dutch Kidney Foundation, Rotterdamse Vereniging Blindenbelangen, Stichting Blindenhulp, Stichting OOG, ANVVB (RR).

REFERENCES

Achenbach, S., Liu, Q. and Pierce E.A. (2004) The RP1 gene and protein in photoreceptor biology. In: Photoreceptor cell biology and inherited retinal degenerations. D.S. Williams, ed. (Singapore: World Scientific Publishing Co. Pte. Ltd.), pp. 223–257.

Adato, A., Michel, V., Kikkawa, Y., Reiners, J., Alagramam, K.N., Weil, D., Yonekawa, H., Wolfrum, U., El Amraoui, A. and Petit, C. (2005) Interactions in the network of Usher syndrome type 1 proteins. Hum. Mol. Genet. *14*, 347–356.

Arshavsky, V.Y., Lamb, T.D. and Pugh, E.N. Jr. (2002) G proteins and phototransduction. Annu. Rev. Physiol. *64*, 153–187.

Baker, S.A., Freeman, K., Luby-Phelps, K., Pazour, G.J. and Besharse, J.C. (2003) IFT20 links kinesin II with a mammalian intraflagellar transport complex that is conserved in motile flagella and sensory cilia. J. Biol. Chem. *278*, 34211–34218.

Baker, S.A., Pazour, G.J., Witman, G.B. and Besharse, J.C. (2004) Photoreceptor and intraflagellar transport. In: Photoreceptor Cell Biology and Inherited Retinal Degenerations. D.S. Williams ed. (Singapore: World Scientific Publishing Co. Pte. Ltd), pp. 109–132.

Beales, P.L. (2005) Lifting the lid on Pandora's box: the Bardet-Biedl syndrome. Curr. Opin. Genet. Dev. *15*, 315–323.

Beales, P.L., Elcioglu, N., Woolf, A.S., Parker, D. and Flinter, F.A. (1999) New criteria for improved diagnosis of Bardet-Biedl syndrome: results of a population survey. J. Med. Genet. *36*, 437–446.

Berson, E.L. (1993) Retinitis pigmentosa. The Friedenwald lecture. Invest. Ophthalmol. Vis. Sci *34*, 1659–1676.

Besharse, J.C., Baker, S.A., Luby-Phelps, K. and Pazour, G.J. (2003a) Photoreceptor intersegmental transport and retinal degeneration: a conserved pathway common to motile and sensory cilia. Adv. Exp. Med. Biol. *533*, 157–164.

Besharse, J.C., Fogerty, J., Baker, S., Link, B., Pazour, G.J. and Luby-Phelps, K. (2003b) Expression of IFT proteins in vertebrate rod photoreceptors. ARVO Abstract 2863/B702.

Besharse, J.C. and Horst, C.J. (1990) The photoreceptor connecting cilium - a model for the transition tone. In: Ciliary and Flagellar Membranes. R.A. Bloodgood, ed. (New York: Plenum), pp. 389–417.

Blacque, O.E. and Leroux, M.R. (2006) Bardet-Biedl syndrome: an emerging pathomechanism of intracellular transport. Cell. Mol. Life Sci. *63*, 2145–2161.

Boeda, B., El-Amraoui, A., Bahloul, A., Goodyear, R., Daviet, L., Blanchard, S., Perfettini, I., Fath, K.R., Shorte, S., Reiners, J., Houdusse, A., Legrain, P., Wolfrum, U., Richardson, G. and Petit, C. (2002) Myosin VIIa, harmonin and cadherin 23, three Usher 1 gene products that cooperate to shape the sensory hair cell bundle. EMBO J. *21*, 6689–6699.

Boylan, J.P. and Wright, A.F. (2000) Identification of a novel protein interacting with RPGR. Hum. Mol. Genet. *9*, 2085–2093.

Chang, B., Khanna, H., Hawes, N., Jimeno, D., He, S., Lillo, C., Parapuram, S.K., Cheng, H., Scott, A., Hurd, R.E., Sayer, J.A., Otto, E.A., Attanasio, M., O'Toole, J.F., Jin, G., Shou, C., Hildebrandt, F., Williams, D.S., Heckenlively, J.R. and Swaroop, A. (2006) In-frame deletion in a novel centrosomal/ciliary protein CEP290/NPHP6 perturbs its interaction with RPGR and results in early-onset retinal degeneration in the rd16 mouse. Hum. Mol. Genet. *15*, 1847–1857.

Chishti, A.H., Kim, A.C., Marfatia, S.M., Lutchman, M., Hanspal, M., Jindal, H., Liu, S.C., Low, P.S., Rouleau, G.A., Mohandas, N., Chasis, J.A., Conboy, J.G., Gascard, P., Takakuwa, Y., Huang, S.C., Benz, E.J., Bretscher, A., Fehon, R.G., Gusella, A.F., Ramesh, V., Solomon, F., Marchesi, V.T., Tsukita, S., Arpin, M.,. Louvard, D., Tonks, N.K., Anderson, J.M., Fanning, A.S., Bryant, P.J., Woods, D.F. and Hoover, K.B. (1998) The FERM domain: a unique module involved in the linkage of cytoplasmic proteins to the membrane. Trends Biochem. Sci. *23*, 281–282.

Cole, D.G., Diener, D.R., Himelblau, A.L., Beech, P.L., Fuster, J.C. and Rosenbaum, J.L. (1998) *Chlamydomonas* kinesin-II-dependent intraflagellar transport (IFT): IFT particles contain proteins required for ciliary assembly in *Caenorhabditis elegans* sensory neurons. J. Cell Biol. *141*, 993–1008.

Cremers, F.P., van den Hurk, J.A. and den Hollander, A.I. (2002) Molecular genetics of Leber congenital amaurosis. Hum. Mol. Genet. *11*, 1169–1176.

den Hollander, A.I., Koenekoop, R.K., Yzer, S., Lopez, I., Arends, M.L., Voesenek, K.E., Zonneveld, M.N., Strom, T.M., Meitinger, T., Brunner, H.G., Hoyng, C.B., van den Born, L.I., Rohrschneider, K. and Cremers, F.P.M. (2006) Mutations in the CEP290 (NPHP6) gene are a frequent cause of Leber Congenital Amaurosis. Am. J. Hum. Genet. *79*, 556–561.

Deretic, D. (2004) From the golgi to the rod outer segment: formation, movement, docking and fusion of rhodopsin transport carriers. In: Photoreceptor Cell Biology and Inherited Retinal Degeneration. D.S. Williams, ed. (Singapore: World Scientific Publishing Co. Pte. Ltd.), pp. 29–64.

Deretic, D., Huber, L.A., Ransom, N., Mancini, M., Simons, K. and Papermaster, D.S. (1995) Rab8 in retinal photoreceptors may participate in rhodopsin transport and in rod outer segment disk morphogenesis. J. Cell Sci. *108*, 215–224.

Ebermann, I., Scholl, H.P., Charbel Issa, P., Becirovic, E., Lamprecht, J., Jurklies, B., Millan, J.M., Aller, E., Mitter, D. and Bolz, H. (2007) A novel gene for Usher syndrome type 2: mutations in the long isoform of whirlin are associated with retinitis pigmentosa and sensorineural hearing loss. Hum. Genet. *121*, 203–211.

Ferreira, P.A. (2005) Insights into X-linked retinitis pigmentosa type 3, allied diseases and underlying pathomechanisms. Hum. Mol. Genet. *14*, R259–R267.

Fliegauf, M. and Omran, H. (2006) Novel tools to unravel molecular mechanisms in cilia-related disorders. Trends Genet. *22*, 241–245.

Fotiadis, D., Jastrzebska, B., Philippsen, A., Muller, D.J., Palczewski, K. and Engel, A. (2006) Structure of the rhodopsin dimer: a working model for G-protein-coupled receptors. Curr. Opin. Struct. Biol. *16*, 252–259.

Fotiadis, D., Liang, Y., Filipek, S., Saperstein, D.A., Engel, A. and Palczewski, K. (2003) Atomic-force microscopy: Rhodopsin dimers in native disc membranes. Nature *421*, 127–128.

Gerber, S., Bonneau, D., Gilbert, B., Munnich, A., Dufier, J.L., Rozet, J.M. and Kaplan, J. (2006) USH1A: Chronicle of a Slow Death. Am. J. Hum. Genet. *78*, 357–359.

Giessl, A., Pulvermüller, A., Trojan, P., Park, J.H., Choe, H.-W., Ernst, O.P., Hofmann, K.P. and Wolfrum, U. (2004a) Differential expression and interaction with the visual G-protein transducin of centrin isoforms in mammalian photoreceptor cells. J. Biol. Chem. *279*, 51472–51481.

Giessl, A., Trojan, P., Pulvermüller, A. and Wolfrum, U. (2004b) Centrins, potential regulators of transducin translocation in photoreceptor cells. In: Photoreceptor Cell Biology and Inherited Retinal Degenerations. D.S. Williams, ed. (Singapore: World Scientific Publishing Co. Pte. Ltd.), pp. 195–222.

Giessl, A., Trojan, P., Rausch, S., Pulvermüller, A. and Wolfrum, U. (2006) Centrins, gatekeepers for the light-dependent translocation of transducin through the photoreceptor cell connecting cilium. Vision Res. *46*, 4502–4509.

Green, J.S., Parfrey, P.S., Harnett, J.D., Farid, N.R., Cramer, B.C., Johnson, G., Heath, O., McManamon, P.J., O'Leary, E. and Pryse-Phillips, W. (1989) The cardinal manifestations of Bardet-Biedl syndrome, a form of Laurence-Moon-Biedl syndrome. N. Engl. J. Med. *321*, 1002–1009.

Hallett, M.A., Delaat, J.L., Arikawa, K., Schlamp, C.L., Kong, F.S. and Williams, D.S. (1996) Distribution of guanylate cyclase within photoreceptor outer segments. J. Cell Sci. *109*, 1803–1812.

Hildebrandt, F and Otto, E. (2005) Cilia and centrosomes: a unifying pathogenic concept for cystic kidney disease? Nat. Rev. Genet. *6*, 928–940.

Hofmann, K.P., Spahn, C.M., Heinrich, R. and Heinemann, U. (2006) Building functional modules from molecular interactions. Trends Biochem. Sci. *31*, 497–508.

Hollander, B.A., Liang, M.-Y. and Besharse, J.C. (1999) Linkage of a nucleolin-related protein and casein kinase II with the detergent-stable photoreceptor cytoskeleton. Cell Motil. Cytoskeleton *43*, 114–127.

Hong, D.H., Pawlyk, B., Sokolov, M., Strissel, K.J., Yang, J., Tulloch, B., Wright, A.F., Arshavsky, V.Y. and Li, T. (2003) RPGR isoforms in photoreceptor connecting cilia and the transitional zone of motile cilia. Invest. Ophthalmol. Vis. Sci. *44*, 2413–2421.

Hong, D.H., Yue, G., Adamian, M. and Li, T. (2001) Retinitis pigmentosa GTPase regulator (RPGRr)-interacting protein is stably associated with the photoreceptor ciliary axoneme and anchors RPGR to the connecting cilium. J. Biol. Chem. *276*, 12091–12099.

Horst, C.J., Forestner, D.M. and Besharse, J.C. (1987) Cytoskeletal-membrane interactions: Between cell surface glycoconjugates and doublet microtubules of the photoreceptor connecting cilium. J. Cell Biol. *105*, 2973–2987

Horst, C.J., Johnson, L.V. and Besharse, J.C. (1990) Transmembrane assemblage of the photoreceptor connecting cilium and motile cilium transition zone contain a common immunologic epitope. Cell Motil. Cytoskeleton *17*, 329–344.

Iannaccone, A., Breuer, D.K., Wang, X.F., Kuo, S.F., Normando, E.M., Filippova, E., Baldi, A., Hiriyanna, S., MacDonald, C.B., Baldi, F., Cosgrove, D., Morton, C.C., Swaroop, A. and Jablonski, M.M. (2003) Clinical and immunohistochemical evidence for an X linked retinitis pigmentosa syndrome with recurrent infections and hearing loss in association with an RPGR mutation. J. Med. Genet. *40*, e118.

Iannaccone, A., Wang, X., Jablonski, M.M., Kuo, S.F., Baldi, A., Cosgrove, D., Morton, C.C. and Swaroop, A. (2004) Increasing evidence for syndromic phenotypes associated with RPGR mutations. Am. J. Ophthalmol. *137*, 785–786.

Joubert, M., Eisenring, J.J., Robb, J.P. and Andermann, F. (1969) Familial agenesis of the cerebellar vermis. A syndrome of episodic hyperpnea, abnormal eye movements, ataxia, and retardation. Neurology *19*, 813–825.

Kajimura, N., Harada, Y. and Usukura, J. (2000) High-resolution freeze-etching replica images of the disk and the plasma membrane surfaces in purified bovine rod outer segments. J. Electron Microsc. *49*, 691–697.

Kang, K., Bauer, P.J., Kinjo, T.G., Szerencsei, R.T., Bonigk, W., Winkfein, R.J. and Schnetkamp, P.P. (2003) Assembly of retinal rod or cone Na(+)/Ca(2+)-K(+) exchanger oligomers with cGMP-gated channel subunits as probed with heterologously expressed cDNAs. Biochemistry *42*, 4593–4600.

Kaplan, M.W., Iwata, R.T. and Sears, R.C. (1987) Lengths of immunolabeled ciliary microtubules in frog photoreceptor outer segments. Exp. Eye Res. *44*, 623–632.

Kaupp, U.B. and Seifert, R. (2002) Cyclic nucleotide-gated ion channels. Physiol. Rev. *82*, 769–824.

Khanna, H., Hurd, T.W., Lillo, C., Shu, X., Parapuram, S.K., He, S., Akimoto, M., Wright, A.F., Margolis, B., Williams, D.S. and Swaroop, A. (2005) RPGR-ORF15, which is mutated in retinitis pigmentosa, associates with SMC1, SMC3, and microtubule transport proteins. J. Biol. Chem. *280*, 33580–33587.

Körschen, H.G., Beyermann, M., Müller, F., Heck, M., Vantler, M., Koch, K.W., Kellner, R., Bode, C., Wolfrum, U., Hofmann, K.P. and Kaupp, U.B. (1999) Interaction of glutamic acid-rich proteins with components of the cGMP-signaling pathway in rod photoreceptors. Nature *400*, 761–766.

Kozminski, K.G., Beech, P.L. and Rosenbaum, J.L. (1995) The Chlamydomonas kinesin-like protein FLA10 is involved in motility associated with the flagellar membrane. J. Cell Biol. *131*, 1517–1527.

Kremer, H., van Wijk, E., Märker, T., Wolfrum, U. and Roepman, R. (2006) Usher syndrome: molecular links of pathogenesis, proteins and pathways. Hum. Mol. Genet. *15*(Supp. 12), R262–R270.

Küssel-Andermann, P., El-Amraoui, A., Safieddine, S., Nouaille, S., Perfettini, I., Lecuit, M., Cossart, P., Wolfrum, U. and Petit, C. (2000) Vezatin, a novel transmembrane protein, bridges myosin VIIA to the cadherin-catenins complex. EMBO J. *19*, 6020–6029.

Li, J.B., Gerdes, J.M., Haycraft, C.J., Fan, Y., Teslovich, T.M., May-Simera, H., Li, H., Blacque, O.E., Li, L., Leitch, C.C., Lewis, R.A., Green, J.S., Parfrey, P.S., Leroux, M.R., Davidson, W.S., Beales, P.L., Guay-Woodford, L.M., Yoder, B.K., Stormo, G.D., Katsanis, N. and Dutcher, S.K. (2004b) Comparative genomics identifies a flagellar and basal body proteome that includes the BBS5 human disease gene. Cell *117*, 541–552.

Li, M., Sun, J., Baker, S., Freeman, K. and Besharse, J.C. (2004a) The intraflagellar transport protein, IFT88, directly interacts with the chaperone protein MRJ in photoreceptors. ARVO Abstract 3650/B111.

Lillo, C., Siemens, J., Kazmierczak, P., Mueller, U. and Williams, D.S. (2005) Roles and interactions of three USH1 proteins in the retina and inner ear. ARVO Abstract 46/5176.

Linari, M., Ueffing, M., Manson, F., Wright, A., Meitinger, T. and Becker, J. (1999) The retinitis pigmentosa GTPase regulator, RPGR, interacts with the delta subunit of rod cyclic GMP phosphodiesterase. Proc. Natl Acad. Sci. U.S.A. *96*, 1315–1320.

Liu, Q., Tan, G., Lenenkova, N., Rux, J., Speicher, D.W. and Pierce, E. (2006) The proteome of mammalian sensory cilium, the mouse photoreceptor outer segment. ARVO Abstract 3725/B368.

Liu, X., Udovichenko, I.P., Brown, S.D., Steel, K.P. and Williams, D.S. (1999) Myosin VIIa participates in opsin transport through the photoreceptor cilium. J. Neurosci. *19*, 6267–6274.

Liu, X.R., Vansant, G., Udovichenko, I.P., Wolfrum, U. and Williams, D.S. (1997) Myosin VIIa, the product of the Usher 1B syndrome gene, is concentrated in the connecting cilia of photoreceptor cells. Cell Motil. Cytoskeleton *37*, 240–252.

Liu, Q., Zhou, J., Daiger, S.P., Farber, D.B., Heckenlively, J.R., Smith, J.E., Sullivan, L.S., Zuo, J., Milam, A.H. and Pierce, E.A. (2002) Identification and subcellular localization of the RP1 protein in human and mouse photoreceptors. Invest. Ophthalmol. Vis. Sci. *43*, 22–32.

Liu, X., Bulgakov, O.V., Darrow, K.N., Pawlyk, B., Adamian, M., Liberman, M.C. and Li, T. (2007) A novel gene for Usher syndrome type 2: mutations in the long isoform of whirlin are associated with retinitis pigmentosa and sensorineural hearing loss. Proc. Natl. Acad. Sci. U.S.A. *104*, 4413–4418.

Marszalek, J.R., Liu, X., Roberts, E.A., Chui, D., Marth, J.D., Williams, D.S., Goldstein and L.S,. (2000) Genetic evidence for selective transport of opsin and arrestin by kinesin-II in mammalian photoreceptors. Cell *102*, 175–187.

McGee, J., Goodyear, R.J., McMillan, D.R., Stauffer, E.A., Holt, J.R., Locke, K.G., Birch, D.G., Legan, P.K., White, P.C., Walsh, E.J. and Richardson, G.P. (2006) The very large G-protein-coupled receptor VLGR1: a component of the ankle link complex required for the normal development of auditory hair bundles. J. Neurosci. *26*, 6543–6553.

Meindl, A., Dry, K., Herrmann, K., Manson, F., Ciccodicola, A., Edgar, A., Carvalho, M.R., Achatz, H., Hellebrand, H., Lennon, A., Migliaccio, C., Porter, K., Zrenner, E., Bird, A., Jay, M., Lorenz, B., Wittwer, B., D'Urso, M., Meitinger, T. and Wright, A. (1996) A gene (RPGR) with homology to the RCC1 guanine nucleotide exchange factor is mutated in X-linked retinitis pigmentosa (RP3). Nat. Genet. *13*, 35–42.

Molday, R.S. (2004) Molecular organization of rod outer segments. In: Photoreceptor Cell Biology and Inherited Retinal Degenerations. D.S. Williams, ed. (Singapore: World Scientific Publishing Co. Pte. Ltd.), pp. 259–300.

Molday, R.S. and Kaupp, U.B. (2000) Ion channels of vertebrate photoreceptors. In: Molecular Mechanism in Visual Transduction. D.G. Stavenga, W.J. Degrip and E.N. Pugh Jr., eds. (Amsterdam: Elsevier Science Publishers B.V.), pp. 143–182.

Mollet, G., Salomon, R., Gribouval, O., Silbermann, F., Bacq, D., Landthaler, G., Milford, D., Nayir, A., Rizzoni, G., Antignac, C. and Saunier, S. (2002) The gene mutated in juvenile nephronophthisis type 4 encodes a novel protein that interacts with nephrocystin. Nat. Genet. *32*, 300–305.

Mollet, G., Silbermann, F., Delous, M., Salomon, R., Antignac, C. and Saunier, S. (2005) Characterization of the nephrocystin/nephrocystin-4 complex and subcellular localization of nephrocystin-4 to primary cilia and centrosomes. Hum. Mol. Genet. *14*, 645–656.

Montell, C. (1999) Visual transduction in *Drosophila*. Annu. Rev. Cell Dev. Biol. *15*, 231–268.

Moore, A., Escudier, E., Roger, G., Tamalet, A., Pelosse, B., Marlin, S., Clement, A., Geremek, M., Delaisi, B., Bridoux, A.M., Coste, A., Witt, M., Duriez, B. and Amselem, S. (2006) RPGR is mutated in patients with a complex X linked phenotype combining primary ciliary dyskinesia and retinitis pigmentosa. J. Med. Genet. *43*, 326–333.

Moore, S.J., Green, J.S., Fan, Y., Bhogal, A.K., Dicks, E., Fernandez, B.A., Stefanelli, M., Murphy, C., Cramer, B.C., Dean, J.C., Beales, P.L., Katsanis, N., Bassett, A.S., Davidson, W.S. and Parfrey, P.S. (2005) Clinical and genetic epidemiology of Bardet-Biedl syndrome in Newfoundland: a 22-year prospective, population-based, cohort study. Am. J. Med. Genet. A *132*, 352–360

Muresan, V., Joshi, H.C. and Besharse, J.C. (1993) Gamma-tubulin in differented cell types: Localization in vicinity of basal bodies in retinal photoreceptors and ciliated epithelia. J. Cell. Sci. 104, 1229–1237.

Nakatani, K., Chen, C., Yau, K.W. and Koutalos, Y. (2002) Calcium and phototransduction. In: Photoreceptors and Calcium. W. Baehr and K. Palczewski, eds. (New York: Kluwer Academic/Plenum Publishers and Landes Biosciences/Eureka.com), pp. 1–20.

Okada, T., Ernst, O.P., Palczewski, K. and Hofmann, K.P. (2001) Activation of rhodopsin: new insights from structural and biochemical studies. Trends Biochem. Sci. *26*, 318–324.

Olbrich, H., Fliegauf, M., Hoefele, J., Kispert, A., Otto, E., Volz, A., Wolf, M.T., Sasmaz, G., Trauer, U., Reinhardt, R., Sudbrak, R., Antignac, C., Gretz, N., Walz, G., Schermer, B., Benzing, T., Hildebrandt, F. and Omran, H. (2003) Mutations in a novel gene, NPHP3, cause adolescent nephronophthisis, tapeto-retinal degeneration and hepatic fibrosis. Nat. Genet. *34*, 455–459.

Orozco, J.T., Wedaman, K.P., Signor, D., Brown, S., Rose, L. and Scholey, J.M, (1999) Movement of motor and cargo along cilia. Nature *398*, 674.

Otto, E., Hoefele, J., Ruf, R., Mueller, A.M., Hiller, K.S., Wolf, M.T., Schuermann, M.J., Becker, A., Birkenhager, R., Sudbrak, R., Hennies, H.C., Nurnberg, P. and Hildebrandt, F. (2002) A gene mutated in nephronophthisis and retinitis pigmentosa encodes a novel protein, nephroretinin, conserved in evolution. Am. J. Hum. Genet. 71, 1161–1167.

Otto, E.A., Loeys, B., Khanna, H., Hellemans, J., Sudbrak, R., Fan, S., Muerb, U., O'Toole, J.F., Helou, J., Attanasio, M., Utsch, B., Sayer, J.A., Lillo, C., Jimeno, D., Coucke, P., de Paepe, A., Reinhardt, R., Klages, S., Tsuda, M., Kawakami, I., Kusakabe, T., Omran, H., Imm, A., Tippens, M., Raymond, P.A., Hill, J., Beales, P., He, S., Kispert, A., Margolis, B., Williams, D.S., Swaroop, A. and Hildebrandt, F. (2005) Nephrocystin-5, a ciliary IQ domain protein, is mutated in Senior-Loken syndrome and interacts with RPGR and calmodulin. Nat. Genet. 37, 282–288.

Otto, E.A., Schermer, B., Obara, T., O'Toole, J.F., Hiller, K.S., Mueller, A.M., Ruf, R.G., Hoefele, J., Beekmann, F., Landau, D., Foreman, J.W., Goodship, J.A., Strachan, T., Kispert, A., Wolf, M.T., Gagnadoux, M.F., Nivet, H., Antignac, C., Walz, G., Drummond, I.A., Benzing, T. and Hildebrandt, F. (2003) Mutations in INVS encoding inversin cause nephronophthisis type 2, linking renal cystic disease to the function of primary cilia and left-right axis determination. Nat. Genet. *34*, 413–420.

Pagh-Roehl, K., Wang, E. and Burnside, B. (1992) Shortening of the calycal process actin cytoskeleton is correlated with myoid elongation in teleost rods. Exp. Eye Res. *55*, 735–746.

Pan, J. and Snell, W.J. (2003) Kinesin II and regulated intraflagellar transport of Chlamydomonas aurora protein kinase. J. Cell. Sci. *116*, 2179–2186.

Pan, J., Wang, Q. and Snell, W.J. (2005) Cilium-generated signaling and cilia-related disorders. Lab. Invest. *85*, 452–463.

Papermaster, D.S. (2002) The birth and death of photoreceptors: the Friedenwald lecture. Invest. Ophthalmol. Vis. Sci. *43*, 1300–1309.

Papermaster, D.S., Schneider, B.G. and Besharse, J.C. (1985) Vesicular transport of newly synthesized opsin from the Golgi apparatus toward the rod outer segment. Invest. Ophthalmol. Vis. Sci. *26*, 1386–1404.

Park, J.H., Pulvermüller, A., Scheerer, P., Rausch, S., Giessl, A., Höhne, W., Wolfrum, U., Hofmann, K.P., Ernst, O.P., Choe, H.-W. and Krauss, N. (2006) Insights into functional aspects of centrins from the structure of N-terminally extended mouse centrin 1. Vision Res. *46*, 4568–4574.

Pazour, G.J., Baker, S.A., Deane, J.A., Cole, D.G., Dickert, B.L., Rosenbaum, J.L., Witman, G.B. and Besharse, J.C. (2002) The intraflagellar transport protein, IFT88, is essential for vertebrate photoreceptor assembly and maintenance. J. Cell Biol. *157*, 103–113.

Pazour, G.J., Dickert, B.L., Vucica, Y., Seeley, E.S., Rosenbaum, J.L., Witman, G.B.and Cole, D.G. (2000) *Chlamydomonas IFT88* and its mouse homologue, polycystic kidney disease *Tg737*, are required for assembly of cilia and flagella. J. Cell Biol. *151*, 709–718.

Pazour, G.J., Dickert, B.L. and Witman, G.B. (1999) The DHC1b (DHC2) isoform of cytoplasmic dynein is required for flagellar assembly. J. Cell Biol. *144*, 473–481.

Pedersen, L.B., Miller, M.S., Geimer, S., Leitch, J.M., Rosenbaum, J.L. and Cole, D.G. (2005) Chlamydomonas IFT172 is encoded by FLA11, interacts with CrEB1, and regulates IFT at the flagellar tip. Curr. Biol. *15*, 262–266.

Piperno, G. and Mead, K. (1997) Transport of a novel complex in the cytoplasmic matrix of *Chlamydomonas* flagella. Proc. Natl Acad. Sci. U.S.A. *94*, 4457–4462.

Poetsch, A., Molday, L.L. and Molday, R.S. (2001) The cGMP-gated channel and related glutamic acid-rich proteins interact with peripherin-2 at the rim region of rod photoreceptor disc membranes. J. Biol. Chem. *276*, 48009–48016.

Pugh, E.N. Jr. and Lamb, T. (2000) Phototransducin in vertebrate rods and cones: Molecular mechanism of amplification, recovery and light adaptation. In: Molecular Mechanism in Visual Transduction. D.G. Stavenga, W.J. Degrip and E.N. Pugh Jr., eds. (Amsterdam, The Netherlands: Elsevier Science Publishers B.V.), pp. 183–255.

Pulvermüller, A., Giessl, A., Heck, M., Wottrich, R., Schmitt, A., Ernst, O.P., Choe, H.-W., Hofmann, K.P. and Wolfrum, U. (2002) Calcium dependent assembly of centrin/G-protein complex in photoreceptor cells. Mol. Cell. Biol. *22*, 2194–2203.

Qin, H., Diener, D.R., Geimer, S., Cole, D.G. and Rosenbaum, J.L. (2004) Intraflagellar transport (IFT) cargo: IFT transports flagellar precursors to the tip and turnover products to the cell body. J. Cell Biol. *164*, 255–266.

Qin, H., Rosenbaum, J.L. and Barr, M.M. (2001) An autosomal recessive polycystic kidney disease gene homolog is involved in intraflagellar transport in C. elegans ciliated sensory neurons. Curr. Biol. *11*, 457–461.

Rattner, A., Chen, J. and Nathans, J. (2004) Proteolytic shedding of the extracellular domain of photoreceptor cadherin. Implications for outer segment assembly. J. Biol. Chem. *279*, 42202–42210.

Rattner, A., Smallwood, P.M., Williams, J., Cooke, C., Savchenko, A., Lyubarsky, A., Pugh, E.N. and Nathans, J. (2001) A photoreceptor-specific cadherin is essential for the structural integrity of the outer segment and for photoreceptor survival. Neuron *32*, 775–786.

Reiners, J., Marker, T., Jurgens, K., Reidel, B. and Wolfrum, U. (2005a) Photoreceptor expression of the Usher syndrome type 1 protein protocadherin 15 (USH1F) and its interaction with the scaffold protein harmonin (USH1C). Mol. Vis. *11*, 347–355.

Reiners, J., Nagel-Wolfrum, K., Jurgens, K., Märker, T. and Wolfrum, U. (2006) Molecular basis of human Usher syndrome: deciphering the meshes of the Usher protein network provides insights into the pathomechanisms of the Usher disease. Exp. Eye Res. *83*, 97–119.

Reiners, J., Reidel, B., El-Amraoui, A., Boeda, B., Huber, I., Petit, C. and Wolfrum, U. (2003) Differential distribution of harmonin isoforms and their possible role in Usher-1 protein complexes in mammalian photoreceptor cells. Invest. Ophthalmol. Vis. Sci. *44*, 5006–5015.

Reiners, J., van Wijk, E., Marker, T., Zimmermann, U., Jurgens, K., te, B.H., Overlack, N., Roepman, R., Knipper, M., Kremer, H. and Wolfrum, U. (2005b) Scaffold protein harmonin (USH1C) provides molecular links between Usher syndrome type 1 and type 2. Hum. Mol. Genet. *14*, 3933–3943.

Roepman, R., Bauer, D., Rosenberg, T., van Duynhoven, G., van de Vosse, E., Platzer, M., Rosenthal, A., Ropers, H.-H., Cremers, F.P.M. and Berger, W. (1996a) Identification of a gene disrupted by a microdeletion in a patient with X-linked retinitis pigmentosa. Hum. Mol. Genet. *5*, 827–833.

Roepman, R., Bernoud-Hubac, N., Schick, D.E., Maugeri, A., Berger, W., Ropers, H.H., Cremers, F.P.M. and Ferreira, P.A. (2000a) The retinitis pigmentosa GTPase regulator (RPGR) interacts with novel transport-like proteins in the outer segments of rod photoreceptors. Hum. Mol. Genet. *9*, 2095–2105.

Roepman, R., Letteboer, S.J.F., Arts, H.H., van, Beersum, S.E.C., Lu, X., Krieger, E., Ferreira, P.A. and Cremers, F.P.M. (2005) Interaction of nephrocystin-4 and RPGRIP1 is disrupted by nephronophthisis or Leber congenital amaurosis-associated mutations. Proc. Natl Acad. Sci. U.S.A. *102*, 18520–18525.

Roepman, R., Schick, D. and Ferreira, P.A. (2000b) Isolation of retinal proteins that interact with retinitis pigmentosa GTPase regulator by interaction trap screen in yeast. Meth. Enzymol. *316*, 688–704.

Roepman, R., van Duynhoven, G., Rosenberg, T., Pinckers, A.J.L.G., Bleeker-Wagemakers, E.M., Bergen, A.A.B., Post, J., Beck, A., Reinhardt, R., Ropers, H.-H., Cremers, F.P.M. and Berger, W. (1996b) Positional cloning of the gene for X-linked retinitis pigmentosa: homology with the guanine-nucleotide-exchange factor RCC1. Hum. Mol. Genet. *5*, 1035–1041.

Rohlich, P. (1975) The sensory cilium of retinal rods is analogous to the transitional zone of motile cilia. Cell Tissue Res. *161*, 421–430.

Rosenbaum, J.L. and Witman, G.B. (2002) Intraflagellar transport. Nat. Rev. Mol. Cell Biol. *3*, 813–825.

Salisbury, J.L. (1995) Centrin, centrosomes, and mitotic spindle poles. Curr. Opin. Cell Biol. *7*, 39–45.

Sayer, J.A., Otto, E.A., O'Toole, J.F., Nurnberg, G., Kennedy, M.A., Becker, C., Hennies, H.C., Helou, J., Attanasio, M., Fausett, B.V., Utsch, B., Khanna, H., Liu. Y., Drummond, I., Kawakami, I., Kusakabe, T., Tsuda, M., Ma, L., Lee, H., Larson, R.G., Allen, S.J., Wilkinson, C.J., Nigg, E.A., Shou, C., Lillo, C., Williams, D.S., Hoppe, B., Kemper, M.J., Neuhaus, T., Parisi, M.A., Glass, I.A., Petry, M., Kispert, A., Gloy, J., Ganner, A., Walz, G., Zhu, X., Goldman, D., Nurnberg, P., Swaroop, A., Leroux, M.R. and Hildebrandt, F. (2006) The centrosomal protein nephrocystin-6 is mutated in Joubert syndrome and activates transcription factor ATF4. Nat. Genet. *38*, 674–681.

Schiebel, E. and Bornens, M. (1995) In search of a function for centrins. Trends Cell Biol. *5*, 197–201.

Schmitt, A. and Wolfrum, U. (2001) Identification of novel molecular components of the photoreceptor connecting cilium by immunoscreens. Exp. Eye Res. *73*, 837–849.

Scholey, J.M. and Anderson, K.V. (2006) Intraflagellar transport and cilium-based signaling. Cell *125*, 439–442.

Schrem, A., Lange, C., Beyermann, M. and Koch, K.W. (1999) Identification of a domain in guanylyl cyclase-activating protein 1 that interacts with a complex of guanylyl cyclase and tubulin in photoreceptors. J. Biol. Chem. *274*, 6244–6249.

Schwarzer, A., Schauf, H. and Bauer, P.J. (2000) Binding of the cGMP-gated channel to the Na/Ca-K exchanger in rod photoreceptors. J. Biol. Chem. *275*, 13448–13454.

Shu, X., Fry, A.M., Tulloch, B., Manson, F.D., Crabb, J.W., Khanna, H., Faragher, A.J., Lennon, A., He, S., Trojan, P., Giessl, A., Wolfrum, U., Vervoort, R., Swaroop, A. and Wright, A.F. (2005) RPGR ORF15 isoform co-localizes with RPGRIP1 at centrioles and basal bodies and interacts with nucleophosmin. Hum. Mol. Genet. *14*, 1183–1197.

Singla, V. and Reiter, J.F. (2006) The primary cilium as the cell's antenna: signaling at a sensory organelle. Science *313*, 629–633.

Smith, C. and Graham, J. (1945) Congenital medullary cysts of the kidneys with severe refractory anemia. Am. J. Dis. Child. 69,369–377.

Sokolov, M., Lyubarsky, A.L., Strissel, K.J., Savchenko, A.B., Govardovskii, V.I., Pugh. E.N. Jr. and Arshavsky, V.Y. (2002) Massive light-driven translocation of transducin between the two major compartments of rod cells: a novel mechanism of light adaptation. Neuron *33*, 95–106.

Spira, A.W. and Milman, G.E. (1979) The structure and distribution of the cross-striated fibril and associated membranes in guinea pig photoreceptors. Am. J. Anat. *155*, 319–337.

Steinberg, R.H., Fisher, S.K. and Anderson, D.H. (1980) Disc Morphogenesis in Vertebrate Photoreceptors. J. Comp. Neurol. *190*, 501–518.

Strissel, K.J., Sokolov, M., Trieu, L.H. and Arshavsky, V.Y. (2006) Arrestin translocation is induced at a critical threshold of visual signaling and is superstoichiometric to bleached rhodopsin. J. Neurosci. *26*, 1146–1153.

Sung, C.H. and Tai, A.W. (2000) Rhodopsin trafficking and its role in retinal dystrophies. Int. Rev. Cytol. *195*, 215–267.

Swiatek-de Lange, M., Beer, M. and Ueffing, M. (2006) Analysis of light-dependent dynamics of protein complexes in the retina. ARVO Abstract 3734/B377.

Thissen, M., Trojan, P., Giessl, A., Wolfrum, U., Krieglstein, J. and Klumpp, S. (2006) Protein phosphatase 2C dephosphorylates centrins. Int. Symposium on Neurodegeneration and Neuroprotection, Münster, Germany, Abstract.

Thompson, J.R., Ryan, Z.C., Salisbury, J.L. and Kumar, R. (2006) The structure of the human centrin 2-xeroderma pigmentosum group C protein complex. J. Biol. Chem. *281*, 18746–18752.

Troutt, L.L., Wang, E., Pagh-Roehl, K. and Burnside, B. (1990) Microtubule nucleation and organization in teleost photoreceptors – microtubule recovery after elimination by cold. J. Neurocyt. *19*, 213–223.

Usukura, J. and Obata, S. (1995) Morphogenesis of photoreceptor outer segments in retinal development. Prog. Retin. Eye Res. *15*, 113–125.

Valente, E.M., Silhavy, J.L., Brancati, F., Barrano, G., Krishnaswami, S.R., Castori, M., Lancaster, M,A., Boltshauser, E., Boccone, L., Al Gazali, L., Fazzi, E., Signorini, S., Louie, C.M., Bellacchio, E., Bertini, E., Dallapiccola, B. and Gleeson, J.G. (2006) Mutations in CEP290, which encodes a centrosomal protein, cause pleiotropic forms of Joubert syndrome. Nat. Genet. *38*, 623–625.

Van Dorp, D.B., Wright, A.F., Carothers, A.D. and Bleeker-Wagemakers, E.M. (1992) A family with RP3 type of X-linked retinitis pigmentosa: an association with ciliary abnormalities. Hum. Genet. *88*, 331–334.

van Wijk, E., van der, Z.B., Peters, T., Zimmermann, U., te, B.H., Kersten, F.F., Märker, T., Aller, E., Hoefsloot, L.H., Cremers, C.W., Cremers, F.P., Wolfrum, U., Knipper, M., Roepman, R. and Kremer, USH2A and VLGR1. Hum. Mol. Genet. *15*, 751–765.

Vervoort, R., Lennon, A., Bird, A.C., Tulloch, B., Axton, R., Miano, M.G., Meindl, A., Meitinger, T., Ciccodicola, A. and Wright, A.F. (2000) Mutational hot spot within a new RPGR exon in X-linked retinitis pigmentosa. Nat. Genet. *25*, 462–466.

Wang, Q., Pan, J. and Snell, W.J. (2006) Intraflagellar transport particles participate directly in cilium-generated signaling in *Chlamydomonas*. Cell *125*, 549–562.

Wolfrum, U., Giessl, A. and Pulvermüller, A. (2002) Centrins, a novel group of Ca^{2+} binding proteins in vertebrate photoreceptor cells. In: Photoreceptors and Calcium. W. Baehr and K. Palczewski, eds. (New York: Landes Bioscience and Kluwer Academic/Plenum Publishers), pp. 155–178.

Wolfrum, U. and Salisbury, J.L. (1998) Expression of centrin isoforms in the mammalian retina. Exp. Cell Res. *242*, 10–17.

Wolfrum, U. and Schmitt, A. (2000) Rhodopsin transport in the membrane of the connecting cilium of mammalian photoreceptor cells. Cell Motil. Cytoskeleton *46*, 95–107.

Wolfrum, U., Trojan, P., Rausch, S., Wünschig, D., Giessl, A. and Pulvermüller, A. (2006) Light-dependent phosphorylation of centrins regulates binding to transducin in mammalian photoreceptors. ARVO Abstract 1091.

Yang, J. and Li, T. (2006) Focus on molecules: rootletin. Exp. Eye Res. *83*, 1–2.

Yang, A., Miron, S., Duchambon, P., Assairi, L,. Blouquit, Y. and Craescu, C.T. (2006) The N-terminal domain of human centrin 2 has a closed structure, binds calcium with a very low affinity, and plays a role in the protein self-assembly. Biochemistry *45*, 880–889.

Young, R.W. (1967) The renewal of photoreceptor cell outer segments. J. Cell. Biol. *33*, 61–72.

Zhao, Y., Hong, D.H., Pawlyk, B., Yue, G., Adamian, M., Grynberg, M., Godzik, A. and Li, T. (2003) The retinitis pigmentosa GTPase regulator (RPGR)- interacting protein: Subserving RPGR function and participating in disk morphogenesis. Proc. Natl. Acad. Sci. U.S.A. *100*, 3965–3970.

Zito, I., Downes, S.M., Patel, R.J., Cheetham, M.E., Ebenezer, N.D., Jenkins, S.A., Bhattacharya, S.S., Webster, A.R., Holder, G.E., Bird, A.C., Bamiou, D.E. and Hardcastle. A.J. (2003) RPGR mutation associated with retinitis pigmentosa, impaired hearing, and sinorespiratory infections. J. Med. Genet. *40*, 609–615.

Zuker, C.S. and Ranganathan. R. (1999) The path to specificity. Science *283*, 650–651.

Corresponding author: Dr. Uwe Wolfrum (wolfrum@uni-mainz.de)

SECTION 4

SUBCELLULAR SYSTEMS BIOLOGY

CHAPTER 11

SYSTEMS BIOLOGY AND THE RECONSTRUCTION OF THE CELL: FROM MOLECULAR COMPONENTS TO INTEGRAL FUNCTION

FRANK J. BRUGGEMAN[1], SERGIO ROSSELL,
KAREN VAN EUNEN, JILDAU BOUWMAN,
HANS V. WESTERHOFF[1] and BARBARA BAKKER
Free University, The Netherlands
[1] *University of Manchester, UK*

Table of Contents

E. Bertrand and M. Faupel (eds.), Subcellular Proteomics, 239–262.

1. SYSTEMS BIOLOGY

1.1. Introduction

Molecular biology was one of the most successful scientific disciplines of the previous century. It matured into an enormous undertaking involving many tens of thousands of scientists all over the world and has spawned a huge industry involved in medicine and (bio-) technology. The discoveries of molecular biology have been driven to a large extent by the development of new technologies enabling the scientists to identify and investigate new molecular phenomena in cells.

Molecular biology began to take shape in the 1940s. In those days the "aim" of the field became transparent: molecular biology was to discover and characterize the molecular constituents of cells (Stent 1968; Hess 1970). And that is certainly what it has been doing very successfully over the last 60 years or so.

We are now at the point in time where our understanding of the properties of the molecular constituents of cells is of sufficient detail to start studying the properties of networks composed of interacting molecules. This requires an approach different form that of molecular biology, involving the application of different techniques and methods. This new field in biology has been emerging in the last few years and is referred to as "Systems Biology" (Ideker et al. 2001a, b; Alberghina and Westerhoff 2005).

Systems biology studies how the functional behaviour of biological networks arises from the interactions between the molecules of which they consist. Accordingly, the introduction of systems biology involved a change in focus: from molecule to network. This has become possible at a large scale only recently, thanks to the enormous knowledge that molecular biology, biochemistry and cell biology have acquired about the molecular organization of cells, and thanks to genome sequencing and a set of techniques enabling functional genomics. There are two forms of systems biology: bottom-up (investigates how the molecular interactions between a set of known components gives rise to functional behaviour) and top-down (uses system level data to identify and characterize underlying patterns of interactions). As we shall see the type of research associated with the former form of systems biology is new (thanks to a number of developments in mathematical biochemistry), as is the applications of top-down systems biology. Systems biology may rightfully be called a new discipline in science by virtue of the fact that it differs essentially from both physical chemistry and biology.

The aim of systems biology is to come to understand general principles underlying the emergence of the behaviour of cells from their molecular organization – it addresses how molecules jointly bring about cellular behaviour. This level of understanding should enable more rational approaches to current challenges encountered in medicine and biotechnology such as drug design and product formation. When systems biology succeeds we should have complete understanding of the molecular functioning of cells. Although systems biology may seem to represent a natural continuation of molecular biology, it requires a transition in biological thinking to enter this new phase towards understanding cells and organisms.

1.2. Two Approaches to Systems Biology

As was mentioned above, systems biology studies how cellular phenomena derive from the interactions between the molecular constituents of cells. This means that studies have to be done on the level of the entire cell, for example, apoptosis, cell division, and adaptation, as well as on the molecular level, for example, studies of protein–protein interactions, cellular location, and molecular transport. Most importantly, and this is perhaps the most challenging aspect of systems biology, the corresponding two datasets have to be integrated to the extent that it allows understanding of how the joint molecular behaviours – the latter dataset – underly the cellular behaviour – the former dataset.

Two strategies may be followed. *Top-down systems biology* starts with the system-level data to identify and characterize underlying molecular mechanisms. *Bottom-up systems biology* starts from the other end, at the level of the molecular mechanisms, and investigates how the latter jointly bring about cellular phenomena. Top-down systems biology is used more often for "discovery" studies, for example, to predict the topology of protein–protein interaction networks, or (presumed) transcription factor networks (target genes for transcription factors) from data concerning all molecules of the corresponding subtype of the entire cell (or a large proportion thereof) (Ideker et al. 2001a, b). This involves more than, for instance, the use of microarrays or yeast two-hybrid screens. It also requires the use of new computational methods, often statistical, to obtain from those datasets network statistically significant topologies that will function as predictions. Top-down systems biology tends to be more associated with studies of poorly-characterized networks of cells. Bottom-up systems biology starts with all the knowledge about one particular subnetwork in a cell and considers the properties of its molecules as well as the processes that underlie their interactions. It then investigates: (i) how its behaviour derives from its internal molecular organization, (ii) how it influences the remainder of the cell and *vice versa* and (iii) what the function(s) are of this subnetwork inside the cell.

Both approaches to systems biology rely on mathematical methods and modelling, as well as on quantitative and precise experimentation. However, the types of mathematical and experimental approaches used differ between the two. Bottom-up systems biology uses either detailed kinetic models describing the processes taking place between molecules in terms of experimentally determined kinetic properties (if available) when it wishes to identify actual mechanisms of emergence of functionality, or "core" models when it examines what new properties might arise from certain interactions (Bakker et al. 1997; Kholodenko et al. 1999; Rohwer et al. 2000; Teusink et al. 2000; Hoefnagel et al. 2002; Hornberg et al. 2005). It relies on knowledge about the topology of the molecular network and the properties of molecules. Top-down systems biology is applied to actually determine topology and occasionally kinetic information possibly in a data driven, hypothesis free fashion. Below we will consider kinetic models, and how they are used in bottom-up systems biology, more elaborately.

1.3. Two Historical Origins of Systems Biology

Systems biology applies sophisticated mathematical and experimental methods to analyse the functioning of molecular networks in cells (Brenner 1999; Kitano 2002; Hood 2003; Bruggeman et al. 2005a; Kriete and Eils 2005). Accordingly, not only developments in molecular biology contributed to the growth of systems biology but also developments in the analysis of nonlinear systems in non-equilibrium physics and mathematics. Whereas developments in the analysis of three-dimensional structures of macromolecules, using crystallography or NMR, were of minor importance for the concepts of system biology, developments such as those of enzyme kinetics, the chemiosmotic theory, allosteric enzyme regulation, covalent modification, regulation of the *lac* operon, were of prime importance. Bioenergetics, enzymology, and biophysics have therefore been influential disciplines for (bottom-up) systems biology. The experimental and theoretical analysis of the stationary and dynamic states of nonlinear systems operating under non-equilibrium conditions is also an important root of systems biology. This includes non-equilibrium thermodynamics, network thermodynamics, the analysis of periodic and stochastic phenomena, control theory, modelling, bifurcation theory, and nonlinear time series analysis. Even though the latter methods have been applied to biology prior to the emergence of systems biology, with exceptions like mosaic non-equilibrium thermodynamics and metabolic and hierarchical control analysis, they have rarely been applied in a way that referred to the macromolecular components of the system, using both experimental data on molecular properties and on the system level. More about the historical roots of systems biology can be found in Westerhoff and Palsson (2004).

2. OVERVIEW OF APPROACHES TO UNDERSTAND CELL FUNCTION BASED ON MOLECULAR INTERACTIONS

2.1. Topological Analysis

Depending on the purpose – "which biological question is to be answered?" – different kinds of modelling can be distinguished (Bruggeman 2005). Very often, the choice of the modelling approach depends also on the quality and extent of molecular information available about the interactions out of which the network is composed. The description that is most qualitative, but nevertheless can answer important biological questions, considers the network as a graph. In this approach, all the molecules in the network represent nodes and their interactions represent edges. This is the only information that is used in this modelling approach. Often such graphs are described in terms of an adjacency matrix denoted by **A**. The rows and columns denote the identity of the interacting molecules and an inhibiting interaction from molecule j to k is denoted by -1 in the j-th row and k-th column of **A**, that is, $a_{jk} = -1$. A "1" would denote an activatory interaction. In some cases, the direction of the interaction is not taken into account, making **A** symmetric, that is, for j and k we have $a_{jk} = a_{kj}$. Research questions concerning the strengths of the interactions or reaction rates (fluxes) cannot be answered with this modelling approach. In the past few years, methods based on graph theory have been used to study the robustness and the modular, scale-free and small-world organization of genetic, protein–protein interaction, signalling and metabolic networks (Watts and Strogatz 1998; Fell and Wagner 2000, 2001; Milo et al. 2002; Ravasz et al. 2002; Spirin and Mirny 2003; Pereira-Leal et al. 2004). Many analytical and numerical tools have been developed to study such networks (Albert and Barabasi 2002; Itzkovitz et al. 2003; Newman 2003).

In this approach only static topological properties of networks can be investigated; such as the number and length of the paths between two molecules, the number and length of cycles, the enumeration of occurrences of particular subnetwork structures, and how those properties scale with the size of the network. This type of modelling we shall refer to as *topological network modelling*.

2.2. Stoichiometric Analysis

The next type of modelling we shall discuss is *stoichiometric modelling* (Schilling et al. 1999; Price et al. 2004). In this approach, it is emphasized that all molecular interactions are in fact reactions having particular stoichiometric coefficients. The latter are the numbers (often integers) preceding the reactants in a reaction, for example, the numbers in: $2A + B + 3C \leftrightarrow 2AC + BC$. This approach appears much more relevant for the analysis of metabolic networks than for genetic and signalling networks where such stoichiometries are absent or ill-defined (for promising exceptions see (Allen and Palsson 2003; Papin and Palsson 2004a, b; Papin et al. 2005)). Again this analysis may be performed in terms of matrices. The stoichiometric matrix **N** has as many rows as there are independent molecule concentrations (or rather variable intermediates engaging in reactions) in the network and as many columns as there are reactions in the network. Then the stoichiometric coefficient of the j-th intermediate in the network, say glutamate, for the k-th reaction is the system, say glutamate synthase, is 2, whereas glutamine, the l-th intermediate, has a -1 in the kth-column of **N**, that is, $n_{jk} = 2$ and $n_{lk} = -1$.

One type of stoichiometric analysis concerns the discovery of moiety-conservation relationships in the network (Reder 1988; Famili and Palsson 2003; Nikolaev et al. 2005; Imielinski et al. 2006). These linear relationships represent sums of concentrations of intermediates in the network that remain constant on the time scale of interest, because there are only reactions that interconvert these particular intermediates. For instance, the following moiety-conservation relationship may pertain: $[P_i] + [AMP] + 2[ADP] + 3[ATP] +$ [other phosphate containing molecules] $= [P]_{total}$ constant.

A more popular form of stoichiometric analysis is the analysis of flux distributions that are consistent with system steady state (Note that in the terminology of metabolic modelling, the rate of a reaction at system steady state is referred to as a flux.) This type of analysis can be done directly on the **N** matrix because of its central role in the description of the mass balances of all the variable intermediates in a network.

A mass balance is an equation for each variable intermediate in the network, say x, that describes at a particular moment in time the rate of change in this variable, that is, dx/dt, in terms of the difference between the rates of all the reactions producing x and all the reactions consuming it. If we denote rates by v then we obtain for each of the intermediates, entries in the state vector $x = \{x_1, \ldots, x_j, \ldots, x_m\}$, the following mass balance:

$$\frac{d}{dt}x_j = \sum_k n_{jk} \cdot v_k \tag{1}$$

In matrix format we obtain (denoting a vector in bold small font and a matrix in bold capital font):

$$\frac{d\mathbf{x}}{dt} = \mathbf{Nv} \tag{2}$$

In steady state, we denote the rate vector **v** by the flux vector **J** (this will be the only vector denoted by a capital letter) to obtain:

$$\mathbf{0} = \mathbf{NJ} \tag{3}$$

Many different flux vectors will satisfy this equation. All of those solutions are linear combinations of the independent fluxes in the network. The coefficients to appear in those linear combinations derive from the nullspace or kernel matrix of **N** denoted by **K** (Schuster et al. 1999). Reordering **J** with the independent fluxes $\mathbf{J}^{\mathrm{I}}$ on top of the dependent fluxes $\mathbf{J}^{\mathrm{D}}$ one obtains:

$$\mathbf{J} = \mathbf{KJ}^{\mathbf{I}} \tag{4}$$

Each column of **K** represents a flux mode, a set of fluxes through the reaction steps in the network that are consistent with a system steady state. The choice of **K** is not unique. To solve this and to make the definition of flux mode biologically more interesting and applicable to bioengineering and pathway analysis, elementary modes and extreme pathways have been developed, both corresponding to choices of **K** (Schilling et al. 2000; Schuster et al. 2000; Papin et al. 2004) that are extreme in terms of simplicity. These choices have been useful in analysing the capabilities of the genome-scale metabolic network, to address characteristics such as possible yields, robustness, the viability of mutants and the redundancy of pathways (Price et al. 2004).

Another application of the analysis of the stoichiometric matrix is flux balance analysis (Edwards et al. 2002). Often the number of fluxes in the system exceeds the number variable metabolites making equation (3) an underdetermined set of linear equations, that is, many different combinations of fluxes are consistent with system steady state. One approach is to measure the fluxes that enter and exit the cell. Because intracellularly there are many redundant pathways, this does not enable one to determine all fluxes. Isotope labelling may help then (Wiechert 2002). Another approach to then find a smaller number of solutions is to postulate that the solution should satisfy an additional objective. This objective is taken to be associated with optimal functioning of the network, for instance maximization of some flux or combination of fluxes (e.g. growth rate or flux ratio). This type of approach is termed flux balance analysis. When investigating genome-scale networks the number of solutions may then still run into the hundreds but at least such numbers are manageable (Mahadevan and Schilling 2003; Reed and Palsson 2004).

2.3. Kinetic Modelling

None of the methods so far were able to deal with dynamics of intracellular networks. They were not able to describe the changes in the concentrations of the network intermediates as function of time upon perturbations made to the network, such as the addition of nutrients, growth factors, or drugs. This is what kinetic modelling does. A kinetic model starts from equation (2) by substituting rate equations into the rate vector. Rate equations describe the dependence of a rate of a reaction in the network with respects to its substrates, products, and effectors by the identification of the enzyme mechanism and the parameterisation of its kinetic constants. An example of a rate equation is the following two substrate (s_1 and s_2) and two product (p_1 and p_2) reaction with the non-competitive inhibitory effect of x:

$$v = \frac{V^{+}_{\mathrm{MAX}} \cdot \frac{s_1 \cdot s_2}{K_{s_1} \cdot K_{s_2}} - V^{-}_{\mathrm{MAX}} \cdot \frac{p_1 \cdot p_2}{K_{p_1} \cdot K_{p_2}}}{\left(1 + \frac{x}{K_x}\right)\left(1 + \frac{s_1}{K_{s_1}} + \frac{s_2}{K_{s_2}}\right)\left(1 + \frac{p_1}{K_{p_1}} + \frac{p_2}{K_{p_2}}\right)} \tag{5}$$

V^{+}_{MAX} and V^{-}_{MAX} denote the maximal rate of the reaction in the forward and the backward directions respectively. The "K" constants reflect the effective affinity of the enzyme for its substrates, products and its one effector. All those constants can be determined experimentally. Most often those experiments are carried in cell free extract or on isolated enzymes. More on enzyme kinetics, including more complicated cases involving multimeric enzymes, can be found in Segel (1993) and Cleland (Cleland 1963a, b, c). This is a discipline in itself, even though much smaller nowadays than in the 1960s and 1970s. Supplemented with initial conditions, the kinetic model, with the parameterized rate equations incorporated, allows the calculation of the temporal behaviour of network.

Not all kinetic models derive their kinetic parameters and the exact enzyme mechanism from experiment. In those cases, the network may still be precise but the descriptions of the enzymes are then phenomenological. On the other end of the scale, "Silicon Cell" models are defined as to contain the experimentally confirmed network structure, enzyme mechanisms, and kinetic parameters (see www.siliconcell.net). Models that serve illustrative purposes or more qualitative purposes are core models and those have simplified network structures, simplified enzyme mechanisms, and guessed or fitted kinetic parameters. It all depends on the purpose of the model whether a model can be judged as good or bad. Silicon Cell models are especially promising because they contain the current status of understanding and should therefore be more easily integrated with experiment. They may assist in pinpointing gaps in our understanding, offering hypotheses, performing *in silico* "experiments," predicting drug targets, and integrating various sources of biological data (fluxes, kinetic constants and concentration of mRNAs, proteins, and metabolites). Silicon cell models come as close as possible to offering a complete description of cells. The Silicon Cell programme will be discussed in more detail in Section 3.

2.4. Metabolic Control Analysis

A powerful tool to understand the behaviour of cells, pathways or models thereof is Metabolic Control Analysis. It was pioneered by Heinrich and Rapoport (1974) and Kacser & Burns (1973) in the seventies for steady-state metabolic networks with early experimental application by Flint et al. (1980, 1981) and Groen et al. (1982). MCA relates changes in systemic properties as quantified by response or control coefficients to properties of enzymes in terms of elasticity coefficients. MCA has later been extended to address branched and cyclic pathways with and without moiety conservation (Fell and Sauro 1985; Hofmeyr et al. 1986; Sauro et al. 1987; Reder 1988; Fell and Sauro 1990; Kacser et al. 1990; Sauro and Kacser 1990; Kholodenko et al. 1994b, 1995b; Sauro 1994, 1995 Westerhoff and Kell 1996), concentration control (Westerhoff and Chen 1984), energy coupling (Westerhoff et al. 1987), control of generalized variables (Schuster 1996), control of transition times (Easterby 1990; Melendezhevia et al. 1990), systems involving quasi-equilibrium reactions and time-scale separation (Delgado and Liao 1995; Kholodenko et al. 1998); oscillatory systems (Kholodenko et al. 1996, 1997a; Ingalls 2004; Sauro and Ingalls 2004), signaling networks (Kholodenko et al. 1997; Bruggeman et al. 2002), channeling (Kholodenko et al. 1994), intra-enzymatic processes (Kholodenko and Westerhoff 1994), hierarchical networks with gene expression, signaling, and metabolism (Westerhoff et al. 1989; Kahn and Westerhoff 1991; Jensen et al. 1999; Hofmeyr and Westerhoff 2001; Bruggeman et al. 2002; Snoep et al. 2002), modular biochemical networks (Schuster et al. 1993), reaction-diffusion systems (Brown and Kholodenko 1999; Peletier et al. 2003), and transient trajectories (Acerenza et al. 1989; Heinrich and Render 1991; Ingalls and Sauro 2003). It has been applied frequently to the experimental analysis of biochemical networks (Groen et al. 1982; Ainscow and Brand 1999a, b). In this section we will outline the theory only briefly. For more comprehensive reviews, see (Westerhoff and Van Dam 1987; Heinrich and Schuster 1996; Fell 1997).

The effect of a change of the activity of an enzyme i on the steady-state flux J is quantified by the corresponding flux control coefficient C_i^J. C_i^J is the percentage change of the flux J upon a 1% change of the enzyme activity v_i causing the change of the flux. Or, in more precise mathematical terms:

$$C_i^J \frac{(\mathrm{d} \ln J/\mathrm{d}p)_{ss}}{(\partial \ln \mathrm{v_i}/\partial \mathrm{p})_{\mathrm{reaction}_i}}$$

in which p is a parameter that only acts on enzyme i, such as an inhibitor concentration or an enzyme concentration. The subscript ss and differential "d" indicate that the change of the flux is analysed after relaxation of all system variables to a new steady-state. The subscript *reaction i* and derivative "∂" indicate that only the change in the local rate of enzyme i is considered, immediately after the change of p at constant values of all system variables, such as the concentrations of substrates, products and effectors of enzyme i.

In simple metabolic pathways, with only one independent flux, the value of C_i^J is usually between zero and 1. Partial inhibition of an enzyme with zero flux control does not affect the flux, while an enzyme with a flux control coefficient of 1 is completely rate-limiting. Flux control coefficients between 0 and 1 have often been found (Groen et al. 1982; Ruijter et al. 1991; Snoep et al. 1996). In ideal metabolic pathways, for example, in the absence of metabolite channelling and coenzyme sequestration (Kholodenko et al. 1995), the sum of the flux control coefficients of all enzymes in the pathway must equal 1:

$$\sum_i C_i^J = 1$$

Control coefficients are properties of a metabolic pathway as a whole. The kinetic properties of individual enzymes that are relevant for control are expressed in terms of their elasticity coefficients. An elasticity coefficient $\varepsilon_{X_j}^i$ quantifies how strongly a metabolite concentration X_j affects an enzyme rate v_i directly, at constant concentrations of all other metabolite concentrations:

$$\varepsilon_{X_j}^i = \frac{\partial \ln v_i}{\partial \ln X_j}$$

Connectivity theorems allow to relate the control coefficients (systemic properties) to the elasticity coefficients (properties of the network's enzymes individually as if in isolation) (Westerhoff and Van Dam 1987; Heinrich and Schuster 1996; Fell 1997). The connectivity theorems have given us a strong insight into the functioning of metabolic pathways. For example, it follows directly from these theorems that enzymes that are very sensitive to the concentrations of metabolites, such as substrates, products and allosteric effectors, tend to have little control over the flux. This is illustrated by overproduction of phosphofructokinase in bakers' yeast, an enzyme often referred to textbooks as "rate-limiting." Yet, overproduction of phosphofructokinase does not lead to a significant flux increase, since the cell compensates by lowering the level of its allosteric effector fructose 2,6-bisphosphate (Schaaff et al. 1989; Davies and Brindle 1992).

2.5. Control versus Regulation

In the biological and biochemical literature the terms *control* and *regulation* are often used loosely as synonyms. In the precise analysis that is applied in systems biology, however, there is a need for a stricter definition of these terms.

We use the term *control* only in the sense in which it is used in Metabolic Control Analysis (Section 2.3). An enzyme is said to control a metabolic flux or a metabolite concentration if activation of this enzyme causes a pronounced effect on this flux or concentration. In this sense questions about control are what-if questions. *If* the cell were to increase the concentration of a controlling enzyme or *if* a scientist were to add an inhibitor of this enzyme, the flux or concentration should change. Control coefficients do not tell us whether the cell uses this potential when it responds to environmental changes.

Regulation indicates which mechanisms the cell actually uses to effectuate changes or robustness of cellular functions. Different aspects of regulation are the (i) tendency of living organisms to respond to their environment, (ii) the internal communication between different parts of cells or organisms and (iii) the maintenance of a relatively constant internal state upon external perturbations (Kahn and Westerhoff 1993). The regulatory strength (Hofmeyr et al. 1993; Kahn and Westerhoff 1993; Hofmeyr 1995) is defined as:

$$ {}^{i}R_{A}^{J} = C_{i}^{J} \cdot \varepsilon_{A}^{1} $$

It quantifies the effect of a perturbation of an external effector A (e.g. the concentration of a nutrient) on the flux J via enzyme i. If A effects different enzymes, perturbations of A can effect J via different routes and the regulatory strengths quantify the relative importance of each of these regulatory routes. A different, but related approach was taken by Westerhoff and Ter Kuile (ter Kuile and Westerhoff 2001). They developed Regulation Analysis, a theory that quantifies the relative importance of flux regulation via metabolism and via gene expression (Section 2.5).

In summary, *control* tells us about the buttons of which the cell disposes and *regulation* tells us which of these buttons the cell actually presses (through the elasticities of the components).

2.6. Regulation Analysis

Because enzymes are catalysts (and not substrates), enzyme rate equations are usually of the shape:

$$ v = v(e, \mathbf{X}, \mathbf{K}) = f(e) \cdot g(\mathbf{X}, \mathbf{K}) $$

in which v is the rate, e is the enzyme concentration, $\mathbf{X}$ is a vector of concentrations of substrates, products and other metabolic effectors, and $\mathbf{K}$ is a vector of constants parameterizing the strength with which the enzymes interact with their substrates, products and allosteric effectors. $f(e)$ describes the dependency of the rate upon the enzyme concentration and can be taken to equal $V_{\max}$, while $g(\mathbf{X}, \mathbf{K})$ describes the interaction of the enzyme with the rest of metabolism through metabolite concentrations and the corresponding affinity constants. The important characteristic of the above equation is that f usually does not depend upon $\mathbf{X}$ and $\mathbf{K}$, and g does not depend upon e.

Considering the transition from one condition to another:

$$1 = \frac{\Delta \log f(e)}{\Delta \log J} \frac{\Delta \log g(\mathbf{X}, \mathbf{K})}{\Delta \log J} = \rho_h + \rho_m$$

ρ_h is called the "hierarchical regulation coefficient" (ter Kuile and Westerhoff 2001), since it quantifies the relative contribution of changes in enzyme capacity (V_{max}) to the regulation of the enzyme's flux and depends on the complete gene-expression cascade of transcription, translation, posttranslational modification, and mRNA and protein degradation. ρ_m is the "metabolic regulation coefficient," quantifying the relative contribution of changes in the interaction of the enzyme with the rest of metabolism, to the regulation of the enzyme's flux. The two regulation coefficients sum up to one, implying that determination of one will yield the other automatically. In practice the hierarchical regulation coefficient is more readily determined, since $f(e)$ usually can be taken to equal V_{max}, and the V_{max} as well as the flux J through the enzyme can be measured or estimated in many cases. Regulation analysis introduces the possibility of making unambiguous and quantitative descriptions of the regulation of fluxes through individual enzymes embedded in biochemical networks of any complexity, in response to any number or kind of simultaneous perturbations.

Regulation Analysis has been applied to understand the strong decrease of the glycolytic flux in the yeast *Saccharomyces cerevisiae* upon nutrient starvation (Rossell et al. 2005, 2006). It turned out that all possible combinations of metabolic and hierarchical regulation that one could possibly envisage, were indeed found *in vivo*. These include pure hierarchical regulation ($\rho_h = 1$), pure metabolic regulation ($\rho_m = 1$), cooperative regulation (both ρ_h and ρ_m positive and between 0 and 1) and antagonistic regulation (one regulation coefficient negative and the other one higher than 1) (Rossell et al. 2005, 2006).

A quantitative theory of regulation allows testing theories of regulation. Among the proposed mechanisms for metabolic flux changes, the two clearest hypotheses are: (i) modulation of single rate-limiting enzymes, and (ii) multi-site modulation, that is, the simultaneous and proportional modulation of all enzymes in the pathway, thus causing a change in flux while leaving metabolite concentrations unchanged (Fell and Thomas 1995). Although examples of multisite modulation exist (e.g. lipogenesis in mice, the urea cycle in rats, and photosynthesis in green plants (Fell and Thomas 1995)), they do not seem to represent a general mode of regulation. Upon a number of different perturbations the glycolytic pathway of bakers' yeast exhibited a complex regulation pattern in which each enzyme played a different role: leading the change, following or pulling back (Daran-Lapujade et al. 2006; Rossell et al. 2006).

When we know how much regulation is in gene expression and how much is in metabolism, it becomes interesting to investigate how this regulation is brought about at the molecular level. This makes Regulation Analysis a tool for making sense of metabolomics, proteomics, and transcriptomics data, or even for preceding such analyses in experiment design. Without such analysis one might get lost in large metabolomics patterns, but if one knows which enzymes are actually regulated at the metabolic level, one need to focus only on the metabolites that affect those enzymes.

Other metabolite levels may change, but these changes are apparently not regulating changes, but cancelled by others or unable to regulate enzyme rates. Similarly, if an enzyme is regulated (partly) by gene expression, we may investigate the quantitative importance of the various processes in the gene-expression cascade. The theory to do so has been developed recently and applied to yeast glycolysis (Daran-Lapujade et al. 2006).

A quantitative analysis such as Regulation Analysis requires highly accurate datasets. If we are to explain a 50% change of flux and the regulation is distributed over various processes, we need numbers with accuracy of less than 10% to identify processes that contribute 20% to the regulation. The analysis of *changes* between conditions is of course central in the study of regulation, but at the same time this makes it error-prone. Thanks to controlled cultivation and highly accurate assay techniques (Piper et al. 2002), transcriptome analysis is adequate to identify the most important regulatory phenomena, but there is a need for more accurate proteomics techniques (Daran-Lapujade et al. 2006).

3. THE SILICON CELL PROGRAMME

Systems Biology combines quantitative experimentation, computer modelling and theory (see Section 2) in order to come to an understanding of cell functioning. The Silicon Cell programme (Snoep et al. 2005, 2006) focuses on the construction of kinetic models of various aspects of cell functioning, such as metabolic, signal transduction and gene-expression models. Central to the Silicon Cell concept is that the models should be as realistic as possible. They are based on experimental data and are validated by comparing the model outcome to the behaviour of real cells. Although the ultimate ambition is to model the entire cell, at present models of subsystems are much more tractable. The idea is therefore to first construct models of subsystems, such as metabolic pathways (Schuster and Holzhutter 1995; Bakker et al. 1999; Teusink et al. 2000; Hoefnagel et al. 2002; Albert et al. 2005), signal transduction cascades (Schoeberl, Eichler-Jonsson et al. 2002; Bruggeman et al. 2005b) or gene-expression (Koster et al. 1988), study their properties, and later connect them to each other.

In view of the enormous effort that is made to make computer models, it is surprising that only recently attention has been paid to their preservation. Biological models were usually published as a set of differential equations. People who wanted to use the models, needed to re-programme them, often to find out that crucial data were missing or incorrect. This made it very difficult to revive a model without the help of the authors. A few years ago, however, Snoep and colleagues started to collect published models in an online database (JWS Online, www.jjj.bio.vu.nl (Snoep 2005)) that can be interrogated via the world-wide web. Today the database consists of some 50 models. Most of these models are realistic kinetic models of the Silicon Cell type, but the database also contains core models and demonstration models that serve to illustrate and investigate general principles and are very useful for teaching of systems biology.

A first attempt to link three models in the database, that is, a detailed glycolysis model of yeast (Teusink et al. 2000), a model of the branch to glycerol (Cronwright et al. 2002) and a model of glyoxylate production (Martins et al. 2001) was successful (Snoep et al. 2006). The behaviour of the glycolysis model was not so much altered by the addition of the two branches. In the combined model, the two branches, however, worked under other conditions than when they were considered in isolation. Their substrates and products were constant in the isolated models, while they were variables in the linked model. Of course various problems may occur when linking models systematically. For example, the models of subsystems may be valid for different experimental conditions or different cell types. Also they may overlap, in which case one has to choose which reactions to take from which model. Or there may be gaps between the models, which requires additional data. This emphasizes the need for standardization in systems biology. Several consortia are active in standardization. Computer scripts from different programmes can be interconverted via the Systems Biology Markup Language that is now supported by over a hundred software systems (Hucka et al. 2003) (http://sbml.org/). For example the Yeast Systems Biology Network (http://www.gmm.gu.se/YSBN/) is active in standardization and integration of Systems Biology activities of yeast, one of the major model organisms in the field. STRENDA (Standards for Reporting Enzymology Data) is active in the standarization of collecting and reporting enzyme data, which are vital for model construction (http://www.strenda.org).

4. ADVANTAGES OF A MODULAR APPROACH

4.1. Modules May Help Us Understand Cells Better

As we have seen above, a good strategy to make realistic kinetic models at a genome-wide scale is to start from smaller modules and to connect these later. There are other reasons to study the cell as if it consists of a number of modules. One is intelligibility. There is an interesting tension in offering quantitative explanations of cellular phenomena. As soon as the explanation becomes more quantitative it will necessarily contain more detail, describing reality more closely. There is a risk that we will miss the general principles by loosing ourselves in the study of details. To keep things understandable, we may temporarily treat parts that are not relevant for our research question at hand, as black boxes. Another reason to modularize models of the cell is that we may be interested in a certain process in the cell as it works *in the cellular context*, but we do not (yet) have the data to model the whole cell. Then we could treat the rest of the cell as a black box, but make sure that the compounds at the boundaries of the module of interest are described well. A final reason to modularize the cell is to determine the control exerted by certain processes in the cell (cf. Section 2.4), when it is not possible to influence these processes directly. In that case one might deduce the control from the elasticity coefficients, using the theorems of metabolic control analysis. Since these theorems only hold for the entire system, this can be done by dissecting the cell in modules of which the elasticity coefficients can be determined experimentally. This is called modular (Schuster et al. 1993) or top-down control analysis.

An intuitive definition of a module would be a subnetwork inside the cell that performs a particular function relatively independently; a module is conceived as a functional unit. Then one could refer to this subnetwork rather than to its internal organization when explaining phenomena occurring at the (higher) cellular level. Examples of such kinds of modules are the ribosome and the degradosome, both macromolecular complexes with a well-defined function. More encompassing modules involving more proteins, involved in metabolism and signalling, and in wel defined pathways. This appears problematic, as sub-networks may be shared between functional modules. On the other hand, the definition of what constitutes a module may also follow other criteria: A more approximate explanation may require a more coarse-grained modularization of the system than a more detailed explanation. All in all, the defining of modules is not unambiguous. An interesting option is to define a module within a biochemical network such that it corresponds with a structural unit of organization, such as an organelle (mitochondrion, lysosome). Because organelles tend to have a limited number of transporters and receptors through which they interact with the rest of the cell, such a definition may indeed identify a functional unit of the network (see below). In hierarchical control analysis (Westerhoff et al. 1989; Kahn and Westerhoff 1991; Hofmeyr and Westerhoff 2001), module definition corresponded with the levels in the gene-expression hierarchy.

Another indication of the problems associated with modularization of complex systems is the small number of formal mathematical methods that allow one to simplify kinetic models. The existing methods are all based on time-scale separation in the system which allows for the decomposition of the system into a module composed of fast processes and one composed of slow processes. Then the fast processes can be considered in the absence of the slow processes. The slow processes are then considered with the fast processes either in steady state or thermodynamic equilibrium (Klonowski 1983; Segel and Slemrod 1989; Schuster and Schuster 1991; Kholodenko et al. 1998; Stiefenhofer 1998; Schneider and Wilhelm 2000). Two successful approaches to modularization of complex networks do not consider dynamics. One is purely structural while the other is applicable only to systems in steady state and concerns the analysis of control.

4.2. Examples of Successful Definitions of Modules

As the first example, we shall discuss two notions of modules often used in topological models of networks (Milo et al. 2002; Ravasz et al. 2002; Itzkovitz et al. 2003; Clauset et al. 2004; Newman 2006). The first notion is that of a "cluster" on the basis of the nodes within the cluster having more interactions among themselves than with nodes outside of the cluster. Clusters can be quantitatively defined on the basis of a cluster coefficient. *Escherichia coli* has been shown to be clustered in a hierarchical fashion (Ravasz et al. 2002). Another definition is that of a network motif (Milo et al. 2002; Shen-Orr et al. 2002; Kashtan et al. 2004; Yeger-Lotem et al. 2004). These authors take a different strategy to define modules, or network motifs as they call them. A network motif is defined as a topological structure that appears with a higher frequency in a biological network than in a random network of the same size. Then the authors reason that this network topology has been selected for and should carry out some particular function contributing to fitness. They indeed found a number of such networks leading to a series of papers about their functional characterization (Mangan and Alon 2003; Mangan et al. 2003; Dekel et al. 2005; Kalir et al. 2005; Mangan et al. 2006). This is a promising approach but the many networks that do not appear significantly often, remain unaddressed. Do those networks then not carry out some function contributing to fitness? For instance, the incoherent feedforward loop network, for example, where X activates Y and inactivates Z and where Z is in addition activated by Y, does occur in *E. coli*'s transcription network, albeit not as frequent as the coherent feedforward loop network (where X activates Z directly rather than inactivating it), but does this mean that it fulfils no function inside *E. coli*? The answer is most certainly "no": it does perform a function – maybe one which is not required so abundantly or one which can also be fulfilled by another network structure occurring more frequently.

The second example of the use of modules derives from metabolic control analysis (Groen et al. 1982; Westerhoff and Van Dam 1987; Kahn and Westerhoff 1991; Schuster et al. 1993; Krab 1995; Brand 1996; van der Gugten and Westerhoff 1997; Ainscow and Brand 1999a, b; Krab et al. 2000; Krauss and Brand 2000; Hofmeyr and Westerhoff 2001; Bruggeman et al. 2002). This branch of metabolic control analysis is often referred to as hierarchical control analysis when the modules interact solely by regulatory influences (e.g. signal transduction cascades) and to modular control analysis if the modules are coupled by way of mass flow. Modular control analysis is sometimes referred to as top-down control analysis (Ainscow and Brand 1995; Brand 1996). In hierarchical control analysis the modules are often referred to as levels. In control analysis a level or a module is defined as a network that behaves the same in isolation of the network upon a parameter perturbation as it would behave in the network when the remainder of the network would be kept fixed upon perturbation of the same parameter. Mathematical analysis then shows that the only conditions for a module to behave sufficiently autonomously is that it should not share any conserved moieties with other modules (Schuster et al. 1993; Heinrich and Schuster 1996). The latter condition is important as it does limit the applicability of modularization to metabolism. On the other hand, there is a reward if modularization is possible: In modular control analysis, control coefficients can be defined for the module as a whole and the module can be considered a supra-enzyme in the entire network (Schuster et al. 1993; Ainscow and Brand 1995; Brand 1996; Ainscow and Brand 1999; Bruggeman et al. 2002). This leads to black-box modules and to an enormous simplification of the analysis of the control properties of the system under study (Schuster et al. 1993; Bruggeman et al. 2002). Importantly, the simplification may or may not correspond to what has been done intuitively for years, that is, discuss cell function in terms of interacting pathways, types of process (e.g. energy metabolism, carbon metabolism, transcription) and organelles.

4.3. Organelles as Natural Modules and Targets for Regulation

Indeed, compartmentation of the cell into various organelles that only communicate via molecules that are outside the organelle membrane or are transported across this membrane by well-defined carrier proteins, suggests a natural way to dissect the cell into modules. The notion that the functioning of the isolated organelle is relevant to the functioning of the whole cell is reflected for example by a long history of research on oxidative phosphorylation in isolated mitochondria, eventually leading to complete analysis of the control distribution (Groen et al. 1982; Ciapaite et al. 2005; Affourtit and Brand 2006) and the construction of detailed models (Demin et al. 1998; Demin et al. 2001).

Since organelles are often specialized to carry out a specific function, it makes sense to regulate the number of organelles depending on the needs of the cell. This is well known for the regulation of the number of mitochondria in yeast as a function of glucose concentration (Cho et al. 2001; Taylor et al. 2005) and in mammals as a function of thyroid hormone (Moraes 2001). Another clear example is the regulation of the number of alcohol oxidase containing peroxisomes in the yeast *Hansenula polymorpha*, depending on the presence of methanol in the medium (Leao-Helder et al. 2003).

In Section 2.3 we discussed the regulation of flux on an enzyme-by-enzyme basis. The overall regulation of flux was dissected in regulation of gene expression and metabolic regulation. However, the organelle abundance represents an additional level of regulation that can be incorporated formally in the theory of regulation analysis. If the organelle copy number per cell is included in the enzyme rate equation, we obtain:

$$v = V^*_{\mathrm{max}} \cdot \{\text{number of organelles per cell}\} \cdot g(\mathbf{X}, \mathbf{K})$$

in which v is again the *in vivo* rate of the enzyme, V^*_{max} is the maximum enzyme capacity *per organelle* (hence the *) and $g(\mathbf{X}, \mathbf{K})$ is again the metabolic part of the rate equation. Since at steady state the rate v equals the flux J, the comparison of the flux under two experimental conditions yields:

$$\frac{\Delta \log v}{\Delta \log J} = 1 = \frac{\Delta \log V_{\mathrm{max}}}{\Delta \log J} + \frac{\Delta \log\{\text{number of organelles per cell}\}}{\Delta \log J} + \frac{\Delta \log g(\mathbf{X}, \mathbf{K})}{\Delta \log J}$$

$$= \rho_h + \rho_{\mathrm{organelle}} + \rho_m$$

in which $\rho_{\mathrm{organelle}}$ now quantifies the importance of organelle abundance in the regulation of flux.

This is a new development for which we do not yet have experimental applications. On the other hand, *T. brucei* with its glycosomes, the concentration of which changes when the parasite shifts between its various stages in its life cycle, seems a most interesting test case.

5. CONCLUDING REMARKS

From the above it is clear that Systems Biology is on its way to reconstruct the living cell and subsequently living organisms in terms of their molecular components. There is much activity at the level of intracellular biochemical reaction, signalling and gene-expression networks. In the context of this book, it is most important to note that there is comparatively little activity in Systems Biology that analyzes the living cell in terms of dynamically interacting structural components. The reason is that Life is fundamentally a non-equilibrium condition that is maintained by a number of physical chemical processes. Much of the existing paradigm of Systems Biology takes the spatial position of all these processes either as homogeneous within the cell, or as well-defined by the position of the enzyme or transporter catalyzing the molecular process. Systems Biology has not yet much been emphasizing that the spatial organization of the molecular components, depends strongly on the presence of intracellular structures.

Making the link from structural organization is sometimes considered trivial. It is not, however. The intracellular structures themselves are not constants but depend on the effect of a multitude of cooperating non-equilibrium processes. This is clear for the supercoiling state of prokaryotic DNA (Snoep et al. 2002), but also for the complex structure of chromatin in higher eukaryotes.

Intriguingly, intracellular structures are the consequence of a number of processes that happen more or less instantaneously (and may be considered Markovian technically), but also of a substantial number of other processes that may be inherited from parental cells. Many of the ribosomes in a cell may in fact have been synthesized in the maternal cell, hence in a completely different context. Most of the plasma membrane of a cell originates in the maternal cell. Consequently, also much of the intracellular dynamic networking of cells may be much more hysteretic than assumed by standard systems biology. Another exciting aspect is the problem of coupling the generation and maintenance of structures that are asymmetrical in space to biochemical reactions that could just as well have been symmetrical. The answer to this issue is known to be the spatial organization of the enzymes, leading to yet another example of the circular or spiralling causality so characteristic of Systems Biology.

The Systems Biology of dynamic intracellular structures and compartmentation has an exciting and bright future. We can only hope that this chapter can help start the new activities.

REFERENCES

Acerenza, L., Sauro, H.M. and Kacser, H. (1989) Control analysis of time-dependent metabolic systems. J. Theor. Biol. *137*, 423–444.

Affourtit, C. and Brand, M.D. (2006) Stronger control of ATP/ADP by proton leak in pancreatic beta-cells than skeletal muscle mitochondria. Biochem. J. *393*, 151–159.

Ainscow, E.K. and Brand, M.D. (1995) Top-down control analysis of systems with more than one common intermediate. Eur. J. Biochem. *231*, 579–586.

Ainscow, E.K. and Brand, M.D. (1999a) Internal regulation of ATP turnover, glycolysis and oxidative phosphorylation in rat hepatocytes. Eur. J. Biochem. *266*, 737–749.

Ainscow, E.K. and Brand, M.D. (1999b) Top-down control analysis of ATP turnover, glycolysis and oxidative phosphorylation in rat hepatocytes. Eur. J. Biochem. *263*, 671–685.

Alberghina, L. and Westerhoff, H.V. eds. (2005) Systems Biology, Definitions and Perspectives (Topics in Current Genetics), (Berlin: Springer-Verlag and Heidelberg: GmbH & Co. K.).

Albert, R. and Barabasi, A.L. (2002) Statistical mechanics of complex networks. Rev. Mod. Phys. *74*, 47–97.

Albert, M.A., Haanstra, J.R, Hannaert, V., Van Roy, J., Opperdoes, F.R., Bakker, B.M. and Michels, P.A. (2005) Experimental and in silico analyses of glycolytic flux control in bloodstream form *Trypanosoma brucei*. J. Biol. Chem. *280*, 28306–28315.

Allen, T.E. and Palsson, B.O. (2003) Sequence-based analysis of metabolic demands for protein synthesis in prokaryotes. J. Theor. Biol. *220*, 1–18.

Bakker, B.M., Michels, P.A.M., Opperdoes, F.R. and Westerhoff, H.V. (1997) Glycolysis in bloodstream form *Trypanosoma brucei* can be understood in terms of the kinetics of the glycolytic enzymes. J. Biol. Chem. *272*, 3207–3215.

Bakker, B.M., Walsh, M.C., ter Kuile, B.H., Mensonides, F.I., Michels, P.A., Opperdoes, F.R. and Westerhoff, H.V. (1999) Contribution of glucose transport to the control of the glycolytic flux in *Trypanosoma brucei*. Proc. Natl Acad. Sci. U.S.A. *96*, 10098–10103.

Brand, M.D. (1996) Top down metabolic control analysis. J. Theor. Biol. *182*, 351–360.

Brenner, S. (1999) Theoretical biology in the third millennium. Philos. Trans. R. Soc. Lond., B, Biol. Sci. *354*, 1963–1965.

Brown, G.C. and Kholodenko, B.N. (1999) Spatial gradients of cellular phospho-proteins. FEBS Lett. *457*, 452–454.

Bruggeman, F.J., Bakker, B.M., Hornberg, J.J. and Westerhoff, H.V. (2005a) Introduction to computational models of biochemical reaction networks. In: Computational Systems Biology. A. Kriete and R. Eils, eds. (London, UK: Elsevier Academic Press), pp. 127–148.

Bruggeman, F.J., Boogerd, F.C. and Westerhoff, H.V. (2005b) The multifarious short-term regulation of ammonium assimilation of *Escherichia coli*, dissection using an in silico replica. Febs. J. *272*, 1965–1985.

Bruggeman, F.J., Westerhoff, H.V., Hoek, J.B. and Kholodenko, B.N. (2002) Modular response analysis of cellular regulatory networks. J. Theor. Biol. *218*, 507–520.

Cho, J.H., Lee, Y.K. and Chae, C.B. (2001) The modulation of the biological activities of mitochondrial histone Abf2p by yeast PKA and its possible role in the regulation of mitochondrial DNA content during glucose repression. Biochim. Biophys. Acta *1522*, 175–186.

Ciapaite, J., Van Eikenhorst, G., Bakker, S.J.L., Diamant, M., Heine, R.J., Wagner, M.J., Westerhoff, H.V. and Krab, K. (2005) Modular kinetic analysis of the adenine nucleotide translocator-mediated effects of palmitoyl-CoA on the oxidative phosphorylation in isolated rat liver mitochondria. Diabetes *54*, 944–951.

Clauset, A., Newman, M.E. and Moore, C. (2004) Finding community structure in very large networks. Phys. Rev. E. Stat. Nonlin. Soft Matter Phys. *70*, 066111.

Cleland, W.W. (1963a). The kinetics of enzyme-catalyzed reactions with two or more substrates or products. I. Nomenclature and rate equations. Biochim. Biophys. Acta *67*, 104–137.

Cleland, W.W. (1963b). The kinetics of enzyme-catalyzed reactions with two or more substrates or products. II. Inhibition: nomenclature and theory. Biochim. Biophys. Acta *67*, 173–187.

Cleland, W.W. (1963c). The kinetics of enzyme-catalyzed reactions with two or more substrates or products. III. Prediction of initial velocity and inhibition patterns by inspection. Biochim. Biophys. Acta *67*, 188–196.

Cronwright, G.R., Rohwer, J.M. and Prior, B.A. (2002) Metabolic Control Analysis of Glycerol Synthesis in *Saccharomyces cerevisiae*. Appl. Environ. Microbiol. *68*, 4448–4456.

Daran-Lapujade, P., Rossell, S.L., Van Gulik, W., Luttik, M.A.H., De Groot, M., Slijper, M., Heck, A.J.R., Daran, J.M., De Winde, J.H., Westerhoff, H.V. et al. (2006) Manuscript in preparation.

Davies, S.E. and Brindle, K.M. (1992) Effects of overexpression of phosphofructokinase on glycolysis in the yeast *Saccharomyces cerevisiae*. Biochemistry *31*, 4729–4735.

Dekel, E., Mangan, S. and Alon, U. (2005) Environmental selection of the feed-forward loop circuit in gene-regulation networks. Phys. Biol. *2*, 81–88.

Delgado, J. and Liao, J.C. (1995) Control of metabolic pathways by time-scale separation. Biosystems *36*, 55–70.

Demin, O.V., Gorianin, I.I., Kholodenko, B.N. and Westerhoff, H.V. (2001) Kinetic modeling of energy metabolism and generation of active forms of oxygen in hepatocyte mitochondria. Mol. Biol. (Mosk.) *35*, 1095–1104.

Demin, O.V., Westerhoff, H.V. and Kholodenko, B.N. (1998) Mathematical modelling of superoxide generation with the bc1 complex of mitochondria. Biochemistry (Mosc.) *63*, 634–649.

Easterby, J.S. (1990) Integration of temporal analysis and control analysis of metabolic systems. Biochem. J. *269*, 255–259.

Edwards, J.S., Covert, M. and Palsson, B. (2002) Metabolic modelling of microbes: the flux-balance approach. Environ. Microbiol. *4*, 133–140.

Famili, I. and Palsson, B.O. (2003) The convex basis of the left null space of the stoichiometric matrix leads to the definition of metabolically meaningful pools. Biophys. J. *85*, 16–26.

Fell, D.A. (1997) Understanding the Control of Metabolism. First edn. (London and Miami: Portland Press).

Fell, D.A. and Sauro, H.M. (1985) Metabolic control and its analysis. Additional relationships between elasticities and control coefficients. Eur. J. Biochem. *148*, 555–561.

Fell, D.A. and Sauro, H.M. (1990) Metabolic control analysis. The effects of high enzyme concentrations. Eur. J. Biochem. *192*, 183–187.

Fell, D.A. and Thomas, S. (1995) Physiological control of metabolic flux: the requirement for multisite modulation. Biochem. J. *311*, 35–39.

Fell, D.A. and Wagner, A. (2000) The small world of metabolism. Nat. Biotechnol. *18*, 1121–1122.

Flint, H.J., Porteous, D.J. and Kacser, H. (1980) Control of the flux in the arginine pathway of *Neurospora crassa*. The flux from citrulline to arginine. Biochem. J. *190*, 1–15.

Flint, H.J., Tateson, R.W., Barthelmess, I.B., Porteous, D.J., Donachie, W.D. and Kacser, H. (1981) Control of the flux in the arginine pathway of *Neurospora crassa*. Modulations of enzyme activity and concentration. Biochem. J. *200*, 231–246.

Groen, A.K., Wanders, R.J., Westerhoff, H.V., van der Meer, R. and Tager, J.M. (1982) Quantification of the contribution of various steps to the control of mitochondrial respiration. J. Biol. Chem. *257*, 2754–2757.

Heinrich, R. and Rapoport, T.A. (1974) A linear steady-state treatment of enzymatic chains. General properties, control and effector strength. Eur. J. Biochem. *42*, 89–95.

Heinrich, R. and Reder, C. (1991) Metabolic control analysis of relaxation processes. J. Theor. Biol. *151*, 57–61.

Heinrich, R. and Schuster, S. (1996) The Regulation of Cellular Systems. First edn. (New York: Chapman & Hall).

Hess, E.L. (1970) Origins of molecular biology. Science *168*, 664–669.

Hoefnagel, M.H.N., Starrenburg, M.J.C., Martens, D.E., Hugenholtz, J., Kleerebezem, M., Van Swam, I.I., Bongers, R., Westerhoff, H.V. and Snoep, J.L. (2002) Metabolic engineering of lactic acid bacteria, the combined approach: kinetic modelling, metabolic control and experimental analysis. Microbiology *148*, 1003–1013.

Hofmeyr, J.H. (1995) Metabolic regulation, a control analytic perspective. J. Bioenerg. Biomembr. *27*, 479–490.

Hofmeyr, J.H., Cornish-Bowden, A. and Rohwer, J.M. (1993) Taking enzyme kinetics out of control; putting control into regulation. Eur. J. Biochem. *212*, 833–837.

Hofmeyr, J.H., Kacser, H. and van der Merwe, K.J. (1986) Metabolic control analysis of moiety-conserved cycles. Eur. J. Biochem. *155*, 631–641.

Hofmeyr, J.H. and Westerhoff, H.V. (2001) Building the cellular puzzle: control in multi-level reaction networks. J. Theor. Biol. *208*, 261–285.

Hood, L. (2003) Systems biology: integrating technology, biology, and computation. Mech. Ageing Dev. *124*, 9–16.

Hornberg, J.J., Bruggeman, F.J., Binder, B., Geest, C.R., de Vaate, A.J., Lankelma, J., Heinrich, R. and Westerhoff, H.V. (2005) Principles behind the multifarious control of signal transduction. ERK phosphorylation and kinase/phosphatase control. FEBS J. *272*, 244–258.

Hucka, M., Finney, A., Sauro, H.M., Bolouri, H., Doyle, J.C., Kitano, H., Arkin, A.P., Bornstein, B.J., Bray, D. and Cornish-Bowden, A. et al. (2003) The Systems Biology Markup Language (SBML): A medium for representation and exchange of biochemical network models. Bioinformatics *19*, 524–531.

Ideker, T., Galitski, T. and Hood, L. (2001a) A new approach to decoding life: systems biology. Annu. Rev. Genomics. Hum. Genet. *2*, 343–372.

Ideker, T., Thorsson, V., Ranish, J.A., Christmas, R., Buhler, J., Eng, J.K., Bumgarner, R., Goodlett, D.R., Aebersold, R. and Hood, L. (2001b) Integrated genomic and proteomic analyses of a systematically perturbed metabolic network. Science *292*, 929–934.

Imielinski, M., Belta, C., Rubin, H. and Halasz, A. (2006) Systematic analysis of conservation relations in *Escherichia coli* genome-scale metabolic network reveals novel growth media. Biophys. J. *90*, 2659–2672.

Ingalls, B. (2004) Autonomously oscillating biochemical systems: parametric sensitivity of extrema and period. Syst. Biol. *1*, 62–70.

Ingalls, B.P. and Sauro, H.M. (2003) Sensitivity analysis of stoichiometric networks: an extension of metabolic control analysis to non-steady state trajectories. J. Theor. Biol. *222*, 23–36.

Itzkovitz, S., Milo, R., Kashtan, N., Ziv, G. and Alon, U. (2003) Subgraphs in random networks. Phys. Rev. E. Stat. Nonlin. Soft Matter Phys. *68*, 026127.

Jensen, P.R., Van Der Weijden, C.C., Jensen, L.B., Westerhoff, H.V. and Snoep, J.L. (1999) Extensive regulation compromises the extent to which DNA gyrase controls DNA supercoiling and growth rate of *Escherichia coli*. Eur. J. Biochem. *266*, 865–877.

Jeong, H., Mason, S.P., Barabasi, A.L. and Oltvai, Z.N. (2001) Lethality and centrality in protein networks. Nature *411*, 41–42.

Jeong, H., Tombor, B., Albert, R., Oltvai, Z.N. and Barabasi, A.L. (2000) The large-scale organization of metabolic networks. Nature *407*, 651–654.

Kacser, H. and Burns, J.A. (1973) The control of flux. Symp. Soc. Exp. Biol. *27*, 65–104.

Kacser, H., Sauro, H.M. and Acerenza, L. (1990) Enzyme-enzyme interactions and control analysis. 1. The case of non-additivity: monomer-oligomer associations. Eur. J. Biochem. *187*, 481–491.

Kahn, D. and Westerhoff, H.V. (1991) Control theory of regulatory cascades. J. Theor. Biol. *153*, 255–285.

Kahn, D. and Westerhoff, H.V. (1993) Regulation and homeostasis in metabolic control theory: interplay between fluctuations of variables and parameter changes. In: Modern Trends in BioThermoKinetics. S. Schuster, M. Rigoulet, R. Ouhabi, J.P. Mazat, ed. (New York: Plenum Press), pp. 199–204.

Kalir, S., Mangan, S. and Alon, U. (2005) A coherent feed-forward loop with a SUM input function prolongs flagella expression in *Escherichia coli*. Mol. Syst. Biol. *1*, 0006.

Kashtan, N., Itzkovitz, S., Milo, R. and Alon, U. (2004) Topological generalizations of network motifs. Phys. Rev. E. Stat. Nonlin. Soft Matter Phys. *70*, 031909.

Kholodenko, B.N., Cascante, M. and Westerhoff, H.V. (1994a) Control theory of metabolic channelling. Mol. Cell. Biochem. *133-134*, 313–331.

Kholodenko, B.N., Demin, O.V. and Westerhoff, H.V. (1996) The metabolic control theory of biochemical oscillating systems. 1. Definitions of the quantitative characteristics and their simplest properties. Biochemistry Mosc. *61*, 423–434.

Kholodenko, B.N., Demin, O.V. and Westerhoff, H.V. (1997a) Control analysis of periodic phenomena in biological systems. J. Phys. Chem. B *101*, 2070–2081.

Kholodenko, B.N., Demin, O.V., Moehren, G. and Hoek, J.B. (1999) Quantification of short term signaling by the epidermal growth factor receptor. J. Biol. Chem. *274*, 30169–30181.

Kholodenko, B.N., Hoek, J.B., Westerhoff, H.V. and Brown, G.C. (1997b) Quantification of information transfer via cellular signal transduction pathways. FEBS Lett. *414*, 430–434.

Kholodenko, B.N., Molenaar, D., Schuster, S., Heinrich, R. and Westerhoff, H.V. (1995a) Defining control coefficients in non-ideal metabolic pathways. Biophys. Chem. *56*, 215–226.

Kholodenko, B.N., Sauro, H.M. and Westerhoff, H.V. (1994b) Control by enzymes, coenzymes and conserved moieties. A generalisation of the connectivity theorem of metabolic control analysis. Eur. J. Biochem. *225*, 179–186.

Kholodenko, B.N., Sauro, H.M., Westerhoff, H.V. and Cascante, M. (1995b) Coenzyme cycles and metabolic control analysis: the determination of the elasticity coefficients from the generalised connectivity theorem. Biochem. Mol. Biol. Int. *35*, 615–625.

Kholodenko, B.N., Schuster, S., Garcia, J., Westerhoff, H.V. and Cascante, M. (1998) Control analysis of metabolic systems involving quasi-equilibrium reactions. Biochim. Biophys. Acta *1379*, 337–352.

Kholodenko, B.N. and Westerhoff, H.V. (1994) Control theory of one enzyme. Biochim. Biophys. Acta *1208*, 294–305.

Kitano, H. (2002) Systems biology: a brief overview. Science *295*, 1662–1664.

Klonowski, W. (1983) Simplifying principles for chemical and enzyme reaction kinetics. Biophys. Chem. *18*, 73–87.

Koster, J.G., Destrée, O.H.J. and Westerhoff, H.V. (1988) Kinetics of histone gene expression during early development of *Xenopus laevis*. J. Theor. Biol. *135*, 139–167.

Krab, K. (1995) Kinetic and regulatory aspects of the function of the alternative oxidase in plant respiration. J. Bioenerg. Biomembr. *27*, 387–396.

Krab, K., Wagner, M.J., Wagner, A.M. and Moller, I.M. (2000) Identification of the site where the electron transfer chain of plant mitochondria is stimulated by electrostatic charge screening. Eur. J. Biochem. *267*, 869–876.

Krauss, S. and Brand, M.D. (2000) Quantitation of signal transduction. FASEB J. *14*, 2581–2588.

Kriete, A. and Eils, R. eds. (2005) Computational Systems Biology. (London, UK: Elsevier Academic press).

Leao-Helder, A.N., Krikken, A.M., Van der Klei, I.J., Kiel, J.A. and Veenhuis, M. (2003) Transcriptional down-regulation of peroxisome numbers affects selective peroxisome degradation in *Hansenula polymorpha*. J. Biol. Chem. *278*, 40749–40756.

Mahadevan, R. and Schilling, C.H. (2003) The effects of alternate optimal solutions in constraint-based genome-scale metabolic models. Metab. Eng. *5*, 264–276.

Mangan, S. and Alon, U. (2003) Structure and function of the feed-forward loop network motif. Proc. Natl Acad. Sci. U.S.A. *100*, 11980–11985.

Mangan, S., Itzkovitz, S., Zaslaver, A. and Alon, U. (2006) The incoherent feed-forward loop accelerates the response-time of the gal system of *Escherichia coli*. J. Mol. Biol. *356*, 1073–1081.

Mangan, S., Zaslaver, A. and Alon, U. (2003) The coherent feedforward loop serves as a sign-sensitive delay element in transcription networks. J. Mol. Biol. *334*, 197–204.

Martins, A.M., Mendes, P., Cordeiro, C. and Freire, A.P. (2001) In situ kinetic analysis of glyoxalase I and glyoxalase II in *Saccharomyces cerevisiae*. Eur. J. Biochem. *268*, 3930–3936.

Melendezhevia, E., Torres, N.V., Sicilia, J. and Kacser, H. (1990) Control analysis of transition times in metabolic systems. Biochem. J. *265*, 195–202.

Milo, R., Shen-Orr, S., Itzkovitz, S., Kashtan, N., Chklovskii, D. and Alon, U. (2002) Network motifs: simple building blocks of complex networks. Science *298*, 824–827.

Moraes, C.T. (2001) What regulates mitochondrial DNA copy number in animal cells? Trends Genet. *17*, 199–205.

Newman, M.E. (2006) From the Cover: modularity and community structure in networks. Proc. Natl Acad. Sci. U.S.A. *103*, 8577–8582.

Newman, M.E.J. (2003) The structure and function of complex networks. SIAM Rev. *45*, 167–256.

Nikolaev, E.V., Burgard, A.P. and Maranas, C.D. (2005) Elucidation and structural analysis of conserved pools for genome-scale metabolic reconstructions. Biophys. J. *88*, 37–49.

Papin, J.A., Hunter, T., Palsson, B.O. and Subramaniam, S. (2005) Reconstruction of cellular signalling networks and analysis of their properties. Nat. Rev. Mol. Cell. Biol. *6*, 99–111.

Papin, J.A. and Palsson, B.O. (2004a). The JAK-STAT signaling network in the human B-cell: an extreme signaling pathway analysis. Biophys. J. *87*, 37–46.

Papin, J.A. and Palsson, B.O. (2004b). Topological analysis of mass-balanced signaling networks: a framework to obtain network properties including crosstalk. J. Theor. Biol. *227*, 283–297.

Papin, J.A., Stelling, J., Price, N.D., Klamt, S., Schuster, S. and Palsson, B.O. (2004) Comparison of network-based pathway analysis methods. Trends Biotechnol. *22*, 400–405.

Peletier, M.A., Westerhoff, H.V. and Kholodenko, B.N. (2003) Control of spatially heterogeneous and time-varying cellular reaction networks: a new summation law. J. Theor. Biol. *225*, 477–487.

Pereira-Leal, J.B., Enright, A.J. and Ouzounis, C.A. (2004) Detection of functional modules from protein interaction networks. Proteins *54*, 49–57.

Piper, M.D., Daran-Lapujade, P., Bro, C., Regenberg, B., Knudsen, S., Nielsen, J. and Pronk, J.T. (2002) Reproducibility of oligonucleotide microarray transcriptome analyses. An interlaboratory comparison using chemostat cultures of *Saccharomyces cerevisiae*. J. Biol. Chem. *277*, 37001–37008.

Price, N.D., Reed, J.L. and Palsson, B.O. (2004) Genome-scale models of microbial cells: evaluating the consequences of constraints. Nat. Rev. Microbiol. *2*, 886–897.

Ravasz, E., Somera, A.L., Mongru, D.A., Oltvai, Z.N. and Barabasi, A.L. (2002) Hierarchical organization of modularity in metabolic networks. Science *297*, 1551–1555.

Reder, C. (1988) Metabolic control theory: a structural approach. J. Theor. Biol. *135*, 175–201.

Reed, J.L. and Palsson, B.O. (2004) Genome-scale in silico models of *E.coli* have multiple equivalent phenotypic states: assessment of correlated reaction subsets that comprise network states. Genome Res. *14*, 1797–1805.

Rohwer, J.M., Meadow, N.D., Roseman, S., Westerhoff, H.V. and Postma, P.W. (2000) Understanding glucose transport by the bacterial phosphoenolpyruvate:glycose phosphotransferase system on the basis of kinetic measurements *in vitro*. J. Biol. Chem. *275*, 34909–34921.

Rossell, S., van der Weijden, C.C., Kruckeberg, A.L., Bakker, B.M. and Westerhoff, H.V. (2005) Hierarchical and metabolic regulation of glucose influx in starved *Saccharomyces cerevisiae*. FEMS Yeast Res. *5*, 611–619.

Rossell, S., van der Weijden, C.C., Lindenbergh, A., van Tuijl, A., Francke, C., Bakker, B.M. and Westerhoff, H.V. (2006) Unraveling the complexity of flux regulation: A new method demonstrated for nutrient starvation in *Saccharomyces cerevisiae*. Proc. Natl Acad. Sci. U.S.A. *103*, 2166–2171.

Ruijter, G.J.G., Postma, P.W. and Van Dam, K. (1991) Control of glucose metabolism by enzyme II^{Glc} of the phosphoenolpyruvate-dependent phosphotransferase system in *Escherichia coli*. J. Bacteriol. *173*, 6184–6191.

Sauro, H.M. (1994) Moiety-conserved cycles and metabolic control analysis: problems in sequestration and metabolic channelling. Biosystems *33*, 55–67.

Sauro, H.M. and Ingalls, B. (2004) Conservation analysis in biochemical networks: computational issues for software writers. Biophys. Chem. *109*, 1–15.

Sauro, H.M. and Kacser, H. (1990) Enzyme-enzyme interactions and control analysis. 2. The case of non-independence: heterologous associations. Eur. J. Biochem. *187*, 493–500.

Sauro, H.M., Small, J.R. and Fell, D.A. (1987) Metabolic control and its analysis. Extensions to the theory and matrix method. Eur. J. Biochem. *165*, 215–221.

Schaaff, I., Heinisch, J. and Zimmermann, F.K. (1989) Overproduction of glycolytic enzymes in yeast. Yeast *5*, 285–290.

Schilling, C.H., Letscher, D. and Palsson, B.O. (2000) Theory for the systemic definition of metabolic pathways and their use in interpreting metabolic function from a pathway-oriented perspective. J. Theor. Biol. *203*, 229–248.

Schilling, C.H., Schuster, S., Palsson, B.O. and Heinrich, R. (1999) Metabolic pathway analysis: basic concepts and scientific applications in the post-genomic era. Biotechnol. Prog. *15*, 296–303.

Schneider, K.R. and Wilhelm, T. (2000) Model reduction by extended quasi-steady-state approximation. J. Math. Biol. *40*, 443–450.

Schoeberl, B., Eichler-Jonsson, C., Gilles, E.D. and Muller, G. (2002) Computational modeling of the dynamics of the MAP kinase cascade activated by surface and internalized EGF receptors. Nat. Biotechnol. *20*, 370–375.

Schuster, S. (1996) Control analysis in terms of generalized variables characterizing metabolic systems. J. Theor. Biol. *182*, 259–268.

Schuster, S., Dandekar, T. and Fell, D.A. (1999) Detection of elementary flux modes in biochemical networks: a promising tool for pathway analysis and metabolic engineering. Trends Biotechnol. *17*, 53–60.

Schuster, S., Fell, D.A. and Dandekar, T. (2000) A general definition of metabolic pathways useful for systematic organization and analysis of complex metabolic networks. Nat. Biotechnol. *18*, 326–332.

Schuster, R. and Holzhutter, H.G. (1995) Use of mathematical models for predicting the metabolic effect of large-scale enzyme activity alterations. Application to enzyme deficiencies of red blood cells. Eur. J. Biochem. *229*, 403–418.

Schuster, S., Kahn, D. and Westerhoff, H.V. (1993) Modular analysis of the control of complex metabolic pathways. Biophys. Chem. *48*, 1–17.

Schuster, R. and Schuster, S. (1991) Relationships between modal-analysis and rapid-equilibrium approximation in the modeling of biochemical networks. Syst. Anal. Model. Simul. *8*, 623–633.

Segel, I.H. (1993) Enzyme kinetics: Behavior and Analysis of Rapid Equilibrium and Steady-state Enzyme Systems. (New York: John Wiley & Sons, Inc.).

Segel, L.A. and Slemrod, M. (1989) The quasi-steady-state assumption - a case-study in perturbation. Siam Rev. *31*, 446–477.

Shen-Orr, S.S., Milo, R., Mangan, S. and Alon, U. (2002) Network motifs in the transcriptional regulation network of *Escherichia coli*. Nat. Genet. *31*, 64–68.

Snoep, J.L. (2005) The Silicon Cell initiative: working towards a detailed kinetic description at the cellular level. Curr. Opin. Biotechnol. *16*, 336–343.

Snoep, J.L., Arfman, N., Yomano, L.P., Westerhoff, H.V., Conway, T. and Ingram, L.O. (1996) Control of glycolytic flux in *Zymomonas mobilis* by glucose-6-phosphate dehydrogenase activity. Biotechnol. Bioeng. *51*, 190–197.

Snoep, J.L., Bruggeman, F., Olivier, B.G. and Westerhoff, H.V. (2006) Towards building the silicon cell: a modular approach. BioSystems *83*, 207–216.

Snoep, J.L., van der Weijden, C.C., Andersen, H.W., Westerhoff, H.V. and Jensen, P.R. (2002) DNA supercoiling in *Escherichia coli* is under tight and subtle homeostatic control, involving gene-expression and metabolic regulation of both topoisomerase I and DNA gyrase. Eur. J. Biochem. *269*, 1662–1669.

Spirin, V. and Mirny, L.A. (2003) Protein complexes and functional modules in molecular networks. Proc. Natl Acad. Sci. U.S.A. *100*, 12123–12128.

Stent, G.S. (1968) That was the molecular biology that was. Science *160*, 390–395.

Stiefenhofer, M. (1998) Quasi-steady-state approximation for chemical reaction networks. J. Math. Biol. *36*, 593–609.

Taylor, S.D., Zhang, H., Eaton, J.S., Rodeheffer, M.S., Lebedeva, M.A., O'rourke, T.W., Siede, W. and Shadel, G.S. (2005) The conserved Mec1/Rad53 nuclear checkpoint pathway regulates mitochondrial DNA copy number in *Saccharomyces cerevisiae*. Mol. Biol. Cell. *16*, 3010–3018.

ter Kuile, B.H. and Westerhoff, H.V. (2001) Transcriptome meets metabolome: hierarchical and metabolic regulation of the glycolytic pathway. FEBS Lett. *500*, 169–171.

Teusink, B., Passarge, J., Reijenga, C.A., Esgalhado, E., van der Weijden, C.C., Schepper, M., Walsh, M.C., Bakker, B.M., van Dam, K., Westerhoff, H.V. and Snoep, J.L. (2000) Can yeast glycolysis be understood in terms of *in vitro* kinetics of the constituent enzymes? Testing biochemistry. Eur. J. Biochem. *267*, 5313–5329.

van der Gugten, A.A. and Westerhoff, H.V. (1997) Internal regulation of a modular system: the different faces of internal control. Biosystems *44*, 79–106.

Watts, D.J. and Strogatz, S.H. (1998) Collective dynamics of "small-world" networks. Nature *393*, 440–442.

Westerhoff, H.V. and Chen, Y.D. (1984) How do enzyme activities control metabolite concentrations? An additional theorem in the theory of metabolic control. Eur. J. Biochem. *142*, 425–430.

Westerhoff, H.V. and Kell, D.B. (1996) What bio technologists knew all along...? J. Theor. Biol. *182*, 411–420.

Westerhoff, H.V., Koster, J.G., van Workum, M. and Rudd, K.E. (1989) On the control of gene expression. In: Control of Metabolic Processes. A. Cornish-Bowden, M.L. Cardenas, eds. (New York: Plenum Press), pp. 399–413, NATO ASI Series.

Westerhoff, H.V. and Palsson, B.O. (2004) The evolution of molecular biology into systems biology. Nat. Biotechnol. *22*, 1249–1252.

Westerhoff, H.V., Plomp, P.J., Groen, A.K. and Wanders, R.J. (1987) Thermodynamics of the control of metabolism. Cell Biophys. *11*, 239–267.

Westerhoff, H.V. and Van Dam, K. (1987) Thermodynamics and Control of Biological Free-energy Transduction. (Amsterdam: Elsevier Science Publishers B.V. (Biomedical Division)).

Wiechert, W. (2002) An introduction to 13C metabolic flux analysis. Genet. Eng. (NY) *24*, 215–238.

Yeger-Lotem, E., Sattath, S., Kashtan, N., Itzkovitz, S., Milo, R., Pinter, R.Y., Alon, U. and Margalit, H. (2004) Network motifs in integrated cellular networks of transcription-regulation and protein-protein interaction. Proc. Natl Acad. Sci. U.S.A. *101*, 5934–5939.

Corresponding author: Dr. Barbara Bakker (barbara.bakker@falw.vu.nl)

CHAPTER 12

AUTOMATED, SYSTEMATIC DETERMINATION OF PROTEIN SUBCELLULAR LOCATION USING FLUORESCENCE MICROSCOPY

ELVIRA GARCÍA OSUNA and ROBERT F. MURPHY
Carnegie Mellon University Pittsburgh, PA

Table of Contents

Abstract: Proteomics is the comprehensive study of all aspects of protein behavior. The subfield of location proteomics is concerned with the systematic analysis of the subcellular location of proteins. In order to perform high-resolution, high-throughput analysis of all protein location patterns, automation is needed both for acquisition and analysis. Automated methods for analyzing subcellular location patterns in fluorescence microscope images have been developed and shown to work well for static 2D and 3D images of single cells. This chapter reviews this work and describes current efforts to extend these approaches, including classification of temporal patterns and building of generative models to represent location patterns.

E. Bertrand and M. Faupel (eds.), Subcellular Proteomics, 263–276.

1. INTRODUCTION

Knowledge of the subcellular location of each protein is critical to understanding protein function and for successful bottom-up systems biology efforts. The advent of high-throughput imaging technology makes large-scale analysis of protein subcellular location feasible, but also creates challenges for the interpretation of the resulting images.

1.1. Large-Scale Protein Subcellular Location Projects

There are two major approaches to the large-scale analysis of subcellular location. One involves *prediction* of subcellular location based on protein sequence. The other involves experimental *determination* of the locations of proteins.

A number of systems for predicting protein localization from sequence have been described (Nakai 2000; Chou and Cai 2003; Park and Kanehisa 2003; Guda et al. 2004; Lu et al. 2004). The limitation of these systems is that they can only assign new proteins to the location categories with which they have been trained. This means that proteins with previously unseen location patterns cannot be recognized. Their predictions are also on a low-resolution basis: they are able to assign proteins to an organelle (or more than one organelle), but not to specific areas within that organelle. Furthermore, prediction systems are not currently able to predict differential localization based on cell type (largely because insufficient training data exist).

Due to the limitations of prediction, there is a need for projects that will collect data on subcellular location for a variety of conditions. These projects *determine* protein location rather than predict it. Although these projects are useful in their own right, they also serve as a way to expand the database for the prediction of protein location. One powerful approach to determination is to isolate specific organelles or structures and then identify their components by expression proteomics methods (Brunet et al. 2003). This provides information rapidly for large numbers of proteins but has limited ability to distinguish different domains within an organelle. This chapter will focus on a complementary approach, automated determination of subcellular location using fluorescence microscopy.

2. AUTOMATED ANALYSIS OF PROTEIN SUBCELLULAR LOCATION IMAGES

Fluorescence microscopy has been used for decades for the analysis of subcellular location patterns. Visual examination has been the primary means of analyzing images from both small and large-scale projects. The volume of images that can be produced by large-scale projects creates a need for computational methods for this task. In the mid-1980s, pattern recognition methods were successfully applied to distinguish cell and nuclear patterns, such as recognizing different disease states in histological preparations (King et al. 1984) or distinguishing chromatin patterns in nuclei (Young et al. 1986). Developing methods that can be widely applied to all subcellular patterns

Table 1. Examples of features used to classify 2D subcellular location patterns

Feature number	Description
SLF1.1	The number of fluorescent objects per cell. An object is defined as a bordering group of above threshold pixels.
SLF1.3	The average number of above-threshold pixels per object (object size). Intended to obtain information about the sizes of the objects.
SLF1.6	The average distance from the center of fluorescence of each object to the center of fluorescence of the whole cell.
SLF2.22	Fraction of above-threshold pixels in the protein image that overlap with the above-threshold pixels in the DNA image.

has been a major focus of the Murphy group for the past decade (Boland et al. 1997; Boland et al. 1998; Boland and Murphy 2001; Murphy et al. 2003; Huang and Murphy 2004a) and more recently of a number of other groups (Danckaert et al. 2002; Conrad et al. 2004; Sigal et al. 2006). Computational methods have the advantage of not only being automated but also providing better sensitivity, accuracy and reproducibility than visual examination.

The starting point for developing these methods is a collection of cell images covering different subcellular patterns. One such collection is a set of 2D images of HeLa cells containing images of all major organelles (Boland and Murphy 2001). Examples of these images are shown in Figure 1. Given such a set of images, the task becomes the development of a system to recognize which subcellular pattern is contained in a previously unseen image. Such a system is referred to as a *classifier,* and the patterns that it is trained to recognize are referred to as *classes*. When given an image, a classifier will output a set of scores or probabilities corresponding to each class. The class with the highest score or probability is considered to be the predicted class. This approach assumes that the unseen image is from one of the classes used in training and that they are collected under similar conditions (i.e. using the same cell type and microscope). Our group has made continuous improvements in such systems for this 2D HeLa collection, from an initial accuracy of 84% (Boland and Murphy 2001) to the most recent system with an accuracy of over 92% (Huang and Murphy 2004b).

At the heart of these systems are sets of numerical features that describe the subcellular pattern. We refer to them as Subcellular Location Features (SLF); detailed descriptions of them have been published (Huang and Murphy 2004b) and are available at http://murphylab.web.cmu.edu/services/SLF. Table 1 gives example of some of these features.

2.1. Classification

Given images for which the class is known, we can estimate the accuracy of a classification approach by training it on some of those images and then evaluating its performance on the remaining images. Repeating this process for different divisions

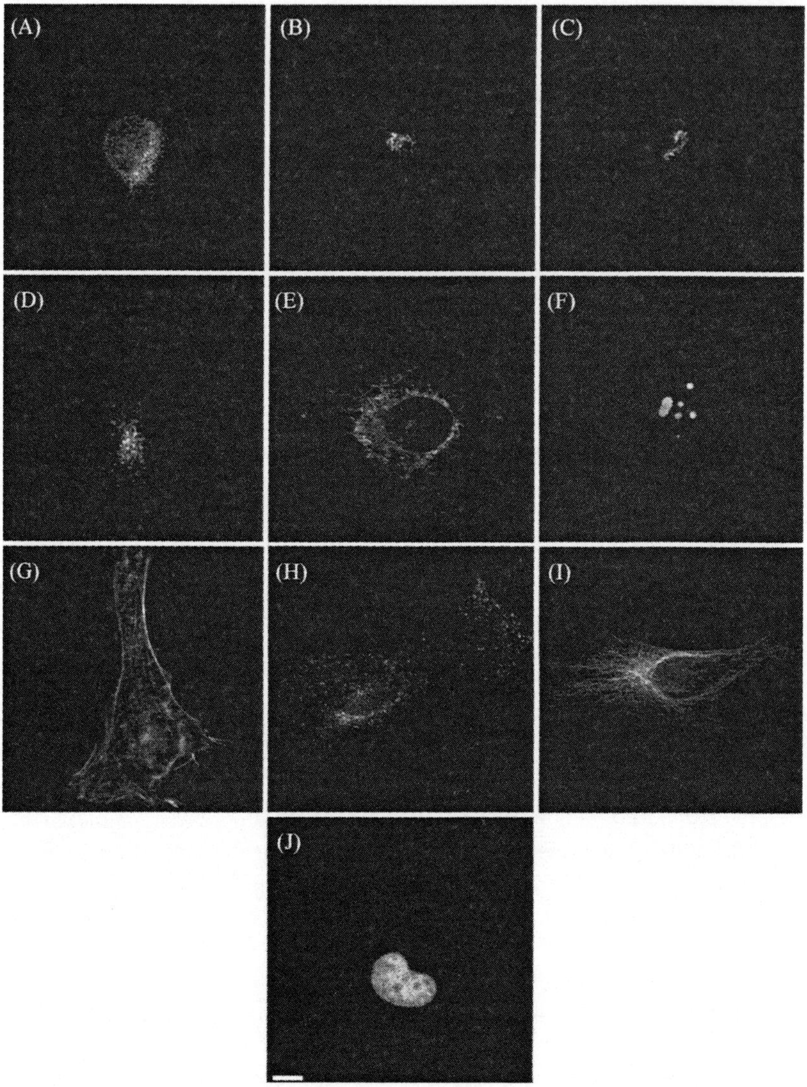

Figure 1. Representative images from the HeLa dataset. The images are (A) an ER protein, (B) Giantin, (C) gpp130, (D) LAMP2, (E) a mitochondrial protein, (F) nucleolin, (G) actin, (H) transferrin receptor, (I) tubulin and (J) DNA. From Boland and Murphy (2001).

into training and test sets is referred to as *cross-validation*, and all accuracies mentioned here for our automated systems were obtained this way. In order to determine how the accuracy of computational analysis compares to visual interpretation, a study was conducted using images from the 2D HeLa set (Murphy et al. 2003). An observer

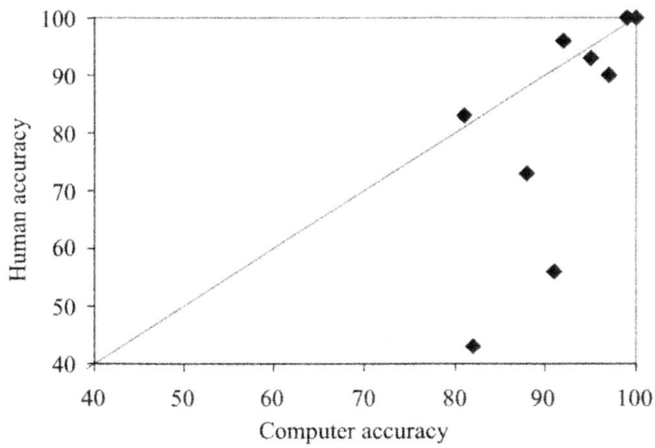

Figure 2. Comparison of classification accuracies from an automated system and from visual examination. Accuracies of automated methods are obtained by using SLF16 and a majority-voting ensemble classifier and are presented versus the average accuracy for the same images obtained by visual examination. Each symbol represents a different pattern class. In increasing order of human accuracy these are: gpp130, Giantin, LAMP2, TfR, ER, Tubulin, Mitochondria, nucleolin and DNA (both at 100% for human and 99% for computer accuracy), and actin (100% for both). From Murphy (2004).

was asked to classify an image into one of the ten classes. The observer was then informed of the correct answer and a new image was presented. This process was repeated for many cycles until the performance did not improve. The results of this study are compared to those for an automated classifier (Huang and Murphy 2004a) in Figure 2, where each symbol represents the relative accuracy of human and computer for each of the patterns shown in Figure 1. The line is a visual aid to indicate exact correspondence between human and computer. Seven of the points are near the line, indicating that both human and machine had similar classification accuracies. However, three of the classes show much higher accuracy for the automated approach than the visual one. This leads us to the important conclusion that automated methods perform better than or equal to visual interpretation for the task of recognizing subcellular patterns. The patterns with the largest difference are giantin and gpp130 (Figure 1B and C), which are both Golgi proteins. They are difficult to distinguish visually (56 and 43%, respectively) (Murphy et al. 2003), but the automated methods are able to correctly classify them with accuracies of at least 90.8 and 82.4%, respectively (Huang and Murphy 2004a).

2.2. Clustering

Of course, classification of images of proteins whose location is already known does not contribute much to our understanding of protein location. We can, however, imagine applying a trained classifier to large collections of images of proteins whose location is *not* known. Automated classification is a powerful approach, but only if its fundamental assumption is correct: that all of the classes are known before starting. Since it is unlikely that a new *major* subcellular location class will be discovered, this may not be a significant hindrance for analyzing microscope images. However, an alternative is to allow the classes to be learned directly from the data (an approached referred to as *unsupervised learning*). This should not only permit previously unknown classes to be discovered, but also to subdivide known classes into finer categories. For this purpose, the SLF features can be combined with *cluster analysis* methods. We have tested this approach on a dataset of images of a number of proteins in NIH 3T3 cells obtained by CD-tagging (Jarvik et al. 1996). This method allows for random tagging of proteins with minimal disruption of protein folding. Tagging proteins with a fluorescent probe is needed so that only that protein can be seen among the tens of thousands of other proteins in a cell.

2.2.1. CD-tagging

CD-tagging inserts a specially designed DNA sequence (CD cassette) into genomic DNA, typically using a modified retrovirus. When the CD cassette is inserted into an intron of a gene, the cassette provides splicing signals that direct the inclusion of a new exon, referred to as the guest exon. Translation of the tagged transcript results in the incorporation of a peptide tag (guest peptide) into the protein (Figure 3). If the coding sequence of Green Fluorescent Protein (GFP) is placed in the CD cassette, the result will be the generation of cells expressing fluorescent chimeras of randomly chosen proteins.

Since retroviral tagging can be done under conditions where only one gene is tagged, and since the guest exon is introduced into the endogenous gene, the natural regulation of the gene and protein is expected to be maintained (Jarvik et al. 2002). In most cases, GFP does not disrupt the natural folding, function and location of the protein because GFP folds itself into a beta-barrel where the N-terminus and C-terminus are close to one another in space. This causes the amino acids in the protein on either side of the guest exon to remain close even after the GFP insertion.

When this method is performed on a population of cells, the cassette will insert itself into different parts of the genomic DNA for each cell. Flow sorting can then be used to sort GFP-positive cells, and separate clones can be established.

To determine which protein has been tagged, RT-PCR (reverse transcription-polymerase chain reaction) can then be performed using a PCR primer from the GFP-coding sequence. A initial collection of over 100 tagged clones has been created (Jarvik et al. 2002) (http://cdtag.bio.cmu.edu).

2.2.2. Imaging and Analysis of Clones

The initial set of CD-tagged clones were imaged using a spinning disk confocal microscope. Single color, single cell images were acquired at 60× magnification. This set contains many 3D images for 90 different clones. Features for each image were calculated and the similarity of the images was determined by calculating the distance

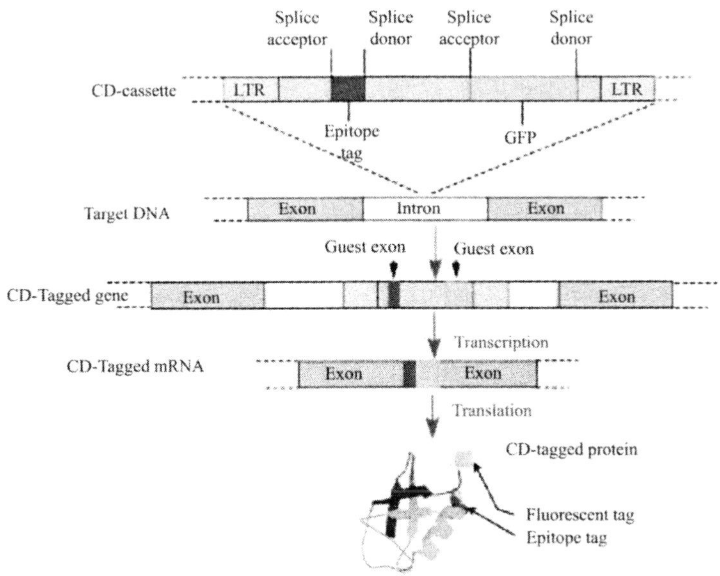

Figure 3. CD-tagging method. When the gene is expressed, the tag will be expressed with it. This ensures that the resulting image is that of only one protein. From http://www.andrew.cmu.edu/user/berget/CDtags/TaggingIntro.html.

between their feature vectors. Two distance functions were tested: z-scored Euclidean and Mahalanobis. The z-scored Euclidean is calculated after normalizing each feature to have zero mean and unit variance. Mahalanobis distance scales the contribution of each feature using the covariance matrix. Using these distance functions, clustering methods were tested on the images of the CD-tagged clones.

In order to evaluate the distance functions, we state that the best distance function should give the most agreement among the different clustering algorithms. To do this, Cohen's κ statistic was used. The larger the κ statistic, the greater the agreement between two partitionings of the data. Table 2 shows the κ statistic for the different clustering methods and distance functions. Figure 4 shows the tree obtained from using the z-scored Euclidean distance function. Upon inspection of the cluster tree, patterns that have similar pattern were, for the most part, grouped together. For example, proteins with predominantly nuclear patterns with some punctate cytoplasmic staining are clustered together (Chen and Murphy 2005). Current work in the Murphy group is extending these methods to images of proteins that have been randomly tagged with GFP. This work will enable us to determine protein location with out previous knowledge of the actual tagged protein.

3. TIME-SERIES ANALYSIS

The patterns of most proteins within cells are dynamic: the proteins themselves may move between/within subcellular structures and the structures themselves may move.

Table 2. Comparison of clustering methods and distance functions. The agreement between the sets of clusters resulting from the four clustering methods was measured using the κ test. The standard deviations of the statistic under the null hypothesis were estimated to range between 0.014 and 0.023 from multiple simulations. From Chen and Murphy (2005).

Clustering approaches compared	z-scored Euclidean distance (κ)	Mahalanobis distance (κ)
k-means/AIC versus consensus	1	0.5397
k-means/AIC versus ConfMat	0.4171	0.3634
Consensus versus ConfMat	0.4171	0.1977
k-means/AIC versus visual	0.2055	0.1854
Consensus versus visual	0.2055	0.1156

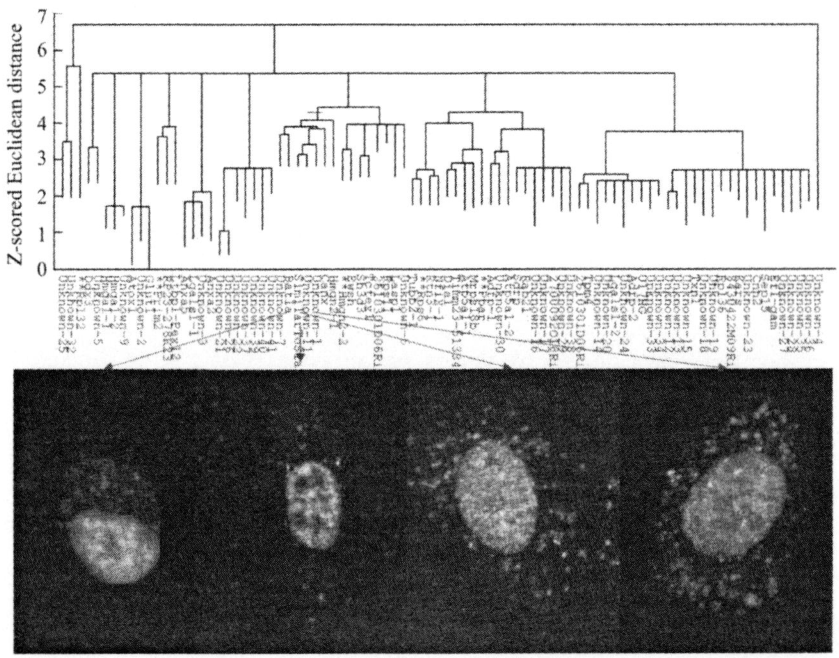

Figure 4. Clustering of subcellular location patterns with example images. From Chen and Murphy (2005).

However, the automated methods that have been described above have been applied to datasets of static images. An area of intensive current work is therefore to extend these approaches to time-series images.

3.1. Object-Tracking

The most powerful approach to characterize the spatiotemporal patterns of proteins in cells would be to track individual proteins. This is difficult under most circumstances since cells express many copies of the same protein. Fortunately, an excellent alternative for proteins that undergo dynamic polymerization and depolymerization is to use speckle microscopy, which takes advantage of the stochastic formation of visible complexes of fluorescently tagged proteins that can be tracked over short time frames (Danuser and Waterman-Storer 2006). Time series images of speckles contain rich information about the dynamics of the tagged protein. For example, this approach has been combined with sophisticated image analysis methods to provide detailed models of the dynamics of the actin cytoskeleton during cell protrusion (Machacek and Danuser 2006).

3.2. Temporal-Texture Features

Sophisticated tracking and modeling methods frequently require extensive initial knowledge of the types of structures and the time scales of motion that a particular protein displays. For location proteomics projects, less powerful but more general methods are needed. One approach to characterizing protein dynamics without tracking is to use temporal-texture features. These features are the temporal equivalents of the spatial texture features that have proven powerful for analyze the static pattern as described above. We have demonstrated that temporal texture features can help to distinguish proteins whose location patterns cannot be distinguished using static images alone. For this purpose, 3D time series images were collected for five CD-tagged clones that were grouped together by the automated clustering described above. When temporal-texture features were combined with 2D and 3D static features and used to train and test a classifier, the five proteins could be distinguished with an accuracy of 85% (compared to 75% using just 2D and 3D static features) (Hu et al. 2006). Clearly, the temporal texture features capture information that is helpful in identifying protein patterns.

3.3. Cell Cycle Analysis

Of particular potential importance for temporal changes in protein patterns are the continuous changes in cells due to the cell cycle. Knowing which proteins change their patterns during the cell cycle will be important not only to identify those that may play a role in the cell cycle process itself, but also so that the mechanisms that cause these changes can be identified. As an illustration of this approach, time series images of twenty cell lines expressing fluorescently-tagged nuclear proteins (generated by CD-tagging) were collected and analyzed for cell cycle dependent changes (Sigal et al. 2006). The time series were synchronized *in silico* so that a complete picture of the expression and localization of each protein during the cell cycle could be obtained.

4. GENERATIVE MODELS

Results from comparing or clustering proteins by their subcellular location can be reported as lists of proteins that share a pattern, or by showing example images from each pattern. However, many examples are required in order to capture the variability within a pattern. We have explored an alternative, building generative models to capture characteristics of an image set that allow for synthesis of new images statistically equivalent to the input images.

The starting point for this work was to develop Subcellular Object Features (SOF). These describe the characteristics of the objects in the pattern and give information regarding how many object types are represented in a dataset. Using these features, a classifier was trained to recognize each object type. This allows for the creation of a vector that describes how many of each object type is present in an image. Using this method, an accuracy of 81–82% was obtained using the 2D HeLa set, where Giantin and gpp130 were merged into one class (Zhao et al. 2005).

The next step is to develop approaches for positioning the objects in the appropriate locations in the cell. Encouraging preliminary work has demonstrated the feasibility of modeling patterns of proteins that are primarily in discrete objects (such as lysosomes) (Zhao and Murphy in preparation). This method begins with synthesizing a nuclear image and a cellular boundary. The protein objects are then positioned in the synthesized cell.

Generative models can provide a high-resolution map for protein location. The availability of these maps should aid systems biology modeling of cell behavior.

5. PROTEIN LOCATION IMAGE DATABASES

A number of databases and image collections have recently been created to contain images of protein patterns in various organisms ranging from yeast to human. Table 3 summarizes these databases. The Yeast GFP Fusion Localization Database was a particularly ambitious project (Huh et al. 2003). Here, homologous recombination was used to attach a GFP-tag to the carboxy terminus of all suspected yeast proteins. A total of 6029 genes were tagged which yielded 4156 expressed proteins. Images were taken at high magnification (100×). The authors assigned each protein one or more labels from a set of 23 labels by visually examining the images.

The GFP-cDNA Localization Project has analyzed a number of human genes by expression of tagged cDNAs in African green monkey cells (Simpson et al. 2000). 107 human open reading frames (ORFs) were used that corresponded to uncharacterized coding regions of the human genome. The cDNA of these ORFs was fused with a GFP coding sequence in a constitutive expression vector. The monkey cells were transfected with the tagged cDNA and then imaged at 63×. Each protein was assigned a location through visual examination. If a location could not be identified, locations were predicted from the sequence.

The CD-tagging database uses internal GFP fusions rather than terminal fusions (Jarvik et al. 1996), as discussed above. The GFP-tagged cells were imaged using a laser spinning disk confocal microscope at 60×. Clustering methods were used to analyze the location patterns of the tagged proteins.

The Protein Atlas project primarily deals with determining protein patterns at the cellular level within tissue (Uhlen et al. 2005). This database contains images of

Table 3. Comparison of subcellular location determination projects

Project	Reference	Tag	Resolution	Analysis	Species
Yeast GFP Fusion Localization Database (yeastgfp.ucsf.edu)	(Ghaemmaghami et al. 2003)	Green fluorescent protein	High (100×)	Visual	Yeast
GFP-cDNA Localization Project (www.gfp-cdna.embl.de)	(Simpson et al. 2000)	Green fluorescent protein	High (63×)	Visual	African green monkey
CD-tagging (www.cdtag.bio.cmu.edu)	(Jarvik et al. 2002; Chen et al. 2003)	Green fluorescent protein (Genomic, c-DNA)	High (60×)	Automated	Mouse
Protein atlas (www.proteinatlas.org)	(Uhlen et al. 2005)	Immunocytochemistry with mono-specific antibodies	Low (20×)	Visual	Human
PSLID (www.pslid.cbi.cmu.edu/PSLID)	(Huang et al. 2002)	GFP and Immunofluorescence with mono-specific antibodies	High (60–100×)	Automated	Mouse and Human

over 700 proteins in 48 normal human tissue and 20 different cancers. The proteins were tagged using antibodies (secondary antibodies conjugated with horseradish peroxidase) in human tissue microarrays. Analysis included five major types of protein families, receptors, kinases, phosphatases, transcription factors and nuclear receptors. Tissue microarrays were imaged using an automated system at 20× magnification, and the resulting images were annotated by pathologists.

With the exception of the CD-tagging database, the databases described above all use visual examination as the primary means of analysis. Therefore, our group has developed the Protein Subcellular Location Image Database (PSLID) which contains images as well as automated methods of analyzing those images (Huang et al. 2002). This database can be accessed at http://pslid.cbi.cmu.edu. This database contains images from various sources and permits the automated methods that have been described in this chapter to be applied to any cell image.

6. CONCLUSION

This chapter has provided a brief review of approaches relevant to the comprehensive determination of subcellular location. These methods are automated, reproducible and more sensitive than visual examination, and their utility has already been demonstrated in pilot projects. However, much work remains to be done, especially to develop automated methods for synthesizing diverse sources information on subcellular location into accurate models of the distribution of all proteins within all cell types.

ACKNOWLEDGEMENTS

The original research from the Murphy group described above was supported in part by NIH grant R01 GM068845 and NSF grant EF-0331657.

REFERENCES

Boland, M.V., Markey, M.K. and Murphy, R.F. (1997) Classification of protein localization patterns obtained via fluorescence light microscopy. 19th Annual International Conference of the IEEE Engineering in Medicine and Biology Society, Chicago, IL, USA.

Boland, M.V., Markey, M.K. and Murphy, R.F. (1998) Automated recognition of patterns characteristic of subcellular structures in fluorescence microscopy images. Cytometry *33*, 366–375.

Boland, M.V. and Murphy, R.F. (2001) A neural network classifier capable of recognizing the patterns of all major subcellular structures in fluorescence microscope images of hela cells. Bioinformatics *17*, 1213–1223.

Brunet, S., Thibault, P., Gagnon, E., Kearney, P., Bergeron, J.J. and Desjardins, M. (2003) Organelle proteomics: looking at less to see more. Trends Cell Biol. *13*, 629–638.

Chen X. and Murphy, R.F. (2005) Objective clustering of proteins based on subcellular location patterns. J. Biomed. Biotechnol. *2005*, 87–95.

Chen, X., Velliste, M., Weinstein, S., Jarvik, J.W. and Murphy, R.F. (2003) Location proteomics – building subcellular location trees from high resolution 3d fluorescence microscope images of randomly-tagged proteins. Proc. SPIE *4962*, 298–306.

Chou, K.C. and Cai, Y.D. (2003) Prediction and classification of protein subcellular location-sequence-order effect and pseudo amino acid composition. J. Cell. Biochem. *90*, 1250–1260.

Conrad, C., Erfle, H., Warnat, P., Daigle, N., Lorch, T., Ellenberg, J., Pepperkok, R. and Eils, R. (2004) Automatic identification of subcellular phenotypes on human cell arrays. Genome Res. *14*, 1130–1136.

Danckaert, A., Gonzalez-Couto, E., Bollondi, L., Thompson, N. and Hayes, B. (2002) Automated recognition of intracellular organelles in confocal microscope images. Traffic *3*, 66–73.

Danuser, G. and Waterman-Storer, C.M. (2006) Quantitative fluorescent speckle microscopy of cytoskeleton dynamics. Annu. Rev. Biophys. Biomol. Struct. *35*, 361–387.

Ghaemmaghami, S., Huh, W.K., Bower, K., Howson, R.W., Belle, A., Dephoure, N., O'Shea, E.K. and Weissman, J.S. (2003) Global analysis of protein expression in yeast. Nature *425*, 737–741.

Guda, C., Fahy, E. and Subramaniam, S. (2004) Mitopred: A genome-scale method for prediction of nucleus-encoded mitochondrial proteins. Bioinformatics *20*, 1785–1794.

Hu, Y., Carmona, J. and Murphy, R.F. (2006) Application of temporal texture features to automated analysis of protein subcellular locations in time series fluorescence microscope images. Proc 2006 IEEE Intl Symp Biomedl Imag 1028–1031.

Huang, K., Lin, J., Gajnak, J.A. and Murphy, R.F. (2002) Image content-based retrieval and automated interpretation of fluorescence microscope images via the protein subcellular location image database. Proc 2002 IEEE Intl Symp Biomedl Imag 325–328.

Huang, K. and Murphy, R.F. (2004a) Boosting accuracy of automated classification of fluorescence microscope images for location proteomics. BMC Bioinformatics *5*, 78.

Huang, K. and Murphy, R.F. (2004b) From quantitative microscopy to automated image understanding. J. Biomed. Opt. *9*, 893–912.

Huh, W.K., Falvo, J.V., Gerke, L.C., Caroll, A.S., Howson, R.W., Weissman, J.S. and O'Shea, E.K. (2003) Global analysis of the protein localization in budding yeast. Nature *425*, 686–691.

Jarvik, J.W., Adler, S.A., Telmer, C.A., Subramaniam, V. and Lopez, A.J. (1996) Cd-tagging: A new approach to gene and protein discovery and analysis. BioTechniques *20*, 896–904.

Jarvik, J.W., Fisher, G.W., Shi,C., Hennen, L., Hauser, C., Adler, S. and Berget, P.B. (2002) *In vivo* functional proteomics: Mammalian genome annotation using cd-tagging. BioTechniques *33*, 852–867.

King, E.B., Kromhout, L.K., Chew, K.L., Mayall, B.H., Petrakis, N.L., Jensen, R.H. and Young, I.T. (1984) Analytic studies of foam cells from breast cancer precursors. Cytometry 5, 124–130.

Lu, Z., Szafron, D., Greiner, R., Lu, P., Wishart, D.S., Poulin, B., Anvik, J., Macdonell, C. and Eisner, R. (2004) Predicting subcellular localization of proteins using machine-learned classifiers. Bioinformatics *20*, 547–556.

Machacek, M. and Danuser, G. (2006) Morphodynamic profiling of protrusion phenotypes. Biophys. J. *90*, 1439–1452.

Murphy, R.F. (2004) Automated interpretation of subcellular location patterns. 2004 IEEE International Symposium on Biomedical Imaging (ISBI-2004),

Murphy, R.F., Velliste, M. and Porreca, G. (2003) Robust numerical features for description and classification of subcellular location patterns in fluorescence microscope images. J. VLSI Sig. Proc. *35*, 311–321.

Nakai, K. (2000) Protein sorting signals and prediction of subcellular localization. Adv. Protein Chem. *54*, 277–344.

Park, K.J. and Kanehisa, M. (2003) Prediction of protein subcellular locations by support vector machines using compositions of amino acids and amino acid pairs. Bioinformatics *19*, 1656–1663.

Sigal, A., Milo, R., Cohen, A., Geva-Zatorsky, N., Klein, Y., Alaluf, I., Swerdlin, N., Perzov, N., Danon, T., Liron, Y., Raveh, T., Carpenter, A.E., Lahav, G. and Alon, U. (2006) Dynamic proteomics in individual human cells uncovers widespread cell-cycle dependence of nuclear proteins. Nat. Methods *3*, 525–531.

Simpson, J.C., Wellenreuther, R., Poustka, A., Pepperkok, R. and Wiemann, S. (2000) Systematic subcellular localization of novel proteins identified by large-scale cdna sequencing. EMBO Rep. *1*, 287–292.

Uhlen, M., Bjorling, E., Agaton, C., Szigyarto, C.A.-K., Amini, B., Andersen, E., Andersson, A.-C., Angelidou, P., Asplund, A., Cerjan, D., Ekstrom, M., Elobeid, A. and Eriksson, C. (2005) A human protein atlas for normal and cancer tissues based on antibody proteomics. Am. Soc. Biochem. Mol. Biol. *4*, 1920–1932.

Young, I.T., Verbeek, P.W. and Mayall, B.H. (1986) Characterization of chromatin distribution in cell nuclei. Cytometry *7*, 467–474.

Zhao, T., Velliste, M., Boland, M.V. and Murphy, R.F. (2005) Object type recognition for automated analysis of protein subcellular location. IEEE Trans. Image Process. *14*, 1351–1359.

Corresponding authors: Dr. Elvira García Osuna (elvira@cmu.edu) and Pr. Robert F. Murphy (murphy@cmu.edu)

CHAPTER 13

SYSTEMS BIOLOGY OF THE ENDOPLASMIC RETICULUM STRESS RESPONSE

MARIE-ELAINE CARUSO[1] and ERIC CHEVET[2]

[1] *Team Avenir, INSERM U889, Bordeaux, France; Université Bordeaux 2, Bordeaux, France*

[2] *McGill University, Montreal, Qc, Canada*

Table of Contents

Abstract: The endoplasmic reticulum (ER) is the first sub-cellular compartment encountered by secretory proteins en route to the plasma membrane. Newly synthesized secretory proteins translocate into the ER lumen and acquire their correct conformation prior to being

E. Bertrand and M. Faupel (eds.), Subcellular Proteomics, 277–298.

exported to later compartments. When folding is not properly achieved, proteins accumulate in the ER due to resident quality control machineries and terminally misfolded proteins are ultimately degraded through the ER-associated degradation pathway. All these molecular machines function in a coordinated fashion to restore and maintain ER homeostasis. A fifth molecular machine plays a coordinating role in the ER. Indeed, the ER stress signaling machinery signals ER dysfunction to the rest of the cell and consequently integrates the functions of the four other molecular machines to improve their operation in stressful conditions.

In this work, we have attempted to define the ER as a molecular biological system regulated by its own specific signaling pathways defined as the Unfolded Protein Response to delineate a systems biology approach of ER stress signaling.

1. INTRODUCTION

Systems-level approaches were defined more than 60 years ago with the application of general systems theory to various scientific fields including chemistry and life sciences with emphasis on morphogenesis (von Bertalanffy 1968; Friboulet and Thomas 2005). In the sixties, concepts derived from Ilya Prigogine's "dissipative structures" theory were also applied to biochemical oscillation and morphogenesis (Friboulet and Thomas 2005). In the seventies, numerical analyses of biochemical systems were developed which could represent a first step into modern systems biology (Hammer et al. 2004; Aderem 2005).

1.1. A Definition of Systems Biology

It is only in 2001 that the foundations of systems biology were set and precisely defined by Kitano (Kitano 2002a,b,c) as follows: systems biology aims at system-level understanding of biological systems. System-level comprehension essentially requires the understanding of (i) system structures, (ii) system behavior, (iii) system control, and finally (iv) how to design the system. This requires research in a broad range of areas such as for instance molecular biology, computer science, control theory and engineering, technologies for quantitative measurements (Kitano 2002c, 2005). The understanding of the structure of the system can be achieved through the characterization of gene regulatory and biochemical network, protein–protein or genetic interaction maps, as well as physical structures. The understanding of the systems dynamics, that is how a system react to an external perturbation or how fast the system returns to its normal state after simulation, can be carried out at both quantitative and qualitative levels. It should lead to the construction of theory/model with powerful prediction capability. Finally, the characterization of system's control machines, and the design of predictive algorithms are essential steps to judge how much we understand the system (Kitano 2002c, 2005). The latter steps are now being dragged into the field of synthetic biology (Benner and Sismour 2005; McDaniel and Weiss 2005; Pawson and Linding 2005) to modify and construct biological systems having desired properties which can be devised based on definite design principles and simulations, instead of blind trial-and-error (Benner and Sismour 2005; McDaniel and Weiss 2005; Pawson and Linding 2005).

Currently a number of issues, such as robustness of biological systems, network structures and dynamics, and applications to drug discovery are actively investigated. This is allowed by significant advances in science and technology to explore how biological component function as a network at sub-cellular, cellular, tissue or organism levels.

1.2. Systems Biology at the Sub-Cellular Level

When systems biology theories meet the more conventional paradigms of cell biology, this leads to the perception of cell biology problematic with a new perspective. Recently, it has been proposed that systems biology could allow the study of complex biological processes, such as intracellular traffic, as a whole (Quenneville and Conibear 2006). In this context, systematic large-scale to genome-wide assays have indeed the potential to identify the transport machinery, delineate pathways and unravel the molecular components of physiological processes that influence trafficking. A goal of this approach is therefore to create predictive models of intracellular trafficking pathways that reflect these relationships. Similarly, reductionist approaches have been spectacularly successful at dissecting at the molecular level many of the key processes that occur within the nucleus, particularly gene expression. At the same time, the limitations of analyzing single nuclear processes in spatial and temporal isolation and the validity of generalizing observations of single gene loci are becoming evident. The next level of understanding of genome function is to integrate our knowledge of their sequences and the molecular mechanisms involved in nuclear processes with our insights into the spatial and temporal organization of the nucleus and to elucidate the interplay between protein and gene networks in regulatory circuits. To this end, catalogues of genomes and proteomes as well as a precise understanding of the behavior of molecules in living cells are required. Converging technological developments in genomics, proteomics, dynamics and computation are now leading towards such an integrated biological understanding of genome biology and nuclear function. Consequently it is clear that systems approaches hold great promise for placing sub-cellular mechanisms/structures in their cellular contexts.

2. CAN THE ENDOPLASMIC RETICULUM BE DEFINED AS A BIOLOGICAL SYSTEM?

Based on the General systems theory (von Bertalanffy 1968), a system is a complex processing of interrelated parts that acquires supplies and turns them into something else. Thus, a living organism can be called a system, as each individual takes in oxygen, nutrients and water, processes them, and turns out movement, growth, and bodily waste.

2.1. General Systems Properties and Definitions

No system can be defined as "closed" indicating that there is no system whose boundary is impermeable to everything. An alternative way of expressing this property is to state that all systems are "open." Thus, all systems can be defined as sub-systems of larger super-systems, and, in particular, the biological system is a subset of the physical system. Systems are dependent upon, and thus controlled by, their various super-systems. This fact becomes readily apparent when a system can no longer obtain needed resources from its environment (super-system) and shuts down. It is also demonstrated when a system's outputs are so excessive or aggravating that its environment can accept no more, and the resulting feedback shuts down the system's operations. We can therefore apply these definitions to biological systems. In particular, each sub-cellular compartment can be considered as a sub-system contained in the super-system defined by the cell. In this context, each sub-system is "open" and regulated by the general condition of its super-system which is itself under the dependence of the sub-systems it includes.

2.2. The ER System

The endoplasmic reticulum (ER) was first identified by Palade and Porter in 1954 when they observed a "lace-like reticulum" in the cytoplasm of avian cells cultured *in vitro* (Palade and Porter 1954). The ER was found to consist in interconnected "strands and vesicles" of small dimensions and relatively low density. During the past 50 years, this sub-cellular compartment has been extensively studied. It has been recognized as a primary site of secretory protein synthesis and post-translational modification (Pryme 1986; Freedman 1989; Price 1992; Bonifacino and Weissman 1998; Kukuruzinska and Lennon 1998; Helenius and Aebi 2001). In addition, following the morphological description of the ER, this compartment was isolated using procedures based on differential centrifugation. This major step has allowed, besides the electron microscope observation of ER microsomes (Figure 1A), the characterization of the import system for newly synthesized secretory proteins through the tranlocation machinery (Campbell and Blobel 1976; Lingappa and Blobel 1980; Blobel 1995). In parallel, yeast genetics experiments have allowed to delineate the ER export machinery to facilitate protein to egress from the ER to the Golgi apparatus (Lyman and Schekman 1996).

Although these two molecular machines are important to maintain a continuous flow of protein through the secretory pathway, they are not sufficient to constitute a system per se since it does not contain the capacity to regulate itself if the above mentioned functions are not properly achieved. This issue was in part addressed by the discovery of (1) ER quality control mechanisms (Ellgaard and Helenius 2003; Sitia and Braakman 2003; Trombetta and Parodi 2003; Helenius and Aebi 2004; Kleizen and Braakman 2004; Vitale and Ceriotti 2004; Bukau et al. 2006; Conn et al. 2006; Wu et al. 2006); (2) the ER-associated degradation machinery (Ellgaard and Helenius 2003; Jarosch et al. 2003; Yoshida 2003; Bar-Nun 2005;

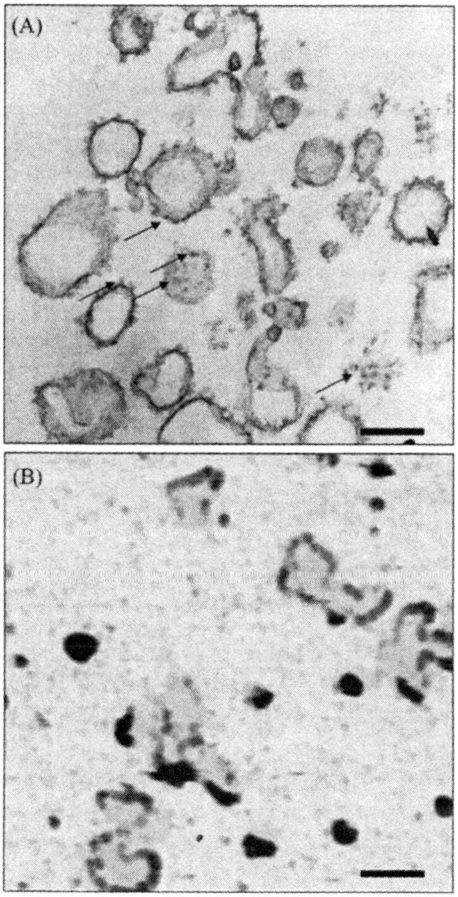

Figure 1. Morphological analysis of dog pancreatic rough ER microsomes. (A) Electron micrograph of dog pancreatic rough microsomes. Arrows indicate ribosomes at the surface of ER microsomal membranes. (B) Contact mode Atomic Force Microscopy micrograph of purified ER microsomes fused on mica. Scale bars are 1 μm.

McCracken and Brodsky 2005; Meusser et al. 2005) and finally (3) the ER signaling machinery (Wu and Kaufman 2006; Wu et al. 2006; Zhang and Kaufman 2006a,b).

2.2.1. The ER quality control machinery

In the past 15 years, the understanding of the mechanisms regulating protein productive folding in the lumen of the ER has considerably increased. Indeed, besides the growing number of chaperones or post-translational modification enzymes acting on newly synthesized proteins prior to their export, several specific mechanisms have been described. Amongst them, the hypothesis of a quality control system, verifying the correct conformation of the proteins to be exported has been raised. This has been specifically confirmed for glycoproteins which constitute the main proportion of secretory proteins (Kukuruzinska and Lennon 1998; Helenius and Aebi 2001).

In particular, the glycoprotein ER quality control machinery includes several proteins whose coordinated and synergistic functions lead to the selective export of correctly folded glycoproteins (Chevet et al. 2001; Jakob et al. 2001; Ellgaard and Helenius 2003; Trombetta and Parodi 2003; Kleizen and Braakman 2004; Bukau et al. 2006; Wu et al. 2006). Amongst these proteins, calnexin, which binds to monoglucosylated glycoporteins, represents a major component of this system. Indeed, the calnexin cycle (Trombetta and Helenius 1998; Schrag et al. 2003; Helenius and Aebi 2004) has been extensively studied to deduce the mechanisms which regulate the selection and sorting of misfolded glycoproteins. This occurs via the prolonged time of association of constituents of the calnexin cycle with misfolded glycoproteins which in turn coincides with their targeting to ER associated degradation (see Section 2.2.2) (Cabral et al. 2001). On other major component of the calnexin cycle is the Uridine diphosphate (UDP)-glucose:glycoprotein glucosyl transferase (UGGT) which is another ER resident protein that has the ability to distinguish between correctly folded and misfolded proteins (Taylor et al. 2003, 2004). UGGT recognizes two features in misfolded proteins, namely the exposure of hydrophobic regions and the oligosaccharide moiety. UGGT is an important component of the ER quality control that recognizes the deepest glucose residue which may only be accessible on denatured protein. By contrast, Sakejima and colleagues (Sekijima et al. 2005) formalized a hypothesis for ER function in which competition of energetically destabilized non-glycoproteins (transthyretin) for the folding machineries of the ER (ER Assisted Folding; ERAF) or the degradation machineries (ERAD) would define its degree of secretion.

Although the "thermodynamic only" hypothesis has not yet been extended to the whole spectrum of secretory proteins, it remains an alternative of choice to the "quality control" model. However, an integrated mechanism combining both hypotheses may represent a third option, then most likely applicable to all secretory proteins.

2.2.2. The ER associated degradation machinery (ERAD)

The ERAD which functions in conjunction with the quality control machinery is an essential pathway for the clearance of proteins accumulated in the lumen of the ER. Indeed, a proteasome dependent machinery has been identified to degrade misfolded proteins accumulated in the ER (Jarosch et al. 2003; McCracken and Brodsky 2005). This can occur after retrotranslocation, eventually deglycosylation, and ubiquitination of misfolded proteins (Jarosch et al. 2003; McCracken and Brodsky 2005). The first step consists in a recognition process which is carried out in the lumen of the ER and again has been very well described for glycoproteins. Indeed, similar to entry into the calnexin cycle for productive folding, entry to ERAD appears to be based on a sugar code for misfolded glycoproteins with the former glucose based and the latter mannose based (Cabral et al. 2001; Schrag et al. 2003). The mannose based code involves at least two resident proteins, ER α1,2-mannosidase I and its paralogue ER degradation enhancing α-mannosidase-like protein (EDEM) (Hebert et al. 2005). The second step includes a retrotranslocation process through the translocon (Bonifacino and Weissman 1998; McCracken and Brodsky 2005; Momoi 2006) while the fourth step occurs in the cytosol (after deglycosylation by N-glycanase, in the cytosol) and consists in the ubiquitination of the retrotranslocated misfolded chains through their association with the AAA ATPase p97/VCP and its partners Derl (Ye et al. 2004; Oda et al. 2006), Dorfin (Huang et al. 2006) and Ataxin3 (Doss-Pepe et al. 2003; Boeddrich et al. 2006). Finally, the polyubiquitinated proteins are recognized and proteolysed by the proteasome machinery.

2.2.3. *The ER sub-systems*

The molecular machines described above constitute therefore the bases for an extended version of the ER system, being itself a sub-system of the cell system and containing several sub-systems represented by the molecular machines described above. At the structural level, it has been possible to visualize the translation component of the ER namely the polysomes associated to the cytosolic surface of the ER (Figure 1, arrows). In addition, the atomic force microscopy analysis of dog pancreatic rough ER microsomes has revealed a structural heterogeneity of the ER membrane, most likely indicating other types of functionally distinct sub-domains of this compartment (Nguyen, Milhiet and Chevet unpublished observations; Figure 1B). Together, these observations lead to the hypothesis that the ER could be constituted of functionally and physically coherent sub-systems functioning in a coordinated fashion to achieve ER functions in the cell and schematized in Figure 2. These observations are compatible with the system representation of the ER and by extension of every subcellular compartment. In addition, as a system formed by several sub-systems, one must predict an additional sub-system in charge of (i) detecting any malfunction of the ER and (ii) coordinating the above-mentioned molecular machines to compensate for these malfunctions. This was indeed the case with the discovery of an ER specific signaling response to the stress generated upon accumulation of misfolded proteins within its lumen.

3. SIGNALING IN THE ENDOPLASMIC RETICULUM – A SYSTEMS APPROACH

Several factors, either endogenous or exogenous, can affect the proper function of the molecular machineries of the ER system described above. These include for instance nutriment privation, pathogen infection, gene mutation, exposure to chemical, oxidative stress and ischemia. Diseases, such as cystic fibrosis (Jarosch et al. 2003; Nanua et al. 2006), emphysema/alpha1-antitrypsin mutations (Chevet et al. 1999a; Nadanaka et al. 2004; Hidvegi et al. 2005; Oda et al. 2006), Charcot-Marie Tooth disease (Dickson et al. 2002), have also been associated with ER dysfunction. These perturbations lead to an accumulation of misfolded protein within the ER which will subsequently results in a general disorder, or stress, of the ER system. This stress has an important impact, not only on the ER sub-system itself, but also on the cell

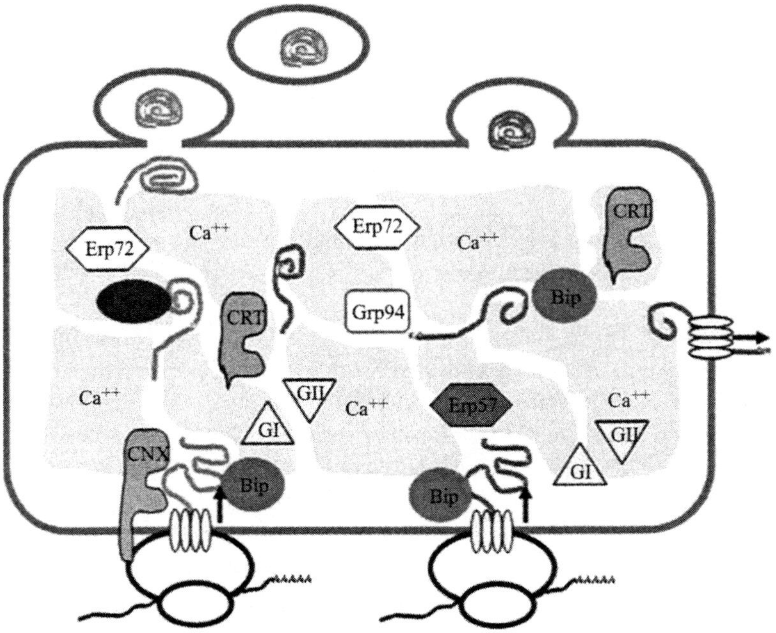

Figure 2. "Pre-system" representation of the ER. Schematic representation of the endoplasmic reticulum including various molecular machines such as protein synthesis and translocation of newly synthesized proteins (from the ribosome through the translocation channel), protein folding and post-translational modification in the lumen of the ER (presence of numerous chaperones and PTM enzymes) and finally either export to later compartment or degradation.

super system because it disrupts important function such as the protein export. If the ER homeostasis is not rapidly restored, dramatic consequences for the cell, such as apoptosis, can occur. Therefore, organisms have developed specific strategies to adapt to this form of stress. This adaptive response, called the Unfolded Protein Response (UPR), has for goal to attenuate the translation of newly synthesized proteins in order to limit the overload of the ER machineries and to up-regulate the expression of genes implicated in ER functions such as protein folding (chaperones, oxidoreductases) and degradation of misfolded proteins (EDEM). The basic components of the UPR were first described in yeast with the identification of IRE, a type-1 transmembrane protein kinase (Mori et al. 1993). This protein senses the stress within the ER and transduces the signal across the ER membrane to initiate the UPR. Whereas the yeast UPR is an *a priori* simple linear pathway, the mammalian UPR constitutes an integrated and complex signaling network. Eukaryotic cell have conserved the essential IRE1 mediated UPR signaling but two additional proximal UPR transducers were also identified: PERK (double-stranded RNA-activated protein kinase like ER kinase) (Shi et al. 1998; Harding et al. 1999) and ATF6 (activating transcription factor 6) (Haze et al. 1999).

3.1. Recognition of Misfolded Proteins

In non stressed cells, the luminal domain of IRE1, PERK and ATF6 is thought to associate with the protein BiP to inhibit their activity by preventing PERK or IRE1 oligomerization (Bertolotti et al. 2000) or maintaining ATF-6 in the ER (Sommer and Jarosch 2002; Shen et al. 2005a). BiP is an abundant ER chaperone with an ATPase activity at its N-terminal. In the ATP-bound form, BiP can bind its substrate with a low affinity. Misfolded proteins accumulating in the ER, stimulate BiP ATPase activity to generate the ADP-bound state of the protein that has a high affinity for the hydrophobic regions exposed in misfolded protein (Hendershot et al. 1996). Under these conditions, BiP preferentially binds to unfolded proteins and is titrated away from IRE1, PERK and ATF6. Dissociation of BiP from the luminal domain of IRE1 and PERK induces their Oligomerization, thus leading to their autophosphorylation and activation. In contrast, ER stress leads to ATF6 transport and processing to the Golgi, thus releasing an active transcription factor. The crystallization of the luminal domain of the yeast IRE1 also revealed that IRE1 (i) exists as dimmers under basal condition and (ii) can directly recognize misfolded proteins peptides in a manner similar to that of histocompatibility complex (MHC; (Credle et al. 2005)). Indeed, the luminal domain of IRE1 resembles to the peptide-binding domain of the MHCs and point mutations in this region of IRE1 impair its activity. It was proposed that direct binding of unfolded proteins changes the conformation of IRE1 oligomers and arranges the kinase domain optimally for autophosphorylation to initiate the UPR (Credle et al. 2005).

After activation of the ER stress sensors, IRE1, PERK and ATF6, distinct signaling pathways are induced to regulate ER chaperones expression, degradation of protein through the ER associated protein degradation (ERAD) and expansion of the ER. If these events fail to restore ER homeostasis, Ca2+ is released to activate apoptotic signaling pathways.

3.2. IRE1/XBP1 Signaling Axis

IRE1 is an 110kDa transmembrane protein containing an ER luminal dimerization domain and cytosolic kinase and RNase domains. The topology and kinase activity of IRE1 suggest that it functions similarly to plasma membrane receptor that respond to external ligands. When misfolded protein accumulates in the ER, IRE1 oligomerizes and undergoes a subsequent trans-autophosphorylation to activate its RNase activity. The major substrate for IRE1 RNase was first identified in yeast as a mRNA encoding the bZIP transcription factor named Hac1p (Sidrauski and Walter, 1997). Activated IRE1 cleaves the 5′ and 3′ exon-intron junction in the HAC-1 mRNA removing a loop that inhibits its translation. This HAC1 mRNA splicing leads to the synthesis of a potent transcriptional factor Hac1p and activate the UPR (Gonzalez et al. 1999; Leber et al. 2004; Niwa et al. 2005). Hac1p binds to the three following promoter elements: (i) unfolded protein response element (UPREs) present in promoters of gene which encode ER chaperones, enzymes involved in maintaining proper oxidizing environment in the ER and components of the ER degradation machinery, (ii) the upstream repressing sequence (URS1) found in the promoters of early meiotic genes and genes involved in carbon and nitrogen utilization and (iii) a subtelomeric ATF/CREB GTA variant element.

In mammals, two yeast IRE1 isoforms were identified: IRE1α and IRE1β (Urano et al. 2000a; Calfon et al. 2002; Back et al. 2005). Whereas IRE1α is ubiquitous, IRE1β expression is restricted to the gut intestinal cells (Bertolotti et al. 2001). The RNAse domain is structurally similar to RNAseL. The functional metazoan homolog of the yeast Hac-1 is the X-box binding protein 1 (Xbp-1) (Urano et al. 2000a; Calfon et al. 2002). In opposite to HAC-1, XBP-1 slicing by active IRE1 removes an unconventional intron on XBP-1 mRNA causing a frame shift which allows the translation of a potent transcriptional activator (Calfon et al. 2002). Xbp-1 is then translocated to the nucleus where it activates various UPR target genes containing an X-box element such as PDI, EDEM, and UGGT (Yoshida et al. 2001). A role for IRE1 has also been proposed in the regulation of protein synthesis via its RNAse activity, by the cleavage of the 28S RNA (Iwawaki et al. 2001).

Moreover, it has been demonstrated that upon activation, IRE1 associates with the adaptor protein TRAF2 (Urano et al. 2000b), which in turn leads to the recruitment and activation of the protein kinases Ask1 and Jnk1 (Urano et al. 2000b; Sekine et al. 2006). In addition, upon activation IRE1 dissociates from the SH2/SH3 containing adaptor Nck thus leading to the activation of the ERK/MAPK pathway (Nguyen et al. 2004).

Finally, the phosphorylation of IRE1 has been shown to be down-regulated in yeast by the protein phosphatase Ptc2 (Welihinda et al. 1998; Valkonen et al. 2004) and more recently, a phospho-tyrosine dependent regulation of IRE1 signaling has been demonstrated through the potentiating role of the phospho-tyrosine phosphatase PTP-1B (Gu et al. 2004).

In summary, besides the IRE1 specific signaling events mediated through the unconventional splicing of XBP-1, this protein is regulated by and also participates to signaling events already described for other pathways (i.e. TRAF2 for TNF signaling (Lee and Lee 2002) and PTP-1B for insulin signaling (Kennedy and Ramachandran 2000; Cheng et al. 2002)), thus demonstrating the integrative IRE1 signaling.

3.3. PERK Signaling

PERK signaling is the most immediate response following disruption of the ER system. Similar to IRE1, PERK is a transmembrane serine/threonine kinase that oligomerizes once activated upon BiP dissociation from its luminal domain (Bertolotti et al. 2000). Activation of PERK has two major consequences. First, it phosphorylates the alpha subunit of eukaryotic translation factor 2 (eIF2a) which shuts down general protein translation (Harding et al. 1999). When eIF2a is phosphorylated, the formation of the ternary translation initiation complex eIF2/GTP/Met-tRNAi is prevented, leading to general attenuation of mRNA translation. However, approximately one-third of UPR-inducible gene transcription also requires phosphorylation of eIF2α. For instance, upon amino acid starvation, phosphorylation of eIF2α by PERK selectively promotes translation of ATF4 mRNA which will activates the transcription of gene involved in amino acid metabolism. Moreover, preferential translation of ATF4 through eIF2α induces CHOP transcription (Ma et al. 2002; Bi et al. 2005), showing that eIF2α has a bimodal function in the UPR: decreasing general translation and increasing expression of URP-inducible gene such as ATF4 (Ma et al. 2002; Bi et al. 2005). It was also demonstrated that the SH2/SH3 containing adaptor protein Nck had a regulatory role on the PERK-mediated phosphorylation of eIF2α (Kebache et al. 2004).

Secondly, active PERK phosphorylates the bZIP Cap'n Collar transcription factor Nrf2 (Cullinan et al. 2003; Cullinan and Diehl 2004, 2006). In unstressed cell, Nrf2 is maintained in the cytoplasm by an interaction with the cytoskeletal anchor Keap1 (Cullinan et al. 2003). Upon ER stress, PERK-dependent phosphorylation of Nrf2 triggers dissociation of the complex and nuclear localization of Nrf2 (Cullinan et al. 2003; Cullinan and Diehl 2004, 2006). Nrf2 activates transcription of gene involved in phase II metabolism of xenobiotics through the antioxidant response element (ARE). ARE is found in the promoter of many genes such as glutathione-S-transferase and NAD(P)H:quinine oxidoreductase involved in protective response against oxidative stresses and which will promote cell survival (Cullinan et al. 2003; Cullinan and Diehl 2004, 2006). Interestingly, both PERK and IRE1 were shown to be processed by presenilin-1 and this processing to be part of a regulation mechanism (Katayama et al. 1999; Niwa et al. 1999; Yasuda et al. 2002) and also to associate with the chaperone Hsp90 as a stabilization process upon stress (Marcu et al. 2002).

3.4. ATF6 Signaling

ATF6 is a type-II transmembrane domain protein with a basic leucine zipper (bZIP) in its cytosolic domain (Haze et al. 1999). Two isoforms exist in mammals, namely ATF6α and ATF6β (Thuerauf et al. 2004). The luminal C-terminal domain of both ATF6 sense stresses occurring within the ER. Upon activation of the UPR, ATF6 translocates in the Golgi where it is processed by site-1 protease (S1P) and site-2 protease (S2P), to generate a cytosolic bZIP fragment (ATF6(p50)) which migrates to the nucleus (Haze et al. 1999). ATF6(p50) then bind to the 3′ half-side of the ER stress element (ERSE) and to the ATF/cAMP response element (CRE) to activate transcription. Binding of ATF6(p50) on ERSE requires NF-Y and regulates the expression of targets such as BiP, XBP 1, CHOP, $P58^{IPK}$ and Herp (Wang et al. 2000; Kokame et al. 2001). Moreover, ATF6 interacts with the transactivation domain of serum response factor (SRF). This interaction leads to the activation of the atrial natriuretic factor (ANF) (Zhu et al. 1997; Thuerauf et al. 1998). Finally, when forming a complex with transcription factor sterol response element binding protein 2 (SREBP2), ATF6 inhibits lipogenesis (Ye et al. 2000; Zeng et al. 2004).

3.5. ER Stress Signaling Proximal Networks and the ER System

Using literature-based knowledge, we have therefore been able to generate a functional interaction network of these three ER stress proximal sensors (Figure 3; **see colour insert**). This network emanating from the three ER stress proximal sensors IRE1, PERK and ATF-6 indicates (i) the nature of the functional partner (kinase, adaptor, transcription factor, protease, chaperone, RNA; inhibitor, activator or target) and (ii) the nature of the interaction (direct or indirect; association or dissociation upon stress).

Thus far no systematic analysis of metazoan ER stress proximal interaction networks has been carried out using large scale experimental approaches. Indeed, although microarray experiments were performed either using mouse embryonic fibroblasts deficient for the ER stress proximal sensors (e.g. Harding et al. 2001; Harding et al. 2003) or the corresponding *C. elegans* strains (either by gene deletion or RNAi) (Shen et al. 2005b). In *S. cerevisiae*, one major attempt was to identify alternative IRE1 substrates besides Hac1 mRNA but this lead to the demonstration that in the experimental conditions used, Hac1 represented the only substrate for IRE1 RNAse activity (Niwa et al. 2005). Recently, Hollien and Weissman reported, using a microarray-based approach, that the murine IRE1 RNAse also mediates the rapid degradation of a specific subset of mRNAs, thus extending the spectrum of IRE1-mediated functions (Hollien and Weissman 2006). In addition, using the systematic analysis of synthetic lethality, Schuldiner and colleagues have proposed the first systems representation of the ER. This study, although providing a systematic genetic interaction map of ER components, was the first to clearly demonstrate an integrated/systems function of the ER (Conibear 2005; Schuldiner et al. 2005).

The functional interaction network presented in Figure 3 brings a new dimension in the understanding of ER stress signaling in the sense that it shows clearly that the three known ER stress proximal sensors are tightly linked. Thus far no direct physical interaction between IRE1, PERK and ATF-6 has been demonstrated, however, a secondary network, including primary targets, activators and inhibitors, can be clearly established (see Figure 3). The existence of a tight functional proximal network can also be confirmed by the experimental evidences suggesting the co-regulation of specific gene sets either in *C. elegans* (Shen et al. 2005b) or in mammals (Harding et al. 2001; Harding et al. 2003). Indeed, numerous gene sets involved in metabolism, calcium homeostasis, oxidative stress, protein degradation, cytoskeleton rearrangement or regulation of the secretory pathway (Shen et al. 2005b) were found to be regulated downstream of at least two of the three ER stress proximal sensors.

Together, the information provided by our representation (Figure 3), although, it does not address time and signal intensity issues, provides a significant piece of information to the integrated and coordinated functions of the three ER stress proximal sensors IRE1, PERK and ATF-6. In addition, as some of the primary, secondary and even tertiary targets of these ER stress proximal sensors were shown to be involved in the regulation of specific ER functions, the impact of ER stress signaling on the ER system can be recognized as an essential component of this system. Indeed, a significant proportion of ER stress target genes are known to encode

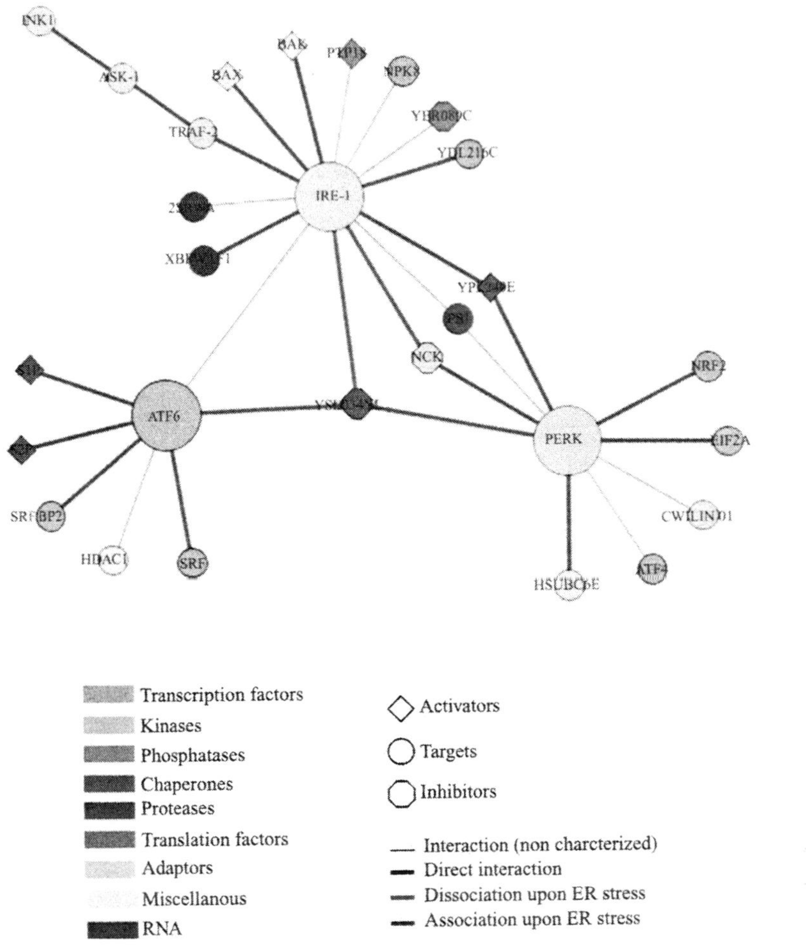

Figure 3. Interaction networks of ER stress proximal sensors. A literature-based reconstruction of IRE1, PERK and ATF6 proximal interaction networks was carried out using Cytoscape v2.0 (www.cytoscape.org/). Nodes represent different functional species based on a color code (including transcription factors, kinases, adaptors, phosphatases, chaperones, proteases, translation factors, RNA and miscellanous), and are classified into three families (activators, targets, inhibitors) of respectively PERK, IRE1 and ATF-6. Edges are characteristic of an interaction, either yet uncharacterized (thin) or characterized (thick). Finally, blue edges are significant of an association upon ER stress and red edges stand for an ER stress induced dissociation.

proteins involved in the regulation of ER function. This is particularly the case for gene products involved in (i) folding with proteins such as BiP, Grp94, or PDI, (ii) ER quality control with proteins such as UGGT or calnexin under certain circumstances (Delom et al. 2006), (iii) ERAD with EDEM or p97 and finally (iv) protein import and export form the ER with genes products such as Sec23 or ERGIC53.

In addition, it was demonstrated that proximal signaling events occurring upon ER stress may also have consequences on either ER stress downstream events such

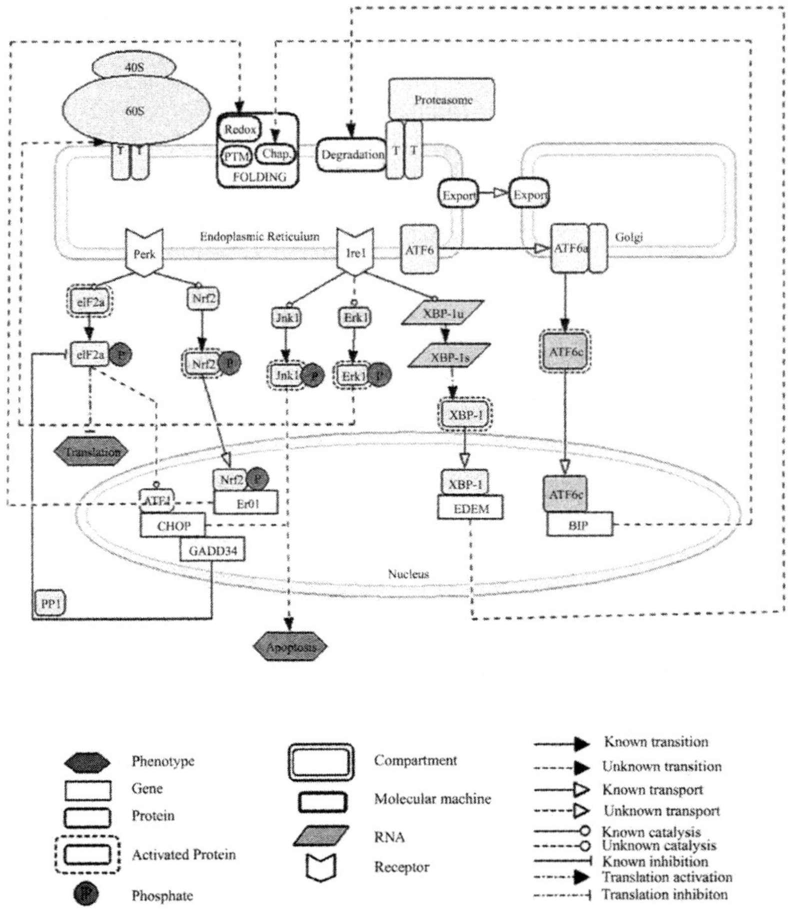

Figure 4. ER stress signaling map. This map was created using CellDesigner ver. 2.0 (http://www.systems-biology.org/002/). A total of 3 compartments, 37 reactions and 37 species were included. 40S: 40S ribosomal subunit, 60S: 60S ribosomal subunit; T: translocon; PTM: post-translational modification; PERK: PKR-like endoplasmic reticulum kinase; IRE1: inositol requiring enzyme; ATF-6: activating transcription factor 6; ATF6a: activating transcription factor 6 activated; ATF-6c: activating transcription factor 6 cytosolic; XBP-1: X-box binding protein 1; Jnk-1: c-Jun N-terminal kinase; Erk-1: Extracellular regulated kinase 1; Bip: Immunoglobulin heavy chain binding-protein ; EDEM: ER degradation enhancer, mannosidase alpha like 1; GADD34: growth arrest and DNA damage 34; CHOP: C/EBP homologous protein; PP1: protein phosphatase 1; ATF-4: activating transcription factor 4; Ero1: Oxidoreductin 1; eIF2alpha: translation elongation initiation factor 2 alpha; Nrf2: Nuclear factor erythroid 2, like 2; P: phosphate.

as the phosphorylation of CHOP by p38^{MAPK} which increases CHOP transcriptional activity (Wang and Ron 1996) or the IRE1 mediated activation of ERK1 (Nguyen et al. 2004) which may have specific substrates in the ER such as calnexin (Chevet et al. 1999b). Finally, ER stress signaling pathways also induce specific regulatory loops to shut down their own activation in a time-dependent fashion. The best example is provided by the PERK signaling-induced transcription of GADD34 which is in turn translated into a cofactor of the protein phosphatase PP1. This complex promotes the dephosphorylation of eIF2α, thus reinitiating protein synthesis (Novoa et al. 2001; Kojima et al. 2003). The integration of the above mentioned information allowed us to constitute a novel schematic representation of the ER system using ER stress signaling as the integration component of all the ER sub-systems. We manually constructed a comprehensive pathway map for ER stress signaling (Figure 4) based on published scientific papers. The map includes IRE1 signaling towards the MAPK/SAPK pathway, IRE1-mediated XBP-1 splicing and the downstream transcriptional activation of specific target genes; PERK signaling towards eIF2α and Nrf2 with the subsequent transcriptional activations and feed-back loops and finally, the transport of ATF-6 to the Golgi apparatus followed by its cleavage, the release of the cytosolic domain which then translocates to the nucleus to regulate the transcription of specific target genes. The map was created using CellDesigner (http://celldesigner.org/), a software package that enables users to describe molecular interactions using a well-defined and consistent graphical notation (Funahashi 2003; Kitano 2003). The map is based on the molecular interactions documented in 99 articles accessible from PubMed (see the list of references for ER stress Pathway Map). It comprises 3 compartments, 37 reactions and 37 species. A "species" is a term defined by SBML as "an entity that takes part in reactions" and it is used to distinguish the different states that are caused by enzymatic modification, association, dissociation, and translocation. The species shown on the ER stress map can be categorized as follows: 18 proteins, 1 ion, 3 oligomers, 6 genes and 2 RNAs.

This scheme represents the regulatory events governed by ER signaling pathways which occur mainly in the ER and the nucleus and, to a lesser extend, in the Golgi apparatus. It indicates also some phenotypic aspects such as the induction of apoptosis under prolonged ER stress which most likely represents another important aspect of ER stress signaling but at the sub-cellular level.

4. ER SIGNALING AND APOPTOSIS: INTEGRATION OF OTHER SUB-CELLULAR SYSTEMS

When cell are exposed to prolonged or excessive ER stress, signaling for the ER also has the potential to produce a cell death signals. Indeed, under those circumstances damaged cells are committed to cell death, which is mediated by ATF4 and ATF6, as well as activation of the JNK/AP-1/Gadd153/CHOP-signaling pathway. Gadd153/CHOP suppresses activation of Bcl-2 and NF-kappaB. UPR-mediated cell survival or cell death is regulated by the balance of GRP78/Bip and Gadd153/CHOP expression, which is co-regulated by NF-kappaB in accordance with the magnitude of ER stress (Paschen 2003; Rutkowski and Kaufman 2004; Schroder and Kaufman 2005; Xu et al. 2005; Lindholm et al. 2006; Zhang and Kaufman 2006b).

In addition, recently IRE1 signaling has been shown to depend on its association with the protein Bax and Bak (Hetz et al. 2006). These two proteins are of particular interest regarding the activation of mitochondrial-dependent pro-apoptotic pathways. In addition recently, it has been demonstrated that ER stress signaling also led to the formation of specific complexes including the ER quality control chaperone calnexin and the caspase-8 substrate Bap31 (Delom et al. 2006). The cleavage product of Bap31 was shown to induce mitochondrial changes leading to the release of cytochrome c and therefore to pro-apoptotic mechanisms (Ng et al. 1997; Nguyen et al. 2000; Breckenridge et al. 2003). Finally, it has been demonstrated that calcium efflux from the ER occurring upon prolonged stress effect the mitochondria to undergo pro-apoptotic changes. Together with the specific activation of signaling pathways regulating the balance of survival versus death which corresponds to the communication between ER and nucleus, the above mentioned results indicate also a strong communication link between ER and mitochondria. Consequently, we may in the future consider not only the ER as a self controlled system (which was the objective of this review) but also the constituent of an organellar communication network itself integrated within the cell system.

5. CONCLUSION AND PERSPECTIVES

The future understanding of ER signaling resides in three major steps. First, the ER resident signaling machines should be exhaustively defined and the identification of proximal regulators carried out. Second, as indicated in Section 4, the communication network between the ER and other organelles such as the nucleus, the mictochondrion, the Golgi apparatus, but also peroxisomes, endosomes and plasma membrane must be clearly established. Third, an increasing number of publications indicates strong biological links between ER stress signaling pathways and known signaling pathways such as TNF Receptor or insulin signaling. In addition, intermediates of these signaling pathways were shown to directly impact on ER stress signaling (e.g. TRAF2 and PTP-1B). This suggests that ER stress signaling pathways may represent significant relays in global cellular signaling and consequently may be essential for appropriate cellular functions. The use of Cell Designer and the generation of standardized signaling map will certainly help in the creation of integrative maps which will on the long term allow the characterization of a new representation of ER stress signaling in the context of the whole cell signaling pathways.

Besides literature searches, another aspect of the analysis of ER stress signaling pathways should also be undertaken. Indeed, approaches combining the use of genetic models with systematic analyses should be developed. Although some of these analyses were already published (in *S. cerevisiae* (Schuldiner et al. 2005) or in *C. elegans* (Shen et al. 2005b)) and some other are currently being carried out (Caruso and Chevet unpublished), these analyses should also bring quantitative information on the ER stress signaling pathways such as time and signal intensity. The integration of these data will then provide solid bases for a better characterization of the ER stress signaling pathways but also of the ER as a biological system per se. Finally, this analytical and integrative strategy could be applied to other cellular sub-systems either independently or in conjunction with the ER system to extend global cellular understanding.

ACKNOWLEDGEMENTS

We thank the Chevet laboratory for critical reading. We apologize for those whose work was not cited due to space limitation. MEC and EC are recipients of salary support from the Fonds de Recherche en Santé du Québec.

REFERENCES

Aderem, A. (2005) Systems biology: its practice and challenges. Cell *121*, 511–513.

Back, S.H., Schroder, M., Lee, K., Zhang, K. and Kaufman, R.J. (2005) ER stress signaling by regulated splicing: IRE1/HAC1/XBP1. Methods *35*, 395–416.

Bar-Nun, S. (2005)The role of p97/Cdc48p in endoplasmic reticulum-associated degradation: from the immune system to yeast. Curr. Top. Microbiol. Immunol. *300*, 95–125.

Benner, S.A. and Sismour, A.M. (2005) Synthetic biology. Nat. Rev. Genet. *6*, 533–543.

Bertolotti, A., Wang, X., Novoa, I., Jungreis, R., Schlessinger, K., Cho, J.H., West, A.B. and Ron, D. (2001) Increased sensitivity to dextran sodium sulfate colitis in IRE1beta-deficient mice. J. Clin. Invest. *107*, 585–593.

Bertolotti, A., Zhang, Y., Hendershot, L.M., Harding, H.P. and Ron, D. (2000) Dynamic interaction of BiP and ER stress transducers in the unfolded-protein response. Nat. Cell Biol. *2*, 326–332.

Bi, M., Naczki, C., Koritzinsky, M., Fels, D., Blais, J., Hu, N., Harding, H., Novoa, I., Varia, M., Raleigh, J., Scheuner, D., Kaufman, R.J., Bell, J., Ron, D., Wouters, B.G. and Koumenis, C. (2005) ER stress-regulated translation increases tolerance to extreme hypoxia and promotes tumor growth. EMBO J. *24*, 3470–3481.

Blobel, G. (1995) Unidirectional and bidirectional protein traffic across membranes. Cold Spring Harb. Symp. Quant. Biol. *60*, 1–10.

Boeddrich, A., Gaumer, S., Haacke, A., Tzvetkov, N., Albrecht, M., Evert, B.O., Muller, E.C., Lurz, R., Breuer, P., Schugardt, N., Plassmann, S., Xu, K., Warrick, J.M., Suopanki, J., Wullner, U., Frank, R., Hartl, U.F., Bonini, N.M. and Wanker, E.E. (2006) An arginine/lysine-rich motif is crucial for VCP/p97-mediated modulation of ataxin-3 fibrillogenesis. EMBO J. *25*, 1547–1558.

Bonifacino, J.S. and Weissman, A.M. (1998) Ubiquitin and the control of protein fate in the secretory and endocytic pathways. Annu. Rev. Cell Dev. Biol. *14*, 19–57.

Breckenridge, D.G., Stojanovic, M., Marcellus, R.C. and Shore, G.C. (2003) Caspase cleavage product of BAP31 induces mitochondrial fission through endoplasmic reticulum calcium signals, enhancing cytochrome c release to the cytosol. J. Cell Biol. *160*, 1115–1127.

Bukau, B., Weissman, J. and Horwich, A. (2006) Molecular chaperones and protein quality control. Cell *125*, 443–451.

Cabral, C.M., Liu, Y. and Sifers, R.N. (2001) Dissecting glycoprotein quality control in the secretory pathway. Trends Biochem. Sci. *26*, 619–624.

Calfon, M., Zeng, H., Urano, F., Till, J.H., Hubbard, S.R., Harding, H.P., Clark, S.G. and Ron, D. (2002) IRE1 couples endoplasmic reticulum load to secretory capacity by processing the XBP-1 mRNA. Nature *415*, 92–96.

Campbell, P.N. and Blobel, G. (1976) The role of organelles in the chemical modification of the primary translation products of secretory proteins. FEBS Lett. *72*, 215–226.

Cheng, A., Dube, N., Gu, F. and Tremblay, M.L. (2002) Coordinated action of protein tyrosine phosphatases in insulin signal transduction. Eur. J. Biochem. *269*, 1050–1059.

Chevet, E., Cameron, P.H., Pelletier, M.F., Thomas, D.Y. and Bergeron, J.J. (2001) The endoplasmic reticulum: integration of protein folding, quality control, signaling and degradation. Curr. Opin. Struct. Biol. *11*, 120–124.

Chevet, E., Jakob, C.A., Thomas, D.Y. and Bergeron, J.J. (1999a). Calnexin family members as modulators of genetic diseases. Semin Cell Dev. Biol. *10*, 473–480.

Chevet, E., Wong, H.N., Gerber, D., Cochet, C., Fazel, A., Cameron, P.H., Gushue, J.N., Thomas, D.Y. and Bergeron, J.J. (1999b). Phosphorylation by CK2 and MAPK enhances calnexin association with ribosomes. EMBO J. *18*, 3655–3666.

Conibear, E. (2005) An E-MAP of the ER. Cell *123*, 366–368.

Conn, P.M., Knollman, P.E., Brothers, S.P. and Janovick, J.A. (2006) Protein folding as post-translational regulation: evolution of a mechanism for controlled plasma membrane expression of a GPCR. Mol. Endocrinol. *12*, 161–171

Credle, J.J., Finer-Moore, J.S., Papa, F.R., Stroud, R.M. and Walter, P. (2005) On the mechanism of sensing unfolded protein in the endoplasmic reticulum. Proc. Natl Acad. Sci. U.S.A. *102*, 18773–18784.

Cullinan, S.B. and Diehl, J.A. (2004) PERK-dependent activation of Nrf2 contributes to redox homeostasis and cell survival following endoplasmic reticulum stress. J. Biol. Chem. *279*, 20108–20117.

Cullinan, S.B. and Diehl, J.A. (2006) Coordination of ER and oxidative stress signaling: the PERK/Nrf2 signaling pathway. Int. J. Biochem. Cell Biol. *38*, 317–332.

Cullinan, S.B., Zhang, D., Hannink, M., Arvisais, E., Kaufman, R.J. and Diehl, J.A. (2003) Nrf^2 is a direct PERK substrate and effector of PERK-dependent cell survival. Mol. Cell. Biol. *23*, 7198–7209.

Delom, F., Emadali, A., Cocolakis, E., Lebrun, J.J., Nantel, A. and Chevet, E. (2007) Calnexin-dependent regulation of tunicamycin-induced apoptosis in breast carcinoma MCF-7 cells. Cell Death Differ. *14*(3), 586–596.

Dickson, K.M., Bergeron, J.J., Shames, I., Colby, J., Nguyen, D.T., Chevet, E., Thomas, D.Y. and Snipes, G.J. (2002) Association of calnexin with mutant peripheral myelin protein-22 *ex vivo*: a basis for "gain-of-function" ER diseases. Proc. Natl Acad. Sci. USA *99*, 9852–9857.

Doss-Pepe, E.W., Stenroos, E.S., Johnson, W.G. and Madura, K. (2003) Ataxin-3 interactions with rad23 and valosin-containing protein and its associations with ubiquitin chains and the proteasome are consistent with a role in ubiquitin-mediated proteolysis. Mol. Cell Biol. *23*, 6469–6483.

Ellgaard, L. and Helenius, A. (2003) Quality control in the endoplasmic reticulum. Nat. Rev. Mol. Cell Biol. *4*, 181–191.

Freedman, R.B. (1989) Protein disulfide isomerase: multiple roles in the modification of nascent secretory proteins. Cell *57*, 1069–1072.

Friboulet, A. and Thomas, D. (2005) Systems Biology-an interdisciplinary approach. Biosens. Bioelectron. *20*, 2404–2407.

Funahashi, A. (2003) [The ERATO Systems Biology Workbench and Systems Biology Markup Language: an integrated environment and standardization for systems biology]. Tanpakushitsu Kakusan Koso *48*, 810–816.

Gonzalez, T.N., Sidrauski, C., Dorfler, S. and Walter, P. (1999) Mechanism of non-spliceosomal mRNA splicing in the unfolded protein response pathway. EMBO J. *18*, 3119–3132.

Gu, F., Nguyen, D.T., Stuible, M., Dube, N., Tremblay, M.L. and Chevet, E. (2004) Protein-tyrosine phosphatase 1B potentiates IRE1 signaling during endoplasmic reticulum stress. J. Biol. Chem. *279*, 49689–49693.

Hammer, G.L., Sinclair, T.R., Chapman, S.C. and van Oosterom, E. (2004) On systems thinking, systems biology, and the in silico plant. Plant Physiol. *134*, 909–911.

Harding, H.P., Zeng, H., Zhang, Y., Jungries, R., Chung, P., Plesken, H., Sabatini, D.D. and Ron, D. (2001) Diabetes mellitus and exocrine pancreatic dysfunction in perk-/- mice reveals a role for translational control in secretory cell survival. Mol. Cell. *7*, 1153–1163.

Harding, H.P., Zhang, Y. and Ron, D. (1999) Protein translation and folding are coupled by an endoplasmic-reticulum-resident kinase. Nature *397*, 271–274.

Harding, H.P., Zhang, Y., Zeng, H., Novoa, I., Lu, P.D., Calfon, M., Sadri, N., Yun, C., Popko, B., Paules, R., Stojdl, D.F., Bell, J.C., Hettmann, T., Leiden, J.M. and Ron, D. (2003) An integrated stress response regulates amino acid metabolism and resistance to oxidative stress. Mol Cell *11*, 619–633.

Haze, K., Yoshida, H., Yanagi, H., Yura, T. and Mori, K. (1999) Mammalian transcription factor ATF6 is synthesized as a transmembrane protein and activated by proteolysis in response to endoplasmic reticulum stress. Mol. Biol. Cell *10*, 3787–3799.

Hebert, D.N., Garman, S.C. and Molinari, M. (2005) The glycan code of the endoplasmic reticulum: asparagine-linked carbohydrates as protein maturation and quality-control tags. Trends Cell. Biol. *15*, 364–370.

Helenius, A. and Aebi, M. (2001) Intracellular functions of N-linked glycans. Science *291*, 2364–2369.

Helenius, A. and Aebi, M. (2004) Roles of N-linked glycans in the endoplasmic reticulum. Annu. Rev. Biochem. *73*, 1019–1049.

Hendershot, L., Wei, J., Gaut, J., Melnick, J., Aviel, S. and Argon, Y. (1996) Inhibition of immunoglobulin folding and secretion by dominant negative BiP ATPase mutants. Proc. Natl Acad. Sci. USA *93*, 5269–5274.

Hetz, C., Bernasconi, P., Fisher, J., Lee, A.H., Bassik, M.C., Antonsson, B., Brandt, G. S., Iwakoshi, N.N., Schinzel, A., Glimcher, L.H. and Korsmeyer, S.J. (2006) Proapoptotic BAX and BAK modulate the unfolded protein response by a direct interaction with IRE1alpha. Science *312*, 572–576.

Hidvegi, T., Schmidt, B.Z., Hale, P. and Perlmutter, D.H. (2005) Accumulation of mutant alpha1-antitrypsin Z in the endoplasmic reticulum activates caspases-4 and -12, NFkappaB, and BAP31 but not the unfolded protein response. J. Biol. Chem. *280*, 39002–39015.

Hollien, J. and Weissman, J.S. (2006) Decay of endoplasmic reticulum-localized mRNAs during the unfolded protein response. Science *313*, 52–53.

Huang, Y., Niwa, J., Sobue, G. and Breitwieser, G.E. (2006) Calcium-sensing receptor ubiquitination and degradation mediated by the E3 ubiquitin ligase dorfin. J. Biol. Chem. *281*, 11610–11617.

Iwawaki, T., Hosoda, A., Okuda, T., Kamigori, Y., Nomura-Furuwatari, C., Kimata, Y., Tsuru, A. and Kohno, K. (2001) Translational control by the ER transmembrane kinase/ribonuclease IRE1 under ER stress. Nat. Cell Biol. *3*, 158–164.

Jakob, C.A., Chevet, E., Thomas, D.Y. and Bergeron, J.J. (2001) Lectins of the ER quality control machinery. Results Probl. Cell Differ. *33*, 1–17.

Jarosch, E., Lenk, U. and Sommer, T. (2003) Endoplasmic reticulum-associated protein degradation. Int. Rev. Cytol. *223*, 39–81.

Katayama, T., Imaizumi, K., Sato, N., Miyoshi, K., Kudo, T., Hitomi, J., Morihara, T., Yoneda, T., Gomi, F., Mori, Y., Nakano, Y., Takeda, J., Tsuda, T., Itoyama, Y., Murayama, O., Takashima, A., St George-Hyslop, P., Takeda, M. and Tohyama, M. (1999) Presenilin-1 mutations downregulate the signalling pathway of the unfolded-protein response. Nat. Cell Biol. *1*, 479–485.

Kebache, S., Cardin, E., Nguyen, D.T., Chevet, E. and Larose, L. (2004) Nck-1 antagonizes the endoplasmic reticulum stress-induced inhibition of translation. J. Biol. Chem. *279*, 9662–9671.

Kennedy, B.P. and Ramachandran, C. (2000) Protein tyrosine phosphatase-1B in diabetes. Biochem. Pharmacol. *60*, 877–883.

Kitano, H. (2002a). Computational systems biology. Nature *420*, 206–210.

Kitano, H. (2002b). Looking beyond the details: a rise in system-oriented approaches in genetics and molecular biology. Curr. Genet. *41*, 1–10.

Kitano, H. (2002c). Systems biology: a brief overview. Science *295*, 1662–1664.

Kitano, H. (2003) [Introductions to systems biology]. Tanpakushitsu Kakusan Koso *48*, 789–793.

Kitano, H. (2005) International alliances for quantitative modeling in systems biology. Mol. Syst. Biol. *1*, 2005 0007.

Kleizen, B. and Braakman, I. (2004) Protein folding and quality control in the endoplasmic reticulum. Curr. Opin. Cell Biol. *16*, 343–349.

Kojima, E., Takeuchi, A., Haneda, M., Yagi, A., Hasegawa, T., Yamaki, K., Takeda, K., Akira, S., Shimokata, K. and Isobe, K. (2003) The function of GADD34 is a recovery from a shutoff of protein synthesis induced by ER stress: elucidation by GADD34-deficient mice. FASEB J. *17*, 1573–1575.

Kokame, K., Kato, H. and Miyata, T. (2001) Identification of ERSE-II, a new cis-acting element responsible for the ATF6-dependent mammalian unfolded protein response. J. Biol. Chem. *276*, 9199–9205.

Kukuruzinska, M.A. and Lennon, K. (1998) Protein N-glycosylation: molecular genetics and functional significance. Crit. Rev. Oral Biol. Med. *9*, 415–448.

Leber, J.H., Bernales, S. and Walter, P. (2004) IRE1-independent gain control of the unfolded protein response. PLoS Biol. *2*, E235.

Lee, N.K. and Lee, S.Y. (2002) Modulation of life and death by the tumor necrosis factor receptor-associated factors (TRAFs). J. Biochem. Mol. Biol. *35*, 61–66.

Lindholm, D., Wootz, H. and Korhonen, L. (2006) ER stress and neurodegenerative diseases. Cell Death Differ. *13*, 385–392.

Lingappa, V.R. and Blobel, G. (1980) Early events in the biosynthesis of secretory and membrane proteins: the signal hypothesis. Recent Prog. Horm. Res. *36*, 451–475.

Lyman, S.K. and Schekman, R. (1996) Polypeptide translocation machinery of the yeast endoplasmic reticulum. Experientia *52*, 1042–1049.

Ma, Y., Brewer, J.W., Diehl, J.A. and Hendershot, L.M. (2002) Two distinct stress signaling pathways converge upon the CHOP promoter during the mammalian unfolded protein response. J. Mol. Biol. *318*, 1351–1365.

Marcu, M.G., Doyle, M., Bertolotti, A., Ron, D., Hendershot, L. and Neckers, L. (2002) Heat shock protein 90 modulates the unfolded protein response by stabilizing IRE1alpha. Mol. Cell. Biol. *22*, 8506–8513.

McCracken, A.A. and Brodsky, J.L. (2005) Recognition and delivery of ERAD substrates to the proteasome and alternative paths for cell survival. Curr. Top Microbiol. Immunol. *300*, 17–40.

McDaniel, R. and Weiss, R. (2005) Advances in synthetic biology: on the path from prototypes to applications. Curr. Opin. Biotechnol. *16*, 476–483.

Meusser, B., Hirsch, C., Jarosch, E. and Sommer, T. (2005) ERAD: the long road to destruction. Nat, Cell, Biol, *7*, 766–772.

Momoi, T. (2006) Conformational diseases and ER stress-mediated cell death: apoptotic cell death and autophagic cell death. Curr. Mol. Med. *6*, 111–118.

Mori, K., Ma, W., Gething, M.J. and Sambrook, J. (1993) A transmembrane protein with a cdc2+/CDC28-related kinase activity is required for signaling from the ER to the nucleus. Cell *74*, 743–756.

Nadanaka, S., Yoshida, H., Kano, F., Murata, M. and Mori, K. (2004) Activation of mammalian unfolded protein response is compatible with the quality control system operating in the endoplasmic reticulum. Mol. Biol. Cell. *15*, 2537–2548.

Nanua, S., Sajjan, U., Keshavjee, S. and Hershenson, M.B. (2006) Absence of typical unfolded protein response in primary cultured cystic fibrosis airway epithelial cells. Biochem. Biophys. Res. Commun. *343*, 135–143.

Ng, F.W., Nguyen, M., Kwan, T., Branton, P.E., Nicholson, D.W., Cromlish, J.A. and Shore, G.C. (1997) p28 Bap31, a Bcl-2/Bcl-XL- and procaspase-8-associated protein in the endoplasmic reticulum. J. Cell Biol. *139*, 327–338.

Nguyen, M., Breckenridge, D.G., Ducret, A. and Shore, G.C. (2000) Caspase-resistant BAP31 inhibits fas-mediated apoptotic membrane fragmentation and release of cytochrome c from mitochondria. Mol. Cell. Biol. *20*, 6731–6740.

Nguyen, D.T., Kebache, S., Fazel, A., Wong, H.N., Jenna, S., Emadali, A., Lee, E.H., Bergeron, J.J., Kaufman, R.J., Larose, L. and Chevet, E. (2004) Nck-dependent activation of extracellular signal-regulated kinase-1 and regulation of cell survival during endoplasmic reticulum stress. Mol. Biol. Cell. *15*, 4248–4260.

Niwa, M., Patil, C.K., DeRisi, J. and Walter, P. (2005) Genome-scale approaches for discovering novel nonconventional splicing substrates of the Ire1 nuclease. Genome Biol. *6*, R3.

Niwa, M., Sidrauski, C., Kaufman, R.J. and Walter, P. (1999) A role for presenilin-1 in nuclear accumulation of Ire1 fragments and induction of the mammalian unfolded protein response. Cell *99*, 691–702.

Novoa, I., Zeng, H., Harding, H.P. and Ron, D. (2001) Feedback inhibition of the unfolded protein response by GADD34-mediated dephosphorylation of eIF2alpha. J. Cell Biol. *153*, 1011–1022.

Oda, Y., Okada, T., Yoshida, H., Kaufman, R.J., Nagata, K. and Mori, K. (2006) Derlin-2 and Derlin-3 are regulated by the mammalian unfolded protein response and are required for ER-associated degradation. J. Cell Biol. *172*, 383–393.

Palade, G.E. and Porter, K.R. (1954) Studies on the endoplasmic reticulum. I. Its identification in cells *in situ*. J. Exp. Med. *100*, 641–656.

Paschen, W. (2003) Shutdown of translation: lethal or protective? Unfolded protein response versus apoptosis. J. Cereb. Blood Flow Metab. *23*, 773–779.

Pawson, T. and Linding, R. (2005) Synthetic modular systems–reverse engineering of signal transduction. FEBS Lett. *579*, 1808–1814.

Price, B.D. (1992) Signalling across the endoplasmic reticulum membrane: potential mechanisms. Cell Signal *4*, 465–470.

Pryme, I.F. (1986) Compartmentation of the rough endoplasmic reticulum. Mol. Cell. Biochem. *71*, 3–18.

Quenneville, N.R. and Conibear, E. (2006) Toward the systems biology of vesicle transport. Traffic *7*, 761–768.

Rutkowski, D.T. and Kaufman, R.J. (2004) A trip to the ER: coping with stress. Trends Cell Biol. *14*, 20–28.

Schrag, J.D., Procopio, D.O., Cygler, M., Thomas, D.Y. and Bergeron, J.J. (2003) Lectin control of protein folding and sorting in the secretory pathway. Trends Biochem. Sci. *28*, 49–57.

Schroder, M. and Kaufman, R.J. (2005) ER stress and the unfolded protein response. Mutat. Res. *569*, 29–63.

Schuldiner, M., Collins, S.R., Thompson, N.J., Denic, V., Bhamidipati, A., Punna, T., Ihmels, J., Andrews, B., Boone, C., Greenblatt, J. F., Weissman, J.S. and Krogan, N.J. (2005) Exploration of the function and organization of the yeast early secretory pathway through an epistatic miniarray profile. Cell *123*, 507–519.

Sekijima, Y., Wiseman, R.L., Matteson, J., Hammarstrom, P., Miller, S.R., Sawkar, A. R., Balch, W.E. and Kelly, J.W. (2005) The biological and chemical basis for tissue-selective amyloid disease. Cell *121*, 73–85.

Sekine, Y., Takeda, K. and Ichijo, H. (2006) The ASK1-MAP kinase signaling in ER stress and neurodegenerative diseases. Curr. Mol. Med. *6*, 87–97.

Shen, X., Ellis, R.E., Sakaki, K. and Kaufman, R.J. (2005b). Genetic interactions due to constitutive and inducible gene regulation mediated by the unfolded protein response in *C. elegans*. PLoS Genet. *1*, e37.

Shen, J., Snapp, E.L., Lippincott-Schwartz, J. and Prywes, R. (2005a). Stable binding of ATF6 to BiP in the endoplasmic reticulum stress response. Mol. Cell Biol. *25*, 921–932.

Shi, Y., Vattem, K.M., Sood, R., An, J., Liang, J., Stramm, L. and Wek, R.C. (1998) Identification and characterization of pancreatic eukaryotic initiation factor 2 alpha-subunit kinase, PEK, involved in translational control. Mol. Cell Biol. *18*, 7499–7509.

Sidrauski, C. and Walter, P. (1997) The transmembrane kinase Ire1p is a site-specific endonuclease that initiates mRNA splicing in the unfolded protein response. Cell *90*, 1031–1039.

Sitia, R. and Braakman, I. (2003) Quality control in the endoplasmic reticulum protein factory. Nature *426*, 891–894.

Sommer, T. and Jarosch, E. (2002) BiP binding keeps ATF6 at bay. Dev. Cell *3*, 1–2.

Taylor, S.C., Ferguson, A.D., Bergeron, J.J. and Thomas, D.Y. (2004) The ER protein folding sensor UDP-glucose glycoprotein-glucosyltransferase modifies substrates distant to local changes in glycoprotein conformation. Nat. Struct. Mol. Biol. *11*, 128–134.

Taylor, S.C., Thibault, P., Tessier, D.C., Bergeron, J.J. and Thomas, D.Y. (2003) Glycopeptide specificity of the secretory protein folding sensor UDP-glucose glycoprotein:glucosyltransferase. EMBO Rep. *4*, 405–411.

Thuerauf, D.J., Arnold, N.D., Zechner, D., Hanford, D.S., DeMartin, K.M., McDonough, P.M., Prywes, R. and Glembotski, C.C. (1998) p38 Mitogen-activated protein kinase mediates the transcriptional induction of the atrial natriuretic factor gene through a serum response element. A potential role for the transcription factor ATF6. J. Biol. Chem. *273*, 20636–20643.

Thuerauf, D.J., Morrison, L. and Glembotski, C.C. (2004) Opposing roles for ATF6alpha and ATF6beta in endoplasmic reticulum stress response gene induction. J. Biol. Chem. *279*, 21078–21084.

Trombetta, E.S. and Helenius, A. (1998) Lectins as chaperones in glycoprotein folding. Curr. Opin. Struct. Biol. *8*, 587–592.

Trombetta, E.S. and Parodi, A.J. (2003) Quality control and protein folding in the secretory pathway. Annu. Rev. Cell Dev. Biol. *19*, 649–676.

Urano, F., Bertolotti, A. and Ron, D. (2000a). IRE1 and efferent signaling from the endoplasmic reticulum. J. Cell. Sci. *113*, 3697–3702.

Urano, F., Wang, X., Bertolotti, A., Zhang, Y., Chung, P., Harding, H.P. and Ron, D. (2000b). Coupling of stress in the ER to activation of JNK protein kinases by transmembrane protein kinase IRE1. Science *287*, 664–666.

Valkonen, M., Penttila, M. and Saloheimo, M. (2004) The ire1 and ptc2 genes involved in the unfolded protein response pathway in the filamentous fungus Trichoderma reesei. Mol Genet Genomics *272*, 443–451.

Vitale, A. and Ceriotti, A. (2004) Protein quality control mechanisms and protein storage in the endoplasmic reticulum. A conflict of interests? Plant Physiol. *136*, 3420–3426.

von Bertalanffy, L. (1968) General System Theory: Foundations, Development, Applications. (New York: George Braziller).

Wang, X.Z. and Ron, D. (1996) Stress-induced phosphorylation and activation of the transcription factor CHOP (GADD153) by p38 MAP Kinase. Science *272*, 1347–1349.

Wang, Y., Shen, J., Arenzana, N., Tirasophon, W., Kaufman, R.J. and Prywes, R. (2000) Activation of ATF6 and an ATF6 DNA binding site by the endoplasmic reticulum stress response. J. Biol. Chem. *275*, 27013–27020.

Welihinda, A.A., Tirasophon, W., Green, S.R. and Kaufman, R.J. (1998) Protein serine/threonine phosphatase Ptc2p negatively regulates the unfolded-protein response by dephosphorylating Ire1p kinase. Mol. Cell Biol. *18*, 1967–1977.

Wu, J. and Kaufman, R.J. (2006) From acute ER stress to physiological roles of the Unfolded Protein Response. Cell Death Differ. *13*, 374–384.

Wu, J.C., Liang, Z.Q. and Qin, Z.H. (2006) Quality control system of the endoplasmic reticulum and related diseases. Acta Biochim. Biophys. Sin. (Shanghai) *38*, 219–226.

Xu, C., Bailly-Maitre, B. and Reed, J.C. (2005) Endoplasmic reticulum stress: cell life and death decisions. J. Clin. Invest.115, 2656–2664.

Yasuda, Y., Kudo, T., Katayama, T., Imaizumi, K., Yatera, M., Okochi, M., Yamamori, H., Matsumoto, N., Kida, T., Fukumori, A., Okumura, M., Tohyama, M. and Takeda, M. (2002) FAD-linked presenilin-1 mutants impede translation regulation under ER stress. Biochem. Biophys. Res. Commun. *296*, 313–318.

Ye, J., Rawson, R.B., Komuro, R., Chen, X., Dave, U.P., Prywes, R., Brown, M.S. and Goldstein, J.L. (2000) ER stress induces cleavage of membrane-bound ATF6 by the same proteases that process SREBPs. Mol. Cell *6*, 1355–1364.

Ye, Y., Shibata, Y., Yun, C., Ron, D. and Rapoport, T.A. (2004) A membrane protein complex mediates retro-translocation from the ER lumen into the cytosol. Nature *429*, 841–847.

Yoshida, Y. (2003) A novel role for N-glycans in the ERAD system. J. Biochem. (Tokyo) *134*, 183–190.

Yoshida, H., Matsui, T., Yamamoto, A., Okada, T. and Mori, K. (2001) XBP1 mRNA is induced by ATF6 and spliced by IRE1 in response to ER stress to produce a highly active transcription factor. Cell *107*, 881–891.

Zeng, L., Lu, M., Mori, K., Luo, S., Lee, A.S., Zhu, Y. and Shyy, J.Y. (2004) ATF6 modulates SREBP2-mediated lipogenesis. EMBO J. *23*, 950–958.

Zhang, K. and Kaufman, R.J. (2006a). Protein folding in the endoplasmic reticulum and the unfolded protein response. Handb. Exp. Pharmacol. *172*, 69–91.

Zhang, K. and Kaufman, R.J. (2006b). The unfolded protein response: a stress signaling pathway critical for health and disease. Neurology *66*, S102–S109.

Zhu, C., Johansen, F.E. and Prywes, R. (1997) Interaction of ATF6 and serum response factor. Mol. Cell Biol. *17*, 4957–4966.

Corresponding author: Dr. Eric Chevet (eric.chevet@mcgill.ca)

SECTION 5

EMERGING TECHNOLOGIES IN PROTEOMICS

CHAPTER 14

SYSTEMS NANOBIOLOGY: FROM QUANTITATIVE SINGLE MOLECULE BIOPHYSICS TO MICROFLUIDIC-BASED SINGLE CELL ANALYSIS

JOERG MARTINI, WIBKE HELLMICH, DOMINIK GREIF, ANKE BECKER, THOMAS MERKLE, ROBERT ROS, ALEXANDRA ROS, KATJA TOENSING, and DARIO ANSELMETTI
Bielefeld University, Germany

Table of Contents

Abstract: Detailed and quantitative information about structure-function relation, concentrations and interaction kinetics of biological molecules and subcellular components is a key prerequisite to understand and model cellular organisation and temporal dynamics. In systems nanobiology, cellular processes are quantitatively investigated at the sensitivity level of single

E. Bertrand and M. Faupel (eds.), Subcellular Proteomics, 301–321.

molecules and cells. This approach provides direct access to biomolecular information without being statistically ensemble-averaged, their associated distribution functions, and possible subpopulations. Moreover at the single cell level, the interplay of regulated genomic information and proteomic variabilities can be investigated and attributed to functional peculiarities. These requirements necessitate the development of novel and ultrasensitive methods and instruments for single molecule detection, microscopy and spectroscopy for analysis without the need of amplification and preconcentration. In this chapter, we present three methodological applications that demonstrate how quantitative informations can be accessed that are representative for cellular processes or single cell analysis like gene expression regulation, intracellular protein translocation dynamics, and single cell protein fingerprinting. First, the interaction kinetics of transcriptionally regulated DNA–protein interaction can be quantitatively investigated with single molecule force spectroscopy allowing a molecular affinity ranking. Second, intracellular protein dynamics for a transcription regulator migrating form the nucleus to the cytoplasm can be quantitatively monitored by photoactivable GFP and two-photon laser scanning microscopy. And third, a microfluidic-based method for label-free single cell proteomics and fingerprinting and first label-free single cell electropherograms are presented which include the manipulation and steering of single cells in a microfluidic device.

1. INTRODUCTION

In systems nanobiology, cellular processes are quantitatively investigated down to the sensitivity level of single molecules and cells. Quantitative information about biological molecules and sub-cellular components, their molecular structure and interaction kinetics is a key prerequisite to understand cellular organisation and spatiotemporal dynamics. Biomolecular information which is not statistically ensemble-averaged gives access to molecular distribution functions, as well as probabilities and variations of subpopulations. There are several reasons to investigate and quantify the physical mechanisms of biological processes, complexes and cells down to the single molecule level. Beside the very pragmatic reason to save expensive material, a more direct and efficient analysis without the need of error inherent amplification and preconcentration steps is possible. Recent developments in ultrasensitive biophysical instrumentation allow optomechanical investigation and characterization of biomolecules, complexes, and cells down to the level of single molecules.

It is hoped, that the interdisciplinary character of systems nanobiology and nanobiotechnology to investigate cellular processes will lead to a better understanding of complex phenomena like gene regulation, molecular signalling and molecular transport in cells and their cellular interplay. Furthermore, new technologies for ultrasensitive biosensors and biodiagnostic devices will contribute to a new quantitative understanding of cell biology in the future and foster research within the framework of synthetic biology (Ball 2005; Benner and Sismour 2005), systems nanobiology (Heath et al. 2003), and single cell analytics (Hellmich et al. 2005a).

In this chapter, three applications how quantitative informations which are representative for cellular processes or single cell analysis like gene expression regulation, intracellular protein translocation dynamics, and single cell protein fingerprinting can be accessed. First, the interaction kinetics of transcriptionally regulated DNA–protein interaction can be quantitatively investigated with single molecule force spectroscopy down to the level of single point mutations. Second, intracellular dynamics for a transcription regulating protein migrating form the nucleus to the cytoplasm was quantitatively monitored by photoactivatable GFP and two-photon laser scanning microscopy. And third, a microfluidic-based method for label-free single cell proteomics and fingerprinting and first label-free single cell electropherograms are presented which include the manipulation and steering of single cells in a microfluidic device.

2. SINGLE MOLECULE BINDING KINETICS OF DNA–PROTEIN INTERACTIONS

Quantitative information about the specific interaction of biological molecules and their related reaction kinetics is a key prerequisite to understand the complex molecular network of cellular organization and its temporal dynamics. Specific protein–DNA interaction is a central and fundamental phenomenon in regulatory processes and covers all aspects of gene transcription and protein expression. In order to understand and model the metabolism of cells and organisms a distinct mathematical framework of the existing protein networks has to be established, which has to be supported with experimental data for spatiotemporal organization models, molecular concentrations and realistic reaction kinetics. Over the last years, we mainly focused on the quantitative investigation of the regulating networks in the Gram-negative soil bacterium *Sinorhizobium meliloti* 2011, which is capable of fixing molecular nitrogen in a symbiotic interaction with alfalfa plants (Glazebrook and Walker 1989). For the understanding of the symbiotic coexistence especially the complex regulation network of the galactoglucan EPS II exopolysaccharide biosynthesis (Figure 1) was investigated in great details (Becker et al. 1997; Rüberg et al. 1999). Here, we focused on the various interactions between a regulatory DNA-binding protein and its specific binding sequence and determined molecular interaction forces, elasticities, reaction lengths and reaction off-rates. In this chapter, we therefore describe the investigation of the molecular binding mechanism of the transcriptional activator ExpG to three associated target sequences in the galactoglucan biosynthesis gene cluster (*exp* genes) with single molecule force spectroscopy based on the atomic force microscope (AFM-FS).

2.1. Single Molecule Dynamic Force Spectroscopy

In AFM single molecule dynamic force spectroscopy (AFM-FS), binding forces between two molecules of interest are measured by functionally immobilizing them on two different surfaces (i.e. the tip of an AFM and an appropriate counter surface) that are brought in mechanical contact under physiological conditions (see Figure 2).

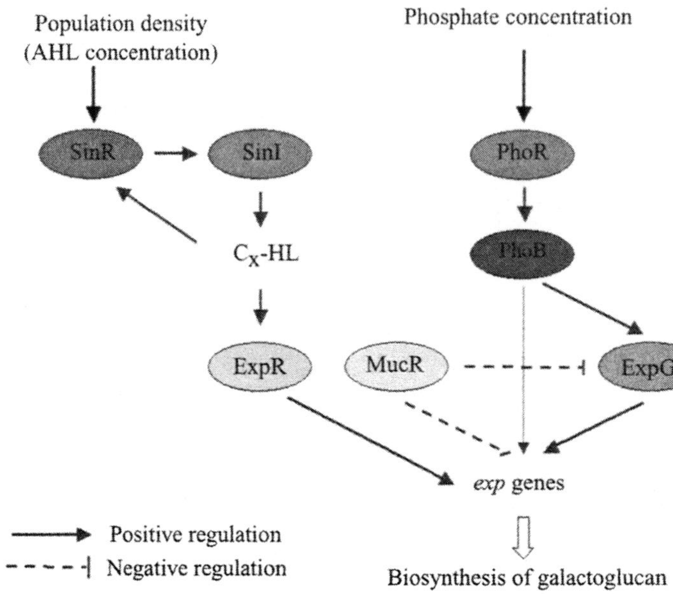

Figure 1. Transcriptionally regulated network of exopolysaccharide galactoglucan biosynthesis in the Gram-negative soil bacterium *S. meliloti* that is affected by extracellular phosphate concentration and bacterial population density (AHL, N-acyl homoserine lactone; HL, homoserine lactone).

The control of the molecular surface density, its appropriate functional immobilization and the introduction of convenient heterobifunctional molecular spacers to reduce unspecific background and to increase steric flexibility, allow functional experiments and measurements with single molecule complexes at a force sensitivity of about 10 pN (Zlatanova et al. 2000). Controlled approaching and withdrawing of the two surfaces allows association and forced dissociation between the molecules in contact upon measuring the molecular dissociation force and elasticity in single molecule force curves. Since molecular dissociation (also under external force) is a stochastic process, many single molecule dissociation events (typically 100 or more) have to be monitored which are commonly plotted in a force histogram (Figure 2). Since its introduction about 10 years ago (Florin et al. 1994; Lee et al. 1994; Dammer et al. 1995), AFM-FS has evolved into a valuable tool for measuring intermolecular binding forces, molecular elasticities, binding specificity and reaction rate constants (Zlatanova et al. 2000). It was found experimentally that the maximum of this force distribution is not constant, however, depends on the experimental approach-retract velocity, and scales with the logarithm of the molecular loading rate (eq. molecular elasticity × experimental velocity) (Bell 1978). The development of an adequate theory for thermal dissociation under an external force in 1997 (Evans and Ritchie 1997) allowed quantitative interpretation of these experimental data with respect of kinetic reaction constants and binding energy landscapes. Over the last decade, many

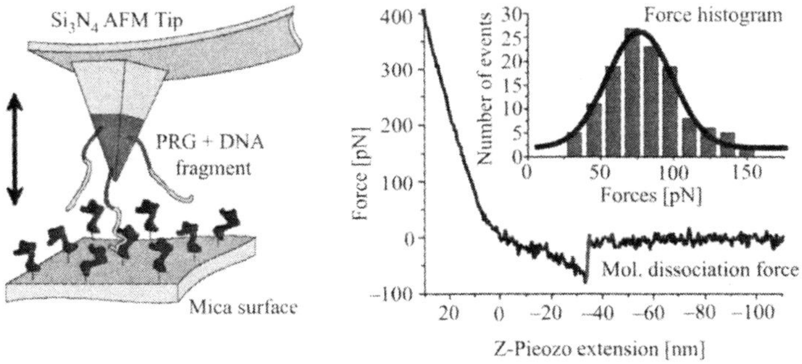

Figure 2. Force spectroscopy measurements. (Left) The experimental setup consists of a Si_3N_4 AFM tip with DNA fragments attached via poly(ethylene glycol) spacer molecules and a flat mica surface on which the (His)6ExpG proteins are immobilized. Cycling the tip and sample between approach and retract with constant velocities results in a series of force-distance measurements. (Right) A typical force-distance curve (only retractive part displayed). An unbinding event can be identified by a certain distance from the point of contact due to the length of the polymer linker and the stretching of this linker directly prior to the point of bond rupture. Rupture forces of a given series are combined to form a histogram (Right, small inset) with an almost Gaussian distribution.

specific receptor-ligand systems have been quantitatively investigated within this theoretical framework from immunological antigen-single chain antibody fragments (Ros et al. 1998; Schwesinger et al. 2000), DNA–protein interactions from bacterial transcription regulation (Bartels et al. 2003; Baumgarth et al. 2005), DNA–peptide interaction in synthetic biology (Eckel et al. 2005b), or synthetic host-guest calixarene receptors from supramolecular chemistry (Eckel et al. 2005a). It was very important to see, that the quantitative binding analysis of single molecules via AFM-FS fully agrees with data from established physicochemical ensemble experiments from free-solution, and allowed a novel and complementary approach of molecular affinity ranking (Schwesinger et al. 2000).

In our AFM-FS experiments of protein–DNA complexes, the respective DNA target sequences were covalently coupled to Si_3N_4 AFM cantilevers via monolayer activation and a functional heterobifunctional spacer molecule. The corresponding $(His)_6$ExpG protein was immobilized on a silanized mica surfaces by a short linker molecule coupled to one of the five ExpG lysines. We used a commercial AFM instrument at 25°C that has been modified with a home-built electronics and data acquisition (Bartels et al. 2003).

2.2. Measuring Intermolecular Forces and Relation to Single Molecule Kinetics

To investigate the specific protein–DNA interaction between the transcriptional activator ExpG to the three promoter DNA-fragments *expA1*, *expG/expD1* and *expE1* on a single molecule basis, the binding partners were covalently bound to the AFM tip and to the sample surface, respectively. A typical force–distance curve from measuring the ExpG binding to the *expG/expD1* sequence is shown in Figure 2. The red part of the force curve corresponds directly to the molecular dissociation force of a single dissociation event. The dissociation forces from multiple approach-retract cycles under a defined retract velocity were combined in a force histogram (Figure 2). The mean value of the nearly Gaussian distribution was taken as the most probable rupture force, its error given by standard deviation. In order to check for the specificity of the measured binding events, always a number of control experiments were conducted. Like in a standard competition assay, the addition of free DNA fragments to the solution which saturates all surface immobilized protein binding sites, leads to a complete loss of activity, visible as a breakdown of the force histogram (data not shown) (Bartels et al. 2003).

To obtain kinetic and structural information concerning the protein–DNA binding, we measured the most probable unbinding forces for the three DNA–protein complexes in dependence on the loading rate by varying the retract velocity. For each DNA target sequence, typically 150–300 unbinding events (from 1000–2000 approach/retract cycles) were recorded at 7–9 different retract velocities ranging from 10 nm/s to 8000 nm/s, which resulted in loading rates ranging from 70 to 79,000 pN/s.

The results for the three DNA target sequences are shown in Figure 3 (**see colour insert**). When analyzing the experimental data according to the standard dissociation theory (Evans and Ritchie 1997) and plotting them against the corresponding loading rates on a logarithmic scale, two regions with different slopes emerge for all DNA fragments. In both regions the experimental data can be fitted to a linear function according to the formula given by:

$$F = \frac{k_B T}{x_\beta} \ln \frac{x_\beta r}{k_B T \, k_{\text{off}}},$$

wherein F is the maximum force of the histogram, $k_B T = 4.114$ pN nm (at 298 K) is the Boltzmann factor, x_β is the molecular reaction length along the reaction coordinate, r is the loading rate, and k_{off} is the thermal off-rate under zero load. In the lower region ($r < 11{,}000$ pN/s), the slopes corresponding to the three fragments do not differ within the experimental error when this fit is applied to each dataset (as shown for the *expG1/G4* fragment in Figure 3d). According to the theoretical model, this slope can be attributed to the last potential barrier in the energy landscape of the system. In this case, the natural thermal off-rate k_{off} can be derived by extrapolating the linear fit to the state of zero external force. We obtain $k_{\text{off}} = (1.45 \pm 1.18) \times 10^{-3}\ \text{s}^{-1}$ for all three DNA target sequences, which corresponds to a mean life time of $\tau = (11.5 \pm 9.4)$ min for the bound protein–DNA complexes. The molecular reaction length x_β defines the distance between the minimum of the potential well of the bound state and the transition state along the reaction coordinate. In the lower region, for all protein–DNA complexes an $x_\beta = (7.2 \pm 1.4)$ Å can be deduced, which is often interpreted as the depth of the binding pocket (p.e. (Merkel et al. 1999)).

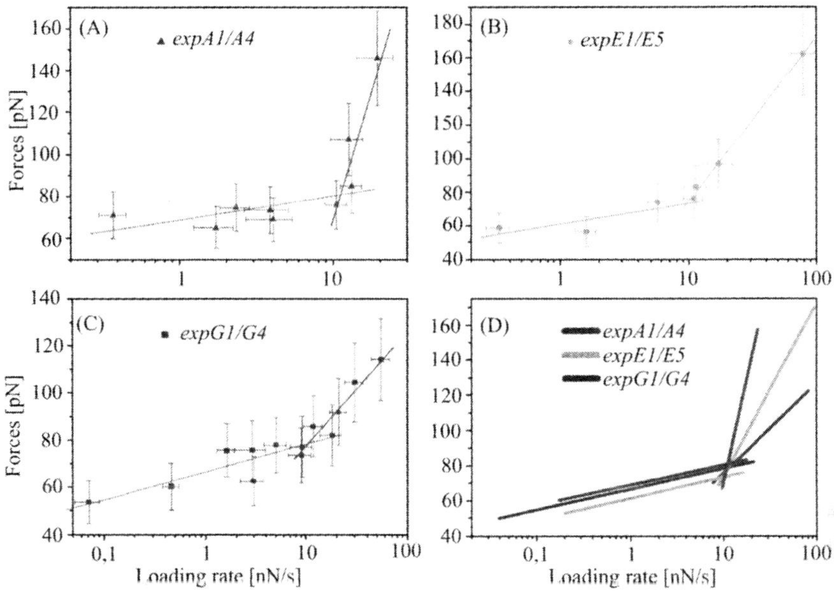

Figure 3. Dynamic force spectroscopy. Loading rate dependent measurements of the unbinding forces are displayed for complexes formed by the $(His)_6$ExpG protein and each of its three DNA target sequences (A–C). Two regions can be distinguished in every dataset. In the higher loading rate region, when each dataset is subjected to a linear fit, the slopes differ widely from each other (see D for comparison of the fits), and the individual DNA fragment in use can be identified by this behavior. In the lower loading rate region, however, the different protein–DNA complexes share a single slope under a linear fit within the error margin. This corresponds to a joint natural thermal off-rate of $k_{\text{off}} = (1.2 \pm 1.0) \times 10^{-3}\text{s}^{-1}$, derived by extrapolating the line fit to the state of zero external force.

In the upper region ($r > 11{,}000$ pN/s), different values for the slopes corresponding to the individual DNA fragments can be found. According to (Merkel et al. 1999), this can be attributed to a second energy barrier in the system, with different properties for the three DNA target sequences. We measure a molecular parameter $x_\beta = (2.1 \pm 0.3)$ Å for the protein–DNA complex with *expG1/G4*, $x_\beta = (0.95 \pm 0.02)$ Å with *expE1/E5*, and $x_\beta = (0.41 \pm 0.14)$ Å with *expA1/A4*, and interpret these differences to originate from the non-identical nucleotides of the three target sequences in the binding region and therefore a locally different molecular binding mechanism.

In summary, the complementary approach of single molecule force spectroscopy allows a novel access to determine and quantifiy molecular mechanisms of binding, reaction off-rates, energy landscapes, at a single molecule level. Since the affinity of an interaction is often commonly dominated by the molecular reaction off-rate (Schwesinger et al. 2000) an affinity ranking with the sensitivity of single point mutations in a broad affinity range of 10^{-4}–10^{-15} M without being essentially affected by solubility effects is possible (Eckel et al. 2005a).

3. MONITORING INTRACELLULAR PROTEIN DYNAMICS IN LIVING CELLS

The understanding of intracellular dynamics and trafficking is the key for a quantitative description of the spatio-temporal dynamics of regulatory networks involving protein interaction. As an example, nucleo-cytoplasmic partitioning of proteins such as transcription factors offers a fast and reversible tool to regulate gene expression, for instance upon extra-cellular and intra-cellular signals (Merkle 2003). Regulated nucleo-cytoplasmic partitioning of transcription factors is just one example of fast protein dynamics that exist in a living cell and that are crucial for signal transduction and cell homeostasis. Unfortunately, conventional 1-photon laser scanning microscopy (1PLSM) reveals only a quasi-stationary picture of protein dynamics at a given steady state. The quantitative real-time analysis of such intracellular protein dynamics therefore requires the combination of a rather fast microscopy technique with high spatial and temporal resolution and the ability for locally confined and selective labeling. Both criteria are met by combination of 2-photon activation (2P-activation) of photo-activatable green-fluorescent protein (pa-GFP) fusion proteins (König 2000; Zipfel et al. 2003b; Post et al. 2005) and 1PLSM or 2PLSM for detection of activated fluorophores. Two-photon excitation (Figure 4; **see colour insert**) depends on extremely high photon densities (Denk et al. 1990; König 2000; Zipfel et al. 2003b), that can be realized by an adequate spatial and temporal focussing of light, that is, diffraction-limited focussing of a femtosecond laser beam. This basic principal however does not only provide the intrinsic sectioning capability of a 2PLSM but it also allows a precise and restrictively confined pa-GFP activation within a defined volume of the cell, the diffraction limited focal volume (<1 fL) of the objective lens, within the cell. Here, we report on the investigation of the protein dynamics due to active translocation of Arabidopsis LCL1 (LHY/CCA1-like 1; (Schmied and Merkle 2006; Martini et al. 2007)), a transcription factor of the MYB1R family with high similarity to the

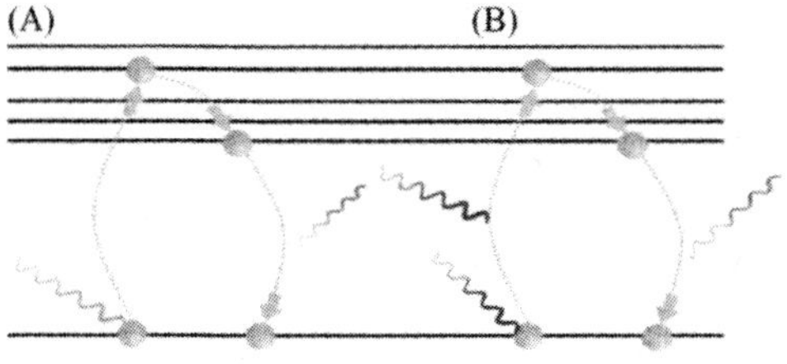

Figure 4. (A) Fluorescence emission after one-photon excitation. (B) Fluorescence emission after two-photon excitation: The sample is illuminated with a wavelength twice the wavelength of the absorption peak of the fluorophore being used. Using a high peak-power, pulsed laser, two-photon absorption will occur in the focal volume of the focussed laser beam. At this point the photon density is sufficiently high that two photons can be absorbed by the fluorophore essentially simultaneously. This is equivalent to a single photon with an energy equal to the sum of the two that are absorbed. In this way, fluorophore excitation will only occur at the point of focus (where it is needed) thereby eliminating excitation of out-of-focus fluorophore and achieving optical sectioning.

Arabidopsis clock proteins LHY and CCA1 (Schaffer et al. 1998; Wang and Tobin 1998). LCL1 contains a nuclear localization signal (NLS) as well as a nuclear export signal (NES) and is therefore an actively translocating nucleo-cytoplasmic shuttle protein. As a control experiment, we monitored the intracellular dynamics of the small protein pa-GFP alone. In contrast to LCL1, the distribution of pa-GFP from the nucleus to the cytoplasm happens by diffusion, only.

3.1. Multifocal Two-Photon Laser Scanning Microscopy

The intracellular dynamics of proteins shuttling between the cell nucleus and the cytoplasm was investigated with a multifocal 2PLSM (Nielsen et al. 2001) that consists of a mode-locked Ti:Sa laser generating 100 fs laser pulses between 760 and 960 nm which is pumped by a solid state laser (Tsunami and Millenia X, both Spectra-Physics, Darmstadt), a multi-focal scanning unit (TriM Scope, LaVision BioTec, Bielefeld) and an inverted microscope (IX 71, Olympus, Hamburg). In Figure 5 the general scheme (A) and an overview photograph (B) of this optical setup are given. The wavelength of the laser can be tuned within 5 s using a home-built motorization. The TriM-Scope laser scanning unit integrates an internal pre-chirp section that compensates for laser pulse dispersion, a beam-multiplexing section and two galvanometric mirror scanners. A set of ten 100% reflective mirrors and one adjustable 50% mirror allows splitting of the exciting NIR laser beam into $1, 2, 4, \ldots, 64$ foci in the sample. This adjustable number of foci is scanned in the focal plane of the microscope's objective lens (UPLAO60XW3/IR, NA $= 1.2$; Olympus) by the two mirror scanners. All measurements were performed in a temperature-controlled environment at 293 ± 1 K. As the multiple foci generate a relatively high two photon-induced fluorescence yield while keeping the energy deposition in each individual focus under the sample's deterioration limit, imaging is possible with 30 ms time resolution. Images are taken by a back-illuminated EMCCD camera (IXON DV887ECS-UVB, Andor Technology). A more detailed description of the instrumental setup has been published recently (Martini et al. 2005; Martini et al. 2006).

3.2. Local Protein Activation

Recently, a photoactivatable variant from *Aequoria victoria* green fluorescent protein (pa-GFP) was reported (Patterson and Lippincott-Schwartz 2002), yielding an increase in fluorescence emission intensity (at $\lambda \sim 520$ nm) by a factor of 100 when excited at $\lambda \sim 488$ nm after spectral activation at $\lambda \sim 408$ nm. This phenomenon is due to an internal photoconversion process in the protein and allows spectral photoactivation of this protein in a very local way such as in the nucleus of a living cell (Post et al. 2005). In tobacco BY-2 protoplasts, we transiently co-expressed pa-GFP or pa-GFP fusion proteins and red-fluorescent protein (DsRed)-tagged prenylated Rab acceptor 1 (Pra1; At2g38360), a membrane protein that localizes in speckles around the nuclear envelope. The DsRed transfection allows proper cell identification and visualization before activation (via Pra1-DsRed fluorescence). After pa-GFP

(A)

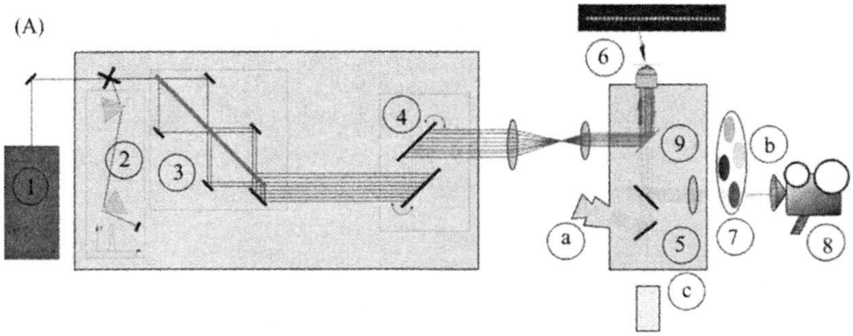

(B)

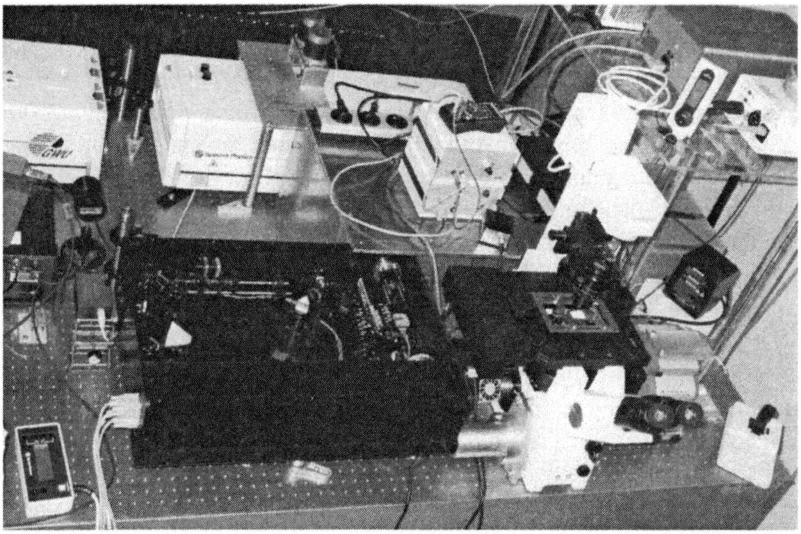

Figure 5. (A) Scheme of two-photon laser scanning microscope: (1) Ti: Sa laser, 100 fs, 80 MHz, 750–980 nm, 1.6W @ 800 nm (TSUNAMI, Spectra Physics), (2) pre-chirp, (3) beam multiplexer, (4) scanning mirrors, (5) microscope (Olympus IX 71, XLUMPLFL20XW, WD = 2 mm, NA = 0.95), (6) fluorescent foci in sample, (7) filter wheel/spectrograph (SpectraPro 2300i, Acton Research Corporation)/spectrometer (home built), (8) back illuminated EMCCD camera (IXON BV887ECS-UVB, Andor Technology), (9) dichroic mirror (2P-Beamsplitter 680 DCSPXR, Chroma). (B) Experimental setup of two-photon laser scanning microscope.

activation with 2PLSM the pa-GFP fluorescence dynamics and the translocation of the pa-GFP from the nucleus to the cytoplasm was monitored and quantitatively determined. Only molecules in the laser focus (activated by a burst of femtosecond 720–840 nm laser light pulses) will be photo-activated and hence will emit fluorescence at $\lambda \sim 520$ nm when excited with light of $\lambda \sim 488$ nm by 1PLSM or $\lambda > 920$ nm by 2PLSM, respectively (Schneider et al. 2005). The use of near-infrared laser light for activation and detection is reasoned by the specific cross sections of pa-GFP, but it also provides the advantage of a higher optical penetration depth, negligible one-photon cross sections, and less potential for cellular damage and fluorescent background (Zipfel et al. 2003a, 2003b). The protein dynamics of the Arabidopsis transcription factor LCL1 (At5g02840) fused to pa-GFP was investigated in this way in co-transfected plant protoplasts. Local activation of pa-GFP-LCL1 in the nucleus was achieved by 2PLSM with a consecutive analysis of the actively translocating nucleo-cytoplasmic shuttling in a spatio-temporal manner. This was compared with a control experiment, where we monitored the localisation and diffusional dynamics of the small protein pa-GFP alone from the nucleus to the cytoplasm.

3.3. Measuring Temporal Fluorescence Dynamics

As a reference experiment, we first recorded the 2P-activation of pa-GFP in the nucleus and the dynamics of its diffusion within the BY-2 protoplast. Five selected images out of a whole series of 1P-transmission-fluorescence images are shown in Figure 6, that have been consecutively taken at the start of recording before photo-activation (0 s), at the start of photo-activation (26 s), and at 33, 180 and 1000 s after start of recording. The cell nucleus, visualized by a red ellipse in the image, has been selected as a region of interest (ROI) for the subsequent determination of the ROI-integrated fluorescence intensity. The decrease of fluorescence intensity over time as seen in the last four images directly displays the diffusion of pa-GFP into the cytoplasm. Simultaneously to the decrease of nuclear fluorescence, an increase of the fluorescence in the cytoplasm can be detected. Furthermore it is worth noting that the rectangular shape of the strong fluorescence signal at time point 26 s (Figure 6b) reflects the fact that pa-GFP activation was accomplished by scanning 4 fs-laser foci for 3 s in the activation area of $7 \times 8\ \mu m^2$ within the nucleus, rather than by single point activation, only. In order to quantify the kinetics of the decrease of pa-GFP fluorescence in the nucleus, we analyzed the fluorescence intensity within the chosen ROI over the whole series of fluorescence images. The spatiotemporal fluorescence dynamics can either be analyzed by 1P-fluorescence or 2P-fluorescence. Our setup allows both possibilities (Martini et al. 2007), however, we focus here on the 2P-epifluorescence results. The quantification of the decrease of nuclear fluorescence was achieved by re-tuning the laser wavelength after 2P-activation of pa-GFP ($\lambda = 800$ nm) to the fluorescence emission wavelength ($\lambda > 920$ nm) which can be achieved within 5 s. The quantification of 2P-fluorescence data within the first 350 s is presented in Figure 7 with a single exponential fit, indicating that photobleaching is not relevant on this time scale. The analysis yielded a diffusion time constant of 123 s for this experiment, in good agreement with the diffusion time constant in a similar experiment measured with 1P-fluorescence (data not shown; Martini et al. 2007).

On the other hand, translocation of the MYB transcription factor LCL1 fused to pa-GFP from the nucleus to the cytoplasm and its intracellular localization dynamics depends on facilitated nuclear export versus facilitated nuclear import. Protoplasts co-transfected with At2g38360-DsRed and pa-GFP-LCL1 were subjected to the 2P-activation procedure and the decrease of nuclear fluorescence intensity was monitored

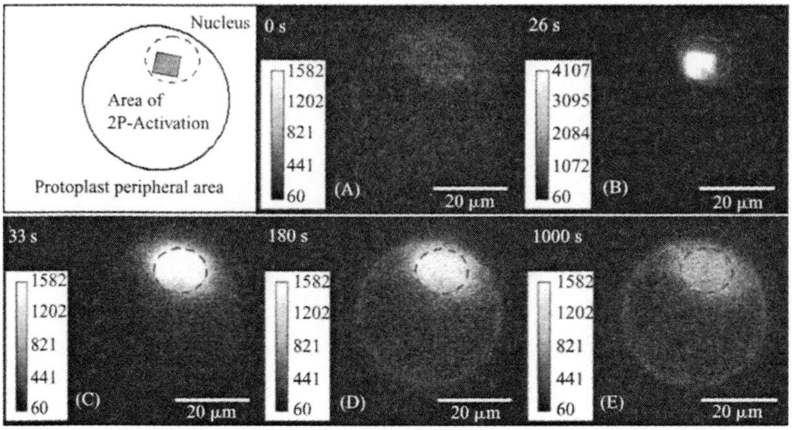

Figure 6. Dynamics of free diffusion of pa-GFP in a live protoplast. Five selected 1P-transmission fluorescence images of a tobacco BY-2 protoplast expressing pa-GFP (A) at the onset of the experiment before activation, (B) during 2P-activation of pa-GFP, and (C–E) after 2P-activation were taken at the time points indicated. (A) Before 2P-activation of pa-GFP in the nucleus (dotted line) the average fluorescence intensity is barely detectable. 2P-activation of pa-GFP was initiated with a fs-laser burst of 3 s covering an area of 7 × 8 μm^2 with 4 parallel laser foci (10 mW at 800 nm per focus). (B) Shortly after photo-activation a strong fluorescence signal was detected and (C–E) the diffusion of photo-activated pa-GFP from the nucleus into the cytoplasm was monitored until equilibrium of partitioning between the two cellular compartments was reached. Fluorescence intensity scales are shown in each panel to the left.

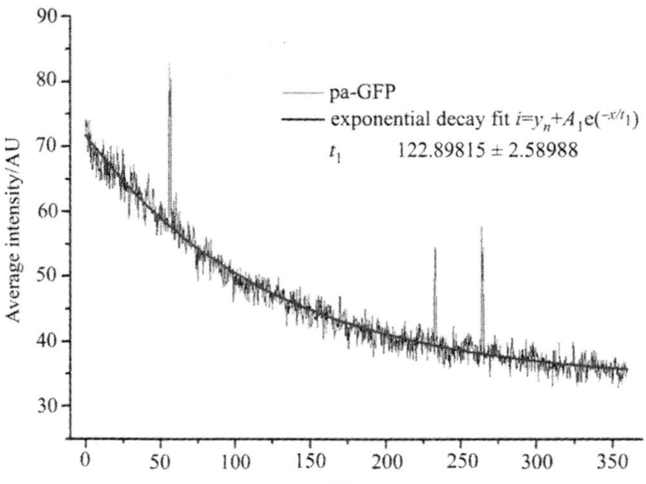

Figure 7. pa-GFP dynamics by parallel 2P-epifluorescence microscopy (64 foci, 920 nm, 240 mW) in a tobacco BY-2 protoplast. Quantitative analysis of the decrease of nuclear pa-GFP 2P-epifluorescence, giving a diffusion time constant of 123 s.

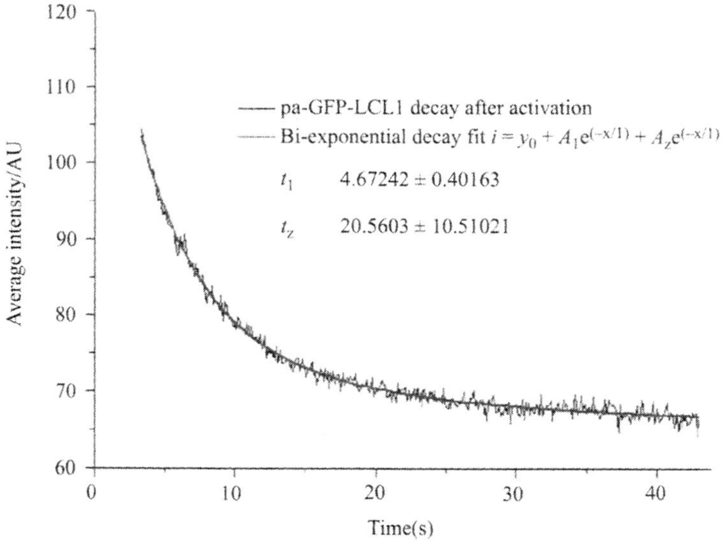

Figure 8. Quantitative analysis of the post-activation decrease of pa-GFP-LCL1 in the nucleus by 2P-epifluorescence with a bi-exponential fit (red line). A time constant of 20 s was calculated for the nuclear fluorescence decrease of pa-GFP-LCL1 due to active transport.

with 2P-fluorescence microscopy (Figure 8). Quantitative analysis of these data revealed an adequate fitting with a bi-exponential function, yielding two time constants of 4.7 and 20.6 s, respectively. While the first time constant can be attributed to non-linear short term photoactivation of DsRed, the time constant of 20.6 s reflects the net translocation of pa-GFP-LCL1 out of the nucleus into the cytoplasm. This time constant is considerably smaller than that of pa-GFP diffusion out of the nucleus (compare Figure 7). In addition, two-color 2PLSM 3D fluorescence microscopy imaging of At2g38360-DsRed and pa-GFP-LCL1, which were taken just before and 45 s after pa-GFP activation allow the visualization of the complex transport process at the level of the cell. This imaging was performed using the fast parallel 64 foci mode of our TriM-Scope that allows sensitive optical sectioning with respect to spectral emission characteristics by using appropriate emission filters. In these 2P-fluorescence images, the DsRed-tagged Pra1 and pa-GFP-LCL1 are colored in red and green, respectively (Figure 9; **see colour insert**). Whereas the transfected protoplast before photo-activation mainly exhibits the red fluorescence of the transfection marker Pra1-DsRed (Figure 9a), the image taken 45 s after start of recording (Figure 9b) additionally exhibits the pa-GFP-LCL1 fluorescence that had been activated in the nucleus and is actively translocated into the cytoplasm.

In conclusion, the combination of localized and selective photoactivation of pa-GFP and 2PLSM allows the quantitative real-time fluorescence monitoring of dynamic intracellular processes *in vivo* that are of crucial importance for a deeper understanding

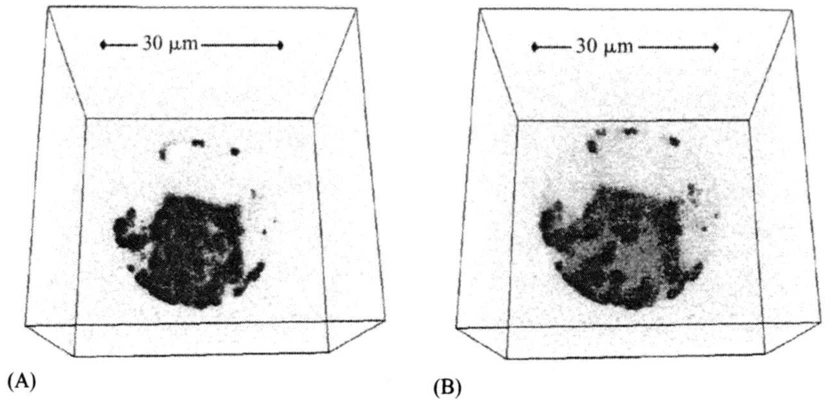

Figure 9. Two-colour-2P-epifluorescence 3D maps of At2g38360-DsRed (transfection marker) and pa-GFP-LCL1 taken (A) just before photo-activation of pa-GFP in the nucleus and (B) after data sampling.

and modeling of regulatory and metabolic processes in living cells (Systems Biology). This method and its application demonstrate, that dynamic cellular processes can be investigated and analyzed *in vivo* by novel biophysical methods that open new possibilities towards a more quantitative biology.

4. TOWARDS MICROFLUIDIC-BASED LABEL-FREE SINGLE CELL PROTEOMICS

Another aspect of systems nanobiology are single cell applications for proteomics that allow expression analyses not subjected to ensemble-averaging, cell-cycle or heterogeneous cell-population effects. To date, proteomes are separated and usually analyzed in two-dimensional electrophoretic gels (Wilkins et al. 1997) at the level of 10^5–10^6 cells accessing functional information only on the basis of that probed cellular ensemble completely neglecting the different and inhomogeneous cellular response to an external stimulus or the introduction of genomic and proteomic variabilities during eucaryotic cell proliferation. The analysis of minute analyte quantities and the hunt for low abundant proteins at the single cell level, however, requires new techniques for efficient and sensitive separation, detection and analysis. The typical protein content of a cell is about 15% proteins (w/w) which equals 75 pg, 2 fmol or 10^9 protein molecules (assuming an average molecular mass of 40 kDa). Low abundant proteins in a cell at μM concentration occur at the amol level which is equivalent to 10^5 molecules. In systems nanobiology, microfabrication and nanotechnology now offer novel tools to detect, measure, analyze, steer, and manipulate individual molecules and cells and will allow more detailed insights in the interplay of genomic information and functional peculiarity at the single molecule or single cell level in the future (Figure 10). For single cell analysis, microfluidic devices and lab-on-a-chip systems (Auroux et al. 2002; Reyes et al. 2002) with characteristic length scales of 10–50 μm

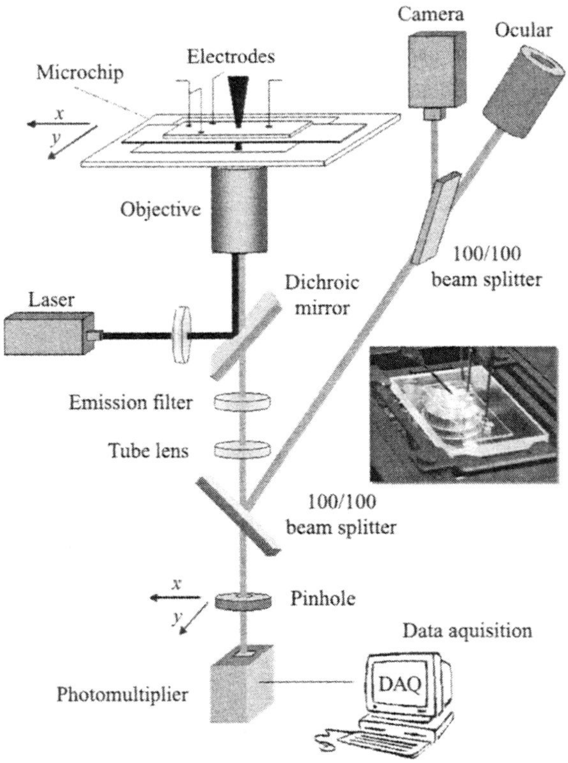

Figure 10. Optical setup used for the detection of native UV-fluorescence for microfluidic single cell analysis. The photograph shows a mounted PDMS microstructure with voltage connections.

are predestined due to typical cell dimensions of several micrometers since a detection level of 10^5 molecules in a cube of 10 μm length (corresponding to the volume of a pL) requires a detection with a sensitivity better than 100 nM. In capillary format, such as capillary electrophoresis, first single cell protein fingerprinting with capillary sieving electrophoresis in one-dimensional (Hu et al. 2003a,b) and two dimensional format (Hu et al. 2004) using fluorescent protein stains and laser-induced fluorescence (LIF) detection in the visible spectral range was pioneered by the group of N. Dovichi.

More recently, we have introduced a method for single cell analysis combining navigation and steering of single cells with optical tweezers, on-chip cell lysis and electrophoretic separation of proteins with subsequent LIF detection in the visible (Hellmich et al. 2005a; Ros et al. 2006) and in the UV spectral range (Hellmich et al. 2006). While LIF in the visible range requires adequate on-chip or off-chip labelling steps for the cell component of interest, LIF detection in the UV spectral range only allows a direct label-free detection of proteins due to the favorable fluorescence properties of the aromatic amino acids Trp, Typ, Phe (Lakowicz 1999). Here, first single cell protein electropherograms taken from single Sf9 insect cells with label-free UV-LIF detection in a microfluidic device format will be presented.

4.1. Manipulation and Handling of Single Cells in the Microfluidic Device

Our microfluidic chip is fabricated by elastomeric poly(dimethylsiloxane) (PDMS) technology (Duong et al. 2003) that consist of two crossing channels which have a typical cross section of 20 × 20 μm^2 (Figure 11a). The channel walls were always coated with a hydrophilic triblock copolymer which significantly reduced unwanted adhesion due to cellular contact during cell steering. The detailed and consecutively optimized fabrication procedure was published in a number of papers (Hellmich et al. 2005a, 2005b, 2006; Ros et al. 2006). At the intersection of the two channels a physical cell trap was realized via several polymer posts (Figure 11b). Individual cells were optically selected and trapped in the reservoir, and injected and steered in the microfluidic channel by a home-made single-beam optical tweezers system. The optical tweezers were incorporated into an inverted optical (fluorescence) microscope (Sischka et al. 2003) and combined with a dedicated x/y-stage allowing long range positioning of individual cells with the optical tweezers within the microchannel for maximal 25 mm with a precision of 1–2 μm. By this means, an individual cell can be captured in the reservoir and navigated to the intersection. Once the cell was navigated into this position the optical trap was switched off and the cell was allowed to adhere to the microchannel wall. Consecutively, cell lysis was performed upon applying a short pulse of high electric field strength (50 ms at 1900 V/cm) to the channel reservoirs. Alternatively, cell lysis can also be accomplished by exposing the cell to a 1% SDS solution. The electrokinetic injection of the cell compounds into the separation channel (channel 2) and separation is performed in separation buffer (100 mM Tris, 100 mM CHES, 4% Pullulan, pH 8.4) at a typical field strength of 200–400 V/cm. The cell lysis was always controlled by optical bright-field microscopy. Figure 11d–g demonstrate a sequence of snapshots from such a single cell lysis at the injection point at the entrance of channel 2, in which subsequent analytics will be performed. We investigated Sf9 (*Spodoptera frugiperda*) insect cells. For the sensitive LIF detection in the UV range (UV-LIF) the inverted microscope had to be optimized for high UV transmission and confocal detection. For excitation of the aromatic amino acids we used a frequency quadrupled Nd:YAG laser (266 nm, 5 mW). A detailed description of the UV-LIF setup was published recently (Hellmich et al. 2006).

4.2. First Label-free Single Cell Electropherograms

First single cell electropherograms from Sf9 cells with label-free UV-LIF detection in microfluidic chip format are presented in Figure 12. In Figure 12a and b electropherograms of one single and of four single cells are shown, respectively. Whereas the single cell electropherogram yields ~10 distinct peaks, in the four-cell-electropherogram ~20 peaks with increased intensities can be resolved that are expected due to an

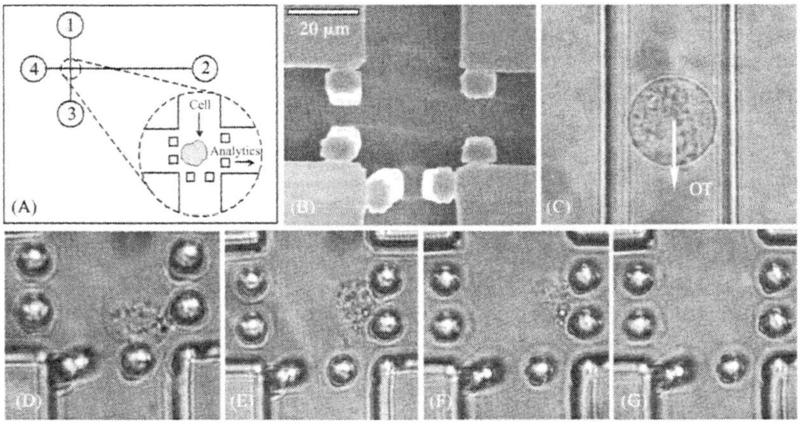

Figure 11. (A) Scheme of the PDMS microfluidic device. Inset: channel crossing with the cell trap composed of microstructured obstacles, (B) Scanning electron micrograph of the cell trap, (C) single cell in a channel navigated by optical tweezers in the microchannel, (D–G) optical micrographs of a single cell at the injection position during SDS lysis. SDS flow is from channel 4 through the cell trap into channel 2.

enhanced protein amount compared to one cell. Although protein expression levels in various cells are expected to vary thus making the interpretation of these electropherograms difficult, we could identify three groups of peaks in both electropherograms labeled I–III. The grouping was achieved by reconciling the peak positions with the corresponding migration time taking into account a recalibration factor of 1.5 from the different separation voltages (360 vs. 240 V/cm). Although the peak number of detected peaks and the separation efficiency are rather low at the moment, we strongly believe that the performance of single cell electropherograms can likely be improved in the future by more appropriate and elaborate separation and detection concepts as it can be seen in figure 12c, where recently 124 peaks in a single cell electropherogram of a Sf9 insect cell could be detected by native UV-LIF detection. Nevertheless, our data represent to our best knowledge the first electropherogram from a single cell with native UV-LIF detection in a microfluidic device opening promising perspectives for an efficient label-free protein fingerprinting from single cells in the future.

5. CONCLUSION

In systems nanobiology, cellular processes are quantitatively investigated at the sensitivity level of single molecules and cells. This approach gives a direct access to biomolecular information which is not statistically ensemble-averaged and makes use of new ultrasensitive biophysical methods that are based on nanotechnological concepts for single molecule detection, microscopy and spectroscopy without the need of amplification and preconcentration. In this chapter, we presented three methodological applications that allow quantitative proteomic analysis of individual molecular complexes and living cells. In the future quantitative investigation of single

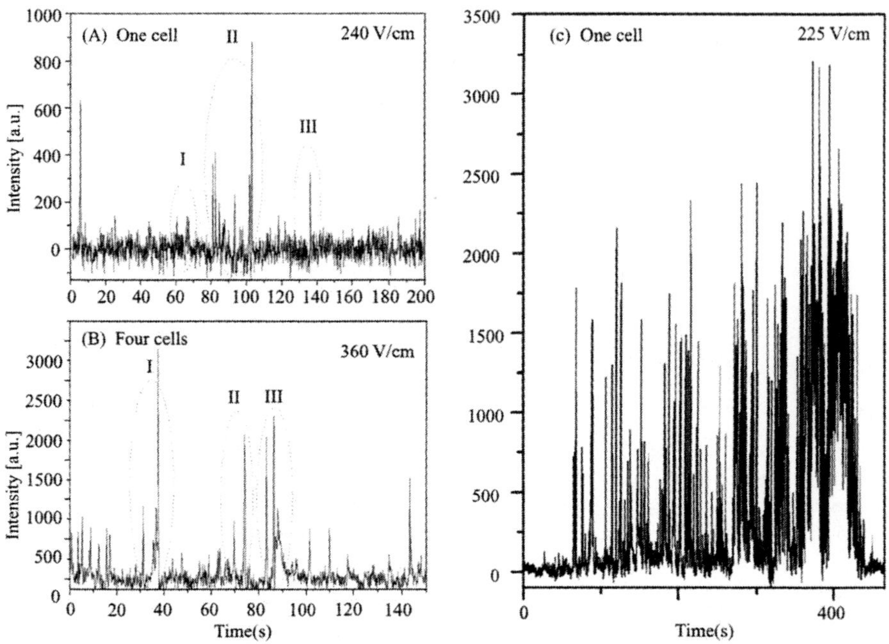

Figure 12. (A) Baseline corrected electropherogram from a single Sf9 cell with native UV-LIF detection at a separation voltage of 240 V/cm. Peak groups I, II and III (dotted ellipses) correspond to 65, 90, and 130 s, respectively. (B) Electropherogram of four Sf9 cells at a separation voltage of 360 V/cm. Peak groups I , II and III (dotted ellipses) correspond to 37, 74, and 87 s, respectively. (C) Baseline corrected single cell electropherogram of Sf9 cell with 124 peaks with native UV-LIF detection.

biological cells at the subcellular level will give access to an efficient and unamplified single cell analysis without being subjected to ensemble-averaging, cell-cycle or cell-population effects and a broad spectrum in applications for systems biology.

ACKNOWLEDGEMENTS

We acknowledge support from Karsten Niehaus, Christoph Pelargus, Kai Leffhalm and Andy Sischka. Financial support from the Deutsche Forschungsgemeinschaft (DFG) within the research grants "Microchip UV-LIF" (An-370/1-2) and from the *Collaborative Research Project SFB 613* – Physics of molecular recognition and single molecule processes in organic systems (D.A.), the Federal Ministry of Education and Research of Germany (BMBF) for the Research Project MEMO – Foundations of laser-based characterization of metabolism and morphology of biological tissue (D.A.) and from the DFG grants ME 1116/4-2 and BIZ 7/1-2 (T.M.) are gratefully acknowledged.

REFERENCES

Auroux, P.A., Iossifidis, D., Reyes, D.R. and Manz, A. (2002) Micro total analysis systems. 2. Analytical standard operations and applications. Anal. Chem. *74*, 2637–2652.

Ball, P. (2005) Synthetic biology for nanotechnology. Nanotechnology *16*, R1–R18.

Bartels, F.W., Baumgarth, B., Anselmetti, D., Ros, R. and Becker, A. (2003) Specific Binding of the Regulatory Protein ExpG to Promoter Regions of the Galactoglucan Biosynthesis Gene Cluster of *Sinorhizobium meliloti* – A Combined Molecular Biology and Force Spectroscopy Investigation. J. Struct. Biol. *143*, 145–152.

Baumgarth, B., Bartels, F.W., Anselmetti, D., Becker, A. and Ros, R. (2005) Detailed studies of the binding mechanism of the *Sinorhizobium meliloti* transcriptional activator ExpG to DNA. Microbiology *151*, 259–268.

Becker, A., Rüberg, S., Küster, H., Roxlau, A., Keller, M., Ivashina, T., Cheng, H.-P, Walker, G.C. and Pühler, A. (1997) The 32-kilobase *exp* gene cluster of *Rhizobium meliloti* directing the biosynthesis of galactoglucan: genetic organization and properties of the encoded gene products. J. Bacteriol. *179*, 1375–1383.

Bell, G.I. (1978) Models for the specific adhesion of cells to cells. Science *200*, 618–627.

Benner, S. and Sismour, A. (2005) Synthetic biology. Nat. Rev. Genet. *6*, 533–543.

Dammer, U., Popescu, O., Wagner, P., Anselmetti, D., Güntherodt, H.-J., and Misevic, G.N. (1995) Binding strength between cell adhesion proteoglycans measured by atomic force microscopy. Science *267*, 1173–1175.

Denk, W., Strickler, J.H. and Webb, W.W. (1990) Two-photon laser scanning fluorescence microscopy. Science *248*, 73–76.

Duong, T., Kim, G., Ros, R., Streek, M., Schmid, F., Brugger, J., Ros, A. and Anselmetti, D. (2003) Size dependent free solution DNA electrophoresis in structured microfluidic systems. Microelectronic Eng. *67–68*, 905–912

Eckel, R., Ros, R., Decker, B., Mattay, J. and Anselmetti, D. (2005a) Supramolecular chemistry at the single molecule level. Angew. Chem. Int. Ed. *44*, 484–488.

Eckel, R., Wilking, S.D., Becker, A., Sewald, N., Ros, R. and Anselmetti, D. (2005b) Single molecule experiments in synthetic biology – A new approach for the affinity ranking of DNA-binding peptides. Angew. Chem. Int. Ed. *44*, 3921–3924.

Evans, E. and Ritchie, K. (1997) Dynamic strength of molecular adhesion bonds. Biophys. J. *72*, 1541–1555.

Florin, E.-L., Moy, V.T. and Gaub, H.E. (1994) Adhesion forces between individual ligand-receptor pairs. Science *264*, 415–417.

Glazebrook, J. and Walker, G.C. (1989) A novel exopolysaccharide can function in place of the calcofluor-binding exopolysaccharide in nodulation of alfalfa by *Rhizobium meliloti*. Cell *56*, 661–672.

Heath, J.R., Phelps, M.E. and Hood, L. (2003) Nanosystems biology. Mol. Imaging Biol. *5*, 312–325.

Hellmich, W., Greif, D., Pelargus, C., Anselmetti, D. and Ros, A. (2006) Improved native laser induced fluorescence detection for single cell analysis in poly(dimethylsiloxane) microfluidic devices. J. Chromatogr. A. *1130*, 195–200.

Hellmich, W., Pelargus, C., Leffhalm, K., Ros, A. and Anselmetti, D. (2005a) Single cell manipulation, analytics and label-free protein detection in microfluidic devices for systems nanobiology. Electrophoresis *26*, 3689–3696.

Hellmich, W., Regtmeier, J., Duong, T., Ros, R., Anselmetti, D. and Ros, A. (2005b) Poly(oxyethylene) based surface coatings for poly(dimethylsiloxane) microchannels. Langmuir *21*, 7551–7557.

Hu, S., Michels, D.A., Abu Fazal, M., Ratisoontorn, Ch., Cunningham, M.L. and Dovichi, N.J. (2004) Capillary sieving electrophoresis/micellar electrokinetic capillary chromatography for two-dimensional protein fingerprinting of single mammalian cells. Anal. Chem. *76*, 4044–4049.

Hu, S., Zhang, L., Krylow, S. and Dovichi, N.J. (2003a) Cell cycle-dependent protein fingerprint form a single cancer cell: image cytometry coupled with single-cell capillary sieving electrophoresis. Anal. Chem. *75*, 3495–3501.

Hu, S., Zhang, L., Newitt, R., Aebersold, R., Kraly, J.R., Jones, M. and Dovichi, N.J. (2003b) Identification of proteins in single-cell capillary electrophoresis fingerprints based on comigration with standard proteins. Anal. Chem. *75*, 3502–3505.

König, K. (2000) Multiphoton microscopy in life sciences. J. Microsc. *200*, 83–104.

Lakowicz, J.R. (1999) Principles of Fluorescence Spectroscopy. (New York: Kluwer Academic/Plenum Publisher).

Lee, G.U., Chrisey, L.A. and Colton, R.J. (1994) Direct measurement of the forces between complementary strands of DNA. Science *266*, 771–773.

Martini, J., Schmied, K., Palmisano, R., Toensing, K., Anselmetti, D. and Merkle, T. (2007) Multifocal two-photon laser scanning microscopy combined with photo-acivatable GFP for in vivo monitoring of intracellular protein dynamics in real time. J. Struct. Biol. in press.

Martini, J., Tönsing, K., Dickob, M. and Anselmetti, D. (2005) 2-photon laser scanning microscopy on native human cartilage. Proc. SPIE *5860*, 16–21.

Martini, J., Tönsing, K., Dickob, M., Schade, R., Liefeith, K. and Anselmetti, D. (2006) 2-photon laser scanning microscopy on native human cartilage and collagen-membranes for tissue engineering. Proc. SPIE *6089*, 274–282.

Merkel, R., Nassoy, P., Leung, A., Ritchie, K. and Evans, E. (1999) Energy landscapes of receptor-ligand bonds explored with dynamic force spectroscopy. Nature *397*, 50–53.

Merkle, T. (2003) Nucleo-cytoplasmic partitioning of proteins in plants: implications for the regulation of environmental and developmental signalling. Curr. Genet. *44*, 231–260.

Nielsen, T., Fricke, M., Hellweg, D. and Andresen, P. (2001) High efficiency beam splitter for multifocal multiphoton microscopy. J. Microsc. *201*, 368–376.

Patterson, G.H. and Lippincott-Schwartz, J. (2002) A photoactivatable GFP for selective photolabeling of proteins and cells. Science *291*, 1873–1877.

Post, J.N., Lidke, K.A., Rieger, B. and Arndt-Jovin, D.J. (2005) One- and two-photon photoactivation of a paGFP-fusion protein in live *Drosophila* embryos. FEBS Lett. *579*, 325–330.

Reyes, D.R., Iossifidis, D., Auroux, P.A. and Manz, A. (2002) Micro total analysis systems. 1. Introduction, theory, and technology. Anal. Chem. *74*, 2623–2636.

Ros, A., Hellmich, W., Regtmeier, J., Duong, T.T. and Anselmetti, D. (2006) Bioanalysis in structured microfluidic systems. Electrophoresis *27*, 2651–2658.

Ros, R., Schwesinger, F., Anselmetti, D., Kubon, M., Schäfer, R., Plückthun, A. and Tiefenauer, L. (1998) Antigen binding forces of individually addressed single-chain Fv antibody molecules. Proc. Natl Acad. Sci. U.S.A. *95*, 7402–7405.

Rüberg, S., Pühler, A. and Becker, A. (1999) Biosynthesis of the exopolysaccharide galactoglucan in *Sinorhizobium meliloti* is subject to a complex control by the phosphate-dependent regulator PhoB and the proteins ExpG and MucR. Microbiology *145*, 603–611.

Schaffer, R., Ramsay, N., Samach, A., Corden, S., Putterill, J., Carre, I.A. and Coupland, G. (1998) The late elongated hypocotyl mutation of *Arabidopsis* disrupts circadian rhythms and the photoperiodic control of flowering. Cell *93*, 1219–1229.

Schmied, K. and Merkle, T. (2006) The *Arabidopsis* LHY/CCA1-like (LCL) protein family: MYB transcription factors containing a nuclear export signal are co-regulators of the circadian clock. submitted.

Schneider, M., Barozzi, S., Testa, I., Faretta, M. and Diaspro, A. (2005) Two-Photon Activation and Excitation Properties of PA-GFP in the 720–920-nm Region. Biophys. J. *89*, 1346–1352.

Schwesinger, F., Ros, R., Strunz, T., Anselmetti, D., Güntherodt, H.-J., Honegger, A., Jermutus, L., Tiefenauer, L. and Plückthun, A. (2000) Unbinding forces of singel antibody-antigen complexes correlate with their thermal dissociation rates. Proc. Natl Acad. Sci. U.S.A. *97*, 9972–9977.

Sischka, A., Eckel, R., Toensing, K., Ros, R. and Anselmetti, D. (2003) Compact, microscope based optical tweezers system for molecular manipulation. Rev. Sci. Instrum. *74*, 4827–4831.

Wang, Z.Y. and Tobin, E.M. (1998) Constitutive expression of the CIRCADIAN CLOCK ASSOCIATED 1 (CCA1) gene disrupts circadian rhythms and suppresses its own expression. Cell *93*, 1207–1217.

Wilkins, M.R., Williams, K.L., Appel, R.D. and Hochstrasser, D.F. (1997). Proteome Research: New Frontiers in Functional Genomics. (Berlin: Springer).

Zipfel, W.R., Williams, R.M., Christie, R., Nikitin, A.Y., Hyman, B.T. and Webb, W.W. (2003a) Live tissue intrinsic emission microscopy using multiphoton-excited native fluorescence and second harmonic generation. Proc. Natl Acad. Sci. USA *100*, 7075–7080.

Zipfel, W.R., Williams, R.M. and Webb, W.W. (2003b) Nonlinear magic: multiphoton microscopy in the biosciences. Nat. Biotechnol. *21*, 1369–1377.

Zlatanova, J., Lindsay, S.M. and Leuba, S.H. (2000) Single molecule force spectroscopy in biology using the atomic force microscope. Prog. Biophys. Mol. Biol. *74*, 37–61.

Corresponding author: Pr. Dario Anselmetti (dario.anselmetti@physik.uni-bielefeld.de)

CHAPTER 15

BIOPHOTONICS APPLIED TO PROTEOMICS

MICHEL FAUPEL, DÉBORA BONENFANT, PATRICK SCHINDLER, ERIC BERTRAND, DIETER MUELLER, MARKUS STOECKLI, FRANCIS BITSCH, TATIANA ROHNER, DIETER STAAB, and JAN VAN OOSTRUM
Novartis Institutes for Biomedical Research, Basel

Table of Contents

Abstract: Since the completion of the human genome sequencing, our understanding of gene and protein function and their involvement in physiopathological states has increased dramatically, partly due to technological developments in photonics. Photonics is a very active area where new developments occur on a weekly basis, while established tools are adapted to fulfill the needs of other disciplines like genomics and proteomics. Biophotonics emerged at the interface of photonics and biology as a very straightforward and efficient approach to observe and manipulate living systems. In this chapter, we review the current applications of photonics and imaging to proteomics from 2D gels analysis to molecular imaging.

1. INTRODUCTION

The complete set of genes of an organism composes its genome, while the complete set of proteins of a cell or tissue constitutes its proteome. Whereas the genome of cells remains stable through their lifetime, the cellular proteome changes in response to the

E. Bertrand and M. Faupel (eds.), Subcellular Proteomics, 323–338.

stimuli from its environment. Since the sequencing of the human genome, we have access to lists and databases of protein sequences encoded by genes. From these, we know the repertoire cells have to build their protein machinery. In most cases, we do not know what drives gene expression or how genes intervene in a physiopathological context. Proteomic analysis makes it possible to follow the quantitative variations of cellular protein expression through processes like tumorigenic transformation or in response to chemical and pharmacological agents. Thereby, proteomic analysis encompasses the whole set of complex interactions that occur within cells or tissues. The analytical instrumentation has to resolve and characterize protein isoforms that are due to genetic variations or post-translational modifications. Post-translational modifications like phosphorylation and proteolytic cleavages are essential to modulate the activity of proteins. One of the purposes of proteomic analysis is precisely to take the post-translational modifications into account. The characterization system relies mainly onto mass spectrometry and biophotonics technologies like MS Imaging, while the proteomic platform is based on two-dimensional electrophoresis techniques which are the only one to possess a sufficient resolution to distinguish post-translational variants.

2. TWO-DIMENSIONAL ELECTROPHORESIS

Two-dimensional electrophoresis was introduced about 30 years ago (O'Farell 1975), at this point, its separation power is still unequaled. Many polypeptidic species observed as distinct spots on the gel are the products of a single gene. This diversity is mainly due to post-translational modifications like cleavages or the addition of some chemical functions on specific amino-acids. As a rule, a chemical modification of an amino-acid is much more likely to affect the isoelectric point of a protein than its observed mass. In the case of phosphorylation and glycosylation, the native form of the protein and its modified form will appear as two separate spots on a two-dimensional gel. Other post-translational modifications like cystein acylation do not affect noticeably the physico-chemical parameters and are not detected by classical electrophoresis. The performance of isoelectric focusing in terms of separation (first dimension of two-dimensional electrophoresis) is constantly improving. Remarkable progresses have been achieved thanks to the introduction of immobilized pH gradients (Bjellqvist et al. 1982; Görg, et al. 2000) and the development of narrow-range high resolution pH gradients (Hoving et al. 2000, 2002). Allelic or post-translational variants that were previously invisible in classical electrophoresis are now easily resolved by isoelectric focusing on narrow-range gradients, even in very basic ranges. In a similar manner, progresses in protein extraction and solubilisation (Rabilloud 2000) have considerably increased the width of 2Dim electrophoresis analysis.

The main challenge in the area is to visualize as many proteins as possible while maintaining the density of polypeptidic spots in two-dimensional gels at a manageable level. We have to take into account that each gel maps the protein content of a single sample and that the goal of most experiment in proteomic analysis is to compare several samples, that is, several gels.

Polypeptidic spots located at the same position across a set of gel also share the same identity, as long as the spot intensity is not so high that it merges with its neighbours. The emergence of immobilized pH gradients allowed the development of overlapping narrow-range pH gradient that have a increased resolution (Hoving et al. 2000). This type of approach allows the visualisation of an increasing number of proteins and preserves the separation power, which brings an additional level of depth to proteomic analysis.

It is also of advantage to use a detection method that is both as sensitive and as linear as possible, in order to achieve a quantitative analysis possible and to limit the amount of sample consumed. In this respect, it is important to underscore that the quantitative precision of proteomic analysis is in the range of ±10%, which permits the detection of small variations in protein amount with a good statistical confidence (usually twofold differences).

In spite of their sensitivity and linearity, methods based on radioactivity are not widely used to detect proteins. The most common methods for protein detection following two-dimensional electrophoresis rely either on colorants, heavy metals (mostly silver salts) or fluorescent probes. An extensive description and comparison of these different methods is out of the scope of the current article. Briefly, the detection sensitivity reaches 10 ng of protein for the methods based on organic colorants, 2–5 ng for those based on fluorescent probes, down to a fraction of a 1 ng for silver staining methods. However, the sensitivity of silver staining methods is variable from one protein to the other, and their dynamic range is quite limited as well. Methods based on organic colorants or fluorescent probes are far more homogeneous and linear: the dynamic range of fluorescence based methods spans more than three orders of magnitude. Moreover, whereas organic colorant or silver staining methods have seemingly reached their maximal sensitivity, fluorescence based methods still have some perspectives of improvement, especially if we consider the detection sensitivity (a few attomoles which corresponds to several femtograms).

3. PROTEOMICS ANALYSIS BY MASS SPECTROMETRY

Proteomics, analysis of a proteome (all proteins expressed by a genome at a given time in a given environment) is constituted of three parts, separation of proteins, enzymatic digestion and analysis of the generated peptide mixtures. The analysis of these peptides is always performed by mass spectrometry (MS). Indeed, two ionization modes, matrix-assisted laser desorption ionization (MALDI) and electrospray (ES) associated respectively to time-of-flight (TOF) and ion trap mass analyzers, are the methods of choice for proteomics analysis because they allowed analysis of peptides rapidly, with high sensibility and resolution. The domain of applications of the proteomics analysis is large and can include for example, the identification of proteins, the quantification of their variable expressions and the localization of their post-translational modifications.

Ionization by MALDI consists in the transfer of energy absorbed by a chemical matrix to sample embedded in this matrix thereby producing desorption of the analyte molecules as ions into the gas phase. Ions formed into the gas phase are then pulsed toward the mass analyzer TOF. The combination of techniques such as two-dimensional polyacrylamide-gel electrophoresis, mass spectrometry with MALDI-TOF coupled with search in protein sequence database allows the identification of a protein (Peptide Mass Fingerprint, PMF) (Henzel et al. 1993). Indeed, the proteins separated by 2D gel are digested with a specific protease such as trypsin to generate peptide mixtures. The masses of the peptides originating from the protein spot are determined by MALDI-TOF. These peptide masses are then compared with those derived from a protein sequence database (such as SwissProt) in order to identify the protein. In some cases, this is not sufficient to identify a protein, therefore, determination of partial amino acids sequence of one or several peptides is necessary for the identification. To obtain these information, analysis by tandem mass spectrometry (MSMS) is mandatory. Generally, electrospray ionization coupled to MSMS analyzer is used. But, since few years, systems like TOFTOF analyzers coupled to MALDI source were also used to analyzed directly the same sample in MS and MSMS modes if the search with PMF does not give reasonable results.

Electrospray is a method which consist in the "desolvatation" and ionization of molecules in solvent under electric field. It takes place in a source at atmospheric pressure. In several proteomics applications, a miniaturize source with a 20–200 nl/min flow can be used, about it is termed nano-electrospray (Mann and Wilm 1995). The electrospray source is coupled to a mass analyzer which enables to obtain amino acids sequence tags of peptides. Determination of a partial amino acids sequence by mass spectrometry coupled with search into database can identified in general without any ambiguity the analyzed protein (Peptide Sequence Tag). In addition, manual interpretation enables *de novo* sequencing of a peptide which could not been matched in the database search. The analyzer used in these cases can be a triple quadrupole, an ionic trap, a hybrid triple quadrupole-TOF or a FT-ICR instrument (Fourier Transform – Ion Cyclotron Resonance Mass Spectrometry). The principle of tandem mass spectrometry is to isolate a peptidic ion, fragment it in general by collision with gas atoms (e.g. N_2 or Ar) and finally analyze the obtained fragment ions. In general, such analyses can be performed with a nano-electrospray source alone (Shevschenko et al. 1996) or coupled with HPLC separation. This technique is particularly used in case of complex peptide mixtures resulting from, for example, digested proteins separated by 1D gel electrophoresis, or originating from protein complexes, organ extracts or total cell extracts. This strategy is often used for the differential analysis of protein complexes (Pandey et al. 2000). In the case of complex peptide mixtures, a bi-dimensional chromatographic separation (cation exchange chromatography followed by C18 reverse phase chromatography) like the Multi-dimensional Protein Identification Technology or MUDPIT method (Link et al. 1999; Washburn et al. 2001) can be used to reduce the complexity of the initial mixture and therefore identify proteins which are present in low abundance and rarely observed in a regular analysis.

The domain of applications in proteomic analysis are abundant and diverse. This method can be used to understand important biological processes involved in cell function and regulation (cell division, immune responses, etc.), to search markers characteristic of diseases for prognostic and diagnostic use, to identify target in drug discovery, etc (Burbaum and Tobal 2002; Petricoin et al. 2002; Seliger and Kellner 2002; Towbin et al. 2003). A comparative and quantitative analysis of proteins expressed, for example, in cells treated versus non-treated cells is often performed. Therefore, 2D-gel electrophoresis followed by proteins identification by mass spectrometry is still a widely used method (Towbin et al. 2003). Recently, an approach based on mass spectrometry and isotopic labeling of specific amino acids such as cysteine has been developed (Gygi et al. 1999). The chemical modification strategy ICAT (Isotope Coded Affinity Tag) (Griffin et al. 2001) uses a reagent which incorporate into a specific amino acid (cysteine) light or heavy isotopes of hydrogen (H/D), or in a more recent version light or heavy isotopes of carbon ($^{12}C/^{13}C$). These isotopic differences enable to compare protein abundance between two samples. The peptides containing such labeled cysteines are separated from the other peptides by affinity chromatography, which results in a reduction of the complexity of peptide mixtures. And finally these set of "marker" peptides are identified and quantified by capLC-ESMSMS or MALDI-TOF. This method reduces drastically the analyzed peptide complexity and enables therefore, identification of low abundance proteins. However, this reduction does not allow to obtain important information like the post-translational modifications. Also, quantitation relies often on one or two peptides per protein, which does not offer a good statistic. To overcome these problems two methods have recently emerged, SILAC and iTRAQ. SILAC (stable isotope labeling with amino acids in cell culture) is based on the introduction of an isotopically labeled amino acid (e.g. 13C6-Lysine) during cell culture resulting in full incorporation of this amino acid in each protein (Ong et al. 2002, 2003). However, the accurate quantitation with this approach is depending of the type of MS instruments used, for example, ion-trap type instruments are less accurate for quantitation purposes in comparison to quadrupole-time-of-flight or FT-ICR type instruments. Recently, a new quantitative method was developed, the isobaric peptide tagging system iTRAQ which label all primary amine, regardless of peptide class (Ross et al. 2004). The labeled isobaric peptides have the same mass and after fragmentation, these peptides generate diagnostic signature ions which enable easy quantitation in the low mass region. Because you do not lose important information with these two approaches, post-translational modifications are observed and accurate quantitation of their relative changes can be performed (Griffin et al. 2001; Ong et al. 2003).

Tandem mass spectrometry coupled to electrospray ion source allowed not only to identify proteins but also to characterize post-translational modifications of a protein (phosphorylation, acetylation, methylation, glycosylation, etc.). Indeed, the presence of such modifications induces an increase of the peptide molecular masses compared to the calculated masses based on the theoretical sequence, which often directly identifies the type of modification. In addition, tandem mass spectrometry allows in general precise localization of the modification at specific residue of the peptide. Analysis of modifications allows to understand biological mechanisms because several processes are controlled and/or induced by such modifications (Mann and Jensen 2003). Being able to quantify post-translational modifications such as phosphorylation, is very important since it is reversible and therefore plays an important role in the regulation of the protein function.

Phosphorylation sites in eukaryotes are present mainly on serine and threonine, tyrosine phosphorylation represents only around 0.1% of all phosphorylated sites. It was estimated that approximately one third of proteins from eukaryotic cells are phosphorylated. To understand the regulation of protein activity by phosphorylation, deep analysis of the phosphorylation on protein are necessary (Loughrey Chen et al. 2002; Bonenfant et al. 2003). Such analysis is shown in Figure 1A. A protein extract from cancer cells was immuno-purified by a specific antibody against phosphotyrosine. The protein mixtures obtained were then enzymatically digested and the peptides generated were analyzed by nanoCalHPLC-MS/MS in automatic mode (reverse phase C18 HPLC column, 75 μm × 15 cm, coupled to a electrospray-quadrupole-TOF mass spectrometer). In this mode, a mass spectrum is first acquired, then the three most intense peptidic ions are selected and fragmented.

The cycle (1 MS scan+3 MSMS scans of selected precursor ions) is repeated during the entire chromatographic separation. For example, in Figure 1B, the peptidic ion at

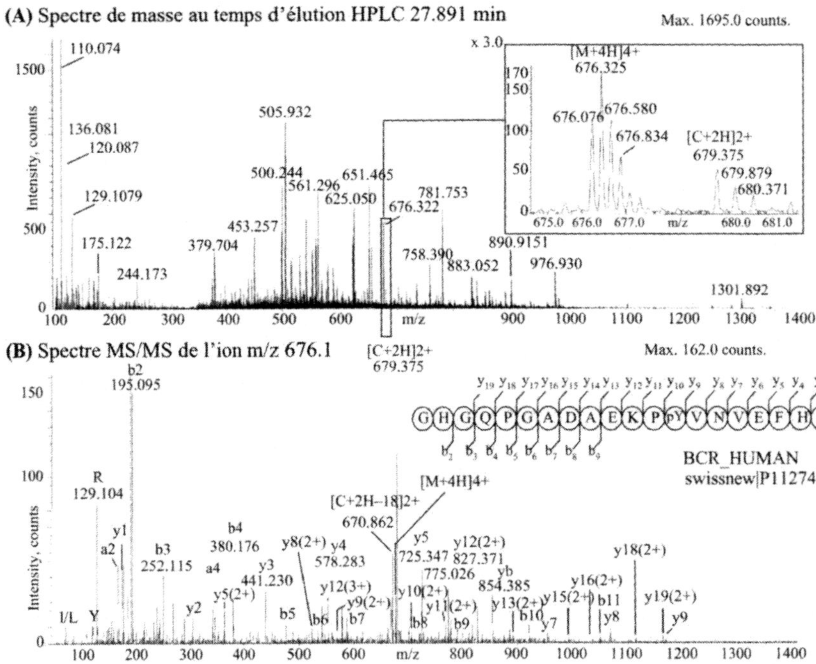

Figure 1. Example of identification and characterization of phosphorylation by mass spectrometry. (A) Mass spectrum of peptides generated after tryptic digestion of a protein extract of cancer cell lines. (B) MS/MS mass spectrum of the ion at *m/z* 676.076 corresponding to peptide 165–185 of BCR protein phosphorylated on tyrosine 177.

m/z 676.076 was selected and fragmented. The mass window for selecting precursor ion was relatively large in automatic mode, therefore in addition the peptidic ion at *m/z* 676.076, a peptidic ion doubly charged at *m/z* 679.375, was also selected. However, the ion fragments observed could almost all be attributed to peptidic ion at *m/z* 676.076 because these two parent ions had very different molecular masses and the fragmentation energy applied depend directly to the mass of the peptide. Finally, total sequencing of the peptide was obtained and the presence of y10 and y11 fragment ions localized the phosphorylation on tyrosine 177 of the BCR protein.

Mass spectrometry is the method of choice for identification of proteins, quantitation of proteins expression at a given time but also for characterization of post-translational modifications. Indeed, allowing direct analysis of peptides and proteins, the MALDI and ES ionization modes have been revolutionary in the field of mass spectrometry applied to biology.

4. MASS SPECTROMETRY IMAGING

A comprehensive analysis of a body fluid or a specific tissue, including target and/or biomarker identification often requires the analysis of only a group of cells or a small region in a tissue. In the drug discovery process, compound and metabolite biodistribution is a key information, which could benefit from the development of sophisticated molecular imaging techniques. *In vivo* methods, such as X-Ray, Magnetic Resonance Imaging (MRI), Positron Emission Tomography (PET), Near-Infrared Fluorescent (NIRF) or bioluminescence imaging, needs reporters to give access to the localization of specific biomolecules in real-time. This requires the development of a tag for each analyte group, and therefore significantly reduces the number of detectable analytes for one experiment set, usually to just one. Besides, the results depend strongly on the specificity of the tag or the label.

In biomedical research, mass spectrometry (MS) is considered as the method of choice for the analysis of metabolites, peptides and proteins, due to its high specificity and unmatched sensitivity (as low as attomol level), especially since the development of soft ionization techniques. Introduced for biopolymer MS analysis in 1987, matrix-assisted laser desorption/ionization (MALDI) applies a weak organic acid (e.g. sinapinic acid, α-cyano-4-hydroxycinnamic acid or 2,5-dihydroxybenzoic acid) to the sample, and the resulting mixture interacts with UV laser pulses. The analyte being embedded in an excess of this so-called matrix, it is ionized by energy and proton transfer from the matrix to the analyte. The molecular weight of theses ions are typically determined using a time-of-flight (TOF) mass analyzer.

The emerging technology MALDI Mass Spectrometry Imaging (MSI) is an extension of MALDI-TOF analysis, since it allows the generation of spatially resolved distribution of analytes, simply by sequentially acquiring a mass spectrum at defined positions. It offers the flexibility of detection without reporter molecules, as well as sensitivity and resolution needed for biological tissues. This *ex-vivo* assay is based on the concept of applying MALDI TOF MS directly on tissue sections. A very important aspect in MALDI MSI on tissue sections is the high number of different analyte classes which can be simultaneously analyzed, including proteins, lipids, carbohydrates and metal ions. This takes benefit of the high dynamic range of the method, even though some analytes are more efficiently ionized during MALDI process (Knochenmuss et al. 1996; Wang and Fitzgerald 2001).

4.1. Principles and Methods

Currently developed for many applications (Stoeckli et al. 1999; Stoeckli et al. 2001, 2002, 2003; Chaurand et al. 2004), MALDI MSI is achieved by rastering sequentially the surface of a defined area while acquiring a mass spectrum from every location (see Figure 2). A typical sample preparation for MSI involves the fixation of the sample, for example, tissue section, on a MALDI plate and the application of the matrix solution over the latter, either as a thin layer or as a spot pattern, to get co-crystallization of analytes with matrix while solvents evaporate. Once dried, the sample is introduced in the mass spectrometer, where, for each defined image position, short UV laser pulses are fired onto the surface to generate ions. Those are analyzed by the TOF instrument and a mass spectrum is acquired.

The acquisition of the mass spectra for peptide and protein imaging is done in our lab on commercial MALDI TOF instruments (Voyager sSTR, 4700 Proteomics Analyzer: Applied Biosystems, Framingham, MA) equipped with a Nd:YAG laser. The acquisition of the data is controlled by an in-house designed software (MALDI MS Imaging Tool, MMSIT). The area to be scanned is specified as a pattern of points, equidistant from each other. The distance is chosen referring to the diameter of the

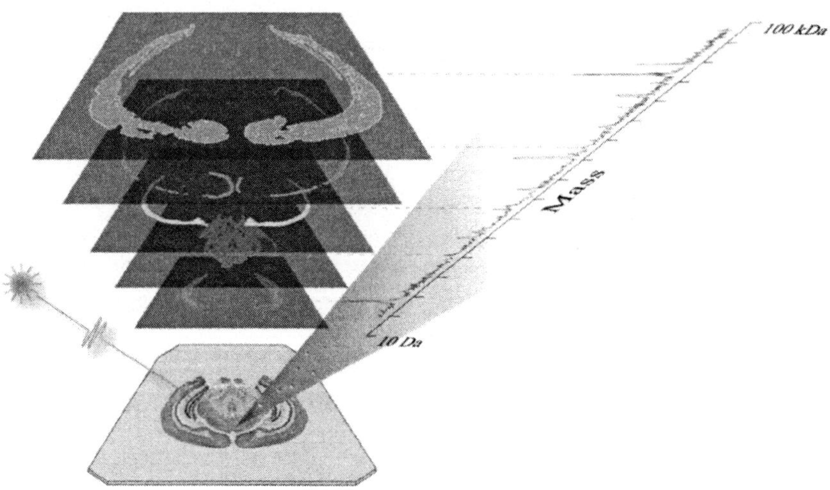

Figure 2. MALDI mass spectrometric imaging. A laser is rastered over a tissue sample while acquiring a complete mass spectrum from each position, allowing to generate molecular images for multiple analytes.

laser spot, here 50 μm – the smallest realistic raster step size – and depending on the lateral resolution required. To limit the data file size, sections or membranes have been imaged with a resolution of 100 μm. Data are recorded without compression by the MMSIT in the Analyze Image 7.5 data format (Mayo Foundation, Rochester, MN) and raw files are transferred to an image processing application named "BioMap" (Stoeckli et al. 2002). This software is based on IDL (Research Systems, Boulder, CO) and provides comprehensive tools for MS image analysis. From each data set, multiple images can be extracted, representing the spatial distribution of the analytes of interest. The same data can be used to compare mass spectra from multiple regions in the same sections. It also allows to select single points or regions of interest (ROIs) on the generated image and to display the corresponding mass spectrum.

4.2. Compound Imaging

Since drug imaging focuses on the low mass range (up to 1000 Da), the resulting MSI distribution can be skewed by an overlapping with other signals from the biological medium. Furthermore, matrix-related signal interference in this mass range is a common shortcoming in MS analysis. To improve MSI analysis accuracy and confidence, a "filtering" of mass signals is then required: for this purpose, specified molecular ions, for example, the drug molecular ion or one of its metabolites, can be fragmented. Fragmentation pattern being highly specific to a defined analyte, the analysis confidence is thus greatly improved. Usually, sections are coated with α-cyano-4-hydroxycinnamic acid (CHCA), known to enhance fragmentation. MSI analysis is so carried out in the so-called MS/MS mode, that is, by the acquisition of mass spectra resulting from the fragmentation of a given parent ion. The distribution is subsequently calculated in Biomap by selecting the mass of one or multiple identified fragment ion of the drug.

The example given in Figure 3 (**see colour insert**) refers to an *in-vitro* experiment studying the uptake of an antiflammatory drug into articular cartilage. For MS/MS drug imaging, two types of joint bovine cartilage have been treated by incubation in an anti-inflammatory drug solution. This procedure mimics the *in-vivo* situation in which the drug is directly injected into the joint synovial fluid. Figure 3 represents MS/MS images from bovine joint cartilage pieces incubated with compound solution. In the control section, that is, coming from an unresponsive cartilage (left panel), no signal related to the compound could be found in the image. Contrary to the control, drug-related signals were observed from cartilage sample (see Figure 3 right panel), indicating that the compound penetrates into the tissue. Signals were obtained from many localized regions, at the edge of the sample but also large areas inside the section. The presence of some isolated signals can be explained by the diffusion that can be induced by matrix application. Matrix coating is a crucial step in MSI: in this sole step, the solution must be able to extract analytes from tissue in order to get them cocrystallized with matrix but without inducing any lateral diffusion. The latter is responsible of the spatial signal artifacts, as illustrated in the right panel of 4.

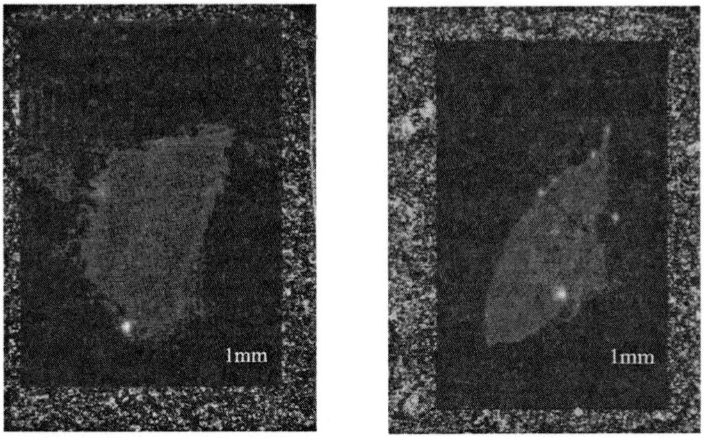

Figure 3. MS/MS molecular images of sections from bovine cartilage immersed in compound solution (left: control; right: drug). The images represent the penetration of the drug into the tissue (signal in red).

4.3. Peptide Imaging

In the mass range from 1 to 10 kD), MSI benefits of the best sensitivity of MALDI TOF instrumentation. Peptide imaging is therefore the optimal application of MSI. To get the maximum information from the tissue sections, sample preparation is a critical step and sections should be carefully processed. First, dissected tissue samples are snap frozen on a metal plate cooled with dry ice and stored at −80°C prior to further processing. Sections are then cut on a cryostat to a thickness of 10 μm and are directly deposited on metal plates compatible with the MALDI MS instrument. The tissue is first fixed by immersing it in 80% ethanol solution for 30 s and dried under vacuum. The matrix coating is achieved by directly depositing 10 $\mu l/cm^2$ of a freshly prepared solution. A slow crystallization is achieved by keeping the plate at 4°C for one hour. This process results in lower noise mass spectra, as already pointed out by Stoeckli and co-workers (Stoeckli et al. 2002). An internal calibrant, typically a peptide in the same mass range as the analyte of interest, can optionally be added to the matrix solution. Hence, the matrix coating can be assessed and a baseline correction can be performed over the scanned area using BioMap.

Coronal mouse brain sections have been processed as abovementioned and mapped by MSI analysis. The most intense MS signals have been selected, the distribution of the corresponding analytes being presented in Figure 4. Depending on the selected mass, different patterns corresponding each to one peptide distribution can be observed. These peptides are either present on the whole area (see Figure 4, A), or specific to an anatomic part of the brain, for example, Figure 4C, D and E. Some distributions are confined to specific brain regions, as pointed out by molecular images B and E or C and F in Figure 4. One key issue for MALDI MSI is the matrix coating. Figure 4c illustrates this problem: the analyte signal, mainly detected in the parietal meninges, is unevenly distributed on the two brain hemispheres. The use of an internal standard for signal normalization is one of the means to overcome this problem. Besides, MS intensities do not necessarily reflect concentration of analytes: the fact that some intensity distortions could occur due to the tissue nature which influences ionization efficiency and could cause ion suppression effect has to be taken into account.

Referring to their specific molecular images, analytes of similar molecular weights can be resolved thanks to mismatched spatial distributions, as illustrated by the peaks A and E of the global mass spectrum in Figure 4. The match of these peptide images to the anatomical images clearly demonstrates that spatial information is preserved during the sample preparation and imaging process and illustrates the relevance of the information provided by MSI: peptide identification and localization on tissue slices is the first step to potentially point out novel disease biomarkers and drug targets.

Once a biomarker has been identified, it can be used to evaluate the action of a drug on a specific pathology. One of the pathological features of Alzheimer's disease (AD) is the presence of amyloid deposits in senile plaques and in blood vessel walls. These deposits are mainly composed of amyloid beta (Aβ) peptide fragments of 4–5 kDa molecular mass, all derived from amyloid precursor protein (APP). The spatial analysis of Aß fragments, which are assumed to be linked to Alzheimer's disease pathogenesis, takes full advantage of MALDI MSI. For analysis, brain sections from APP23 transgenic mice – featured by an enhanced enzymatic degradation of APP were prepared as described above. These mice develop massive Aβ deposits in cortical and hippocampal structures upon aging.

A single scan of a brain section allows simultaneous localization of the different peptides, which were identified based on their masstocharge (m/z) ratio. As illustrated in Figure 5 (**see colour insert**), the Aβ(1–40) is by far the most abundant amyloid peptide. Two very intense regions are located in the parietal and the occipital cortical lobe and the third one close to the low part of Sylvian fissure, that is, in the hippocampus region. Referring to Figure 5, the normalized distributions of Aβ-(1–40) and Aβ-(1–42) show that they are the most abundant amyloid peptides. In agreement with these results, it has been reported that vascular amyloids were essentially composed of Aβ(1–40) and Aβ(1–42), the major peptide in aqueous cerebral cortical extracts from AD brains being the Aβ(1–40) proteolytic fragment of APP (Mori et al. 1992), while the insoluble amyloid Aβ(1–42) peptide was primarily in senile plaque cores (Miller et al. 1993; Roher et al. 1993). Thus, MSI gives access to the levels of known targets but also allows the mapping of the different targets with accuracy, which is not possible when whole-brain extracts are analyzed.

4.4. Protein Imaging

As mentioned above, MSI for tissue section analysis is based on MALDI-TOF technology. This latter allows very sensitive detection of a wide variety of compounds, with masses up to 10 kDa. For biomolecules of larger molecular weights, MSI is no longer very sensitive due to detection limitations, but also due to the MALDI process itself as well as the sample complexity, that is, tissue section. However, a major part of interesting proteins has their molecular weight well above 20 kDa, which means that this target mass range is currently out of reach for MSI tissue section analysis. To overcome this limitation, the molecular scanner, introduced by Bienvenut and coworkers (Bienvenut et al. 1999) for 2D gel electrophoresis post-processing, has

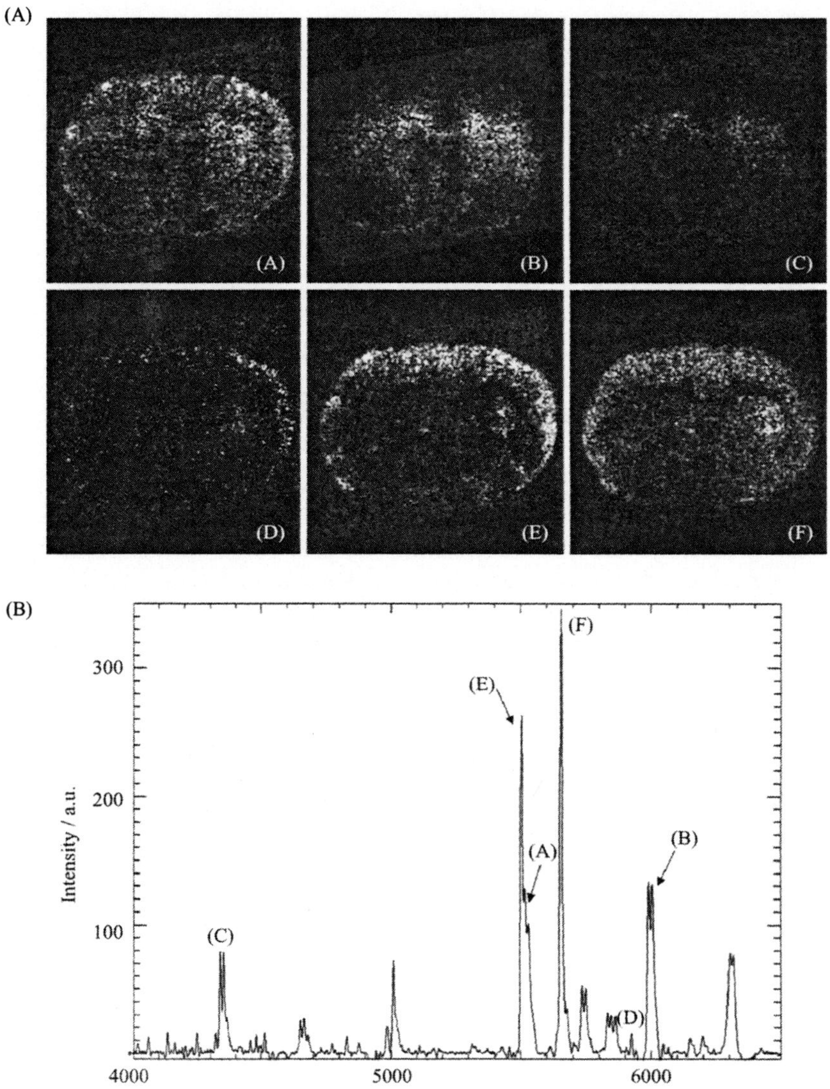

Figure 4. MSI of mouse brain coronal section. Molecular images for (A) *m/z* 5506; (B) *m/z* 6006; (C) *m/z* 4358; (D) *m/z* 5923; (E) *m/z* 5498; (F) *m/z* 5656; Average mass spectrum calculated over the mouse brain section area.

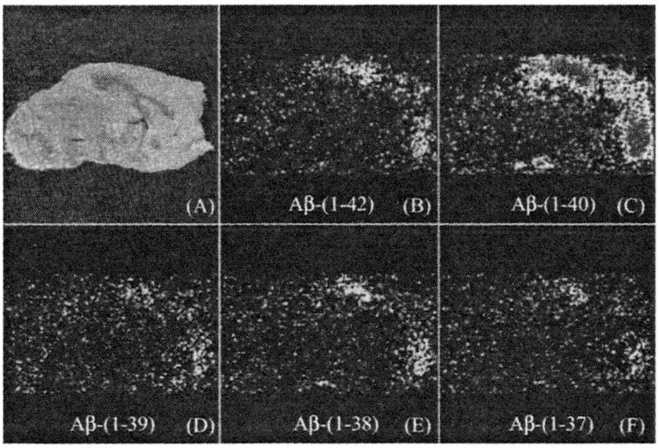

Figure 5. MSI on sagittal AD brain section. (A) Optical image of the sagittal AD brain section; (B) Aβ-(1–42) molecular image (*m/z* 4515.1); (C) Aβ-(1–40) molecular image (*m/z* 4330.9); (D) Aβ-(1–39) molecular image (*m/z* 4231.7); (E) Aβ-(1–38) molecular image (*m/z* 4132.6); (F) Aβ-(1–37) molecular image (*m/z* 4075.5).

been evaluated and further developed for tissue sections. The molecular scanner is based on the principle of protein identification by peptide mass fingerprint, that is, the analysis of the digestion of one protein. The protocol is very similar to a semi-dry electroblotting transfer, except that the starting material is a tissue section and that a so-called trypsin membrane is implemented in the sandwich between the section and the capture membrane. This membrane, in which trypsin has been covalently bound, allows the digestion of proteins while they are migrating towards the capture membrane. Once tryptic peptides are generated, they will also migrate under the action of the electric field and will be entrapped on the capture membrane through hydrophobic interactions. Thus, this latter is afterwards analyzed by MSI, resulting in molecular images of tryptic peptides, with lower mass, and so allows identification of proteins through peptide mass fingerprint analysis.

The molecular scanner approach benefits from MS/MS imaging similar to drug imaging. It helps to reduce analysis complexity while keeping a high level of confidence in the identification of the proteins. Indeed, the transblotting imposes the separation between proteins and the other components of the tissue. However, some proteins are more easily extracted from the tissue section than others, even with the use of detergent. Therefore, only a part of the whole protein set within a tissue preparation can be effectively analyzed. Nevertheless, as illustrated for a mouse brain slice in Figure 6, distributions of a large number of peptides can be measured while preserving the spatial information. Even if the explored mass range stays below 4 kDa, it should be kept in mind that these peptides are digest products of proteins. In Figure 6, the distributions corresponding to m/z 2324 and 3099 are similar, and may come from the same protein. To perform protein identification by database searching with a sufficiently high degree of confidence, at least four peptides issued from one single

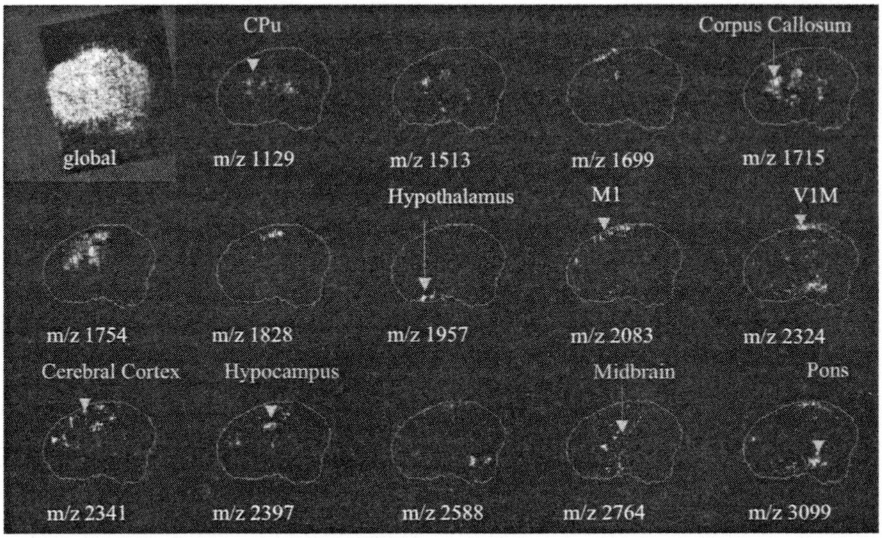

Figure 6. Mass spectrometry imaging of a capture membrane processed with molecular scanner on a mouse brain (CPu = caudate putamen (striatum); M1 = primary motor cortex; V1M = primary visual cortex, monocular region).

protein digestion are needed. If this requirement is not fulfilled, subsequent peptide analysis by MS/MS provides significant amino acid sequence information, allowing unambiguous protein identification.

5. CONCLUSIONS

MALDI MS and MALDI MSI takes full advantage of the high sensitivity of Biophotonics and mass spectrometry together with its ability to detect simultaneously a wide range of compounds, regardless from their nature and mass. Key issues in MALDI MSI that need to be overcome are analyte dependant ionization efficiency, suppression effects and cross-linking, and all of them are more or less related to sample processing and matrix coating. This step is indeed crucial since tissue sections or membranes should be wet enough to extract a part of the biological material and make it available to be co-crystallized with matrix. However, if the support becomes too wet when matrix solution is applied, biomolecules can be delocalized or even stripped off. As highlighted by the studies presented here, MALDI MSI can be used to track biomarkers involved in disease states, such as peptides or proteins, but also to map drug/tissue interactions, even though only a fraction of the whole set of biomolecules will actually be analyzed.

REFERENCES

Bienvenut, W.V., Sanchez, J.-C., Karmime, A., Rouge, V., Rose, K., Binz, P.A. and Hochstrasser, D.F. (1999) Toward a clinical molecular scanner for proteome research: Parallel protein chemical processing before and during werstern blot. Anal. Chem. *71*, 4800–4807.

Bjellqvist, B., Ek, K., Righetti, P.G., Gianazza, E., Görg, A., Westermeier, R. and Postel, W. (1982) Isoelectric focusing in immobilized pH gradients: principle, methodology and some applications. J. Biochem. Biophys. Methods *6*, 317–339.

Bonenfant, D., Schmelzle, T., Jacinto, E., Crespo, J.L., Mini, T., Hall, M.N. and Jenoe, P. (2003) Quantitation of changes in protein phosphorylation: a simple method based on stable isotope labeling and mass spectrometry. Proc. Natl Acad. Sci. U.S.A. *100*, 880–885.

Burbaum, J. and Tobal, G.M. (2002) Proteomics in drug discovery. Curr. Opin. Chem. Biol. *6*, 427–433.

Chaurand, P., Schwartz, S.A., Billheimer, D., Xu, B.J., Crecelius, A. and Caprioli, R.M. (2004) Integrating histology and imaging mass spectrometry. Anal. Chem. *76*, 1145–1155.

Görg, A., Obermaier, C., Boguth, G., Harder, A., Scheibe, B. Wildgruber, R. and Weiss, W. (2000) The current state of two-dimensional electrophoresis with immobilized pH gradients. Electrophoresis *21*, 1037–1053.

Griffin, T.J., Han, D.K., Gygi, S.P., Rist, B., Lee, H., Aebersold, R. and Parker, K.O. (2001) Toward a high-throughput approach to quantitative proteome analysis: expression-dependent protein identification by mass spectrometry. J. Am. Mass Spectom. *12*, 1238–1248.

Gygi, S.P., Rist, B., Gerber, S.A., Turecek, F., Gelb, M.H. and Aebersold, R. (1999) Quantitative analysis of complex protein mixtures using isotope-coded affinity tags. Nat. Biotechnol. *17*, 994–997.

Henzel, W.J., Billeci, T.M., Stults, J.T., Wong ,S.C., Grimley, C. and Wanatabe, C. (1993) Identifying proteins from two-dimensional gels by molecular mass searching of peptide fragments in protein sequence databases. Proc. Natl Acad. Sci. U.S.A. *90*, 5011–5015.

Hoving, S., Gerrits, B., Voshol, H., Müller, D., Roberts, R.C. and van Oostrum, J. (2002) Preparative two-dimensional gel electrophoresis at alkaline pH using narrow range immobilized pH gradients. Proteomics *2*, 127–134.

Hoving, S., Voshol, H. and van Oostrum, J. (2000) Toward high performance two-dimensional gel electrophoresis using ultrazoomgels. Electrophoresis *21*, 2617–2621.

Knochenmuss, R., Dubois, F., Dale, M.J. and Zenobi, R. (1996) The matrix suppression effect and ionization mechanisms in matrix-assisted laser desorption/ionization. Rapid Commun. Mass Spectrum. *10*, 871–877.

Link, A.J., Enj, J., Scjhieltz, D.M., Carmack, E., Mize, G.J., Morris, D.R., Gar, B.M. and Yates, J.R. 3rd (1999) Direct analysis of protein complexes using mass spectrometry. Nat. Biotechnol. *17*, 676–682.

Loughrey C.S., Huddleston, M.J., Shou, W., Deshaies, R.J., Annan, R.S. and Carr, S. (2002) Mass spectrometry-based methods for phosphorylation site mapping of hyperphosphorylated proteins applied to Net1, a regulator of exit from mitosis in yeast. Mol. Cell. Proteomics *1*, 186–196.

Mann, M. and Jensen, O.N. (2003) Proteomic analysis of post-translational modification. Nat. Biotechnol. *21*, 255–261.

Mann, M. and Wilm, M. (1995) Electrospray mass spectrometry for protein characterization. Trends Biochem. Sci. *20*, 219–224.

Miller, D.L., Papayannopoulos, I.A., Styles, J., Bobin, S.A., Lin, Y.Y., Biemann, K. and Iqbal, K. (1993) Peptide compositions of the cerebrovascular and senile plaque core amyloid deposits of Alzheimer's disease. Arch. Biochem. Biophys. *301*, 41–52.

Mori, H., Takio, K., Ogawara, M. and Selkoe, D.J. (1992) Mass spectrometry of purified amyloid beta protein in Alzheimer's disease. J. Biol. Chem. *267*, 17082–17086.

O'Farell, P.H. (1975) High resolution two-dimensional electrophoresis of proteins. J. Biol. Chem. *250*, 4007–4021.

Ong, S.E., Blagoev, B., Kratchmarova, I., Kristensen, D.B., Stenn, H., Pandey, M. and Mann, M. (2002) Stable isotope labeling by amino acids in cell culture, SILAC, as a simple and accurate approach to expression proteomics. Mol. Cell. Proteomics *1*, 376–386.

Ong, S.E., Foster, L.J. and Mann, M. (2003) Mass spectrometric-based approaches in quantitative proteomics. Methods *29*, 124–130.

Pandey, A., Podtelejnikov, A.V., Blagoev, B., Bustelo, X.R. and Mann, M. (2000) Analysis of receptor signalling pathways by mass spectrometry:identification of Vav-2 as a substrate of the epidermal and platelet-derived growth factor receptors. Proc. Natl Acad. Sci. U.S.A. *97*, 179–184.

Petricoin, E.F., Zoon, K.C., Kohn, E.C., Barrett, J.C. and Liotta, L.A. (2002) Clinical proteomics: translating benchside promise into bedside reality. Nat. Rev. Drug Discov. *1*, 683–695.

Rabilloud, T. (2000) Proteome Research: Two-dimensional Gel Electrophoresis and Identification Methods. (Berlin: Springer).

Roher, A.E., Lowenson, J.D., Clarke, S., Woods, A.S., Cotter, R.J., Gowing, E. and Ball, M.J. (1993) Beta-Amyloid-(1–42) is a major component of cerebrovascular amyloid deposits: Implications for the pathology of Alzheimer disease. Proc. Natl Acad. Sci. U.S.A. *90*, 10836–10840.

Ross, P.L., Huang, Y.N., Marchese, J.N., Williamson, B., Parker, K., Hattan, S., Khainovski, N., Pillai, S., Dey, S., Daniels, S., Purkayastha, S., Juhasz, P., Martin, S., Bartlet-Jones, M., He, F., Jacobson, A. and Pappin, D.J. (2004) Multiplexed protein quantitation in Saccharomyces cerevisiae using amine-reactive isobaric tagging reagents. Mol. Cell. Proteomics *12*, 1154–1169.

Schwartz, S.A., Reyzer, M.L. and Caprioli, R.M. (2003) Direct tissue analysis using matrix-assisted laser desorption/ionization mass spectrometry: Practical aspects of sample preparation. J. Mass Spectrom. *38*, 699–708.

Seliger, B. and Kellner, R. (2002) Design of proteome-based studies in combination with serology for the identification of biomarkers and novel targets. Proteomics *2*, 1641–1651.

Shevschenko, A., Wilm, M., Vorm, O. and Mann, M. (1996) Mass spectrometric sequencing of proteins silver-stained polyacrylamide gels. Anal. Chem. *68*, 850–858.

Stoeckli, M., Chaurand, P., Hallahan, D.E. and Caprioli, R.M. (2001) Imaging mass spectrometry: A new technology for the analysis of protein expression in mammalian tissues. Nat. Med. *7*, 493–496.

Stoeckli, M., Farmer, T.B. and Caprioli, R.M. (1999) Automated mass spectrometry imaging with a matrix-assisted laser desorption ionization time-of-flight instrument. J. Am. Soc. Mass Spectrom. *10*, 67–71.

Stoeckli, M., Staab, D., Staufenbiel, M., Wiederhold, K.H. and Signor, L. (2002) Molecular imaging of amyloid beta. peptides in mouse brain sections using mass spectrometry. Analy. Biochem. *311*, 33–39.

Towbin, H., Bair, K.W., DeCaprio, J.A., Eck, M.J., Kim, S., Kinder, F.R., Morollo, A., Mueller, D.R., Schindler, P., Song, H.K., Van Oostrum, J., Versace, R.W., Voshol, H., Wood, J., Zabludoff, S. and Phillips, P.E. (2003) Proteomics-based target identification. J. Biol. Chem. 278, 52964–52971.

Wang, M.Z. and Fitzgerald, M.C. (2001) A solid sample preparation method that reduces signal suppression effects in the MALDI analysis of peptides. Anal. Chem. *73*, 625–631.

Washburn, M.P., Wolters, D. and Yates, J.R. 3rd (2001) Large-scale analysis of the yeast proteome by multidimentional protein identification technology. Nat. Biotechnol. *19*, 242–247.

Corresponding author: Michel Faupel (michel_daniel.faupel@novartis.com)

CHAPTER 16

DIFFERENTIAL EPITOPE IDENTIFICATION OF ANTIBODIES AGAINST INTRACELLULAR DOMAINS OF ALZHEIMER'S AMYLOID PRECURSOR PROTEIN USING HIGH RESOLUTION AFFINITY-MASS SPECTROMETRY

XIAODAN TIAN[1], MADALINA MAFTEI[1], MARKUS KOHLMANN[1], BERNADETTE ALLINQUANT[2], and MICHAEL PRZYBYLSKI[1]
[1] *University of Konstanz, 78457 Konstanz*
[2] *Inserm U573, France*

Table of Contents

E. Bertrand and M. Faupel (eds.), Subcellular Proteomics, 339–354.

Abstract: Several polypeptides comprising the carboxy-terminal domain of the ß-amyloid precursor protein (cAPP) were prepared by solid phase peptide synthesis, and employed as antigens for the determination of the epitopes recognised by anti-cAPP antibodies. Selective proteolytic epitope-excision and -extraction on the immobilised immune complexes, in combination with high resolution Fourier transform ion cyclotron resonance mass spectrometry (FTICR-MS) were used as major methods for epitope identification. The epitope recognised by a polyclonal anti-cAPP antibody (36-BO) was identified as APP(727–737), a sequence close to the APP transmembrane region. In contrast, the epitope recognised by a monoclonal anti-cAPP antibody (Jonas-mAb) was identified at APP(740–747) to be located more remote from the transmembrane region. The two adjacent, yet distinct epitopes recognised by two different antibodies should provide efficient tools for (i), molecular diagnostic applications, and (ii), the study of intracellular processing pathways of APP relevant to Alzheimer's disease, utilising suitable mass spectrometric and molecular imaging approaches.

1. INTRODUCTION

Amyloid-ß-peptide (Aß) is the primary peptide constituent of senile amyloid plaques in brain, a hallmark of Alzheimer's disease (AD), and has been implicated in the pathogenesis of the disease (Selkoe 1999). The 39–43-residue Aß peptide is a proteolytic product of the ß-amyloid precursor protein (APP). The proteolytic processing of APP leads to the release of a carboxy-terminal cytoplasmic domain (cAPP) which may be targeted to the nucleus and/or degraded in the cytoplasm (Selkoe et al. 1996; De Strooper and Annaert 2000; Cao and Sudhof 2001; Cupers et al. 2001). Thus, knowledge of the molecular degradation pathways of APP, cAPP, Aß and its aggregation products is essential for understanding the pathogenic mechanism(s) of AD at the molecular level. Since Aß-containing plaques are a defining pathological feature of AD, several approaches have been pursued for blocking Aß production, or increasing its clearance (Bard et al. 2000; Hong et al. 2000; Janus et al. 2000; Morgan et al. 2000; Seiffert et al. 2000; De Mattos et al. 2001; Doerfler et al. 2001; Dovey et al. 2001; Ghosh et al. 2001; Hadland 2001; Kitazume et al. 2001; Luo et al. 2001; McLaurin et al. 2002; Manea et al. 2004; Tian et al. 2004, 2005). Blocking proteolytic enzymes involved in the formation of Aß has the principal disadvantage that the same enzymes are involved in physiological processes (Hong et al. 2000; Seiffert et al. 2000; Doerfler et al. 2001; Dovey et al. 2001; Ghosh et al. 2001; Hadland et al. 2001; Kitazume et al. 2001; Luo et al. 2001). A second general strategy for preventing Aβß-aggregation is to enhance the Aß clearance (Bard et al. 2000; Janus et al. 2000; Morgan et al. 2000; Mclaurin et al. 2002; Tian et al. 2005). Recently, novel approaches are being pursued by immunisation, either actively using Aß-peptide derivatives (19, 20), or passively with anti-Aß antibodies to clear soluble or "pre-aggregates" of Aß in brain (De Mattos et al. 2001; Dodart et al. 2002; Du et al. 2003). In transgenic mice as AD models, immunisation with Aß(1–42) has been shown to produce antibodies that cleared the Aß plaques and reversed AD symptoms (Bard et al. 2000; Janus et al. 2000; Morgan et al. 2000; Mclaurin et al. 2002). However, a first clinical trial was withdrawn because of severe toxic side effects observed (Grubeck-Lobenstein et al. 2000; Monsonego 2002; Schenk 2002; Senior 2002). In previous studies, we have identified an Aß-plaque specific epitope recognised by antisera isolated from transgenic AD mice immunised with Aß(1–42), using specific proteolytic epitope excision of immobilised immune complexes and high resolution mass spectrometry (McLaurin et al. 2002; Manea et al. 2004; Tian et al. 2004; Zirah et al. 2004; Tian et al. 2005). The epitope recognised by these therapeutically active antibodies was identified at the sequence Aß(4–10) (McLaurin et al. 2002; Manea et al. 2004; Tian et al. 2004; Zirah et al. 2004; Tian et al. 2005); its specificity was ascertained by identification of the identical epitope recognised in Aß-plaques and -extracts from Aß-protofibrils, synthetic Aß(1–42), and polypeptides of the APP-ectodomain comprising the Aß(4–10) sequence (Manea et al. 2004; Tian et al. 2004; Zirah et al. 2004; Tian et al. 2005). These results raise the possibility that a vaccine comprising this epitope may be effective without severe side effects, and might lead to the development of new molecular vaccine mimics (Manea et al. 2004; Tian et al. 2004; Zirah et al. 2004; Tian et al. 2005).

Based on the previous results and due to several unknown features of the APP processing, we are interested in the detailed evaluation of the biochemical degradation pathway of APP, particularly that of intracellular APP following proteolytic formation of Aß. The cAPP domain consists of 47 amino acids and contains several potential signal sequences for trafficking, processing and binding of specific target molecules, including G0, Disabled-1, adapter proteins, and Fe65 (De Strooper and Annert 2000). In addition, a caspase cleavage site has been described leading to neurotoxic fragments in primary neurons and in neuroblastoma cell lines (Gervais et al. 1999; Pellegrini et al. 1999; Weidemann et al. 1999; Lu et al. 2000; Passer et al. 2000; Bertrand et al. 2001; Dumanchin-Njock et al. 2001). The presence of caspase-cleaved APP fragments has been reported in AD brain (Lu et al. 2000; Zhao et al. 2003). Binding of these proteins to cytoplasmic APP may play an important role in the pathogenesis of AD, thus necessitating sensitive and specific tools for analysis of cytoplasmic APP fragments.

A rabbit polyclonal antibody against the C-terminal region of APP (36BO) was generated using synthetic APP(724–770), and has been successfully used to detect C-terminal APP fragments in granules (Langui et al. 2004). Furthermore, a commercially available monoclonal antibody (Jonas-mAb) directed against cytosolic APP has been shown to be a valuable tool for studying the membrane location of APP in brain cells and for purifying APP from human brain tissue (De Strooper and Annert 2000). A crucial precondition for using anti-cAPP antibodies in evaluating biochemical pathways of APP is the knowledge of their molecular recognition specificity. For the identification of epitope structures to both monoclonal and polyclonal antibodies, mass spectrometric methods by selective, limited proteolytic digestion of intact antigen-antibody complexes (epitope excision), and by antibody selection of antigen digestion mixtures (epitope extraction) have been previously developed in our laboratory (Suckau et al. 1992), and successful applications to linear and assembled epitope determinations have been reported (Suckau et al. 1992; Macht et al. 1996). In the present study, the epitopes of both antibodies was identified using epitope-excision and-extraction and high resolution mass spectrometric methods (Tian et al. 2005), and were ascertained by immunoanalytical characterisation of the residues essential for antibody binding.

2. MATERIALS AND METHODS

2.1. Antibody Production and Immobilisation

A recombinant GST-C-APP fusion protein (724–770) was expressed in *E. coli* BL21 and purified using GSH-Sepharose (Amersham Biosciences). Immunisation of rabbits was performed by Animal Pharma Services, Inc (Healdsburg, CA). A first injection of 150 μg was followed 30 days later by five injections of 100 μg at one week intervals. After 15 days, additional injections of 100 μg were continued at one week intervals. Prior to all injections blood samples were taken to check the production of antibodies.

Purification of antisera (10 ml) for mass spectrometric epitope analysis was performed on Protein A-Sepharose (Amersham Biosciences) according to manufacturer's instructions. Protein quantification was performed using the micro-BCA kit (Pierce, Rockford); positive elution fractions were concentrated to 7 μg/ml.

2.2. Immunoprecipitation

Primary neurons at 5DIV (10^6) were metabolically labelled for 12 h with 10 μCi ^{35}S-methionine. Cells were then processed as previously described (Rousselet et al. 1998). Briefly, cells were lysed in 200 μl 2% SDS-Tris buffer saline containing protease inhibitors. After boiling for 5 min, the extract was added to 800 μl of Tris-buffer saline containing 2.5% Triton X100. A solution of 10 μl primary cAPP polyclonal antibody (36BO) was then added and incubated overnight at 4°C under slight agitation. Protein A sepharose was then added and incubated for 3 h at 20°C under agitation. The beads were washed 4 times in Tris- buffered saline containing 0.005% SDS and 0.1% Triton X100. A final washing step was performed with Tris-EDTA and 500 mM NaCl, and beads were eluted in Laemmli buffer.

2.3. Peptide Synthesis

The C-terminal APP peptides were synthesised on an Abimed EPS 221 semi-automatic peptide synthesiser using standard SPPS Fmoc chemistry. All chemicals were of analytical grade or higher purity. Fmoc amino acids (Novabiochem; Bad Soden, Germany) were coupled using Benzotriazole-1-yl-oxy-tris-pyrrolidino-phosphonium hexafluorophosphate (PyBOP)/N-methylmorpholine (NMM) in dimethylformamide. Coupling times of 40 min and 5 min deprotection times in 20% (v/v) piperidine in DMF were employed for all syntheses. Final cleavage of peptides from the resin was performed with TFA-H_2O-triethylsilane (90:5:5) for 2 h. Peptides were precipitated by adding a 7-fold molar excess of *tert*-butyl methyl ether, and kept overnight at −20°C. The crude peptide was then purified by reversed-phase C_{18} HPLC on a Waters -Bondapak column (250 × 20 mm). A linear gradient of 0.1% aqueous TFA (solvent A) and acetonitrile containing 0. 1% TFA (solvent B) was used. Sequences and homogeneities of all peptides were confirmed by mass spectrometry.

2.4. Circular Dichroism

CD spectroscopy was carried out with a JASCO J-715 spectropolarimeter equipped with a Xe-lamp for recording spectra in the UV region. The spectra were recorded at 25°C with a scan speed of 1 nm/min in 0.1 cm cuvettes. The spectra were averages of six scans in a wavelength range between 190 and 260 nm. Before each measurement a blank spectrum of solvent was recorded. Molar ellipticities were calculated using the Jasco-700 Software.

2.5. Proteolytic Digestion

Proteolytic enzymes used were of analytical grade or highest available purity. L-(Tosylamido-2-phenyl) ethyl chloromethyl ketone (TPCK)-treated trypsin and endoproteinase Glu-C were obtained from Sigma (St Louis, USA). Tryptic digestion of cAPP peptides was performed at pH 8 at 37°C for 4 h. Synthetic peptides were dissolved in 50 mM NH_4HCO_3 buffer, and a 2% protease solution was added. Rection mixtures were quenched by freezing, and proteolytic fragment mixtures were directly analysed by MALDI-FTICR-MS. For digestion with GluC-protease, identical digestion conditions and times were employed.

2.6. Antibody Immobilisation

A solution of 100 μg IgG in 500 μl 0.2 M $NaHCO_3$/0.5 M NaCl (pH 8) was added to n-hydroxysuccinimidyl (NHS)-activated 6-aminohexanoic acid-coupled sepharose (Sigma, St Louis, USA), and allowed to bind for 60 min at 20°C before transferring onto a microcapillary (MoBiTec, Göttingen, Germany). Columns were washed alternately with blocking (ethanolamine/NaCl) and washing buffer (NaAc/NaCl).

2.7. Epitope-excision and -extraction

Epitope excision was performed by application of 2–5 μg antigen to the antibody microcolumn (Tian et al. 2005). Binding was performed for 60 min at 20°C. After washing with binding buffer, proteolytic digestion was carried out on the column with 0.2 μg protease in 200 μl PBS for 2 h at 37°C. Supernatant unbound peptides were removed using washing buffer and the epitope dissociated by addition of 500 μl 0.1% trifluoroacetic acid (TFA). After incubation for 15 min at 20°C, the epitope eluate was lyophilised and reconstituted in 10 μl NaAc buffer.

2.8. FTICR Mass Spectrometry

FTICR-MS was performed with a Bruker Apex II 7T FT-ICR mass spectrometer (Bruker Daltonics, Bremen, Germany) equipped with an Apollo II electrospray/nanoelectrospray multiport ion source, and an external Scout 100 fully automated X-Y target stage MALDI source with pulsed collision gas (Tian et al. 2005). A 100 mg/ml solution of 2,5-dihydroxybenzoic acid (DHB, Aldrich, Germany) in acetonitrile: 0.1% TFA in water (2:1) was used as the MALDI matrix. The pulsed UV-nitrogen laser is operated at 337 nm. Ions generated by the UV-laser shots were accumulated in the hexapole for 0.5–1 s at 15 V and extracted at −7 V into the analyser cell. Aliquots of 0.5 μl of sample solution were mixed on the stainless-steel MALDI sample target and allowed to dry. Typical ESI conditions were ∼2 kV needle voltage and 100 nA spray current. Ions were accumulated in the hexapole for 2 s and then transferred into the cylindrical ICR cell.

3. RESULTS

3.1. Synthesis and Structural Characterisation of Carboxy-terminal APP Polypeptides

C-terminal APP polypeptides were synthesised on a semi-automated peptide synthesiser using the Fmoc strategy as described in Materials and Methods. For the direct synthesis of large polypeptides such as the complete cAPP domain of 47 amino acid residues, SPPS protocols using improved capping and double-coupling procedures were employed (Manea et al. 2004). The structures and purities of all synthesized polypeptides were ascertained by MALDI-FTICR-MS; molecular ions observed (monoisotopic molecular masses) are summarised in Table 1. Both MALDI- and ESI-FTICR-MS provided high accuracies of mass determinations of approximately 1 ppm (Table 1 and Figure 1).

Fragment ions obtained by MS/MS experiments covered almost the complete sequences and provided detailed information for structural characterisation. Sequences of the synthesized peptides were further confirmed using capillary/skimmer (or nozzle/skimmer) fragmentation. Figure 1A shows the ESI-FTICR mass spectrum of cAPP(723–767). The experimentally determined and calculated molecular masses, as well as sequence-specific fragments are in complete agreement. The $[M+7H]^{7+}$ and $[M+6H]^{6+}$ ions were isolated and dissociated with a capillary/ skimmer potential of 130 V, as illustrated in Figure 1B by the corresponding ESI-FTICR fragment ion spectrum of the 6+ charged ion. The ions obtained in this experiment were mostly b- and y-fragments, which covered almost the entire sequence of the peptide.

To obtain secondary structure information of the cAPP peptides, CD spectra were recorded in different solvents. These data (not shown) provided strongly negative ellipticities around 200 nm in both phosphate buffer (pH 7) and deionised water, indicating flexible, unstructured states.

3.2. Epitope Identification of a Polyclonal Anti-cAPP Antibody (36BO)

The polyclonal antiserum, obtained by immunisation of rabbits as described in Materials and Methods was initially used for immunoprecipitation of metabolically labelled primary cortical neurons, providing a major band at 110 kDa corresponding to intact neuronal APP (Figure 2A). The antibody was also tested by western analysis yielding

Table 1. Monoisotopic molecular ions [M + H]+ cAPP polypeptides, and their proteolytic fragments by trypsin and GluC-protease digestion determined by MALDI-FTICR-MS

Peptide	Sequence	$[M+H]^+_{calc.}$	$[M+H]^+_{exp.}$	Δm(ppm)
APP(724–770)	KKKQYTSIHHGVVEVDAAVTPEERHLSKMQQNGYENPTYKFFEQMQN-$CONH_2$	5563.7274	5563.7719	8.0±0.5
APP(723–767)	LKKKQYTSIHHGVVEVDAAVTPEERHLSKMQQNGYENPTYKFFEQ-$CONH_2$	5303.6695	5303.7001	5.9±0.3
APP(724–737)	KKKQYTSIHHGVVE-COOH	1653.9077	1653.9012	−3.9±0.6
APP(747–766)	RHLSKMQQNGYENPTYKFFE-COOH	2517.1985	2517.2141	6.2±0.8
APP(747–770)	RHLSKMQQNGYENPTYKFFEQMQN-$CONH_2$	3017.4145	3017.4385	−7.9±0.8
APP(738–766)	VDAAVTPEERHLSKMQQNGYENPTYKFFE-COOH	3428.6221	3428.6002	−6.4±0.6
APP(738–770)	VDAAVTPEERHLSKMQQNGYENPTYKFFEQMQN-$CONH_2$	3928.8381	3928.8498	3.0±0.4
APP(752–763)	MQQNGYENPTYK-COOH	1472.6480	1472.6552	4.9±0.4
PP(727–747)	QYTSIHHGVVEVDAAVTPEER-COOH	2337.1475	2337.1373	−4.4±0.3

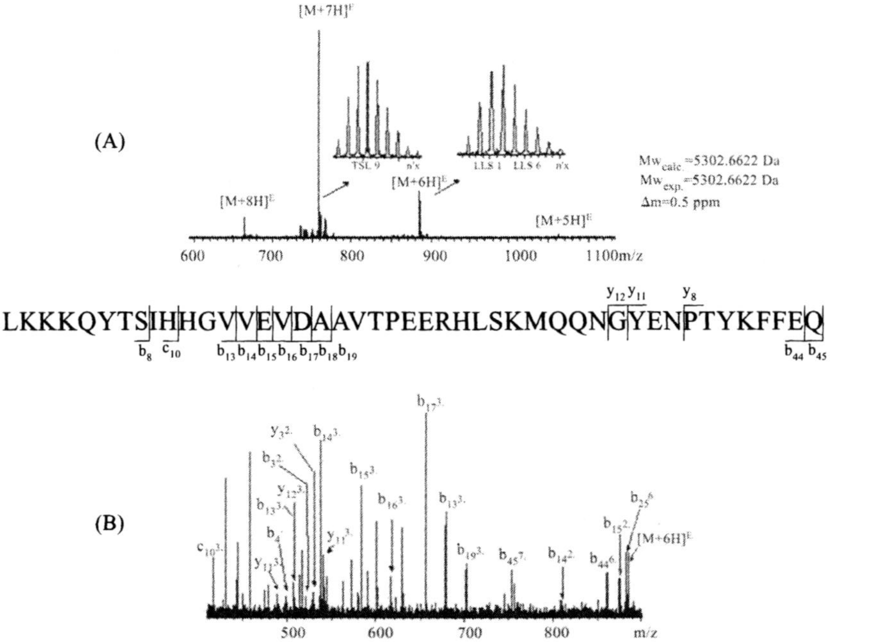

Figure 1. Characterisation of synthetic cAPP polypeptides by FTICR-MS and FTICR-MS/MS. (A) ESI-FTICR spectrum of APP(723–767) (0. 1 μg/μl) showing molecular ions with 4+ to 8+ charge states; (B) ESI-FT ICR-MS/MS of 6+ charged ion (0.01 μg/μl) at capillary exit voltage of 130 V.

a band at 110 kDa similar to that observed previously with an anti-APP monoclonal antibody (22C11) and with other polyclonal antibodies (Tian et al. 2005) (Figure 2B). Both in immunoprecipitation and in Western Blot analysis only full length APP was detected. Furthermore, this antibody was successfully employed for immunocytochemistry detection yielding a similar pattern of that previously observed using the 22C11 antibody (Figure 2C) (Allinquant et al. 1994).

For epitope elucidation of the polyclonal anti-cAPP antibody, the analytical scheme summarised in Figure 3 was employed (Macht et al. 1996; Tian et al. 2005). The antibody was first immobilised on Sepharose embedded in a microcolumn (s. Materials and Methods). The synthetic peptide cAPP(724–770) was used as antigen and initially cleaved by trypsin and Glu-C protease, yielding the proteolytic fragments summarised in Table 1.

In the epitope extraction experiment, cAPP(724–770) was cleaved by trypsin in solution. Three proteolytic fragments were identified by MALDI-FTICR-MS, APP(727–747), APP(752–763) and APP(764–770) (Figure 4A). The same fragments were also detected in the supernatant fraction after the affinity column was incubated with the tryptic peptide mixture. After washing with 30 ml PBS buffer, no fragments were detected in the last 1 ml of the washing fraction (data not shown). In the elution

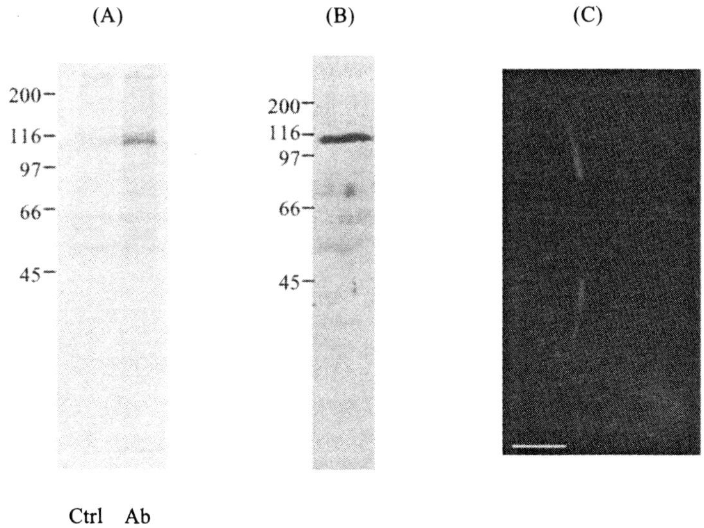

Figure 2. The polyclonal antibody to C-terminal APP 36BO recognises a band at 110 kDa in neuronal extracts and labels membranous APP. (A), APP immunoprecipitation after metabolic labelling of neurons. Ctrl, immunoprecipitation in the absence of cAPP antibody. The antibody immunoprecipitates a band at 110 kDa corresponding to the neuronal isoform of APP. (B) the same band is detected by immunoblot after western blotting of a neuronal extract (equivalent of 10^6 neurons). (C) the antibody labels the surface of primary neurons fixed with 4% paraformaldehyde. Note that stretches of brightly labelled neurites interrupted by APP negative regions are observed as previously described using the 22C11 recognizing the N-terminal part of APP (Bertrand et al. 2001); scale bar, 10 μm.

fraction, a single peak was present in the FTICR spectrum which could be identified as fragment cAPP(727–747), indicating a strong binding of this peptide to the antibody (Figure 4B). The same fragment was also identified in the epitope excision experiment. Therefore, there was no shielding of the cleavage sites at lysine and arginine residues, K727 and R747 due to interaction between the peptide and antibody.

Further epitope excision/extraction experiments of the polyclonal antibody were performed using GluC-endopeptidase. Five potential cleavage sites of cAPP(724–770) were efficiently cleaved in solution by GluC (Figure 4C), however only fragment APP(724–737) was identified in the elution fraction (Figure 4D). Since trypsin digestion generated the epitope-elution peptide cAPP(727–747), while GluC digestion produced the peptide cAPP(724–737), the epitope sequence, cAPP(727–737) could be derived. In order to ascertain this assignment, a two-step proteolytic cleavage was performed by combined epitope-extraction and -excision, in which the antigen was first cleaved by trypsin followed by endoprotease GluC digestion (s. Figure 3). The tryptic digest mixture of the peptide was first applied onto the affinity column; after 1 h the unbound fragments were removed by washing, and the remaining antibody-bound peptide APP(727–747) was cleaved further with GluC protease. After removal of the supernatant fragments, the immune complex was dissociated with 0.1% TFA.

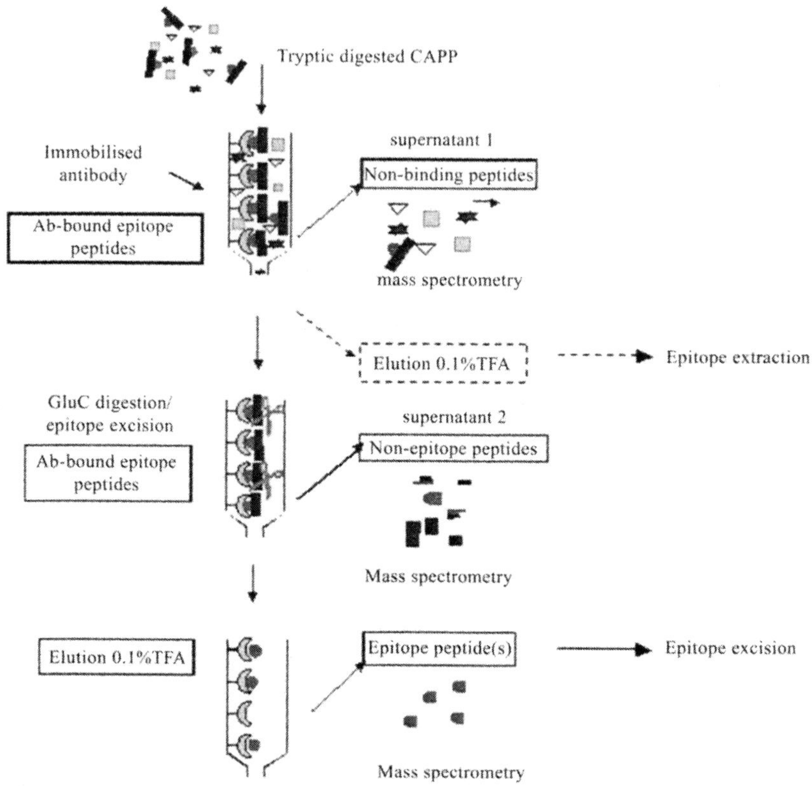

Figure 3. Analytical scheme of epitope-extraction and -excision. The antigen is first digested in solution and the fragment mixture added onto the antibody affinity column (epitope extraction); in the epitope excision procedure the antigen is first bound to the antibody affinity column and then the antibody-antigen complex is proteolytically digested. In this study, a two-step cleavage was performed consisting of (i) epitope extraction using trypsin, and (ii) epitope excision using GluC-protease digestion of the tryptic epitope-fragment, cAPP(727–747) that remained bound to the affinity column.

Only the epitope peptide APP(727–737) was identified in the epitope elution fraction (Figure 4). Interestingly, this epitope sequence is just adjacent to the epitope APP(740–747) recognised by the monoclonal anti-cAPP antibody (Jonas-mAb) (Tian et al. 2005).

3.3. Epitope Identification and Immunological Characterisation of a Monoclonal Anti-C-APP Antibody (Jonas-mAb)

The cAPP polypeptide APP(724–770) was also used as antigen for elucidation of the epitope recognised by a monoclonal anti-cAPP antibody (Jonas-mAb), which has been

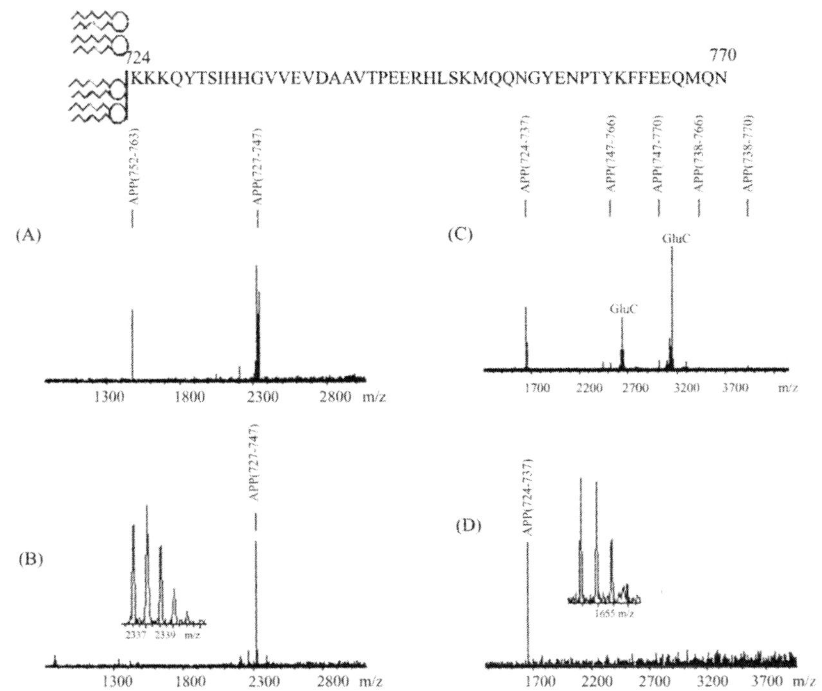

Figure 4. MALDI-FTICR-MS of trypsin and GluC epitope-extraction experiments of polyclonal anti-cAPP antibody using APP(724–770) as antigen. (A) Trypsin digestion of APP(724–770) in solution; (B) Elution fraction (epitope peptide) after trypsin digestion; (C) GluC digestion of APP(724–770) in solution; (D) Elution fraction (epitope peptide) after GluC digestion of APP(724–770).

previously used in studies of several APP polypeptides (Tian et al. 2005). An epitope extraction experiment using tryptic digestion provided the peptide APP(727–747) in the elution fraction, which was confirmed by an epitope excision- mass spectrometry experiment. When this antibody-bound epitope peptide, cAPP(727–747) was further cleaved with GluC protease, the peptide cAPP(740–747) was identified as the minimal epitope in the elution fraction (Figure 5).

Using synthetic epitope- containing peptides, immunoanalytical binding studies were carried out to further characterise the essential residues for the antibody binding. The biotinylated epitope peptides cAPP(733–751), cAPP(735–751), cAPP(737–751) and two mutant peptides in which residues Val-742 and Glu-745 were substituted by Lys were synthesised and examined by Dot-blot binding. In these experiments all three epitope peptides were found to bind to the Jonas mAb. In contrast, the mutant peptides, cAPP(737–751)Val742/Lys and cAPP(737–751)Glu745/Lys showed completely abolished binding to the antibody (Figure 6), indicating that these residues Val-742 and Glu-745 are an essential part of the core epitope.

APP-(737-751)		dot-blot Jonas mAb (10 mg/ml) 1 mg/ml	0,01 mg/ml
EVDA AVTPEERHLSK	15	+ + +	+ +
EVDA AKTPEERHLSK $V^{742}K$	15	–	–
EVDA AVTPKERHLSK $E^{745}K$	15	–	–

Figure 5. Epitope characterisation of the Jonas-mAb by Dot blot using synthetic mutant peptides. Wild type peptides were found to bind to the antibody, while mutants cAPP(737–751)Val742/Lys and APP(737–751)Glu745/Lys were devoid of binding affinity.

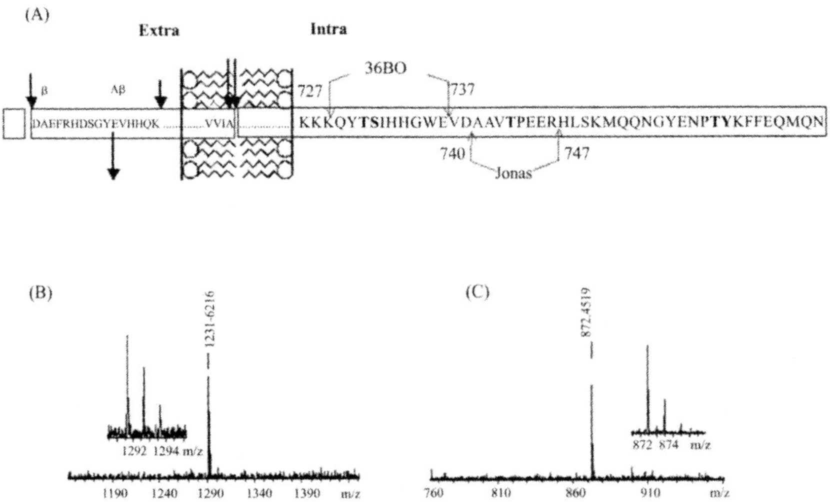

Figure 6. (A) Schematic representation of epitope sequences in cAPP identified for polyclonal (36BO) and monoclonal (Jonas-mAb) anti-cAPP antibodies. (B) MALDI-FTICR-MS of epitope peptide APP(727–737) recognised by polyclonal anti-cAPP 36BO antibody. (C) MALDI-FTICR-MS of epitope peptide APP(740–747) recognised by Jonas-mAb.

4. DISCUSSION AND PERSPECTIVES

In this study, a number of C-terminal APP polypeptides were synthesised and efficiently employed as antigens in the structural determination of epitopes recognised by monoclonal and polyclonal antibodies specific for the Carboxy-terminal, intracellular domain of APP, using proteolytic epitope-excision and -extraction in combination with high resolution FTICR-mass spectrometry. The results of these experiments specifically define the epitope of the cAPP polyclonal antibody (36BO) as the sequence, APP(727–737), while the epitope recognised by the Jonas-mAb was defined at the amino acids APP(740–747) including the potential Caspase-3 cleavage site of APP. The direct, chemical epitope identication approach using epitope excision/extraction -mass spectrometry provided differential information of epitope structures by comparison of shielding of specific proteolytic cleavage sites; this information is not directly amenable by affinity-binding such as quantitative ELISA, but can be readily ascertained and verified by binding studies using synthetic peptides. Hence, the polyclonal cAPP antibody 36BO can be used for detecting full length APP and large C-terminal APP fragments, while the monoclonal Jonas-mAb should be a useful tool for the study of APP processing in apoptotic neuronal cells and in brain extracts. Furthermore, the differential molecular detection of intact and cleaved APP could be a useful tool in the future development of molecular imaging techniques applied to the *in vivo* analysis of APP processing and Aß aggregation.

ACKNOWLEDGEMENTS

This work has been supported by the Deutsche Forschungsgemeinschaft, Bonn, Germany (Transregio-SFB Konstanz – Zuerich; and Biopolymer-Mass Spectrometry; MP). The expert assistance by Andreas Marquardt and Dmitry Galetskiy with the FTICR-mass spectrometry is gratefully acknowledged.

REFERENCES

Allinquant, B., Moya, K. L., Bouillot, C. and Prochiantz, A. (1994) Amyloid precursor protein in cortical neurons: coexistence of two pools differentially distributed in axons and dendrites and association with cytoskeleton. J. Neuroscience *14*, 6842.

Bard, F., Cannon, C., Barbour, R., Burke, R.L., Games, D., Grajeda, H., Guido, T., Hu, K., Huang, J., Johnson-Wood, K., Khan, K., Kholodenko, D., Lee, M., Lieberburg, I., Motter, R., Nguyen, M., Soriano, F., Vasquez, N., Weiss, K., Welch, B., Seubert, P., Schenk, D. and Yednock, T. (2000) Peripherally administered antibodies against amyloid-ß peptide enter the central nervous system and reduce pathology in a mouse model of Alzheimer disease. Nature Nat. Med. *6*, 916.

Bertrand, E., Brouillet, E., Caillé, I., Bouillot, C., Cole, G.M., Prochiantz, A. and Allinquant, B. (2001) A short cytoplasmic domain of the amyloid precursor protein induces apoptosis *in vitro* and *in vivo*, Mol. Cell. Neuroscience. *18*, 503–511.

Cao, X. and Sudhof, T.C. (2001) A transcriptionnally active complex of APP with Fe65 and histone acetyltransferase Tip60, Science *293*, 115.

Cupers, P., Orlans, I., Craessaerts, K., Annaert, W. and de Strooper, B. (2001) The amyloid precursor protein (APP)-cytoplasmic fragment generated by gamma-secretase is rapidly degraded but distributes partially in a nuclear fraction of neurones in culture. J. Neurochem. *78*, 1168.

De Strooper, B. and Annaert, W. (2000) Proteolytic processing and cell biological functions of the amyloid precursor protein, J. Cell Sci. *113*, 1857.

DeMattos, R.B., Bales, K.R., Cummins, D.J., Dodart, J.C., Paul, S. M. and Holtzman, D.M. (2001) Peripheral anti-Aß antibody alters CNS and plasma Aß clearance and decreases brain Aß burden in a mouse model of Alzheimer's disease, . Proc. Natl. Acad. Sci. U.S.A. *98*, 8850.

Dodart, J.C., Bales, K.R., Gannon, K.S., Greene, S.J., DeMattos, R. B., Mathis, C., DeLong, C.A., Wu, S., Wu, X., Holtzman, D.M. and Paul, S.M. (2002) Immunization reverses memory deficits without reducing brain Aß burden in Alzheimer's disease model, Nature Nat. Neurosci. *5*, 452.

Doerfler, P., Shearman, M.S. and Perlmutter, R.M. (2001) Presenilin-dependent Gamma-secretase activity modulates thymocyte development. Proc. Natl Acad. Sci. U.S.A. *98*, 9312.

Dovey, H.F., John, V., Anderson, J.P., Chen, L.Z., de Saint Andrieu, P., Fang, L.Y., Freedman, S.B., Folmer, B., Goldbach, E., Holsztynska, E.J., Hu, K.L., Johnson-Wood, K.L., Kennedy, S.L., Kholodenko, D., Knops, J.E., Latimer, L.H., Lee, M., Liao, Z., Lieberburg, I.M., Motter, R.N., Mutter, L.C., Nietz, J., Quinn, K.P., Sacchi, K.L., Seubert, P.A., Shopp, G.M., Thorsett, E.D., Tung, J.S., Wu, J., Yang, S., Yin, C.T., Schenk, D.B., May, P.C., Altstiel, L.D., Bender, M.H., Boggs, L.N., Britton, T.C., Clemens, J.C., Czilli, D.L., Dieckman-McGinty, D.K., Droste, J.J., Fuson, K.S., Gitter, B.D., Hyslop, P.A., Johnstone, E.M., Li, W.Y., Little, S.P., Mabry, T.E., Miller, F.D. and Audia, J.E. (2001) Functional gamma-secretase inhibitors reduce ß-amyloid peptide levels in brain. J. Neurochem. *76*, 173.

Du, Y., Wei, X., Dodel, R., Sommer, N., Hampel, H., Gao, F., Ma, Z., Zhao, L., Oertel, W.H. and Farlow, M. (2003) Human anti-b-amyloid antibodies block ß-amyloid fibril formation and prevent b-amyloid-induced neurotoxicity, Brain *126*, 1935.

Dumanchin-Njock, C., Alves da Costa, C., Mercken, L., Pradier, L., Checler, F. (2001) The caspase-derived C-terminal fragment of ßAPP induces caspase-independent toxicity and triggers selective increase of Aß42 in mammalian cells J. Neurochem. *78*, 1153.

Gervais, F. G., Xu, D., Robertson, G. S., Vaillancourt, J. P., Zhu, Y., Huang, J., LeBlanc, A., Smith, D., Rigby, M., Shearman, M. S., Clarke, E. E., Zheng, H., Van der Ploeg, L. H., Ruffolo, S. C., Thornberry, N. A., Xanthoudakis, S., Zamboni, R. J., Roy, S. and Nicholson, D. W. (1999) Involvement of caspases in proteolytic cleavage of Alzheimer's amyloid-beta precursor protein and amyloidogenic A beta peptide formation, Cell *97*, 395.

Ghosh, A.K., Bilcer, G., Harwood, C., Kawahama, R., Shin, D., Hussain, K.A., Hong, L., Loy, J. A., Nguyen, C., Koelsch G., Ermolieff, J. and Tang, J. (2001) Structure-based design: potent inhibitors of human brain memapsin 2 (β-secretase), J. Med. Chem. *44*, 2865.

Grubeck-Loebenstein, B., Blasko, I., Marx, F.K. and Trieb, I. (2000) Immunization with beta-amyloid: could T-cell activation have a harmful effect?, Trends Neurosci. *23*, 114.

Hadland, B.K., Manley, N.R., Su, D.M., Longmore, G.D., Moore, C.L., Wolfe, M.S., Schroeter, E.H. and Kopan, R. (2001) gamma-Secretase inhibitors repress thymocyte development. Proc. Natl Acad. Sci. U.S.A. *98*, 7487.

Hong, L., Koelsch, G., Lin, X., Wu, S., Terzyan, S., Ghosh, A. K., Zhang, X. C. and Tang, J. (2000) Structure of the protease domain of memapsin 2 (β-secretase) complexed with inhibitor, Science *290*, 150.

Janus, C., Pearson, J., Mclaurin, J., Mathews, P.M., Jiaaang, Y., Schmidt, S.D., Chishti, M.A., Horne, P., Heslin, D., French, F., Mount, H.T.J., Nixon, R.A., Merken, M., Bergeron, C., Fraser, P.E., St George-Hyslop, P. and Westaway, D., (2000) Aß-eptide immunization reduces behavioural impairment and plaques in a model of Alzheimer's disease. Nature *408*, 979.

Kitazume, S., Tachida, Y., Oka, R., Shirotani, K., Saido, T.C. and Hashimoto, Y., (2001) Alzheimer's β-secretase, β-site amyloid precursor protein-cleaving enzyme, is responsible for cleavage secretion of a Golgi-resident sialyltransferase. Proc. Natl Acad. Sci. U.S.A. *98*, 13554.

Langui, D., Girardot, N., El Hachimi, K.H., Allinquant, B., Blanchard, V., Pradier, L. and Duyckaerts, C. (2004) Subcellular topography of neuronal Abeta peptide in APPXPS1 transgenic mice, Am. J. Pathol. *165*, 1465.

Lu, D.C., Rabizadeh, S., Chandra, S., Shayya, R.F., Ellerby, L.M., Salvesen, G.S., Koo, E.H. and Bredesen, D.E. (2000) A second cytotoxic proteolytic peptide derived from amyloid ß-protein precursor, Nature Nat. Med. *6*, 397.

Luo, Y., Bolon, B., Kahn, S., Bennett, B.D., Babu-Khan, S., Denis, P., Fan, W., Kha, H., Zhang, J., Gong, Y., Martin, L., Louis, J.C., Yan, Q., Richards, W.G., Citron, M. and Vassar, R. (2001) Mice deficient in BACE1, the Alzheimer's β-secretase, have normal phenotype and abolished β-amyloid generation, Nature Nat. Neurosci. *4*, 231.

Macht, M., Fiedler, W., Kürzinger, K. and Przybylski, M. (1996) Mass spectrometric mapping of protein epitope structures of myocardial infarct markers myoglobin and troponin T. Biochemistry *35*, 15633.

Manea, M., Horvati, K., Mezo, G., Cecal, R., Tian, X., Stefanescu, R., Hudecz, F. and Przybylski, M. (2004) Synthesis, antibody recognition and enzymatic stability of linear, cyclic and branched polypeptides containing a ß-amyloid plaques specific epitope, J. Peptide Pept. Sci. *10* (2004) 291.

McLaurin, J., Cecal, R., Kierstead, M., Tian, X., Phinney, A., Manea, M., French, J., Lambermon, M., Darabie, A., Brown, M., Janus, C., Chishti, M., Horne, P., Westaway, D., Fraser, P., Mount, H., Przybylski, M. and St George-Hyslop, P. (2002) Sera generated by Aß42-vaccination recognise a specific epitope, Inhibit Aß-cytotoxiticity and fibrillogenesis, and induce fibril disaggregation. Nature Nat. Med. *8*, 1263.

Monsonego, A. (2002) Immunogenic aspects of amyloid beta peptide: implications for pathogenesis and treatment of Alzheimer's disease, Neurobiol. Aging *23*, S112.

Morgan, D., Diamond, D.M., Gottschall, P.E. Ugen, K.E., Dickey, C., Hardy, J., Duff, K., Jantzen, P., DiCarlo, G., Wilcock, D., Connor, K., Hatcher, J., Hope, C., Gordon, M. and Arendash, G.W. (2000) Aß-peptide vaccination prevents memory loss in an animal model of Alzheimer's disease. Nature *408*, 982.

Passer, B., Pellegrini, L., Russo, C., Siegel, R.M., Lenardo, M.J., Schettini, G., Bachmann, M., Tabaton, M. and d'Adamio, L. (2000) Generation of an apoptotic intracellular peptide by g-secretase of Alzheimer's ßamyloid precursor protein. J. Alzheimer's Disease. *2*, 289.

Pellegrini, L., Passer, B.J., Tabaton, M., Ganjei, K. and d'Adamio, L., (1999) Alternative, non-secretase processing of Alzheimer's ß-amyloid precursor during apoptosis by caspase 6 and 8. J. Biol. Chem. *274*, 21011.

Rousselet, A., Fetler, L., Chamak, B. and Prochiantz, A., (1988) Rat mesencephalic neurons in culture exhibit different morphological traits in the presence of media conditioned on mesencephalic or striatal astroglia., Dev. Biol. *129*, 495–504.

Schenk, D. (2002) Amyloid-beta immunotherapy for Alzheimer's disease: the end of the beginning, Nat. Rev. Neurosci. *3*, 824.

Seiffert, D., Bradley, J.D., Rominger, C.M., Rominger, D.H., Yang, F., Meredith, Jr., J. E. Jr., Wang, Q., Roach, A.H., Thompson, L.A., Spitz, S.M., Higaki, J.N., Prakash, S.R., Combs, A.P., Copeland, R.A., Arneric, S.P., Hartig, P.R., Robertson, D.W., Cordell, B., Stern, A.M., Olson, R.E. and Zaczek, R. (2000) Presenilin-1 and -2 are molecular targets for gamma-secretase inhibitors. J. Biol. Chem *275*, 34086.

Selkoe, D.J. (1999) Translating cell biology into therapeutic advances in Alzheimer's disease. Nature *299*, A23.

Selkoe, D.J., Yamazaki, T., Citron, M., Podlisny, M. B., Koo, E. H., Teplow, D.B. and Haass, C. (1996) The role of APP processing and trafficking pathways in the formation of amyloid beta-protein. Ann. N.Y. Acad. Sci. *777*, 57.

Senior, K. (2002) Dosing in phase II trial of Alzheimer's vaccine suspended, . Lancet Neurol. *1*, 3.

Suckau, D., Mak, M. and Przybylski, M. (1992) Protein surface topology probing by selective chemical modificaton and mass spectrometric peptide mapping, Proc. Natl Acad. Sci. U.S.A. *89*, 5630.

Tian, X., Cecal, R., McLaurin, J., Manea, M., Stefanescu, R., Grau, S., Harnasch, M., Amir, S., Ehrmann, M., St George-Hyslop, P., Kohlmann, M. and Przybylski, M., (2005) Characterisation of target polypeptides of Alzheimer's Disease and specific antibody epitopes using high resolution FTICR-MS Eur. J. Mass Spectrom. *10*, 1255.

Tian, X., Cecal, R., Mezo, G., Manea, M., Stefanescu, R., McLaurin, J., St George-Hyslop, P., Hudecz, F. and Przybylski, M. (2004) New amyloid-derived vaccine lead structures against Alzheimer's disease discovered by affinity proteomics and high resolution mass spectrometry. In: Innovation and Perspectives in Antibody and Peptide Vaccine Development, R. Epton, ed. (Kingswinford, UK: Mayflower Worldwide Ltd), p. 79.

Weidemann, A., Paliga, K., Durrwang, U., Reinhard, F.M., Schuckert, O., Evin, G. and Masters, C.L. (1999) Proteolytic processing of the Alzheimer's Disease amyloid precursor protein within its cytoplasmic domain by caspase-like proteases, J. Biol. Chem. *274*, 5823.

Zhao, M., Su, J., Head, E. and Cotman, C.W. (2003) Accumulation of caspase cleaved amyloid precursor protein represents an early neurodegenerative event in aging and in Alzheimer's disease Neurobiology of Disease. *14*, 391.

Zirah, S., Stefanescu, R., Manea, M., Tian, X., Cecal, R., Kozin, S., Debey, P., Rebuffat, S. and Przybylski, M. (2004) Zinc binding agonist effect on the recognition of the ß-amyloid (4–10) epitope by anti-ß-amyloid antibodies. Biochem. Biophys. Res. Comm. *321*, 324.

Corresponding author: Pr. Michael Przybylski (Michael.Przybylski@uni-konstanz.de)

CHAPTER 17

LC-MALDI MS AND MS/MS – AN EFFICIENT TOOL IN PROTEOME ANALYSIS

DIETER R. MUELLER[1], HANS VOSHOL[1], ANNICK WALDT[1], BRIGITTE WIEDMANN[2] and JAN VAN OOSTRUM[1]
[1] *Novartis Institutes for BioMedical Research, Switzerland*
[2] *Infectious Disease Area, USA*

Table of Contents

Abstract: Liquid chromatography-matrix-assisted laser desorption/ionization mass spectrometry represents a sensitive, hyphenated MS- and MS/MS-technique with a broad range of

E. Bertrand and M. Faupel (eds.), Subcellular Proteomics, 355–380.

applications in all areas of proteome analysis. Whereas a number of interface types have been developed for coupling MALDI MS and liquid chromatography, in this chapter selected on-line and off-line types and techniques will be discussed with respect to their individual properties and performance. The technique is especially attractive in off-line mode where LC-separation and MS analyses are decoupled and each step can be performed at its individual optimum. Different speed of chromatographic separation and achievement of S/N criteria in MS or MS/MS mode can be optimized independently by individual adjustment of specific operating parameters. This flexibility makes LC-MALDI MS attractive for the analysis of peptide mixtures from low to medium complexity. Using sequential MS analysis of parallel LC runs (multiplexing), even highly complex samples can be handled. Quantitation at the MS and MS/MS level can be accomplished by a variety of labeling techniques, where the predominant formation of singly charged ions in MALDI alleviates the assignment of isotopomers. After discussing the level of complementarity between LC-MALDI and LC-ESI MS, selected applications of LC-MALDI MS are presented. Examples of membrane protein analysis applying 1D SDS PAGE are discussed in detail as well as applications in protein interaction analysis. These application examples clearly show that in all respects LC-MALDI MS and MS/MS are flexible and sensitive techniques which can be adapted to a wide range of different workflows.

1. INTRODUCTION

In proteome analysis the application of multidimensional separation methods to deconvolute the highly complex analyte mixtures is a prerequisite for the identification and characterization of individual components by mass spectrometric techniques. To achieve these objectives two main, but quite different workflows are currently pursued. While in the protein-centered approach extensive separation is performed at the protein level and peptides are generated only in the last step, the peptide-centered approach starts with the conversion of protein mixtures into highly complex peptide mixtures, which are then separated accordingly. The classical 2-dimensional gel-based approach combines protein separation according to pI (isoelectric focusing, IEF) and molecular mass (polyacrylamide gel-electrophoresis, SDS-PAGE) and the protein mixtures are generally well resolved. The resulting individual protein spots are of low complexity and the corresponding peptide mixtures can be analyzed directly by MALDI peptide mass fingerprinting (PMF) and MALDI MS/MS partial sequencing. By reducing protein separation to one dimension, the remaining complexity of the mixture requires at least one additional high resolution separation step at the protein or peptide level (e.g. after SDS-PAGE). In the latter case, reversed-phase microbore and nano-HPLC are the methods of choice because they provide high separation power and reasonable loading capacity. In the purely peptide-centered approach to proteomics, RP-HPLC is combined as second step with ion exchange chromatography as the first dimension of separation. In this workflow peptides are traditionally analyzed by on-line electrospray LC-MS and -MS/MS. State-of-the-art equipment allows to perform peptide mass detection and partial sequencing of a small number of precursor ions in seconds but depth of analysis is of course restricted by the number of experiments which can be performed on-line during elution of a chromatographic peak. Already on this basis, decoupling of liquid chromatography and MS analysis would be advantageous. This can be achieved by depositing LC-effluents as traces or spots together with matrix on a suitable carrier material before they are analyzed off-line by MALDI mass spectrometry. The chromatogram is thus frozen and with minimal precautions (exclusion of humidity and light) stable for at least days. In a first step MS analysis of the complete trace or spotset provides now time-resolved masses and intensities of all peptide ions, and candidate precursor ions for MS/MS can be selected from positions with highest intensity. This is not only important for protein identification but also for optimal quantitation in MS as well as in MS/MS mode. With current instrumentation (tandem time-of flight instruments, MALDI-ion traps and other MS combinations) partial sequence information can thus be obtained for selected peptide ions in the higher attomole range.

In this chapter, the interfacing of LC and MALDI-MS and -MS/MS will be described, performance will be discussed and selected applications in the proteomics field including analyses of membrane proteins will be presented.

2. INTERFACING LC AND MALDI MS

For interfacing liquid chromatography and matrix-assisted laser desorption-ionization mass spectrometry several MALDI-specific criteria have to be considered (for reviews see Murray 1997; Gusev 2000; Orsnes and Zenobi 2001). In contrast to electrospray, the desolvation process of LC-effluents should not just provide analyte ions in the gas phase but, ideally, analytes fully incorporated in matrix crystals suitable for MALDI ionization. Since standard MALDI matrices can not be directly added to LC solvents without compromising chromatographic performance, special techniques were developed to achieve this goal. For on-line LC-MALDI, desolvation and matrix/sample crystallization need to be perfectly time-controlled and have to proceed without interference with the high vacuum (and in selected cases high voltages) in the MS systems. Therefore LC methods with low flow rates like microbore- or nano-HPLC are most suitable. Moreover, they also provide the most concentrated analytes, which is of prime importance for detection sensitivity in ESI-MS as well as LC-MALDI MS techniques.

2.1. On-line Interfaces

As a result of the technical constraints summarized above, only a small number of on-line LC-MALDI interfaces have been described in the literature. In aerosol LC-MALDI (Murray et al. 1994) column effluent was mixed with matrix solution and passed through a heated nebulizer; the resulting aerosol beam was further desolvated in a heated tube and laser desorption was performed "on the fly" from the airborne particles. Because of limited focusing of the neutral particles and the high flow rates (0.5–1 mL/min) sensitivity was only in the nanomole range. In continous flow MALDI (Nagra and Li 1995) a low flow (5 μL/min) of a mixture of 3-nitrobenzyl alcohol as liquid UV adsorbing matrix, ethylene glycol and TFA was admixed to 1–5 μL of LC effluent. At the tip of the capillary located in front of the repeller of the MS ion source, laser desorption was performed and ions were extracted perpendicularly. With microbore HPLC and reduced split ratios (ca 1:100) small proteins could be detected in low picomole amounts. In a modified version with a porous stainless steel frit at the capillary tip and well balanced matrix and LC solvent conditions continous analyte and matrix co-crystallisation was achieved (Zhan et al. 1999). Solvent flushing and laser ablation were applied for interface regeneration. In two interfaces primarily developed for CE-MALDI, the area of sample and matrix crystallisation was separated from the point of laser desorption. Matrix was added to the sheath flow used in capillary electrophoresis and the combined eluent mixture was then vacuum deposited on a rotating quartz wheel located directly behind the repeller of the MALDI ion source (Figure 1, Preisler et al. 1998). The quartz wheel was later replaced by a 80 m long Mylar® tape, which allowed up to 24 h of continuous operation (Preisler et al. 2000). Vacuum deposition provides very homogenous amorphous crystalline matrix layers resulting in attomole MS sensitivity for test peptides and in a significant reduction of suppression effects in peptide mixtures. To prevent freezing and concomitant formation of ice crystals, solvents had to contain at least 30% methanol. This interface-type was later modified to accept up to eight parallel traces which could be interrogated by a 5 kHz laser positioned by a fast-scanning mirror (Preisler et al. 2002). Good performance was also achieved with an interface using a rotating stainless steel ball (Orsnes et al. 2000).

Despite the high performance of selected interface types, analyte deposition and MS analysis have to be synchronized which restricts the available time window, for

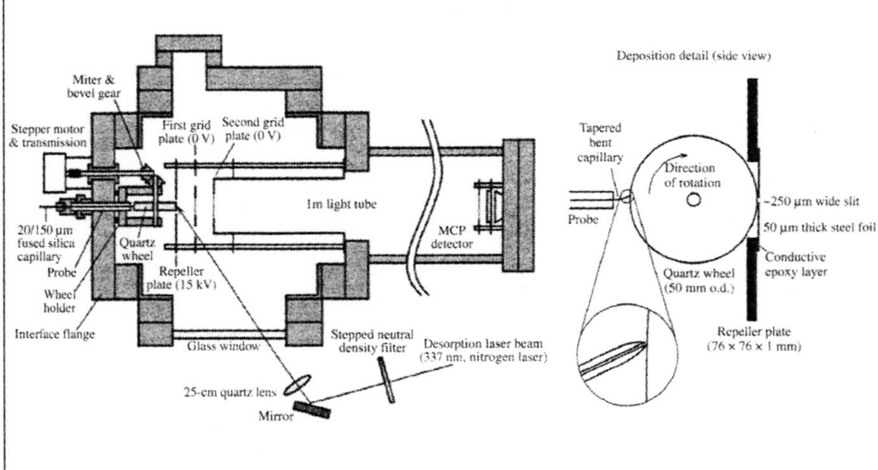

Figure 1. Rotating Wheel Vacuum Deposition Interface. Schematic of the on-line MALDI-TOF MS instrument (left) and details of the deposition process (right). Reprinted with permission from Preissler et al. (1998). Copyright ©1998 American Chemical Society.

example for MS/MS experiments in protein identification. In addition, all on-line approaches are technically demanding and were – to the best of our knowledge – not commercialized.

2.2. Off-line Interfaces

In the off-line approach, analyte and matrix deposition are decoupled from MS analysis and only instrument-specific plate formats have to be considered in interface design. In this case interface design focuses on the development of techniques for a perfect transfer of LC-effluent and matrix to the surface from which laser desorption takes place at a later stage. The development of suitable techniques started very early in conjunction with capillary electrophoresis (CE) and some of the methods which are applied nowadays for low-flow LC-MALDI are included in the following compilation.

2.2.1. Continous deposition

Continous deposition of a CE trace on a MALDI probetip using post-column addition of 2,5-DHB matrix (T-piece) was already reported in 1992 (Castoro et al. 1992) and detection limits in the picomole range were obtained for small proteins. By addition of matrix to the CE sheath flow selected peptides deposited on a moving target could be detected down to high femtomole amounts (Van Veelen 1993). When the CE trace was deposited on a conducting cellulose membrane precoated with alpha-cyano-4-hydroxycinnamic acid, detection limits for peptides like neurotensin could be extended to the medium attomole range using linear UV-MALDI MS (Zhang and Caprioli 1996). For LC traces blotted onto PVDF membranes, infrared MALDI with succinic acid as matrix could be successfully applied for molecular mass determination of peptides at the low picomole level (Eckerskorn et al. 1997). Continous vacuum deposition, first described in on-line mode (cf. Section 2.1.) is also applicable off-line in a separate vacuum chamber; trace deposition on metal MALDI targets allowed to use low percentage of organic solvents owing to better heat dissipation (Rejtar et al. 2002). In combination with CE, electrophoretic resolution was perfectly preserved and not only MALDI MS but also MS/MS spectra were obtained at the 200 femtomole level. For higher flow rates (up to 3 μL/min) like in nanoLC or fast LC with monolithic columns, the MALDI plate was coated with nitrocellulose to prevent uncontrolled droplet formation (Chen et al. 2005b). In order to reduce fast depletion from amorphous thin matrix layers especially in MS/MS mode (elevated laser power), deposition conditions (solvent, pressure) were adjusted in such a way that fine micro-crystalline layers were formed (Chen et al. 2005b). Another method for solvent removal is the application of heat. In a heated nebulizer interface nitrogen gas and LC-gradient-dependent heating was applied to the tip of the interface capillary and a fine mist of droplets was sprayed onto targets precoated with alpha-cyano-4-hydroxycinnamic acid (Wall et al. 2002). For flow rates up to 10 μL/min, MS sensitivity in the high attomole range was reported for testpeptides and, for protein digests, femtomole detection limits were achieved. A similar approach was applied for spot deposition and is discussed in the next section. Aerosol deposition can also be achieved without heating for flow rates up to 3 μL/min with an oscillating capillary nebulizer (Fung et al. 2004). A high gas flow between two coaxial capillaries leads to the oscillation of the inner capillary and, with optimized dimensions, an aerosol is formed.

2.2.2. Spot deposition

In spot deposition the continuous LC-trace is divided into fractions which are collected on a MALDI sample plate. Continous information is thus "digitized" and, in order to preserve the chromatographic resolution, spotting intervals have to be adjusted accordingly. With too many sample points the analyte is dispersed over several fractions and as a consequence, analyte to matrix ratio and sensitivity are reduced. In the opposite case peaks start to merge and suppression effects become operative. Consequently, deposition of three to five fractions across a peak profile seems to be a reasonable compromise. To maintain chromatographic resolution apart from spotting frequency, any sample dissipation during droplet formation has to be suppressed. This is especially important for interface types which avoid contact of capillary and MALDI plate and which do not actively support droplet transfer. With a high content of organic solvent and a small outer diameter of the spotting capillary, liquid can easily be retracted at the outside of the tip by capillary forces. Apart from these technical details, fraction deposition in a standardized pattern presents a clear advantage compared to continous techniques with respect to ease of data management and data acquisition.

Spot deposition on a computer controlled x-y table in a pattern of 100, 400 or 1024 spots was already demonstrated in 1994 (Blakley et al. 1994) and by MALDI MS, picomole level sensitivity was achieved for a LC separation of peptides at high flow rates (2.1 mm column). For CE a similar procedure allowed to detect high femtomole amounts of smaller proteins (Walker et al. 1995). Matrix was either spotted onto dried fractions or was admixed to the sheath flow. Modified silica sleeves have been especially developed for CE interfacing (Tegeler et al. 2004). Parallel deposition from two neighbouring capillaries is also feasible; this method was applied in MS and MS/MS (PSD) analysis of peptides from a single neuron (Hsieh et al. 1998). An elegant way to transfer liquid as droplets from a continous flow to a carrier is piezoelectric droplet dispensing (Oennerfjord et al. 1998). By the action of a piezoelectric element, the volume of a flow channel is reduced and droplets are ejected from a nozzle (Figure 2). With MALDI targets precoated with a alpha-cyano-4-hydroxycinnamic acid/nitrocellulose matrix, peptide detection at the higher attomole level could be demonstrated (Miliotis et al. 2000). In a similar workflow MALDI plates with prespotted matrix/calibrant layers in hydrophilic surface areas were prepared in a first step and CE eluate was spotted in a separate second step. After recrystallisation, low femtomole amounts of standard peptides could be detected (Johnson et al. 2001). A very effective alternative for droplet transfer is the application of a short electric pulse (−2 kV, 20 ms) between the droplet hanging at a metal capillary tip and the metal target plate located at a distance of 2–5 mm (Ericson et al. 2003). Droplets are first polarized and then attracted

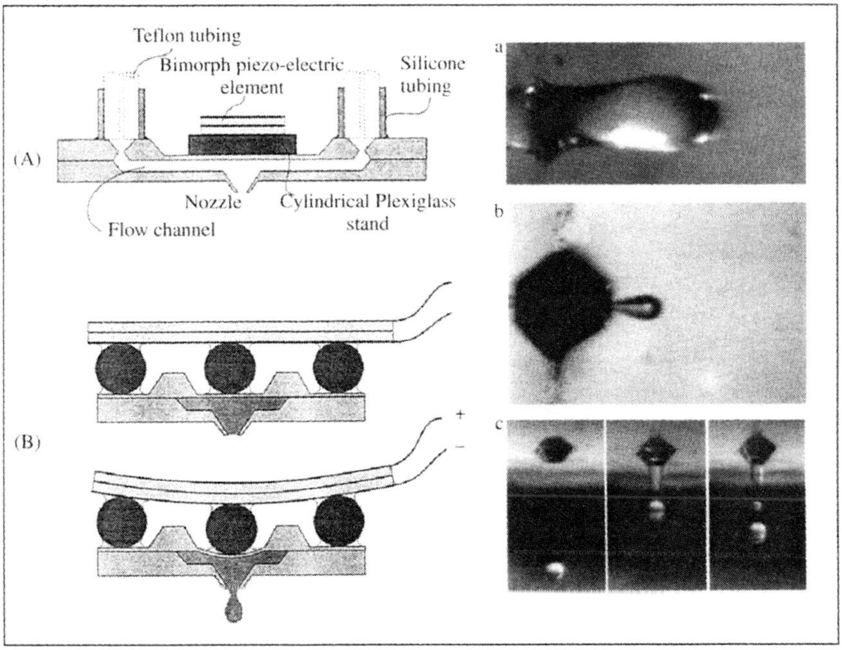

Figure 2. Piezoelectric Droplet Dispenser. Schematic of the dispenser construction (panel A, left) and actuation principle (panel B, left). Droplet formation at the dispenser tip (right, a and b) and droplet appearance at different voltages (right, c). Optimal voltage (right, c, left panel) and voltage too high (right, c, middle and right panel). Reprinted with permission from P. Oennerfjord et al. (1998). Copyright ©1998 American Chemical Society.

to the MALDI plate. This principle was implemented in a multichannel interface which allowed parallel deposition of four LC-traces. Matrix was added via T-piece or was prespotted onto plates with 384 or 1536 hydrophilic surface areas (Figure 3, Ericson et al. 2003). With prespotted alpha-cyano-4-hydroxycinnamic acid matrices (1 mg/mL) and recrystallisation after spot deposition, 50 attomoles of a standard peptide could be detected. An interface type which is adapted to a wider range of flow rates uses partial droplet desolvation. This was achieved by applying LC-gradient-dependent heating to a flow of nitrogen at the capillary tip. Aerosol formation was thus prevented and the remaining solvent was evaporated by heating of the MALDI plate to 110°C (Zhang et al. 2004). Flows up to 200 μL/min could be handled and standard peptides were detected at the low femtomole level. Only nitrogen flow assistance and careful capillary selection were used in an interface for low flow nano-LC; with precoated MALDI plates containing hydrophilic anchors and a TFA washing step after sample deposition, nine peptides of a 1 fmol BSA digest could be detected (Mirgorodskaya et al. 2005a). Another means to transfer droplets is a mechanical impulse; this principle was realized by applying a pulsed voltage to a selenoid surrounding the capillary outlet (Young and Li 2006). By careful adjustment of plate distance low femtomole sensitivity was achieved.

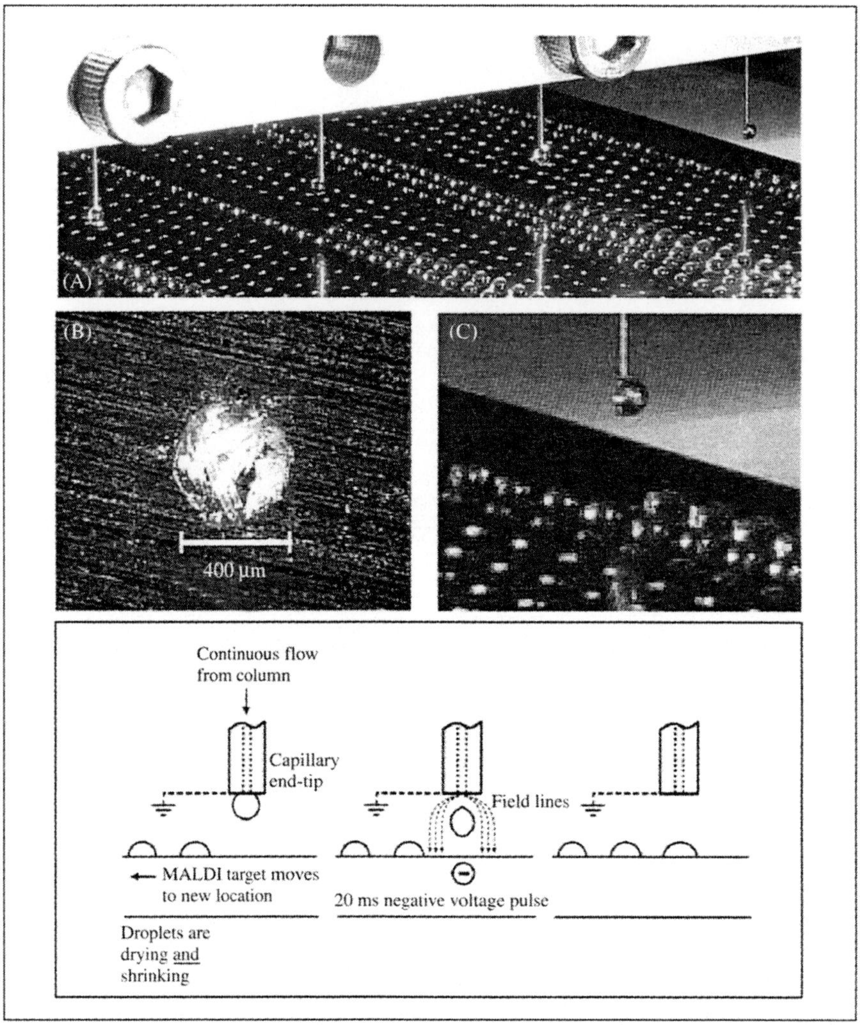

Figure 3. Interface with pulsed field droplet deposition. Principle of the electrically mediated liquid deposition (lower panel). Deposition of the eluents from 4 parallel LC separations (upper panel, A). Samples cocrystallized with matrix on hydrophilic islands with 400 μm diameter (upper panel, B) and droplets just before application of electrical field (upper panel, C). Reprinted with permission from Ericson et al. 2003. Copyright ©2003 American Chemical Society.

Most of the currently applied LC-MALDI interfaces use off-line spot deposition and a number of commercial instruments are available (Mukhopadhyay 2005). Compared to on-line techniques, this interface type is technically less demanding but only careful adjustment guarantees optimal performance.

3. LC-MALDI MS IN PROTEOMICS

3.1. Introduction

In the early days of proteomics, MALDI MS was mainly applied for peptide mass fingerprinting using time-of-flight (TOF) mass spectrometers as the major instrument type. MALDI MS/MS on commercial TOF-instruments was only available with post-source decay analysis (PSD, Kaufmann et al. 1994) and MS/MS spectra had to be assembled from 10 to 20 subspectra acquired within a long time frame. Therefore partial MS/MS sequencing of peptide ions was mainly performed on instruments fitted with electrospray ionization (ESI), for example on triple quadrupole (QQQ), quadrupole-TOF (Q-TOF) or ion trap instruments (iTRAP). The situation started to change with the introduction of (vacuum) MALDI ion sources which could be fitted for example to Q-TOF instruments (Harvey et al. 2000; Krutchinsky et al. 2000; Loboda et al. 2000; Wattenberg et al. 2002). A major breakthrough for MALDI MS/MS was however achieved with the introduction of tandem time-of-flight instrumentation (Medzihradszky et al. 2000) and the implementation of LIFT-TOF/TOF technology (Suckau et al. 2003). Interfacing of MALDI to ion trap-TOF (vacuum MALDI, Martin and Brancia 2003) and ion trap instruments (vacuum and atmospheric pressure MALDI, Krutchinsky et al. 2001, Tan et al. 2004) completed this development. With this set of technologies MALDI MS and MS/MS could be performed on the same instrument and sample without splitting. In the proteomics field peptide mass fingerprinting and MS/MS could now be merged and, for the analysis of complex peptide mixtures, MALDI MS and MS/MS could be directly combined with LC separation. Figure 4 shows a standard LC-MALDI setup as used in our laboratory; peptide mixtures are desalted and concentrated on a short trap column, separated by capillary LC and after UV detection and matrix admixture, fractions are directly spotted on a MALDI target. In the standard workflow (Figure 5) MS spectra are first acquired from all spots and subsequently precursor ions for MS/MS are selected. After MS/MS data acquisition, database seach is performed with the complete set of MS/MS spectra.

3.2. General Aspects

3.2.1. Adaptation of LC-MALDI MS workflow to sample complexity

To achieve the best performance in protein identification or quantitation, the extent of protein or peptide separation should match the capabilities of the technique applied in the identification or quantitation step. With LC-MALDI MS and MS/MS as analysis technique, the number of good-quality MS/MS spectra, which can be acquired from one sample spot (LC fraction) represents one limitation; the number of components in one fraction should therefore not exceed this maximum. By increasing matrix concentration more laser shots could be acquired from one spot, but analyte concentration in the crystals and detection sensitivity would

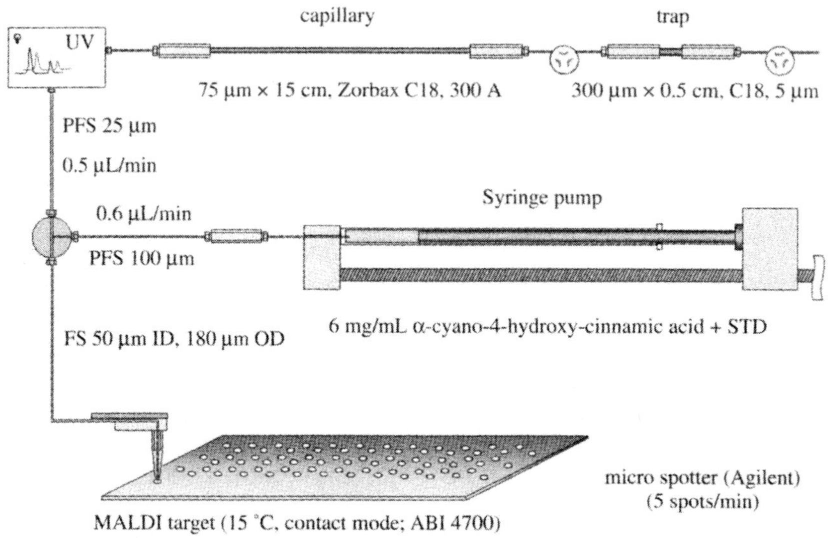

Figure 4. Standard LC-MALDI setup with sample concentration/desalting step, nanocolumn LC separation and off-line deposition of LC-eluent after admixture of matrix solution.

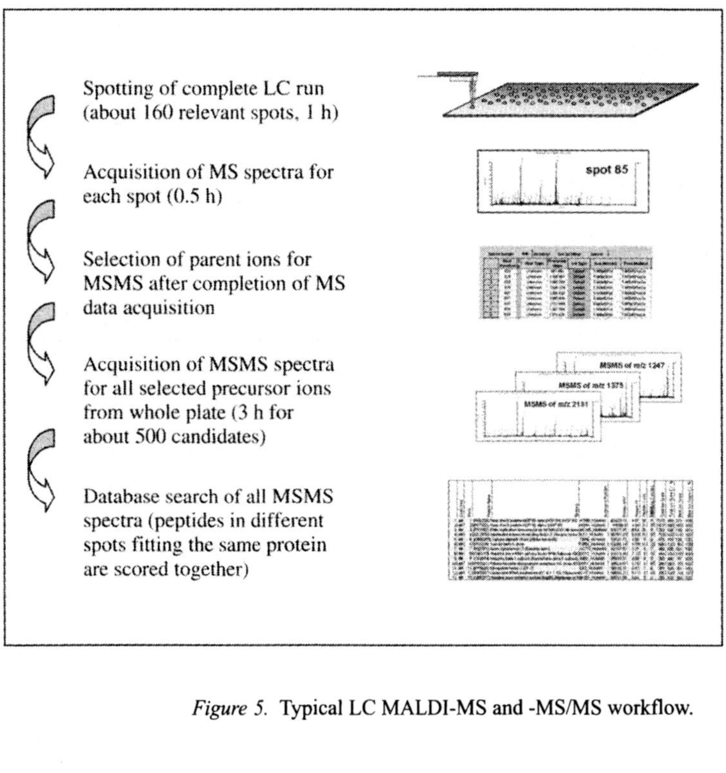

Figure 5. Typical LC MALDI-MS and -MS/MS workflow.

thus be reduced. The same dilution effect reduces detection sensitivity when analytes are distributed over too many fractions. In our setup with a LC-gradient of 50 min, a chromatographic peak is dispersed over three to four spots, tailing effects are small and with a matrix concentration of 3 mg/mL MS/MS sensitivity is good (Figure 6). The other limitation for MS and MS/MS detection is the suppression of components in peptide mixtures; this effect can be caused for example by preferential ionization of components in MALDI or by differences in the crystallisation step (e.g. Rejtar et al. 2002). Suppression cannot be quantitated in general (independent of individual peptide properties) but in a peptide mixture with more than 20 components present in different concentrations, suppression has to be anticipated. Therefore a smaller average number of mixture components would be of advantage. Depending on sample concentration and instrument performance, either the suppression effect or the maximum number of good-quality MS/MS spectra which can be acquired from one fraction is the limiting factor. On this basis the extent of separation should be adjusted to analyte complexity.

For the analysis of a complex sample (yeast whole cell lysate), three different separation schemes combined with LC-MALDI have been evaluated for identical sample amounts (Hattan et al 2005; cf. Figure 7); (i) fractionation at the protein level by RP-HPLC (12 fractions) followed by tryptic digestion and cation exchange (4 elution steps; 48 LC-MALDI runs in total); (ii) separation at the peptide level by strong cation exchange (SCX, 15 fractions, gradient elution; 10 used) or (iii) strong anion exchange (SAX, 14 fractions, gradient elution; 12 used). In the second dimension, peptides were separated by a dual solvent gradient within 40 min and spots were deposited in 5 s intervals. About 12,000 peptides were identified by approach (i) using the most extensive separation, 4000 by SCX (ii) and 9100 by SAX (iii). The number of unique peptides was 4871 for (i) 3350 for (ii) and 5700 for (iii) indicating a high redundancy for the extensive separation (4871 of 12,000), a low value for SCX (3350 of 4000) and a medium value for SAX (5700 of 9100). The most homogenous peptide distribution over the fractions was achieved by SAX in contrast to SCX, were peptides were concentrated in a few fractions. While distribution over too many fractions in the extensive separation (mostly in the protein chromatography step) reduced detection efficiency, in this example SAX provided the most homogenous flow of peptides. 1215 proteins were identified by LC-MALDI analysis of the 12 SAX fractions (iii) which is about 80% of the number of proteins identified in a similar preparation by ESI-LCMS/MS analysis of 80 SCX fractions (Peng et al. 2003).

Compared to yeast with about 5800 ORFs which code for proteins, samples from higher organisms can be much more complex and the separation scheme has to be adjusted accordingly. For very simple species as the other extreme, already subcellular fractionation can provide the appropriate pre separation. For example only one LC-MALDI MS/MS run of a peptide mixture derived from the separated E-coli 50S ribosomal subunit allowed to identify 30 of the 33 expected proteins (Mirgorodskaya et al. 2005a).

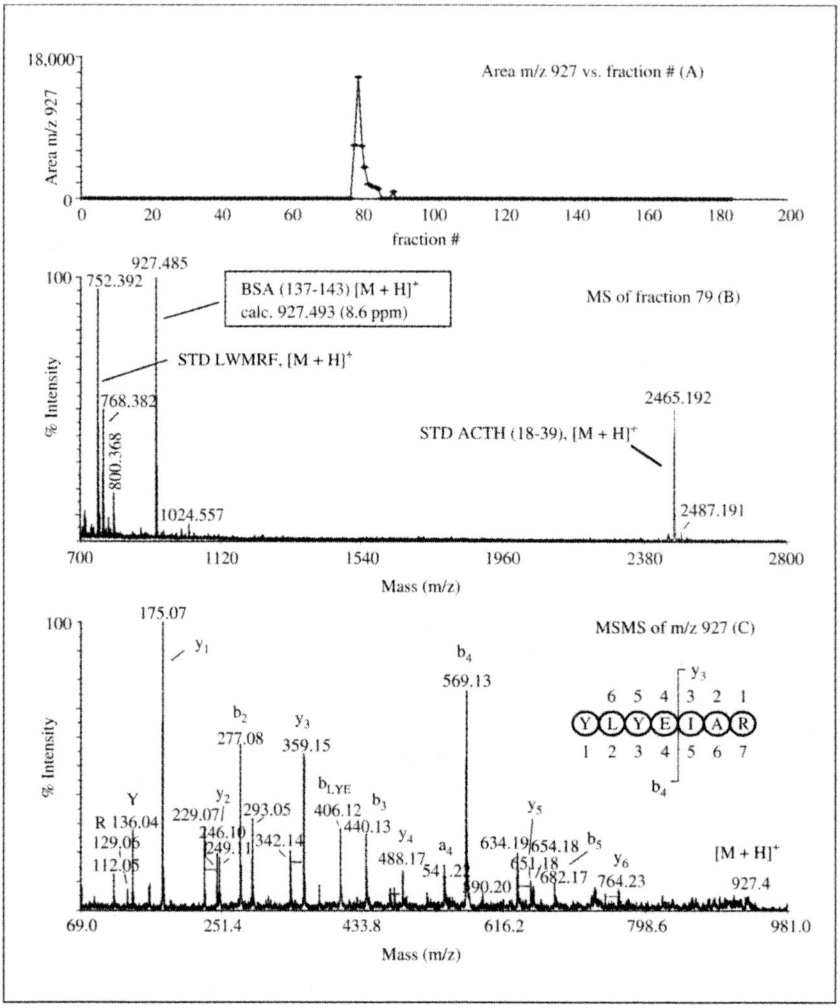

Figure 6. Chromatographic resolution, MS- and MS/MS sensitivity achieved with the LC-MALDI setup described in Fig. 4. Trace of m/z 927 ($[M + H]^+$ of peptide 137–143) from 10 fmol of a tryptic BSA hydrolysate using 12 sec wide fractions and a matrix concentration of 3 mg/mL (A). MALDI MS (B) and MS/MS spectrum (C) of the fraction with the highest intensity of m/z 927 (fraction #79).

3.2.2. *Multiplexing*

To cope with the high number of RP-HPLC runs required to analyze fractions in the two-dimensional LC-MS approach, multiplexed LC-separation and spotting for LC-MALDI analyses was applied very early. In quantitative analysis of differential protein expression by ICAT technology, total analysis time could be significantly reduced with four parallel separation lines (Lee et al. 2002). Prerequisite is of course

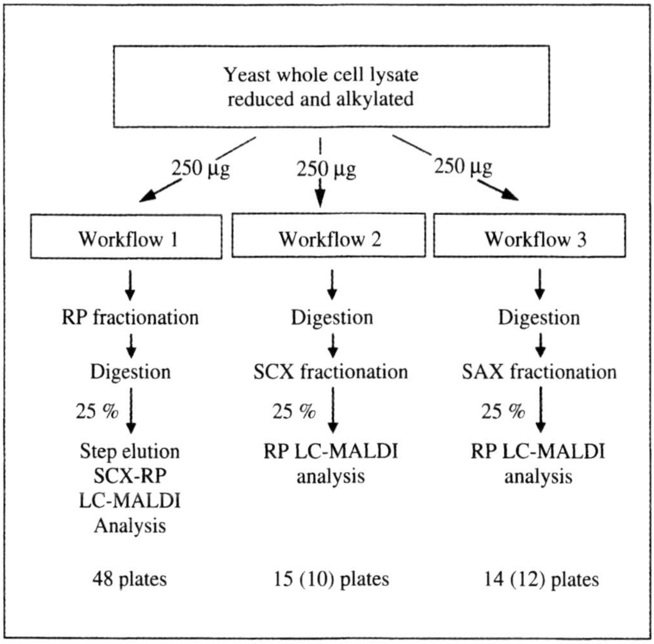

Figure 7. Comparison of different prefractionation steps in typical LC-MALDI workflows for complex mixture analysis. Adapted from Hattan et al. (2005). Copyright ©2005 American Chemical Society.

that MS and MS/MS capacity matches the output from multiplexed LC and MALDI spotting. Especially for the combination with fast LC-chromatography on monolithic columns, the implementation of faster lasers and the development of faster data systems would be of high value. Multiplexing is also advantageous when parallel LC-MALDI MS and MS/MS analyses are performed on complex (split) samples (Chen et al. 2005a). In this case peptides reliably identified by MS and MS/MS in the first LC-MALDI run were excluded from MS/MS experiments on corresponding spots of the next LC-trace (spotted at the same time). With this strategy the number of peptide identifications from an *E. coli* total cell lysate improved from 1042 to 1418 already by using two steps (1 exclusion list); by four steps 1735 unique peptides were identified in total. Proper alignment of the four LC-separations was performed by a linear correction applied to 20 chromatographic segments.

3.2.3. Quantitation and LC-MALDI MS

Relative protein quantitation is the basis of all types of differential proteome analyses. In the 2D-gel approach protein staining with either visible or fluorescent dyes provides a reliable and sensitive method to detect changes in protein expression or isoform abundance. In the multidimensional LC approach quantitation relies mostly on stable isotope labeling and ratios between light and heavy isotopomers are determined by MS or MS/MS at the peptide level. Labeling can be performed on the protein level by incorporation of labeled amino acids in cell cultures (Ong et al. 2002), by feeding with isotope-labeled nutrients (e.g. ^{15}N labeled yeast; Krijgsveld et al. 2003) or by chemical derivatization with suitable reagents (e.g. Schmidt et al. 2005). At the peptide level a variety of derivatization methods are available which allow to introduce ^{2}H, ^{13}C, ^{18}O, ^{15}N or combinations thereof. Since the methodology was recently reviewed (Moritz and Meyer 2003), only selected examples important in the context with LC-MALDI will be presented.

3.2.3.1. Quantitation at the Protein Level In differential protein labeling with isotope-coded affinity tags (ICAT) a biotin-containing iodoacetyl derivative is used for cystein alkylation either as d_0- or d_8-isotopomer. Mixture complexity is significantly reduced by avidin affinity enrichment of cysteine-containing peptides which are then selected for relative quantitation (Gygi et al. 1999; Gygi et al. 2002). In the differential analysis of yeast grown on different media (Lee et al. 2002), separation was performed by 2D LC and a multiplexed version of LC-MALDI (cf. Section 3.2.2) was evaluated. Multiplexing reduced total analysis time but in addition, MS analysis after completed LC separation allowed to select only peptide ions displaying changes in expression level for MS/MS (expression-dependent MS/MS). A second generation of reagents with acid-cleavable biotin-moiety and ^{13}C-incorporation for quantitation provided improved MS/MS fragmentation behavior and identical chromatographic retention of the isotopomers. 2D-LC-MALDI and -ESI were assessed in a study evaluating this improved ICAT methodology (cICAT) for differential analysis of proteins attached to the prion protein (pull downs) and of low μg amounts of tracheal epithelial gland secretions (Hansen et al. 2003). In this study standard deviations below 10% were observed for quantitation of protein-specific cICAT-labeled peptides resulting in the identification of a number of proteins upregulated in cystic fibrosis (cf. also Section 3.2.5.). Compared to ICAT, labeling of lysines with differently isotope-labeled nicotinic acid derivatives increased the number of peptides which can be used for protein quantitation (Schmidt et al. 2005). This method which was originally developed for combination with the 2D gel approach is also of value for LC-MALDI MS and MS/MS, especially with expression-dependent precursor ion selection (Tebbe et al. 2005).

3.2.3.2. Quantitation at the Peptide Level At the peptide level, methylesterification (d_0/d_3; e.g. Goodlett et al. 2001) or acetylation (d_0/d_3; e.g. Hunt et al. 1980) are simple means for the introduction of different isotope labels and also trypsin-catalyzed ^{18}O incorporation has a long tradition (Rose et al. 1983). Another alternative is reductive methylation with CH_2O or CD_2O (Melanson et al. 2006). In MS/MS the N,N-dimethylated peptides show a strong transition from $[M+H]^+$ and $[M(d_4)+H]^+$ to the fragment a_1 which can be selected for quantitation by multiple reaction monitoring MS/MS (MRM). For a combination of MRM with LC-MALDI on a prototype triple quadrupole instrument with a MALDI ion source, quantitation of testpeptides was reported at the attomole level. The same labeling method (Ji et al. 2005) was previously applied for quantitation and identification of differentially expressed proteins from E-Cadherin deficient SCC9 cells and SCC9 transfectants expressing E-cadherin. With a combination of SCX fractionation and microbore RP-HPLC-MALDI, 49 proteins were identified by MS/MS analysis of only those 320 peptide ion pairs exhibiting a more than 2-fold change. In this expression-dependent MS/MS experiment the other 5160 peptide ion pairs detected could be excluded from MS/MS analysis. Other derivatizing agents which improve detection sensitivity are reagents that convert lysines into their 4,5-dihydro-1H-imidazol-2-yl derivatives. They not only improved MS/MS-based identification but also allowed the introduction of four ^{2}H atoms for quantitation. (Peters et al. 2001). In all these methods quantitation is based on the ratio of different isotopomers detected in MS mode or in MRM. In a new class of reagents (iTRAQ) this step is performed on the MS/MS level (Ross et al. 2004). Four different isotopomers of a N-methylpiperazine reporter group provide four different MS/MS signatures for quantitation. Their mass differences are counterbalanced by a CH_2CO-linker with corresponding opposite isotope labels incorporated in such a way that in MS mode the overall mass increment of all four reagents is the same. Since in MS mode signals are not split into isotopomers and in MS/MS mode only those of the reporter ions, sensitivity is improved in both modes and quantitation is simplified. Multiplexed analyses can be performed by applying all four labels for example in time-course experiments. Quantitation of protein standards and protein-spiked depleted plasma by 2D LC-MALDI combined with cleavable ICAT (cICAT) or iTRAQ labeling as well as DIGE technology were recently assessed and all three techniques were subsequently applied to the differential analysis of HCT-116 cell lysates (Wu et al. 2006). While all methods provided similar accuracy for standards (90–110%; standard deviations at least $<15\%$) and spiked plasma (80–120% ; standard deviations $<30\%$), the best sensitivity was achieved with iTRAQ followed by cICAT and DIGE. The reduced cICAT sensitivity was attributed to losses during avidin affinity purification and the fact that proteins with few cysteines are less likely be detected. However, ICAT is a technique designed to reduce sample complexity and it has its inherent drawbacks like any other method. For the MS/MS-based iTRAQ quantitation for example, the purity of precursor ions is of high importance.

3.2.4. Derivatization and LC-MALDI MS

For non-quantitative LC-MALDI applications, derivatization chemistry is not restricted to compounds which allow a convenient incorporation of heavy isotopes. For example for improved MS/MS detection sensitivity, tryptic peptides were labeled with sulfonated coumarin-tags at the N-termini after guanidylation of lysines (Pashkova et al. 2005). Despite reduced MS sensitivity for arginine-terminated peptides (in alpha cyano-4-hydroxycinnamic acid matrix), formation of y-ions was enhanced in MS/MS by the second mobile proton provided from the sulfonic acid group. For a SCX fraction from a *E.coli* hydrolysate 50% more peptides and 30% more proteins could be identified by multiplexed LC-MALDI MS and MS/MS after derivatization.

3.2.5. Complementarity of LC-MALDI and LC-ESI MS techniques

Since the ionization in MALDI and in ESI is based on different mechanisms, peptides and proteins are protonated in a method-specific form with different efficiencies. While in ESI multiply charged ions are the preferentially observed species for peptides, MALDI leads predominantly to singly protonated molecules. Their fragmentation behavior in MS/MS is different to that of multiply charged ions and, depending on peptide sequence, one or the other method can provide more information for protein identification. Similarly, response factors in MALDI and in electrospray have different method-specific contributions, for example matrix incorporation of analyte or surface properties in solution. All these factors together lead to a different peptide-specific response in LC-MALDI MS/MS and LC-ESI MS/MS. Parallel application of both techniques has therefore a considerable potential for increasing proteome coverage and the complementarity of both methods was evaluated.

A direct comparison of both techniques was performed on the same instrument with different ion sources for the analysis of proteins derived from the 39S and 28S subunit of mitochondrial ribosomes (Bodnar et al. 2003). 21% of the proteins were identified by LC-MALDI only, 16% by LC-ESI only and 63% by both methods. In a comparison of LC-MALDI MS/MS analysis on a TOFTOF instrument with LC-ESI MS/MS on a Q-TOF, values for method-unique identifications ranged from 20 to 50% (Hansen et al. 2003). Similar proportions were obtained in membrane protein analyses on a LC-ESI ion trap instrument compared to LC-MALDI on a MALDI-Q-TOF (Zhang et al. 2004b). LC-ESI-only identified 27%, LC-MALDI-only 38% and both methods together 35% of all proteins. While in ESI more shorter peptides were observed, MALDI MS/MS preferentially allowed to identify higher mass peptides (Hansen et al. 2003; Zhang et al. 2004b). Although different instrument types and samples of different complexity were applied in the various comparisons, all studies revealed a substantial complementarity of both methods at least for protein identification. A parallel application of LC-MALDI MS/MS and LC-ESI MS/MS is therefore essential to achieve optimal coverage in proteome analysis.

3.3. Selected Application Areas of LC-MALDI MS in Proteomics

3.3.1. Membrane protein analysis

Membrane proteins play a critical role in many biological processes especially in the transfer of information and material between different intra- and/or extracellular compartments (e.g. receptors, ion channels). Their analysis represents a special challenge because of their strong aggregation behavior and the poor solubility of hydrophobic sequence stretches. These characteristics are the reason for their underrepresentation in protein identification based on 2D PAGE, most likely due to solubility problems in the isoelectric focusing step. One promising alternative method for appropriate reduction of mixture complexity is the combination of 1D SDS PAGE with LC-MALDI MS and MS/MS, where prefractionation is achieved by dividing the gel into slices.

In the course of a proteomics study of Haemophilus influenzae (1740 open reading frames) we evaluated this approach for establishing a map of membrane and membrane-associated proteins. In the first step proteins from a carbonate-washed membrane fraction (Fujiki et al. 1982) were separated by SDS PAGE on a (4–20%) gradient gel, which was subsequently divided into 65 slices covering the complete separation range. After in-gel digestion the 65 peptide mixtures were analyzed with our standard LC-MALDI setup (Figure 4) using a 40 min. LC-gradient. From the about 190 spots deposited in each run on average 120 contained peptides. A maximum of 6 MS/MS spectra were acquired from each individual spot (average of 4000 shots). Confidence levels for peptide identification using Mascot/GPS Explorer were determined by manual examination of a substantial number of MS/MS spectra in the critical confidence range. Based on 22067 MS/MS spectra, 1763 confident protein hits were identified in all 65 runs with 47% of all MS/MS spectra contributing to these hits (Figure 8). On average, proteins were distributed over 3.7 bands. After removal of duplicates from adjacent bands 480 unique protein identifications could be established. From these, 426 proteins (88%) were identified by more than one peptide which indicates good reliability. In general the molecular masses of the identified proteins fitted well with those extrapolated from the standard calibrants on the gel (Figure 9). From the 480 identified proteins 189 are annotated as membrane proteins in the SwissProt database, representing 42% of the 443 membrane proteins in H. flu. (SwissProt). From the 376 proteins with predicted alpha-helical transmembrane domains (SwissProt) 40% (149) could be identified. The distribution of the number of membrane domains of these 149 identified proteins matched that of the predicted 376 proteins with TM-domains (Figure 10). Only proteins with

Summary of 1D SDS-PAGE/LC-MALDI analysis

65	LC-MALDI MS- and MSMS runs
22067	MSMS spectra (mean 340/run)
10268	MSMS spectra contribute to confident hits (47 %)
1763	"confident" protein hits (BIS >= 37)
480	nonredundant proteins
3.7	"ID's" per protein indicates good separation by 1D SDS PAGE
189	"membrane proteins"
42 %	of 443 proteins with SwissProt annotation "membrane protein"
149	with predicted alpha-helical TM domains
40 %	of 376 proteins with predicted alpha-helical TM domains

Figure 8. Summary of 1D SDS-PAGE/LC-MALDI analysis of a membrane preparation derived from Haemophilus Influenzae.

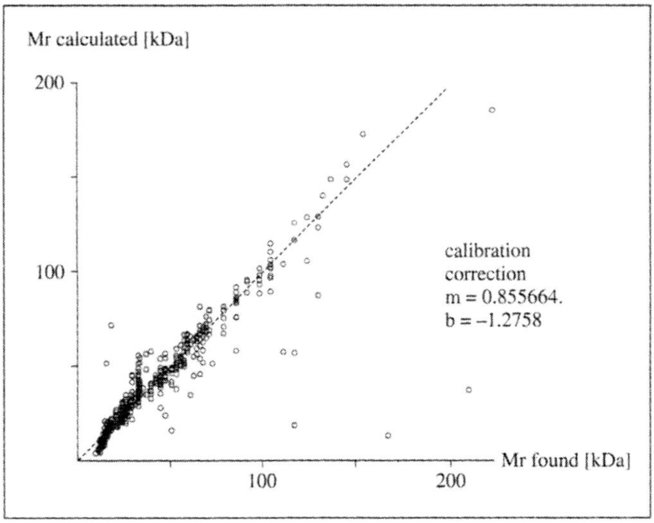

Figure 9. Comparison of calculated and observed molecular masses of all proteins identified by LC-MALDI in a membrane preparation derived from Haemophilus Influenzae.

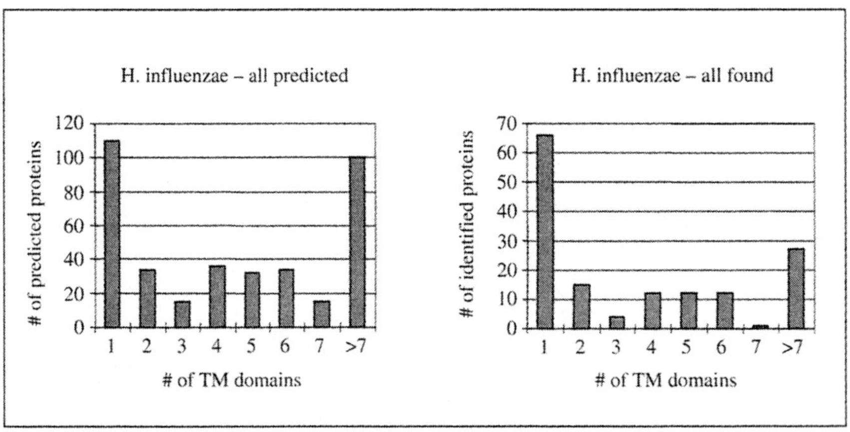

Figure 10. Comparison of the transmembrane domain distribution of all proteins identified by LC-MALDI in a membrane preparation derived from Haemophilus Influenzae with that of all H. flu. proteins predicted to contain alpha-helical transmembrane domains.

more than seven transmembrane domains were still underrepresented. The study presents currently one of the most comprehensive maps of membrane proteins of Haemophilus Influenzae and it serves as a basis for further quantitative differential analysis.

1D PAGE combined with LC-MALDI was also applied for membrane protein analysis of staphylococcus aureus (strain N315, 2596 ORFs) in a study correlating transcriptomic and proteomic expression profiles (Scherl et al. 2005). Total protein extracts were analyzed by 2D LC-MALDI (SXC as first step) and the standard 2D PAGE approach; with these methods 245 and 211 proteins were identified, respectively. After phase partitioning, the two different membrane fractions were first separated by 1D SDS-PAGE, gels were divided into 30 slices, and after in-gel digestion hydrolysates of each slice were subjected to LC-MALDI MS and MS/MS (75μ × 150 mm columns). 291 proteins were identified in the soluble part and 269 in the pellet. 24% of the expected proteins with one transmembrane (TM) domain were detected and 5% or less of the classes containing 2–5, 6–10 and >10 TM-domains. With 2D gels only 1% of the expected membrane proteins could be identified.

As an alternative to the application of protein separation methods which are compatible with membrane protein characteristics, direct in-solution digestion of SDS-containing extracts was applied in the analysis of lipid rafts of THP-1 monocytes (Li et al. 2004). A SDS-guard column was used to reduce the SDS concentration and with LC-MALDI, which is less sensitive to trace amounts of detergents, 71 proteins were identified. By adding cation exchange chromatography as pre-separation step, the number of proteins identified with high confidence could be increased to 126. For the analysis of membrane proteins of human HT 29 cells, in-solution digestion and direct LC-MALDI was even superior to 2D LC-ESI MS and MS/MS; 50 membrane or membrane-associated proteins were identified compared to 24 (Zhang et al. 2004a). A possibility to avoid the detergents used in solubilization is to apply microwave-assisted TFA hydrolysis in suspension (25% TFA) instead of enzymatic digestion (Zhong et al. 2005). When this approach was combined with LC-MALDI MS/MS analysis, 119 proteins (including 41 membrane proteins) with gravy indices from −1.135 to 0.992 were identified in a membrane preparation derived from human MCF7 cells. Most enrichment techniques for membrane proteins rely on several stages of differential centrifugation and washing. As an elegant alternative for mapping of membrane proteins cell surface proteins were derivatized with different biotin-containing sulfo-succinimidyl derivatives and captured on streptavidin beads (Scheurer et al. 2005). Best overall performance was achieved with a SS-containing biotin-linker; after reductive cleavage from beads, in-solution digestion of the released derivatized proteins and LC-MALDI MS/MS analyses, 549 proteins could be identified. 28% of these 549 were either membrane or extracellular matrix proteins.

3.3.2. Protein interaction analysis (pull downs)

For the understanding of protein function in a complex cellular environment the analysis of interaction partners is of high importance. In many essential protein complexes these partners are proteins, mRNA or DNA. Another class of interacting entities is represented by small molecules. They are very often involved in regulatory events or in cell signalling (e.g. via receptors) and many pharmacologically active compounds belong to this class. Knowledge about their interacting partners is essential to understand their mode of action. For interaction analysis, the protein (or small molecule) of interest is normally attached (as bait) with a linking element to a carrier (e.g. beads) and the potential complex partners are selected from a suitable protein mixture, for example from a cell lysate. Low affinity binders are removed by washing steps and only the remaining complex partners are analyzed. They can form relative complex mixtures and in general, at least one separation step (1D PAGE or LC) is applied before MS and MS/MS analysis. To assess the potential of LC-MALDI in this application area, ampicillin was attached as bait to a carrier and treated with solubilized membranes of *E. coli* with the aim to detect penicillin receptors (Zhen et al. 2004). All known Penicillin-binding proteins were identified despite the fact that they were expressed in the range between 5–10 copies/cell up to thousands of copies/cell. This wide dynamic range could only be covered by incorporating LC-separation. For more complex pulldowns (e.g. double-tagged GRB2) 2D LC separation (SCX, RP-HPLC) was necessary to reduce mixture complexity. LC-MALDI analysis required significantly fewer SCX-fractions compared to LC-ESI mainly because of the much longer analysis time available for MS/MS experiments (Zhen et al. 2004). Differentiation of true and unspecific binders in affinity pulldowns was achieved by differential ^{18}O-labeling of tryptic peptides and quantitative analysis by LC-MALDI (Figure 11) (Mirgorodskaya et al. 2005b). Proteins attached to beads with GST tag but without bait were trypsinized with $H_2^{16}O$, beads with GST tag and bait (but without protein mixture) with $H_2^{18}O$ and both mixed in a 1:2 ratio. The pulldown was trypsinized with $H_2^{16}O$ and added to give a 3:1:2 ratio of all three samples. In MS analyses tryptic peptides from true interactors with the bait appear as singlets, unspecific binders as doublets with changed ^{16}O:^{18}O ratio and peptides observed only in controls present the original 1:2 ratio. Nontryptic peptides can be sorted out by reversed labeling (Wang et al. 2001).

3.3.3. Other applications areas

With the high performance of LC-MALDI, application areas come now into reach which were formerly a domain of on-line ESI-MS/MS. One such area is the analysis of major histocompatibility complex (MHC) Class I-associated peptides. After prefractionation on RP-HPLC, MHC-associated peptides eluated from human leukocyte antigene complexes were recently identified by LC-MALDI (Hofmann et al. 2005). The prefractionation step allowed to reduce (adjust) sample complexity and, with MALDI MS screening, MHC-peptide-containing fractions were specifically selected. By a combination of automated LC-MALDI MS and MS/MS with standard database search and de novo sequencing, 30 novel HLA-presented peptide ligands were identified including species presented at very low levels. Another application with high sensitivity requirements is the analysis of tear-derived proteins (Li et al. 2005). In a study of such proteins from small amounts of clinical samples (5 μL), LC-MALDI analysis was applied to three ZipTip derived peptide fractions and 44 proteins could be identified. Especially ease of detection and in-depth analysis of phosphorylated

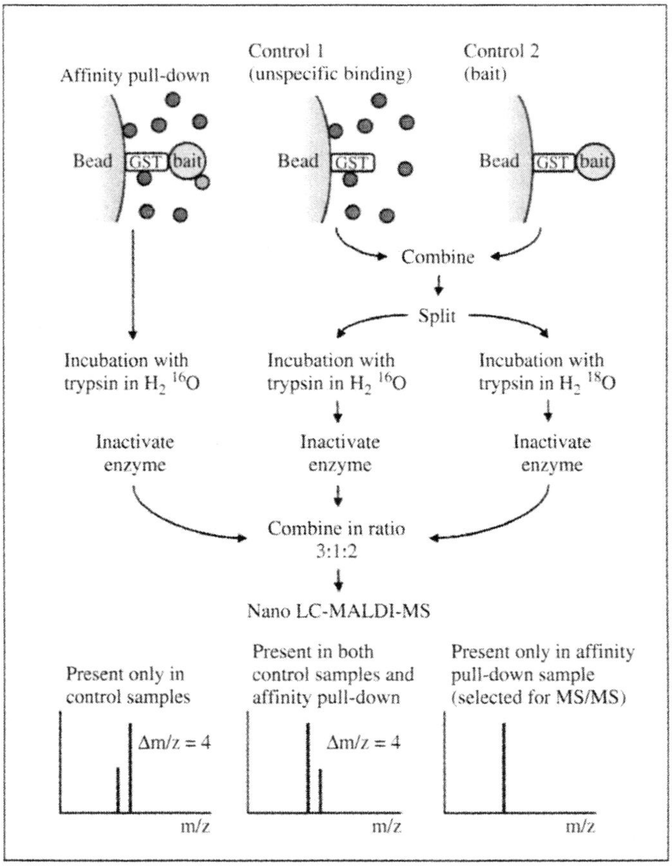

Figure 11. Strategy for Affinity-Pulldown. Differentiation of true binding proteins and unspecific binding background by ^{18}O-labeling (for details cf. text). Reprinted with permission from Mirgorodskaya et al. (2005b). Copyright ©2005 American Chemical Society.

and glycosylated peptides was reported as advantage in the LC-MALDI setup. Phosphopeptide detection in LC-MALDI and MALDI in general could be significantly improved by using 2,5-DHB matrix containing 1% phosphoric acid (Kjellström and Jensen 2004).

4. SUMMARY

In the last years off-line LC-MALDI has developed into an independent and sensitive technique which is now applied in all areas of proteome analysis. By decoupling separation, MS and MS/MS analysis, each step can be performed at its individual optimum. For example, different speed of chromatographic separation can be adapted by different spotting frequencies and spectral quality in MS and MS/MS mode can be optimized without time constraints by setting signal to noise criteria. Since MS/MS is performed only after MS data acquisition for all fractions is completed, precursor ions can be selected from spots where they show maximum intensity. The number of precursor ions is only limited by sample depletion or overall analysis time, which can probably be reduced by higher frequency lasers and faster data acquisition in the future. All parts of the spots which remain after one round of analysis are still available for further experiments, which is a major advantage for the characterization of post-translational modifications. In all respects, LC-MALDI is a very flexible technique which can be adapted to a wide range of different workflows.

REFERENCES

Blakley, C.R., Vestal, M.L., Verentchikov, A. and Wang, Y.P. (1994) Liquid Chromatograph and Capillary Electrophoresis Interfaces for Matrix Assisted Laser Desorption Time-of-flight Mass Spectrometry. Proceedings of the 42nd ASMS Conference on Mass Spectrometry and Allied Topics, Chicago, IL, p. 1045.

Bodnar, W.M., Blackburn, R.K., Krie, J.M. and Moseley, M.A. (2003) Exploiting the complementary nature of LC/MALDI/MS/MS and LC/ESI/MS/MS for increased proteome coverage. J. Am. Soc. Mass Spectrom. *14*, 971–979.

Castoro, J.A., Chiu, R.W., Monnig, C.A. and Wilkins, C.L. (1992) Matrix-assisted laser desorption/ionization of capillary electrophoresis effluents by fourier transform mass spectrometry. J. Am. Chem. Soc. *114*, 7571–7572.

Chen, H., Rejtar, T., Andreev, V., Moskovets, E. and Karger, B.L. (2005a) Enhanced characterization of complex proteomic samples using LC-MALDI MS/MS: exclusion of redundant peptides from MS/MS analysis in replicate runs. Anal. Chem. *77*, 7816–7825.

Chen, H., Rejtar, T., Andreev, V., Moskovets, E. and Karger, B.L. (2005b) High-speed, high-resolution monolithic capillary LC-MALDI MS using an off-line continous deposition interface for proteomic analysis. Anal. Chem. *77*, 2323–2331.

Eckerskorn, C., Strupat, K., Kellermann, J., Lottspeich, F. and Hillenkamp, F. (1997) High-sensitivity peptide mapping by micro-LC with on-line membrane blotting and subsequent detection by scanning-IR-MALDI mass spectrometry. J. Protein Chem. *16*, 349–362.

Ericson, C., Phung, Q.T., Horn, D.M., Peters, E.C., Fitchett, J.R., Ficarro, S.B., Salomon, A.R., Brill, L.M. and Brock, A. (2003) An automated noncontact deposition interface for liquid chromatography matrix-assisted laser desorption/ionization mass spectrometry. Anal. Chem. *75*, 2309–2315.

Fujiki, Y., Hubbard, A.L., Fowler, S. and Lazarow, P.B. (1982) Isolation of intracellular membranes by means of sodium carbonate treatment: application to endoplasmic reticulum. J. Cell Biol. *93*, 97–102.

Fung, K.Y.C., Askovic, S., Basile, F. and Duncan, M.W. (2004) A simple and inexpensive approach to interfacing high-performance liquid chromatography and matrix-assisted laser desorption/ionization-time of flight-mass spectrometry. Proteomics *4*, 3121–3127.

Goodlett, D.R., Keller, A., Watts, J.D., Newitt, R., Yi, E.C., Purvine, S., Eng, J.K., Von Haller, P., Aebersold, R. and Kolker, E. (2001) Differential stable isotope labeling of peptides for quantitation and de novo sequence derivation. Rapid Commun. Mass Spectrom. *15*, 1214–1221.

Gusev, A.I. (2000) Interfacing matrix-assisted laser desorption/ionization mass spectrometry with column and planar separations. Fresenius J. Anal. Chem. *366*, 691–700.

Gygi, S.P., Rist, B., Gerber, S.A., Turecek, F., Gelb, M.H. and Aebersold, R. (1999) Quantitative analysis of complex protein mixtures using isotope-coded affinity tags. Nat. Biotechnol. *17*, 994–999.

Gygi, S.P., Rist, B., Griffin, T.J., Eng, J. and Aebersold, R. (2002) Proteome analysis of low-abundance proteins using multidimensional chromatography and isotope-coded affinity tags. J. Proteome Res. *1*, 47–54.

Hansen, K.C., Schmitt-Ulms, G., Chalkley, R.J., Hirsch, J., Baldwin, M.A., and Burlingame, A.L. (2003) Mass spectrometric analysis of protein mixtures at low levels using cleavable 13C-isotope-coded affinity tag and multidimensional chromatography. Mol. Cell. Proteomics *2*, 299–314.

Harvey, D.J., Bateman, R.H., Bordoli, R.S. and Tyldesley, R. (2000) Ionisation and fragmentation of complex glycans with a quadrupole time-of-flight mass spectrometer fitted with a matrix-assisted laser desorption/ionisation ion source. Rapid Commun. Mass Spectrom. *14*, 2135–2142.

Hattan, J.H., Marchese, J., Khainovski, N., Martin, S. and Juhasz, P. (2005) Comparative study of [Three] LC-MALDI workflows for the analysis of complex proteomic samples. J. Proteome Res. *4*, 1931–1941.

Hofmann, S., Glückmann, M., Kausche, S., Schmidt, A., Corvey, C., Lichtenfels, R., Huber, C., Albrecht, C., Karas, M. and Herr, W. (2005) Rapid and sensitive identification of major histocompatibility complex class I-associated tumor peptides by nano-LC MALDI MS/MS. Mol. Cell. Proteomics *4*, 1888–1897.

Hsieh, S., Dreisewerd, K., Van der Schoors, R.C., Jimenez, C.R., Stahl-Zeng, J., Hillenkamp, F., Jorgenson, J.W., Geraerts, W.P.M. and Li, K.W. (1998) Separation and identification of peptides in single neurons by microcolumn liquid chromatography-matrix-assisted laser desorption/ionization time-of-flight mass spectrometry and postsource decay analysis. Anal. Chem. *70*, 1847–1852.

Hunt, D.F., Buko, A.M., Ballard, J.M., Shabanowitz, J. and Giordani, A.B. (1980) Sequence analysis of polypeptides by collision activated dissociation on a triple quadrupole mass spectrometer. In: Soft Ionization Biological Mass Spectrometry. H. Morris, ed. (London: Heyden & Sons Ltd.), pp. 85–109.

Ji, C., Li, L., Gebre, M., Pasdar, M. and Li, L. (2005) Identification and quantitation of differentially expressed proteins in E-cadherin deficient SCC9 cells and SCC9 transfectants expressing E-cadherin by dimethyl isotope labeling, LC-MALDI MS and MS/MS. J. Proteome Res. *4*, 1419–1426.

Johnson, T., Bergquist, J., Ekman, R., Nordhoff, E., Schürenberg, M., Klöppel, K.D., Müller, M., Lehrach, H. and Gobom, J. (2001) A CE-MALDI interface based on the use of prestructered sample supports. Anal. Chem. *73*, 1670–1675.

Kaufmann, R., Kirsch, D. and Spengler, B. (1994) Sequencing of peptides in a time-of-flight mass spectrometer: evaluation of postsource decay following matrix-assisted laser desorption ionisation (MALDI). Int. J. Mass Spectrom. Ion Process. *131*, 355–385.

Kjellström, S. and Jensen, O.N. (2004) Phosphoric acid as a matrix additive for MALDI MS analysis of phosphopeptides and phosphoproteins. Anal. Chem. *76*, 5109–5117.

Krijgsveld, J., Ketting, R.F., Mahmoudi, T., Johansen, J., Artal-Sanz, M., Verrijzer, C.P., Plasterk, R.H.A. and Heck, A.J.R. (2003) Metabolic labeling of *C.elegans* and *D.melanogaster* for quantitative proteomics. Nat. Biotechnol. *21*, 927–931.

Krutchinsky, A.N., Kalkum, M. and Chait, B.T. (2001) Automatic identification of proteins with a MALDI-quadrupole ion trap mass spectrometer. Anal. Chem. *73*, 5066–5077.

Krutchinsky, A.N., Zhang, W. and Chait, B.T. (2000) Rapidly switchable matrix-assisted laser desorption/ionization and electrospray quadrupole-time-of-flight mass spectrometry for protein identification. J. Am. Soc. Mass Spectrom. *11*:493–504.

Lee, H., Griffin, T.J., Gygi, S.P., Rist, B. and Aebersold, R. (2002) Development of a multiplexed microcapillary liquid chromatography system for high-throughput proteome analysis. Anal. Chem. *74*, 4353–4360.

Li, N., Shaw, A.R.E., Zhang, N., Mak, A. and Li, L. (2004) Lipid raft proteomics: analysis of in-solution digest of sodium dodecyl sulfate-solubilised lipid raft proteins by liquid chromatography-matrix-assisted laser desorption/ionization tandem mass spectrometry. Proteomics *4*, 3156–3166.

Li, N., Wang, N., Liu, X.M., Lever, O.W., Erickson, P.M. and Li, L. (2005) Characterization of human tear proteome using multiple proteomic analysis techniques. J. Proteome Res. *4*, 2052–2061.

Loboda, A.V., Krutchinsky, A.N., Bromirski, M., Ens, W. and Standing, K.G. (2000) A tandem quadrupole/time-of-flight mass spectrometer with a matrix-assisted laser desorption/ionization source: design and performance. Rapid Commun. Mass Spectrom. *14*, 1047–1057.

Martin, R.L. and Brancia, F.L. (2003) Analysis of high mass peptides using a novel matrix-assisted laser desorption/ionisation quadrupole ion trap time-of-flight mass spectrometer. Rapid Commun. Mass Spectrom. *17*, 1358–1365.

Medzihradszky, K.F., Campbell, J.M., Baldwin, M.A., Falick, A.M., Juhasz, P., Vestal, M.L. and Burlingame, A.L. (2000) The characteristics of peptide collision-induced dissociation using a high-performance MALDI-TOF/TOF tandem mass spectrometer. Anal. Chem. *72*, 552–558.

Melanson, J.E., Chisholm, K.A. and Pinto, D.M. (2006) Targeted comparative proteomics by liquid chromatography/matrix-assisted laser desorption/ionization triple-quadrupole mass spectrometry. Rapid Commun. Mass Spectrom. *20*, 904–910.

Miliotis, T., Kjellström, S., Oennerfjord, P., Nilsson, J., Laurell, T., Edholm, L.E. and Marko-Varga, G. (2000) Protein identification platform utilizing micro dispensing technology interfaced to matrix-assisted laser desorption ionization time-of-flight mass spectrometry. J. Chromatogr. A *886*, 99–110.

Mirgorodskaya, E., Braeuer, C., Fucini, P., Lehrach, H. and Gobom, J. (2005a) Nanoflow liquid chromatography coupled to matrix-assisted laser desorption/ionization mass spectrometry: Sample preparation, data anlysis, and application to the analysis of complex peptide mixtures. Proteomics *5*, 399–408.

Mirgorodskaya, E., Wanker, E., Otto, A., Lehrach, H. and Gobom, J. (2005b) Method for qualitative comparisons of protein mixtures based on enzyme-catalyzed stable-isotope incorporation. J. Proteome Res. *4*, 2109–2116.

Moritz, B. and Meyer, H.E. (2003) Approaches for the quantification of protein concentration ratios. Proteomics *3*, 2208–2220.

Mukhopadhyay, R. (2005) The automated union of LC and MALDI MS. Anal. Chem. 77:150A–152A.

Murray, K.K. (1997) Coupling matrix-assisted laser desorption/ionization to liquid separations. Mass Spectrom. Rev. 16:283–299.

Murray, K.K., Lewis, T.M., Beeson, M.D. and Russell, D.H. (1994) Aerosol matrix-assisted laser desorption ionization for liquid chromatography/time-of-flight mass spectrometry. Anal. Chem. *66*, 1601–1609.

Nagra, D.S. and Li, L. (1995) Liquid chromatography-time-of-flight mass spectrometry with continous-flow matrix-assisted laser desorption ionization. J. Chromatogr. A *711*, 235–245.

Oennerfjord, P., Nilsson, J., Wallmann, L., Laurell, T. and Marko-Varga, G. (1998) Picoliter sample preparation in MALDI-TOF MS using a micromachined silicon flow-through dispenser. Anal. Chem. *70*, 4755–4760.

Ong, S.E., Blagoev, B., Kratchmarova, I., Kristensen, D.B., Steen, H., Pandey, A. and Mann, M. (2002) Stable isotope labeling by amino acids in cell culture, SILAC, as a simple and accurate approach to expression proteomics. Mol. Cell. Proteomics *1*, 376–386.

Orsnes, H., Graf, T., Degn, H. and Murray, K.K. (2000) A rotating ball inlet for on-line MALDI mass spectrometry. Anal. Chem. *72*, 251–254.

Orsnes, H. and Zenobi, R. (2001) Interfaces for on-line sample delivery for matrix-assisted laser desorption ionisation mass spectrometry. Chem. Soc. Rev. *30*, 104–112.

Pashkova, A., Chen, H.S., Rejtar, T., Zang, X., Giese, R., Andreev, V., Moskovets, E. and Karger, B.L. (2005) Coumarin tags for analysis of peptides by MALDI-TOF MS and MS/MS. 2. Alexa fluor 350 tag for increased peptide and protein identification by LC-MALDI-TOF/TOF MS. Anal. Chem. *77*, 2085–2096.

Peng, J., Elias, J.E., Thoreen, C.C., Licklider, L.J. and Gygi, S.P. (2003) Evaluation of multidimensional chromatography coupled with tandem mass spectrometry (LC/LC-MS/MS) for large-scale protein analysis: the yeast proteome. J. Proteome Res. *2*, 43–50.

Peters, E.C., Horn, D.M., Tully, D.C., Brock, A. (2001) A novel multifunctional labeling reagent for enhanced protein characterization with mass spectrometry. Rapid Commun. Mass Spectrom. *15*, 2387–2392.

Preisler, J., Foret, F. and Karger, B.L. (1998) On-line MALDI-TOF MS using a continous vacuum deposition interface. Anal. Chem. *70*, 5278–5287.

Preisler, J., Hu, P., Rejtar, T. and Karger, B.L. (2000) Capillary-electrophoresis-matrix-assisted laser desorption/ionization time-of-flight mass spectrometry using a vacuum deposition interface. Anal. Chem. *72*, 4785–4795.

Preisler, J., Hu, P., Rejtar, T., Moskovets, E. and Karger, B.L. (2002) Capillary array electrophoresis-MALDI mass spectrometry using a vacuum deposition interface. Anal. Chem. *74*, 17–25.

Rejtar, T., Hu, P., Juhasz, P., Campbell, J.M., Vestal, M.L., Preisler, J. and Karger, B.L. (2002) Off-line coupling of high-resolution capillary electrophoresis to MALDI-TOF and TOF/TOF MS. J. Proteome Res. *1*, 171–179.

Rose, K., Simona, M.G., Offord, R.E., Prior, C.P., Otto, B. and Thatcher, D.R. (1983) A new mass spectrometric C-terminal sequencing technique finds a similarity between g-interferon and a2-interferon and identifies a proteolytically clipped g-interferon that retains full antiviral activity. Biochem. J. *215*, 273–277.

Ross, P.L., Huang, Y.N., Marchese, J.N., Williamson, B., Parker, K., Hattan, S., Khainovski, N., Pillai, S., Dey, S., Daniels, S., Purkayastha, S., Juhasz, P., Martin, S., Bartlet-Jones, M., He, F., Jacobson, A. and Pappin, D.J. (2004) Multiplexed protein quantitation in saccharomyces cerevisiae using amine-reactive isobaric tagging reagents. Mol. Cell. Proteomics *3*, 1154–1169.

Scherl, A., Francois, P., Bento, M., Deshusses, J.M., Charbonnier, I., Converset, V., Huyghe, A., Walter, N., Hoogland, C., Appel, R.D., Sanchez, J.C., Zimmermann-Ivol, C.G., Corthals, G.L., Hochstrasser, D.F. and Schrenzel, J. (2005) Correlation of proteomic and transcriptomic profiles of Staphylococcus aureus during the post-exponential phase of growth. J. Microbiol. Methods *60*, 247–257.

Scheurer, S.B., Roesli, C., Neri, D. and Elia, G. (2005) A comparison of different biotinylation reagents, tryptic digestion procedures, and mass spectrometric techniques for 2-D peptide mapping of membrane proteins. Proteomics *5*, 3035–3039.

Schmidt, A., Kellermann, J. and Lottspeich, F. (2005) A novel strategy for quantitative proteomics using isotope-coded protein labels. Proteomics *5*, 4–15.

Suckau, D., Resemann, A., Schuerenberg, M., Hufnagel, P., Franzen, J. and Holle, A. (2003) A novel MALDI LIFT TOF/TOF mass spectrometer for proteomics. Anal. Bioanal. Chem. *376*, 952–965.

Tan, P.V., Laiko, V.V., Doroshenko, V.M. (2004) Atmospheric pressure MALDI with pulsed dynamic focusing for high-efficiency transmission of ions into a mass spectrometer. Anal. Chem. *76*, 2462–2469.

Tebbe, A., Klein, C., Schmidt, A., Bisle, B., Konstandinidis, K., Scheffer, B., Lottspeich, F., Siedler, F., Pfeiffer, F. and Oesterhelt, D. (2005) Analysis of the cytosolic proteome of halobacterium salinarum – implications for genome annotation and differential expression. Proceedings of the 53rd ASMS Conference on Mass Spectrometry and Allied Topics, San Antonio, TX, June5-June9, 2005.

Tegeler, T.J., Mechref, Y., Boraas, K., Reilly, J.P. and Novotny, M.V. (2004) Microdeposition device interfacing capillary electrochromatography and microcolumn liquid chromatography with matrix-assisted laser desorption/ionization mass spectrometry. Anal. Chem. *76*, 6698–6706.

Van Veelen, P.A., Tjaden, U.R., Van der Greef, J., Ingendoh, A.and Hillenkamp, F. (1993) Off-line coupling of capillary electrophoresis with matrix-assisted laser desorption mass spectrometry. J. Chromatogr. *647*, 367–374.

Walker, K.L., Chiu, R.W., Monnig, C.A. and Wilkins, C.L. (1995) Off-line coupling of capillary electrophoresis and matrix-assisted laser desorption/ionization time-of-flight mass spectrometry. Anal. Chem. *67*, 4197–4204.

Wall, D.B., Berger, S.J., Finch, J.W., Cohen, S.A., Richardson, K., Chapman, R., Drabble, D., Brown, J. and Gostick, D. (2002) Continous sample deposition from reversed-phase liquid chromatography to tracks on a matrix-assisted laser desorption/ionization precoated target for the analysis of protein digests. Electrophoresis *23*, 3193–3204.

Wang, Y.K., Ma, Z., Quinn, D.F. and Fu, E.W. (2001) Inverse 18O labeling mass spectrometry for the rapid identifcation of marker/target proteins. Anal. Chem. *73*, 3742–3750.

Wattenberg, A.,Organ, A.J., Schneider, K., Tyldesley, R., Bordoli, R. and Bateman, R.H. (2002) Sequence dependent fragmentation of peptides generated by MALDI quadrupole time-of-flight (MALDI Q-TOF) Mass spectrometry and its implications for protein identification. J. Am. Soc. Mass Spectrom. *13*, 772–783.

Wu, W.W., Wang, G., Baek, S.J. and Shen, R.F. (2006) Comparative study of three proteomic quantitative methods, DIGE, cICAT, and iTRAQ, Using 2D Gel- or LC-MALDI TOF/TOF. J. Proteome Res. *5*, 651–658.

Young, J.B. and Li, L. (2006) An impulse-driven liquid-droplet deposition interface for combining LC with MALDI MS and MS/MS. J. Am. Soc. Mass Spectrom. *17*, 325–334.

Zhan, Q., Gusev, A. and Hercules, D.M. (1999) A novel interface for on-line coupling of liquid capillary chromatography with matrix-assisted laser desorption/ionization detection. Rapid Commun. Mass Spectrom. *13*, 2278–2283.

Zhang, H. and Caprioli, R.M. (1996) Capillary electrophoresis combined with matrix-assisted laser desorption/ionization mass spectrometry; continous sample deposition on a matrix-precoated membrane target. J. Mass Spectrom. *31*, 1039–1046.

Zhang, N., Li, N. and Li, L. (2004a) Liquid chromatography MALDI MS/MS for membrane proteome analysis. J. Proteome Res. *3*, 719–727.

Zhang, N., Li, N., Mak, A., Shaw, A.R. and Li, L. (2004b) Lipid raft proteome analysis by 2D LC/ESI and 2D LC/MALDI MS/MS mass spectrometry. Proceedings of the 52nd ASMS Conference on Mass Spectrometry and Allied Topics, Nashville, TN, May23-May27, 2004.

Zhang, B., McDonald, C. and Li, L. (2004) Combining liquid chromatography with MALDI mass spectrometry using a heated droplet interface. Anal. Chem. *76*, 992–1001.

Zhen, Y., Xu, N., Richardson, B., Becklin, R., Savage, J.R., Blake, K. and Peltier, J.M. (2004) Development of an LC-MALDI method for the analysis of protein complexes. J. Am. Soc. Mass Spectrom. *15*, 803–822.

Zhong, H., Marcus, S.L. and Li, L. (2005) Microwave-assisted acid hydrolysis of proteins combined with liquid chromatography MALDI MS/MS for protein identification. J. Am. Soc. Mass Spectrom. *16*, 471–481.

Corresponding author: Dieter Mueller (dieter.mueller@novartis.com)

INDEX

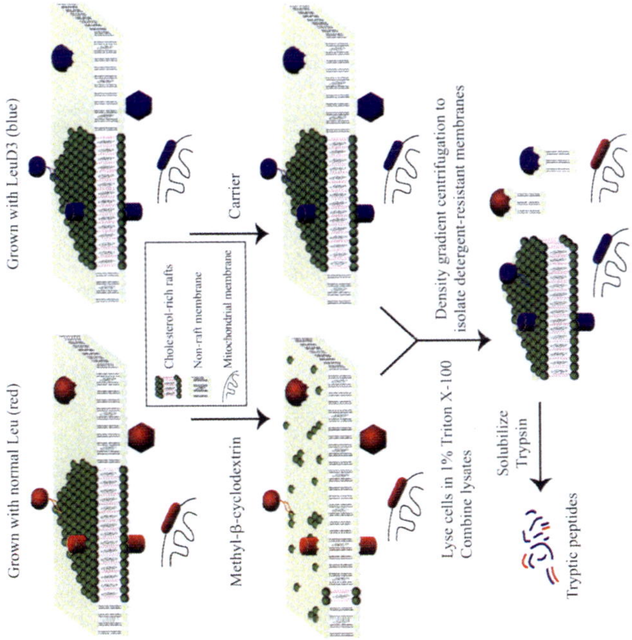

Figure 4.1.

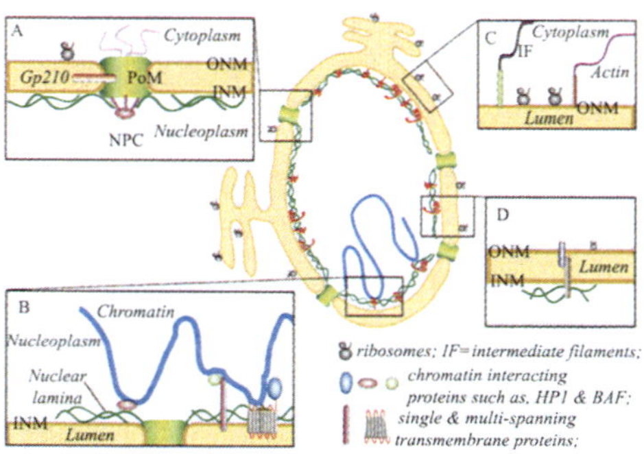

Figure 5.1.

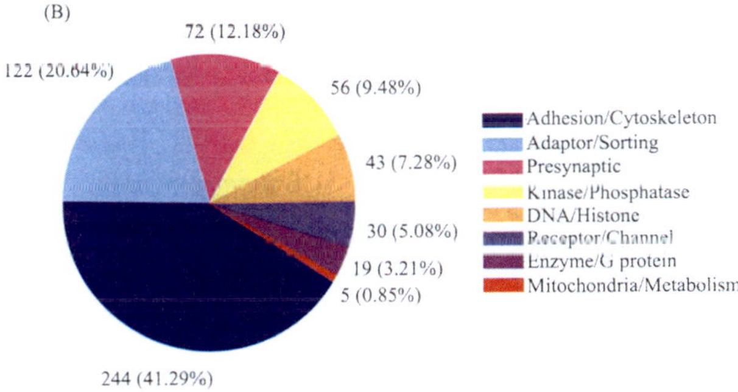

Figure 6.5.

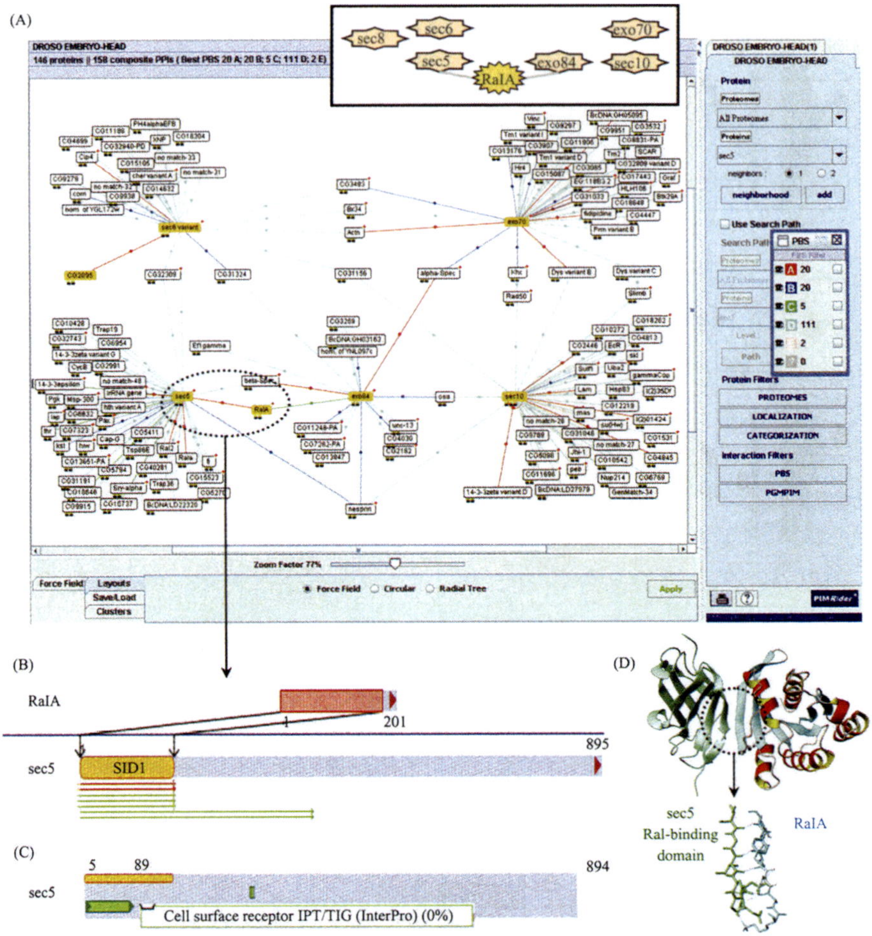

Figure 8.4. Typical dissection of a supramolecular complex (the exocyst, see Figure 1) carried out with PIMRider platform tools (pim.hybrigenics.com). (A) *PIM viewer* tool is used to display the partial PIM arising from the yeast two-hybrid high-throughput screening of *Drosophila* (Formstecher et al. 2005) focused on some components of Sec6/8 secretory machine (i.e. exocyst): Sec5, RalA, Exo84, Sec10, Exo70, Sec6, Sec8 (CG2095) are orange-boxed. The resulting graph (proteins are nodes, interactions are edges, baits have typical radiating-hub display) may be used to browse and navigate through a pathway or assembly subunits. Interaction scores follow a coloring scheme from red (PBS A) to pink (PBS E) transcribed on PPI linkage and may be used to filter and decrease PIM densification. (B) Screen copy of the *Interaction Viewer* tool showing the highly confident (PBS A) PPI between RalA and sec5 and the SID (Selected Interacting Domain) resulting from common part of experimental prey fragments (arrows). (C) Snapshot of the *Domain Viewer* tool that allows to map sec5 SID to IPT, a Ral-binding domain annoted within InterPro database. (D) X-ray crystallographic structure of the complex between the 9-183 fragment of RalA and the 120 residues N-terminal fragment of Sec5, that adopts an original immunoglobulin-like β-sandwich fold (Fukai et al. 2003; pdb entry 1UAD). These authors have well-depicted the hydrogen bonds networks formed at the junction of the intermolecular antiparallel β-sheet formed by sec5 β1 and RalA β2. 3D pictures performed with TRIPOS 3D-Explorer.

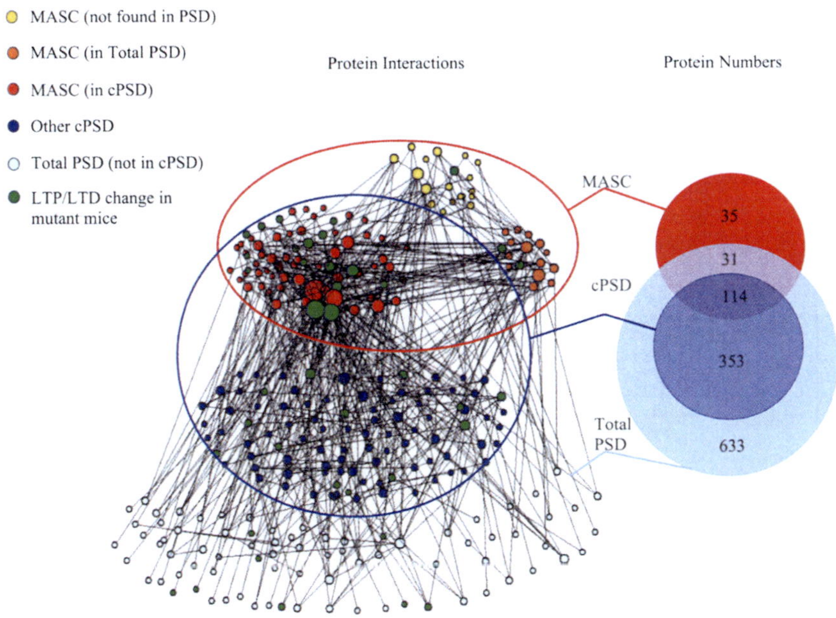

Figure 9.5.

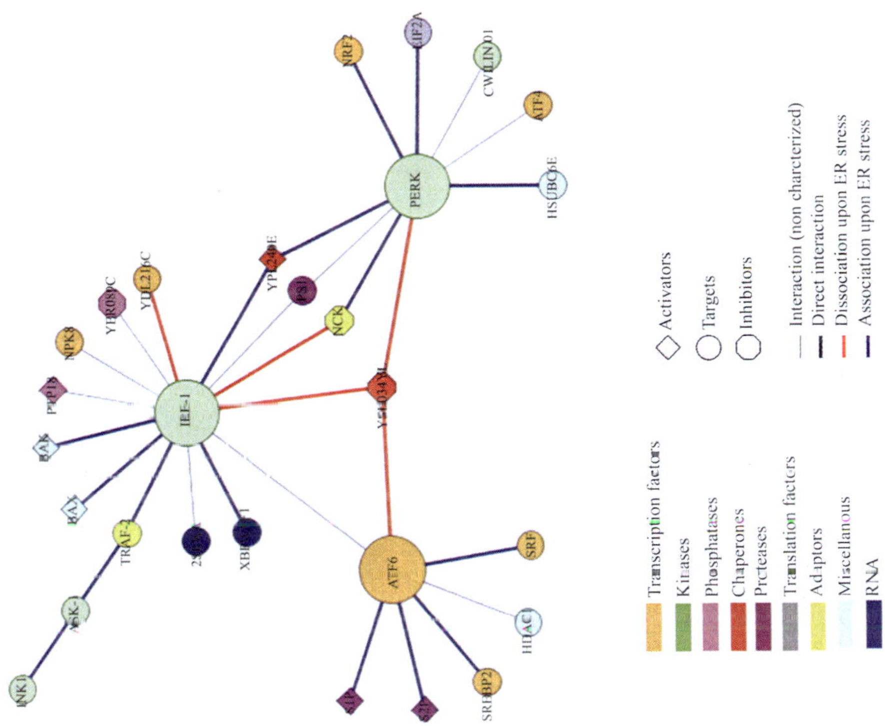

Figure 13.3.

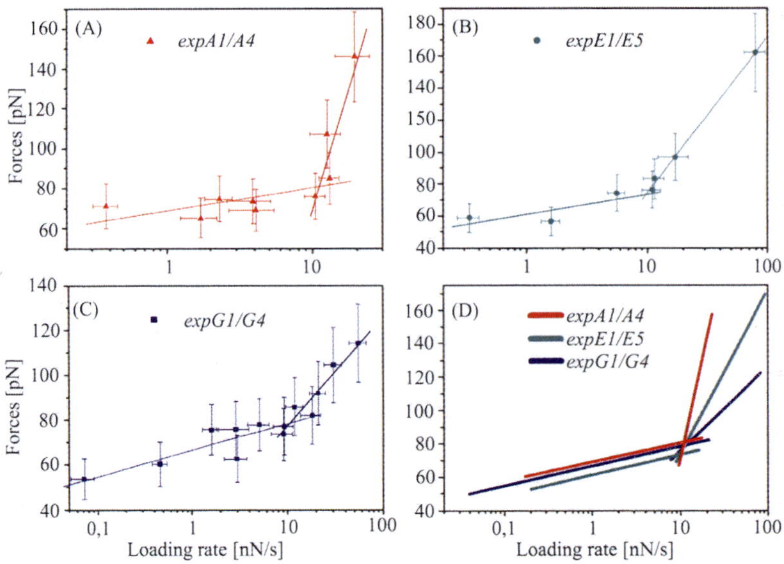

Figure 14.3.

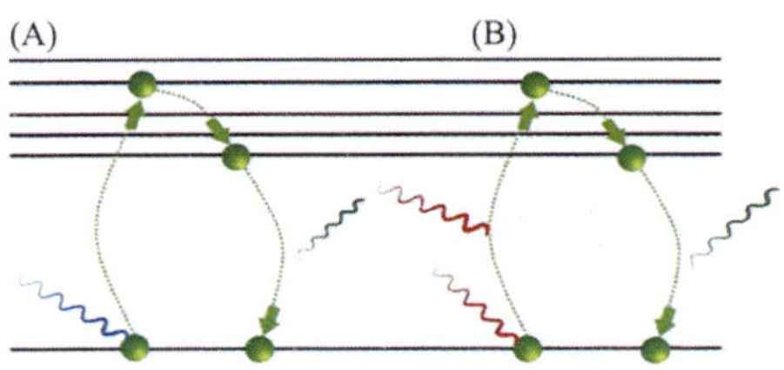

Figure 14.4.

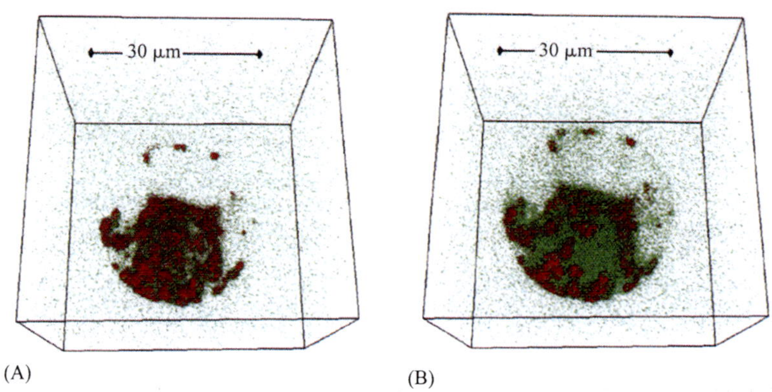

Figure 14.9.

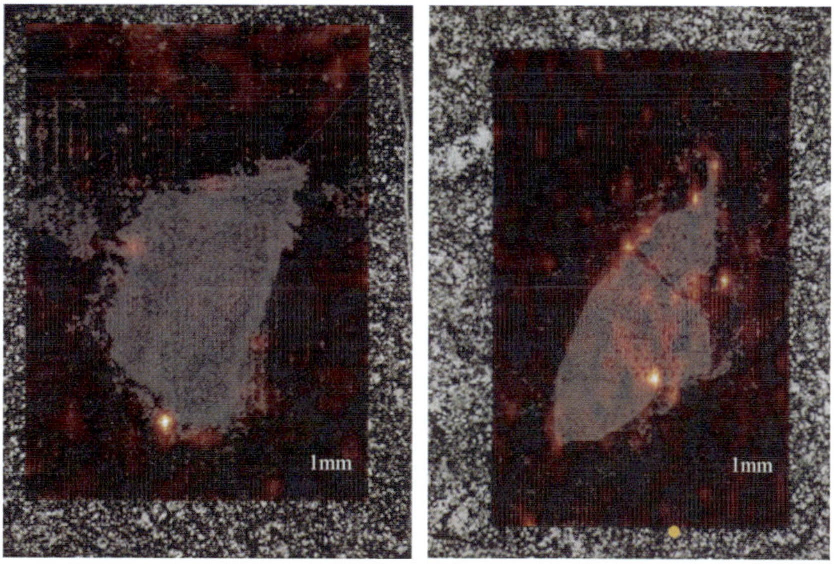

Figure 15.3. MS/MS molecular images of sections from bovine cartilage immersed in compound solution (left: control; right: drug). The images represent the penetration of the drug into the tissue (signal in red).

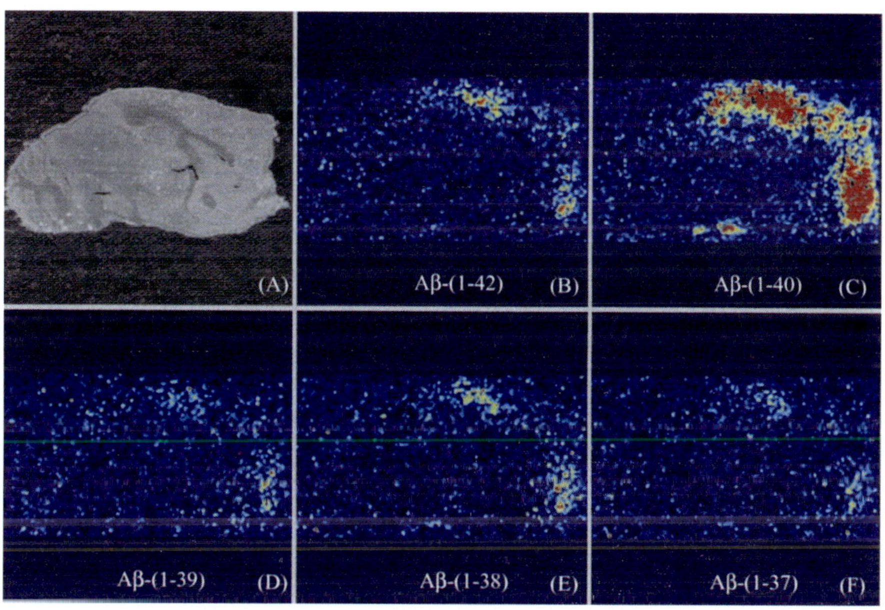

Figure 15.5. MSI on sagittal AD brain section. (A) optical image of the sagittal AD brain section; (B) Aβ-(1–42) molecular image (*m*/*z* 4515.1); (C) Aβ-(1–40) molecular image (*m*/*z* 4330.9); (D) Aβ-(1–39) molecular image (*m*/*z* 4231.7); (E) Aβ-(1–38) molecular image (*m*/*z* 4132.6); (F) Aβ-(1–37) molecular image (*m*/*z* 4075.5).